Plant Biotechnology: Genetic Modification of Plants

Plant Biotechnology: Genetic Modification of Plants

Editor: Isabelle Nickel

R CALLISTO REFERENCE

www.callistoreference.com

Callisto Reference,
118-35 Queens Blvd., Suite 400,
Forest Hills, NY 11375, USA

Visit us on the World Wide Web at:
www.callistoreference.com

ISBN: 978-1-63239-910-6 (Hardback)

Cataloging-in-Publication Data

Plant biotechnology : genetic modification of plants / edited by Isabelle Nickel.
    p. cm.
Includes bibliographical references and index.
ISBN 978-1-63239-910-6
1. Plant biotechnology. 2. Plant genetic engineering. 3. Plant genetics.
I. Nickel, Isabelle.
TP248.27.P55 P53 2018
631.523 3--dc23

Table of Contents

Preface

Plant biotechnology has been practiced for decades. In present times it studies the cross-breeding, texture and color of plants and other mechanisms. It can be accomplished through different techniques from simply selecting plants with desirable characteristics for propagation to more complex techniques. A number of latest researches have been included herein to keep the readers up-to-date with the global concepts in this area of study. This book, with its detailed analyses and data, will prove immensely beneficial to professionals and students involved in this area at various levels.

Various studies have approached the subject by analyzing it with a single perspective, but the present book provides diverse methodologies and techniques to address this field. This book contains theories and applications needed for understanding the subject from different perspectives. The aim is to keep the readers informed about the progresses in the field; therefore, the contributions were carefully examined to compile novel researches by specialists from across the globe.

Indeed, the job of the editor is the most crucial and challenging in compiling all chapters into a single book. In the end, I would extend my sincere thanks to the chapter authors for their profound work. I am also thankful for the support provided by my family and colleagues during the compilation of this book.

Editor

The Complete Chloroplast Genome of Banana (*Musa acuminata*, Zingiberales): Insight into Plastid Monocotyledon Evolution

Guillaume Martin[1], Franc-Christophe Baurens[1], Céline Cardi[1], Jean-Marc Aury[2], Angélique D'Hont[1]*

1 CIRAD (Centre de coopération Internationale en Recherche Agronomique pour le Développement), UMR AGAP, Montpellier, France, **2** Genoscope, Evry, France

Abstract

Background: Banana (genus *Musa*) is a crop of major economic importance worldwide. It is a monocotyledonous member of the Zingiberales, a sister group of the widely studied Poales. Most cultivated bananas are natural *Musa* inter-(sub-)specific triploid hybrids. A *Musa acuminata* reference nuclear genome sequence was recently produced based on sequencing of genomic DNA enriched in nucleus.

Methodology/Principal Findings: The *Musa acuminata* chloroplast genome was assembled with chloroplast reads extracted from whole-genome-shotgun sequence data. The *Musa* chloroplast genome is a circular molecule of 169,972 bp with a quadripartite structure containing two single copy regions, a Large Single Copy region (LSC, 88,338 bp) and a Small Single Copy region (SSC, 10,768 bp) separated by Inverted Repeat regions (IRs, 35,433 bp). Two forms of the chloroplast genome relative to the orientation of SSC versus LSC were found. The *Musa* chloroplast genome shows an extreme IR expansion at the IR/SSC boundary relative to the most common structures found in angiosperms. This expansion consists of the integration of three additional complete genes (*rps15*, *ndhH* and *ycf1*) and part of the *ndhA* gene. No such expansion has been observed in monocots so far. Simple Sequence Repeats were identified in the *Musa* chloroplast genome and a new set of *Musa* chloroplastic markers was designed.

Conclusion: The complete sequence of *M. acuminata* ssp *malaccensis* chloroplast we reported here is the first one for the Zingiberales order. As such it provides new insight in the evolution of the chloroplast of monocotyledons. In particular, it reinforces that IR/SSC expansion has occurred independently several times within monocotyledons. The discovery of new polymorphic markers within *Musa* chloroplast opens new perspectives to better understand the origin of cultivated triploid bananas.

Editor: James G. Umen, Donald Danforth Plant Science Center, United States of America

Funding: Centre de coopération Internationale en Recherche Agronomique pour le Développement (CIRAD), French National Research Agency (ANR) and Commissariat à l'Energie Atomique (CEA). The funders had no role in study design, data collection and analysis, decision to publish, or preparation of the manuscript.

Competing Interests: The authors have declared that no competing interests exist.

* E-mail: angelique.d'hont@cirad.fr

Introduction

Chloroplasts are the photosynthetic organelles that provide energy for plants and algae. They are also involved in major functions such as sugar synthesis, starch storage, the production of several amino acids, lipids, vitamins and pigments and also in key sulfur and nitrogen metabolic pathways. In angiosperms, chloroplastic (cp) genomes exist at least in part as a circular DNA molecule [1] ranging from 120 to 160 kb in length. Most cp genomes have a quadripartite organization comprising two copies of 20 to 28 kb Inverted Repeats (IRs) which separate the rest of the genome into a 80–90 kb Large Single Copy region (LSC) and a 16–27 kb Small Single Copy region (SSC) [2]. In angiosperms, the cp genome usually encodes 4 rRNAs, 30 tRNAs, and about 80 unique proteins. Earlier studies, using restriction site mapping, have demonstrated that gene content, gene order, and genome organization are largely conserved within land plants [3,4]. However, with the increasing number of whole cp genome available, many structural rearrangements, large IR expansion and gene loss have been reported [2,5,6]. These events can be used for the reconstruction of plant phylogeny [7]. Besides, the availability of whole chloroplast genomes or complete sets of cp genes have helped resolving relationships among major clades of angiosperms [8,9] with more accuracy than even well-chosen "Lucky Genes" [10]. Most of the reported complete monocotyledons chloroplast genomes are from the Poales group (so far 31 of the 46 complete chloroplast genomes deposited in Genbank). It is thus important to have more representatives of other clades to better understand the evolution of cp genome within monocots.

Bananas (genus *Musa*, family Musaceae) are monocotyledons from the Zingiberales, a sister group of the Poales. Banana is of major economic importance in many tropical and subtropical countries where it is vital for food security and also a major source of incomes. Bananas are widely exported to industrialized countries where they represent the most popular fruit. A reference sequence of *Musa acuminata* nuclear genome has recently been published based on sequencing a DNA extract enriched in nucleus [11], yet providing additional sequence data to assemble a chloroplastic genome. In banana, the peculiar paternal inheritance of the mitochondrial genome associated to the classic maternal inheritance of the chloroplast genome [12] make cytoplasmic markers potentially very useful for analyzing the origin of cultivars, most of which are spontaneous triploid inter-(sub)-specifc hybrids [13–15]. In previous studies based on RFLP [16] or PCR-RFLP [17] a total of nine different chloroplastic patterns have been identified among cultivated bananas and related wild species. However, most *M. acuminata* sub-species and cultivars had identical pattern restraining the identification of cultivars progenitors [18].

In this study, we report the assembly, annotation and structure analysis of the complete cp genome of banana. We compare its organization (gene content, IR expansion/contraction, structural rearrangement) with the complete genome of 34 monocots and 10 more basal angiosperms. We also provided new cp markers designed from Simple Sequence Repeats (SSR).

Materials and Methods

Sequence Data

A reference nuclear genome sequence of the doubled-haploid Pahang accession (DH-Pahang) was produced based on DNA extraction enriched in nuclear content. A total of 27,495,411 reads were generated using Roche/454 GSFLX pyrosequencing platform. An addition of 1,069,954 paired-Sanger 10 kb insert-size reads and 49,216 paired-Sanger BAC-ends sequenced on two BAC libraries generated with *Hind*III and *Bam*HI restriction enzymes were produced [11].

The plastid reads were extracted from the total using blast similarity search against *Phoenix dactylifera* whole chloroplast genome (NC_013991). The 454 filtered reads were then assembled into sequence contigs using de novo assembly with Newbler. A total of six contigs were obtained. Using a python script, an iterative elongation for both ends of each contig using the total 454 reads was applied to ensure that contribution of *Musa* specific sequences was taken into account. The resulting four contigs (one contig for each region except two for the IRs) were then ordered based on *P. dactylifera* chloroplast structure. A mapping step using the paired-Sanger 10 kb insert-size reads was then applied to confirm and correct contig junctions. A total of 1,800,008 GS FLX Titanium reads and 33,583 paired-Sanger reads were mapped to the assembled plastid genome representing 6.5% and 3.1% of the total 454 and Sanger reads for an average coverage of 5,341 X (sd = 2,048), the large standard deviation mainly due to the doubling of coverage in the IRs. The minimum coverage was 619 X and the maximum coverage was reached in the IRs with a value of 9,500 X. The junction between the two contigs corresponding to the IRs was confirmed with the Sanger reads. The four junctions between the single-copy regions and IRs were confirmed by PCR.

LSC Orientation Relative to SSC

In order to verify the orientation of the SSC region relative to the LSC region, paired-Sanger BAC-end reads were mapped on the assembled *Musa* chloroplast genome using BLAST. Only pairs presenting more than 90% identity on more than 60% of their length were conserved. An additional filter was applied to conserve only pairs having a mate on the SSC region while the other was on the LSC. A total of 180 paired-Sanger BAC-ends were retained. Orientation visualization of the different paired BAC-ends reads was performed using CIRCOS [19] and was used to infer LSC orientation relative to SSC.

Genome Annotation

The genome was annotated by using DOGMA [20], followed with manual corrections for start codons. Intron positions were determined based on those of *P. dactylifera* [21] and *Elaeis guineensis* [22]. The transfer RNA genes were annotated using DOGMA and tRNAscan-SE (version1.23) [23]. Some intron-containing genes in which exons are too short to be detected were identified based on comparisons to corresponding exons in *P. dactylifera* and *E. guineensis*. The resulting annotated sequence has been deposited at the European Nucleotide Archive under accession number HF677508.

Codon Usage

Codon usage frequencies and the relative synonymous codon usage (RSCU) was calculated from coding sequences (CDS) of all different protein coding genes in the *M. acuminata* chloroplast genome using seqinr R-cran package [24].

Cp DNA Transfers to the Nucleus

Chloroplast DNA transfers to the nucleus were detected using Blast based approach. The assembled *M. acuminata* chloroplast genome, with one of its IR removed, was compared to the 11 chromosomes of *Musa* nuclear reference genome with high stringency blast parameter (e-value $<10^{-5}$, hit length >100 bp). A *per base* insertion value of each plastid base has been calculated as described in The Tomato Genome Consortium [25].

Phylogenetic Analysis

The phylogeny was performed using 79 plastid protein-coding genes derived from 48 plant species (Table S1) with complete chloroplast sequence, most belonging to monocotyledons. A codon based alignment was performed for each gene using homemade scripts that grouped together homologous genes and then converted them into proteins. An alignment was then applied to the protein sequence using MAFFT [26] and this protein alignment was then used to make the codon based aligment. Each aligned gene was then concatenated into a single matrix. Missing genes were replaced by Ns. A nucleotide matrix of 76,524 sites was then constituted. Evolutionary model choice was performed using jModelTest 2.0.2 software [27]. A maximum likelihood (ML) phylogenetic analysis was then performed using GTR+G+I model of sequence evolution using PhyML v3.0 [28]. Branch support was estimated based on aLRT statistics.

Musa Chloroplast Structure Comparison with others Whole cp Genomes

Gene positions of the different cp genomes were collected from the Genbank file and ordered based on their positions within the genome. Gene order and composition were then compared between the different species. Large events, *e.g.* gene loss, IR gene gain/loss, large structural rearrangement, relative to the basal angiosperm *Amborella trichopoda* [29] were recorded and used to infer scenarios in the different monocot lineages.

Short Tandem Repeats

Microsatellites (mono-, di-, tri-, tetra-, penta-, and hexanucleotide repeats) detection was performed using MISA [30] with minimum number of repeats of 10, 5, 4, 3, 3, 3 for 1, 2, 3, 4, 5, 6 unit size respectively (Table S2). Minisatellites (unit size ≥10) were detected manually using dot plot with Gepard software [31] with the *Musa* chloroplast sequence plotted against itself. Sequences with unit repeat equal to or higher than 10 bp repeated tandemly at least twice were conserved. The dot plot was

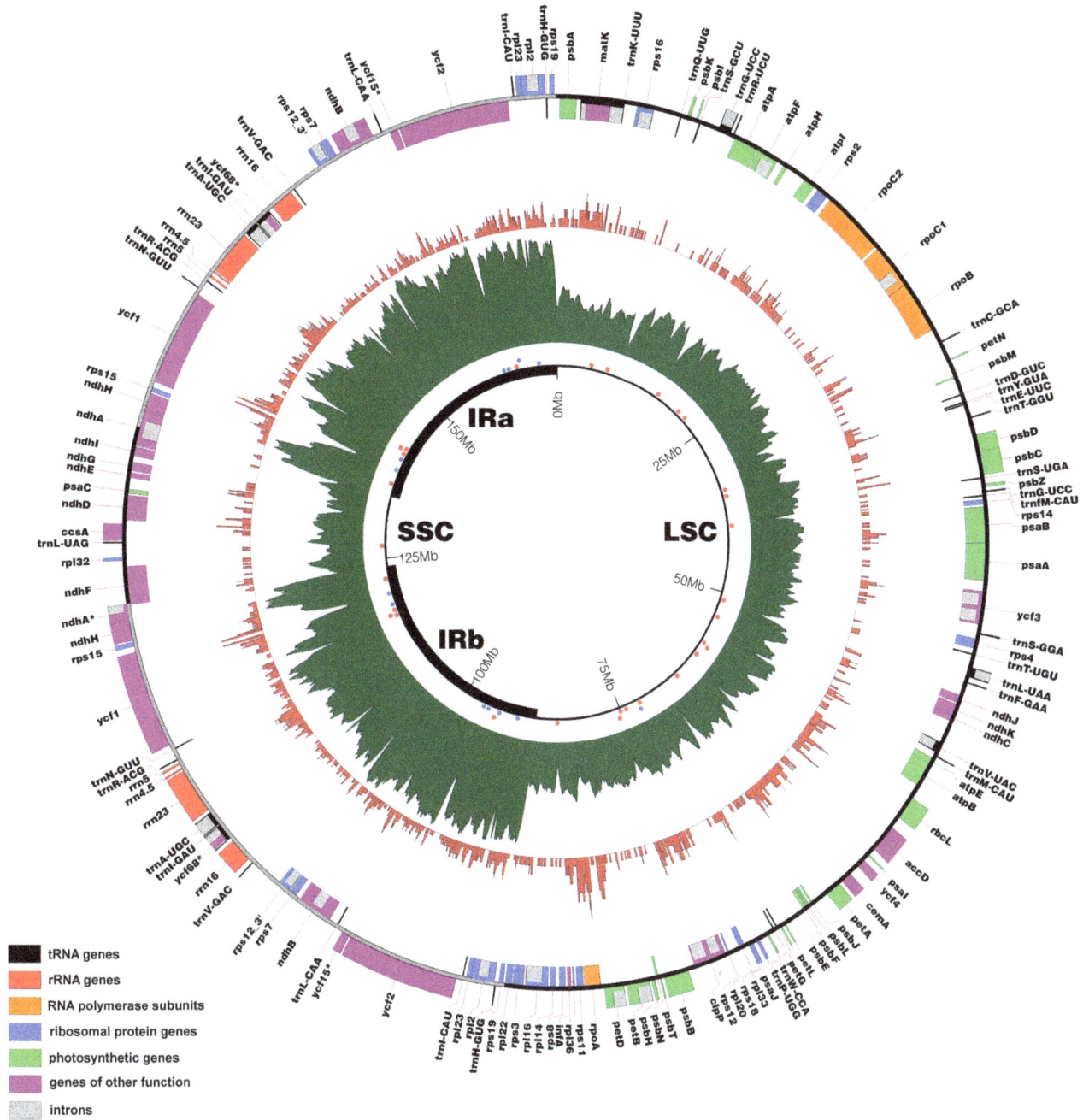

Figure 1. Circular *Musa acuminata* chloroplast map. Genes are represented with boxes inside or outside the circle to indicate clockwise or counterclockwise transcription direction respectively. The color of the gene boxes indicates the functional group to which the gene belongs. Read depth of the genome is represented in the inner green circle. The locations of the short tandem repeats, tested for their polymorphism, are represented with red and blue dots for microsatellites and minisatellites respectively. The *per base* insertion value in the nucleus is drawn in the red circle. The *per base* insertion value of the IR analyzed has been divided by two and applied to both IR. Pseudogenes are marked with asterisks.

inspected on overlapping windows of 5 kbp with an overlap of 1 kbp. Mini- and microsatellites located in the IR regions were only counted once. Primer design was performed using Primer3 software [32] (Table S3).

A total of 32 SSR located all over the plastid genome (Figure 1) were tested for their polymorphism in a testing panel comprising *M. boman*, *M. balbisiana* and three *M. acuminata* spp., *M.a.* ssp *banksii*, *M.a.* ssp *zebrina* and *M.a.* ssp *malaccensis* (DH-Pahang) with the

Applied Biosystems® 3500×L Genetic Analyzer. This set included 7 minisatellites and 26 microsatellites. The 12 most polymorphic markers were evaluated onto 5 additional cultivated accessions, including the triploid accession Cavendish Grande Naine, all belonging to the *Musa* chloroplastic group II [16].

Table 1. *Musa acuminata* plastome characteristics.

Plastome characteristics	
Size (bp)	169,972
LSC size in bp (%)	88,338 (52.0)
SSC size in bp (%)	10,768 (6.3)
IR length in bp	35,433
Size in bp (%) coding regions	100,277 (59.0)
Size in bp (%) of protein-encoding regions	88,336 (52.0)
Size in bp (%) of introns	19,312 (11.4)
Size in bp (%) of rRNA	9,056 (5.3)
Size in bp (%) of tRNA	2,885 (1.7)
Size in bp (%) of IGS	50,389 (29.6)
Number of different genes	113
Number of different protein-encoding genes	79
Number of different tRNA genes	30
Number of different rRNA genes	4
Number of different genes duplicated by IR	24
Number of different genes with introns	18
Overall % GC content	36.8
% GC content in protein-encoding regions	37.3
% GC content in introns	37.8
% GC content in IGS	31.6
% GC content in rRNA	55.2
% GC content in tRNA	53.1

Results and Discussion

General Feature of *Musa* Acuminata cp Genome

The *M. acuminata* chloroplast genome is a DNA molecule of 169,972 bp in length. Similar to most other angiosperms, the chloroplast genome of *M. acuminata* is circular with a quadripartite structure: a pair of Inverted Repeats (IRs) (35,433 bp) separated by the Single Copy region (SSC) (10,768 bp) and Large Single Copy region (LSC) (88,338 bp) (Figure 1). A total of 136 functional genes were predicted, including 113 distinct genes comprising 79 protein-coding genes, 30 transfer RNA (tRNA) genes and 4 ribosomal RNA (rRNA) genes (Table 1). All 4 rRNA genes, 8 tRNA and 10 protein-coding genes are repeated in the IR. Protein-coding genes, tRNA and rRNA represent respectively 52.0%, 1.7% and 5.3% of the plastid genome. Non-coding DNA, including intergenic spacers (IGSs) and introns represent 41.0% of the genome. Similar to other plastid genomes, the overall GC content of the *M. acuminata* plastid genome is 36.8%. This value is slightly higher in protein coding genes (37.3%) and introns (37.8%), slightly lower in IGS (31.6%) while tRNA and rRNA show higher GC value with 53.1% and 55.2% respectively.

A total of 23,199 codons represent the 79 different protein-coding genes of the *M. acuminata* chloroplast genome. Among these, 2,350 (10.6%) code for leucine and 269 (1.2%) for cysteine, which are the most frequent and the least frequent amino acids, respectively (Table 2). The 30 different tRNA found in the chloroplast genome correspond to 28 different codons, at least one for each amino acid. Only 7 of the 28 different anticodon tRNAs encoded in the *Musa* plastid genome correspond to the most common codon (where synonymous codons exist). The codon usage is biased towards a high representation of A and T at the

third position, as observed in most land plant chloroplast genomes [33].

The *M. acuminata* chloroplast genome has 18 different intron-containing genes, six of which are tRNA. Most have a single intron except two genes, *clpP* and *ycf3*, which contain two introns. The gene *rps12* is trans-spliced and has the 5′exon in the LSC and two exons in the IR. The *ycf15* and *ycf68* genes were found to have 5 and 7 internal stop codons respectively. This suggests that *ycf15* and *ycf68* have become pseudogenes in *M. acuminata* chloroplast genome. These two pseudogenes were mentioned in very few chloroplast studies and thus were not used in our phylogenetic study. The incomplete duplication of the 5′ end of *ndhA* at the IRa and SSC boundary resulted in two ndhA gene copies: a pseudogene at the boundary of IRb and SSC and a complete copy at the IRa and SSC boundary.

LSC Orientation Relative to SSC

Due to the inverted repeated regions it was not possible to conclude on the orientation of the SSC relative to the LSC using the 454 and Sanger reads 10 kb paired reads. BAC-end-sequences (BES) were used to orient the SSC relative to LSC. A total of 77 BES, 29 and 48 in the Forward/Reverse and Reverse/Forward orientation respectively, support the orientation presented in this paper (Figure 2A). Another set of 103 BES, 29 and 74 in the Forward/Forward and Reverse/Reverse support a SSC in the reverse complement order (Figure 2B). These results imply that the two forms co-exist in the *M. acuminata* chloroplast genome. This coexistence of two orientation-forms has previously been reported in *Phaseolus vulgaris* [34] and *Zea mays* [35] using RFLP analysis.

DNA Transfer to the Nucleus

A total of 563 hits (Table 3) of more than 100 bp were found on the eleven chromosomes of *M. acuminata* for a cumulative length of 134,491 bp of the *Musa* nuclear genome (0.41 ‰). This value is situated between those of *Arabidopsis thaliana* (0.17‰) [36] and tomato (0.75‰) [25]. A much higher value (1.85‰) had been found in the rice genome [37]. Matsuo et al. [38], using the rice nuclear genome, reported that the plant nuclear genome is in equilibrium between integration and elimination of the chloroplast genome. The various proportions of inserted chloroplast genome observed in plant species reveal different levels of equilibrium. These variations may result from distinct speed of cp DNA transfer flow to the nucleus or a distinct speed of elimination of the inserted cp DNA or a combination of these two processes.

Based on a *per base* insertion value calculated for each plastid base, we showed that the cp DNA inserted in the *Musa* nuclear DNA originate from every part of the chloroplast genome and covers 57.4% of the chloroplast (without IRa) (Figure 1). The highest per base insertion values appeared around the regions carrying the *rpoA* gene and to a further extent in a region containing the *ycf1* gene. In the tomato genome, the *per base* insertion value was also higher in two regions carrying *ycf* genes [25].

Unlike tomato and rice that countain numerous large insertions, only 6 hits of more than 1 kb but not exceeding 2 kb were found on the *M. acuminata* nuclear genome. Chloroplast insertions were found on all chromosomes (Table 3) with a relatively homogeneous distribution unlike the uneven distribution observed in rice [37]. Chromosome 2, with 2.43% of all chloroplast genome insertions, was the chromosome with the least plastid insertion while chromosome 6 was the one having the most abundant plastid insertion (11.08%). The cp DNA insertions into the nuclear genome of *Musa* were evenly distributed over the chromosomes with a reduced number of insertions in pericentromeric regions

Table 2. Codon usage and codon-anticodon recognition pattern of the *Musa acuminata* chloroplast genome.

Amino acid	Codon	Number	RSCU[a]	Frequency[b]	Amino acid	Codon	Number	RSCU[a]	Frequency[b]
F	TTT	**839**	1.28	**64.19**	A	**GCA**	353	1.13	28.35
F	**TTC**	468	0.72	35.81	A	GCG	122	0.39	9.80
L	**TTA**	**750**	1.91	**31.91**	Y	TAT	**682**	1.58	**78.75**
L	**TTG**	497	1.27	21.15	Y	**TAC**	184	0.42	21.25
L	CTT	465	1.19	19.79	H	CAT	**432**	1.55	**77.56**
L	CTC	158	0.40	6.72	H	**CAC**	125	0.45	22.44
L	**CTA**	324	0.83	13.79	Q	**CAA**	**640**	1.54	**77.11**
L	CTG	156	0.40	6.64	Q	CAG	190	0.46	22.89
I	ATT	**980**	1.47	**48.98**	N	AAT	**850**	1.55	**77.27**
I	**ATC**	384	0.58	19.19	N	**AAC**	250	0.45	22.73
I	ATA	637	0.96	31.83	K	**AAA**	**907**	1.51	**75.52**
M	**ATG**	546	1.00	100.00	K	AAG	294	0.49	24.48
V	GTT	468	1.42	35.51	D	GAT	**763**	1.61	**80.32**
V	**GTC**	168	0.51	12.75	D	**GAC**	187	0.39	19.68
V	**GTA**	**505**	1.53	**38.32**	E	**GAA**	**976**	1.50	**74.90**
V	GTG	177	0.54	13.43	E	GAG	327	0.50	25.10
S	TCT	**512**	1.72	**28.72**	C	TGT	**200**	1.49	**74.35**
S	**TCC**	290	0.98	16.26	C	**TGC**	69	0.51	25.65
S	**TCA**	342	1.15	19.18	W	**TGG**	394	1.00	100.00
S	TCG	161	0.54	9.03	R	**CGT**	326	1.39	23.24
S	AGT	390	1.31	21.87	R	CGC	76	0.33	5.42
S	**AGC**	88	0.30	4.94	R	CGA	312	1.33	22.24
P	CCT	**374**	1.57	**39.33**	R	CGG	105	0.45	7.48
P	CCC	196	0.82	20.61	R	**AGA**	441	1.89	**31.43**
P	**CCA**	280	1.18	29.44	R	AGG	143	0.61	10.19
P	CCG	101	0.42	10.62	G	GGT	539	1.38	34.46
T	ACT	**454**	1.54	**38.41**	G	**GGC**	154	0.39	9.85
T	**ACC**	235	0.80	19.88	G	**GGA**	643	1.64	**41.11**
T	**ACA**	366	1.24	30.96	G	**GGG**	228	0.58	14.58
T	ACG	127	0.43	10.74	*	TAA	**41**	1.56	**51.90**
A	GCT	**575**	1.85	**46.18**	*	TAG	20	0.76	25.32
A	GCC	195	0.63	15.66	*	TGA	18	0.68	22.78

[a]: relative synonymous codon usage.
[b]: codon frequency relative to each amino acid.
Codons shown in bold complement the anticodons of the tRNAs encoded in the chloroplast genome. Frequencies shown in bold indicate the most common codon (where synonymous codons exist for that amino acid or termination).

(Figure S2). However, this may be due to lower assembly quality of this type of regions.

Phylogenetic Analysis

A ML phylogenetic analysis was conducted based on 79 protein coding gene from 48 plant taxa. The resulting topology is presented in Figure 3 and Figure S1. All except two nodes are well supported with aLRT statistics higher than 0.98. The first not well supported node with a value of 0.972 is in the Basal Angiosperms group and is positioning *Chloranthus spicatus* at a basal position to the group constituted of *Drimys granadensis*, *Piper cenocladum*, *Calycanthus floridus*, *Magnolia kwangsiensis* and *Liriodendron tulipifera*. The second ambiguous node, with an aLRT value of 0.396, is the relative position of the Bamboo species *Ferrocalamus rimosivaginus* and *Acidosasa purpurea* at the basal position of the others

Arundinarieae included in the analysis. Speciation of Zingiberales, Arecales and Poales has long been difficult to resolve and conflicting results have been reported [9]. Our *M. acuminata* chloroplast data positions Zingiberales as sister to the Poales. The Arecales is positioned as sister group to the Poales and Zingiberales in agreement with previous study based on chloroplast genes [6,8,9]. However, these results differ from the phylogenetic trees obtained with 93 nuclear single genes, that regroup Zingiberales and Arecales in a sister group to the Poales [11]. Similar incongruence between analyses of single-copy nuclear genes and the chloroplast genes has been observed in the phylogenetic placement of the Malpighiales within the Rosids [39]. These incongruences may be caused by incomplete lineage sorting [40], long-branch attraction phenomenon [41,42] or chloroplast introgressions between Musaceae and Poales ancestors

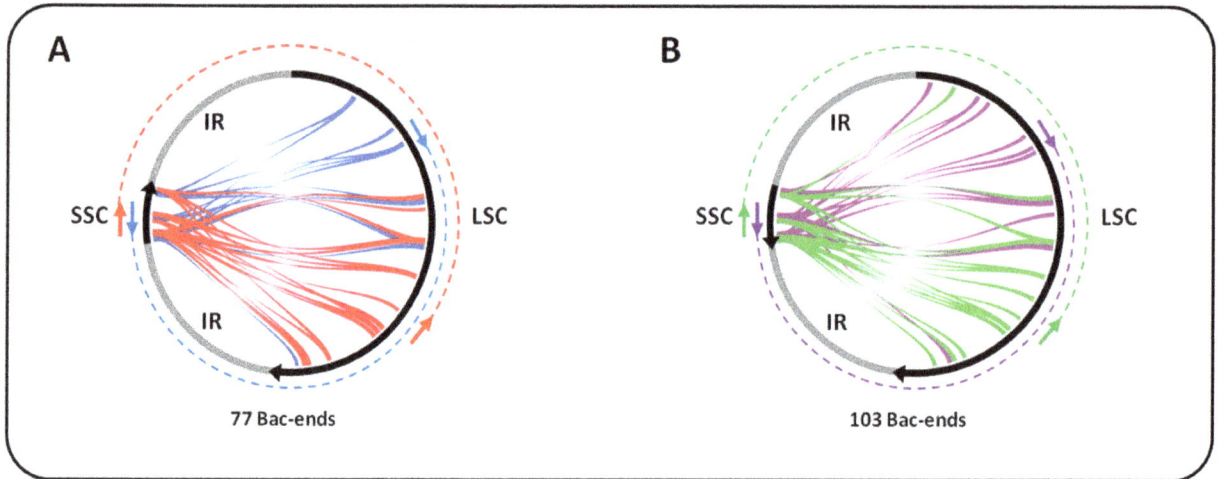

Figure 2. BAC-end-sequences (BESs) mapped on the LSC and SSC *Musa* chloroplast genome. A, BESs mapping with the Forward/Reverse (FR), and Reverse/Forward (RF) orientations respectively in blue and red, supporting the SSC orientation relative to the LSC as displayed in the assembled *Musa* chloroplast sequence. **B**, BESs mapping with the Forward/Forward (FF), and Reverse/Reverse (RR) orientations respectively in purple and green, supporting the presence of another form relative to the orientation of the SSC vs LSC in *M. acuminata*.

(see [43] for example). Additional taxa sampling and coalescence-based analyses will be required to resolve this conflict.

Structural Comparison within Angiosperms

The *M. acuminata* chloroplast genome structure was compared to other angiosperms. Major chloroplast genome structural events (gene losses, IR expansion/contraction and structural rearrangements) and inferred scenarios impacting several Monocotyledons clades are reported on the phylogenetic analysis in Figure 3 (for details on each species and basal angiosperms see Figure S1).

Gene content. The *infA* gene has been lost through multiple independent events from at least 24 Angiosperm lineage chloroplast genomes [5]. It is present in the *Musa acuminata* chloroplast genome as well as other Monocotyledons studied so far to the exception of the Alismatale lineage (Figure 3 and [44]). The *accD*, *ycf1* and *ycf2* genes are annotated as functional genes in the *Musa* chloroplastic genome while they have been lost in Poaceae cp genomes [6]. In addition *clpP* and *rpoC1* introns, found in the *Musa*

chloroplast genome, have been lost in Poaceae with the exception of the basal Poales *Anomochloa* [45]. The *AccD* and *ycf1* genes have also been lost in the Acoraceae and Orchidaceae. The *ndhB*, *ndhJ*, *ndhC*, *ndhK*, *ndhD*, *ndhF*, *ndhA*, *ndhH*, *ndhG* genes lost in all Orchidaceae genomes sequenced [46–48] are all annotated functional in the *Musa* chloroplastic genome as for the *rps16* gene lost in *Dioscorea elephantipes* [49].

Structural rearrangements. Dot plot analysis showed that *Musa* chloroplastic genome organization is similar to those found within most angiosperms. The *M. acuminata* chloroplast genome does not present the major structural rearrangement of 30 kb found in Poaceae [6,50–53]. Relative to the *M.* chloroplast genome, this rearrangement consisted in two inversions of 25 and 1 kb respectively and a translocation of about 5 kb all located in the same region.

IR expansion/contraction. The most derived chloroplast genome sequenced of Araceae, Bambusoideae, Poideae, Ehrhartoideae, Panicoideae and Anomochlooideae show events of IR/

Table 3. Chloroplast genome insertion into the nuclear genome in *Musa acuminata*.

Chromosomes	Nb_hits	Nb_bases	Proportion (%) of all cp insertions	Proportion (%) of the chromosome relative to the total nuclear DNA
chr01	57	10,468	5.49	8.32
chr02	24	4,637	2.43	6.65
chr03	44	9,629	5.05	9.19
chr04	63	17,752	9.31	9.06
chr05	39	6,993	3.67	8.86
chr06	83	21,129	11.08	10.53
chr07	42	10,838	5.68	8.63
chr08	44	10,482	5.50	10.69
chr09	48	11,125	5.83	10.30
chr10	60	13,659	7.16	10.16
chr11	59	17,779	9.32	7.70

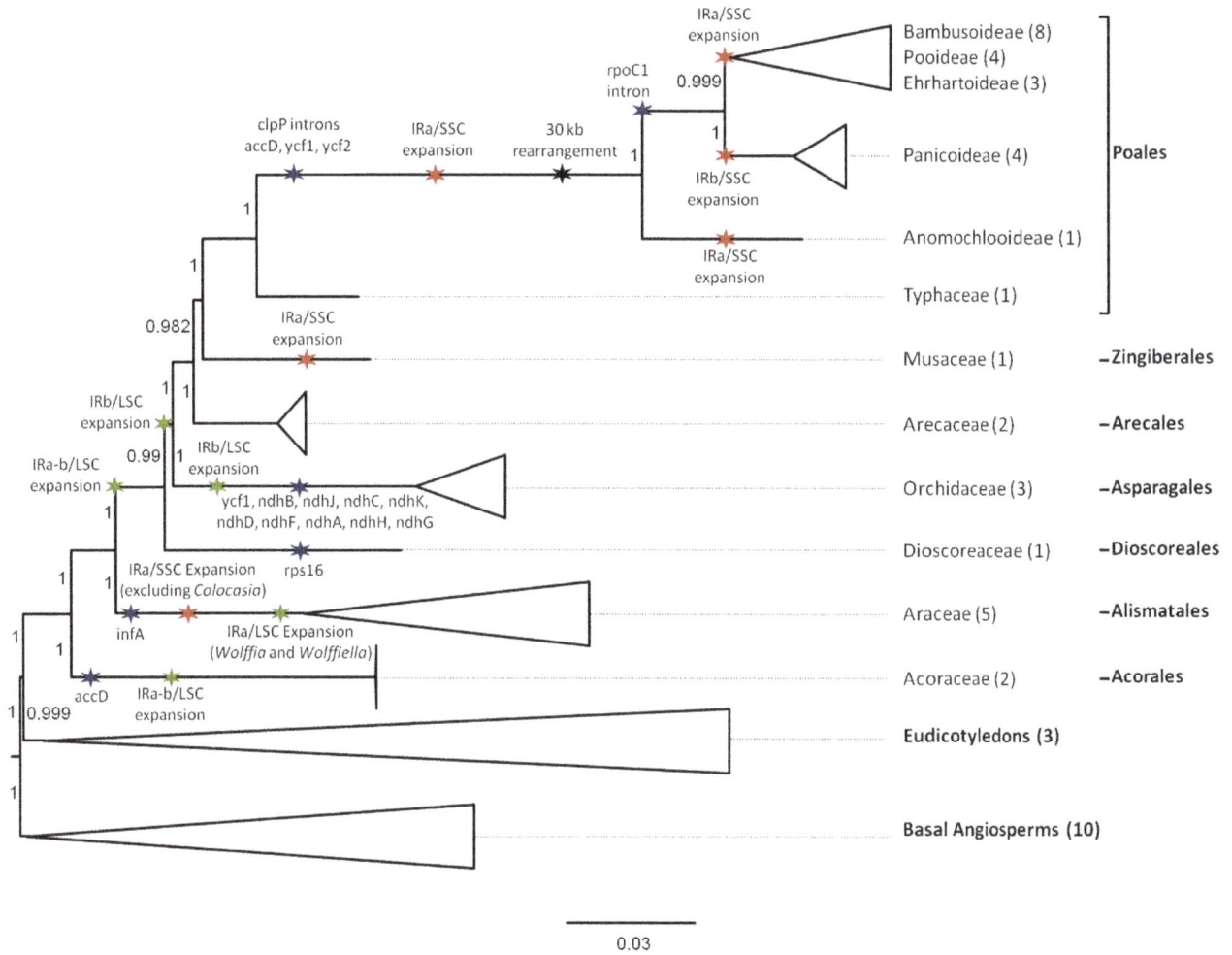

Figure 3. Condensed tree based on the maximum likelihood phylogenetic analysis constructed on 79 chloroplast protein coding genes of 10 basal angiosperms, 35 monocotyledons and 3 dicotyledons. The tree has a -lnL of −527912.066159. Support values for ML are provided at the nodes. Gene losses in all members of the different clades are indicated with blue stars. Putative events of IR expansions/contractions in the monocots are indicated with red and green stars for IR/SSC and IR/LSC boundaries respectively. Major structural rearrangements are indicated with black stars. Numbers indicate aLRT branches support.

SSC expansion relative to *Amborella trichopoda*. IRs of the *Musa acuminata* chloroplast genome show an extreme extension that includes two additional genes (*rps15* and *ndhH*) plus the full sequence of *ycf1* and 1030 bp of the *ndhA* gene relative to the IR structure of *Amborella trichopoda*. The expansion is made at the IRa/SSC junction and is the largest observed in monocots. In all other monocot groups where IR/SSC expansion is observed, except for the Panicoideae group, the expansion has occurred only at the IRa/SSC junction. The result of these expansions is the inclusion of the whole sequences of *ycf1* and *rps15* in the IRs and a part of the *ndhH* gene except for the Araceae group where *ndhH* is not always included. In the Panicoideae, the IRs contain *rsp15*, and a part of the *ndhF* gene suggesting that the IR/SSC extension has been made in two steps: first an IRa/SSC extension that has included *rps15* in the IR and a second step with an IRb/SSC extension including a part of the *ndhF* gene in the IR. These results suggest that an event of IRa/SSC extension has occurred prior to the divergence of Anomochlooideae, Bambusoideae, Poideae, Ehrhartoideae and Panicoideae including *rps15* gene in the IRs. After the divergence, Anomoclooideae group and Bambusoideae/Poideae/Ehrhartoideae group have been subjected to independent

additional IRa/SSC extension to include a part of *ndhH* in the IRs while Panicoideae have been subjected to IRb/SSC extension to include a part of the *ndhF* gene in the IRs. This scenario is similar to the one proposed by Guisinger et al. [6] but it adds the independents events of secondary IRa/SSC expansion in Anomochlooideae and the Bambusoideae/Pooideae/Ehrhatoideae group. This secondary IRa/SSC expansion provides further support for the sister relationship between Bambusoideae, Pooideae and Ehrhatoidea. The phylogenetic position of *M. acuminata* relative to Typhaceae at the basis of Poales and the chloroplastic structure of *Typha* showing no event of IR/SSC expansion suggest that *M. acuminata* has been subjected to an independent event of IRa/SSC expansion relative to Poales. Further investigation should be conducted to determine if this event is common to other Musaceae and the Zingiberales. In the Araceae the most derived species show an IRa/SSC expansion while the basal species *Colocasia esculenta* and the sister group Acoraceae and the Dioscoreaceae, Orchidaceae and Arecaceae do not show IR/SSC expansion. This suggests that this event of IRa/SSC expansion in the Araceae is another independent event. To summarize, three major IRa/SSC expansions may have occurred

Table 4. Number of alleles detected within the Musaceae, Eumusa, *M. acuminata* ssp (M. a.) and within the chloroplastic group II samples.

Markers	Musaceae (10)	Eumusa (9)	M. a (8)	cp group II (6)
mMaClRcp01	4	4	4	3
mMaClRcp02	2	2	2	2
mMaClRcp19	4	3	2	1
mMaClRcp20	5	4	3	3
mMaClRcp25	4	4	4	4
mMaClRcp27	4	3	3	2
mMaClRcp29	4	3	2	1
mMaClRcp30	5	4	3	2
mMaClRcp31	3	3	2	1
mMaClRcp32	4	3	3	3
mMaClRcp33	5	5	5	4
mMaClRcp34	4	3	2	2
Average per marker	4.00	3.42	2.92	2.33

The number of accession tested for each group is in parenthesis.

Overview of the Short Tandem Repeats Landscape

Short tandem repeats (also named Simple sequence repeats (SSR)) can exhibit high variation within the same species and are thus considered valuable markers for population genetics [55,56] and phylogenetic analyses [57]. A total of 112 SSRs were detected in the *Musa* chloroplast genome. Among them, 54 are microsatellites (mono-, di-, tri-, tetra-, penta-, and hexanucleotide repeats) and 58 are minisatellites (unit size ≥10). Minisatellites detected have a unit repeat mean length of 20.8 bp with a minimum of 11 bp and a maximum of 43 bp. The most repeated minisatellite has 14 units of 30 bp repeated tandemly. Among the microsatellites, 39 are exclusively constituted of A/T nucleotides while only one microsatellite is exclusively constituted of C/G nucleotides. Fourteen microsatellites are a mixture of puric and pyrimidic bases.

Sixteen of the homopolymer loci contain multiple A or T nucleotides while only one contains multiple G or C nucleotides. This higher proportion of poly(A)/(T) relative to poly(G)/(C) has also been reported in Poceae [57] and more divergent species such as *Panax ginseng* and *Nicotiana tabacum* [58], *Cucumis sativus* [59], *Magnolia kwangsiensis* [60], *Megaleranthis saniculifolia* [61] or *Sesamum indicum* [62]. However *P. ginseng* and *S. indicum* showed a slightly higher proportion of poly(G)/(C). In *Musa*, among the 10 dinucleotide repeat loci found, 8 are multiple AT or TA and 2 are multiple GA or AG. In Poaceae and *M. saniculifolia*, AT and TA repeats are the most common but others forms are found while only multiple AT or TA are reported in *S. indicum*. In *Musa*, seven trinucleotide repeat loci, fourteen tetranucleotide, five pentanucleotide and one hexanucleotide are found. While in Poaceae tri-, tetra-, penta-, and hexanucleotide repeats are reported [57], no tetra-, penta- and hexanucleotide are reported in the eudicotyledon *S. indicum* and no hexanucleotide are reported in the eudicotyledon *M. saniculifolia*.

Musa Chloroplast PCR Markers

A total of 32 SSR (Table S3) were tested for their polymorphism within a sample of *Musa*. Seven markers appeared monomorphic and 25, 21 and 15 were polymorphic in Musaceae, Eumusa and within the *M. acuminata* sub-species, respectively. The 12 most polymorphic markers were further tested in a sample of six accessions belonging to the chloroplastic group II defined by Carreel et al. [16]. The number of haplotypes detected within our panel is presented in Table 4 for the 12 SSR markers. The average polymorphism level was 4.00, 3.42, 2.92 and 2.33 alleles per marker respectively in Musaceae, Eumusa, *M. acuminata* and within the chloroplastic group II. Among these 12 markers, 9 revealed polymorphism within the chloroplastic group II and showed from 2 to 4 alleles. This new set of chloroplastic PCR markers represents a new, fast and efficient tool for studying the diversity of bananas and the origin of cultivars. Most cultivated bananas are triploids derived from spontaneous hybridization between *M. acuminata* sub-species and a few other *Musa* species but their exact origin is still not completely understood [15,63]. Their high level of sterility complicates their use in breeding programs. In this context the identification of their fertile progenitors would be very useful for breeders.

Conclusion

We assembled, annotated and analyzed the complete chloroplast sequence of banana (*Musa acuminata* ssp *malaccensis*). This first Zingiberale chloroplast (cp) genome was compared to other available monocotyledon cp genomes, providing new insight in their evolution. IR/SSC expansion is particularly pronounced in banana and has occurred independently several times within

independently in monocotyledons, one in the Araceae, one in Musaceae and one in the Poaceae. Three secondary independent events of IR/SSC expansion in the Poaceae have occurred, an IRa/SSC expansion in Anomochlooideae, an IRa/SSC expansion in Bambusoideae/Pooideae/Ehrhartoideae group and an IRb/SSC expansion in Panicoideae.

All Monocots sequenced except the most basal Araceae show events of IR/LSC expansion relative to *A. trichopoda* (Figure 3). The Acoraceae and Dioscoreacea display the insertion of the *trnH-GUG* gene at the IRa/LSC boundary and a partial copy of the *rps19* gene at the IRb/LSC boundary. The most derived plastid genomes sequenced of Araceae (*Wolffia australiana* and *Wolffiella lingulata*) only display a partial expansion of the IR including a partial copy of the *rps19* gene at the IRb/LSC boundary. All sequenced plastid genomes belonging to the sister group of the Dioscoreacea (Poales, Zingiberales, Arecales and Asparagales) present the insertion of complete *trnH-GUG* and *rps19* genes located in the LSC of *Amborella* at the IRa/LSC and IRb/LSC boundaries respectively. Asparagales show an additional IRb/LSC expansion as all their whole cp genome sequenced includes a partial copy of the *rpl22* gene. The relative order of *trnH-GUG* and *rps19* genes in IR suggests that in Acoraceae, Dioscoreaceae, Poales, Zingiberales, Arecales and Asparagales the expansion has been made in two steps as proposed by Mardanov et al. [44]: an IRa/LSC expansion leading to the inclusion of the *trnH-GUG* gene in the IR followed with an IRb/LSC expansion leading to the total or partial inclusion of the *rps19* gene in the IR, depending of the group. The structure of the IR/LSC boundary observed in the different clades can be explained by three independent events of IRa-b/LSC expansion, one in Acoraceae, one in the most derived Araceae and one at the basis of the Dioscoreaceae, Poales, Zingiberales, Arecales and Asparagales group. A second round of IRb/LSC expansion has taken place in the last group excluding the Dioscoreaceae leading to the complete inclusion of the *rps19* gene in the IR. A third round of expansion of IRb/LSC expansion has taken place in Asparagales leading to the partial inclusion of *rpl22* gene in the IR as it has been proposed in Wang et al. [54].

monocotyledons. The availability of new chloroplast markers within *Musa* opens new perspective to refine the phylogeny of *Musa* and the origin of cultivated triploid bananas.

Supporting Information

Figure S1   Maximum likelihood phylogenetic analysis based on 79 chloroplast protein coding genes of 45 basal angiosperms and monocotyledons and 3 Dicotyledons. The tree has a -lnL of -527912.066159. Support values for ML are provided at the nodes. Gene losses in chloroplast genomes are indicated with red triangles. Green and red stars represent partial or total IR gain of genes belonging respectively to LSC or SSC relative to *A. trichopoda* structure. Green and red minus signs represent loss of one of the two partial or complete gene copies belonging to IR respectively to become member of LSC or SSC relative to *A. trichopoda* structure.

Acknowledgments

We thank the SouthGreen Bioinformatics Platform – UMR AGAP - CIRAD (http://southgreen.cirad.fr) for providing us with computational resources. We thank Dr Jim Leebens-Mack for critical reading of the manuscript.

Author Contributions

Conceived and designed the experiments: GM FCB AD. Performed the experiments: GM FCB CC JMA. Analyzed the data: GM FCB. Wrote the paper: GM FCB AD.

References

1. Bendich AJ (2004) Circular Chloroplast Chromosomes: The Grand Illusion. The Plant Cell Online 16: 1661–1666. doi:10.1105/tpc.160771.
2. Chumley TW, Palmer JD, Mower JP, Fourcade HM, Calie PJ, et al. (2006) The Complete Chloroplast Genome Sequence of Pelargonium × hortorum: Organization and Evolution of the Largest and Most Highly Rearranged Chloroplast Genome of Land Plants. Molecular Biology and Evolution 23: 2175–2190. doi:10.1093/molbev/msl089.
3. Palmer JD (1991) Plastid chromosomes: structure and evolution. In: Bogorad L, Vasil I, editors. Cell Culture and Somatic Cell Genetics of Plants. San Diego: Academic Press. 5–53.
4. Raubeson LA, Jansen RK (2005) Chloroplast genomes of plants. In: Henry RJ, editor. Plant diversity and evolution: genotypic and phenotypic variation in higher plants. Cambridge: CAB International. 45–68.
5. Millen RS, Olmstead RG, Adams KL, Palmer JD, Lao NT, et al. (2001) Many Parallel Losses of infA from Chloroplast DNA during Angiosperm Evolution with Multiple Independent Transfers to the Nucleus. The Plant Cell Online 13: 645–658. doi:10.1105/tpc.13.3.645.
6. Guisinger M, Chumley T, Kuehl J, Boore J, Jansen R (2010) Implications of the Plastid Genome Sequence of Typha (Typhaceae, Poales) for Understanding Genome Evolution in Poaceae. J Mol Evol 70: 149–166. doi:10.1007/s00239-009-9317-3.
7. Downie SR, Palmer JD (1992) Use of chloroplast DNA rearrangements in reconstructing plant phylogeny. In: Soltis PS, Soltis DE, Doyle JJ, editors. Molecular systematics of plants. New York: Chapman and Hall. 14–35.
8. Jansen RK, Cai Z, Raubeson LA, Daniell H, dePamphilis CW, et al. (2007) Analysis of 81 genes from 64 plastid genomes resolves relationships in angiosperms and identifies genome-scale evolutionary patterns. Proceedings of the National Academy of Sciences 104: 19369–19374. doi:10.1073/pnas.0709121104.
9. Givnish TJ, Ames M, McNeal JR, McKain MR, Steele PR, et al. (2010) Assembling the Tree of the Monocotyledons: Plastome Phylogeny and Evolution of Poales1. Annals of the Missouri Botanical Garden 97: 584–616. doi:10.3417/2010023.
10. Logacheva MD, Penin AA, Samigullin TH, Vallejo-Roman CM, Antonov AS (2007) Phylogeny of flowering plants by the chloroplast genome sequences: in search of a "lucky gene." Biochemistry Moscow 72: 1324–1330. doi:10.1134/S0006297907120061.
11. D'Hont A, Denoeud F, Aury J-M, Baurens F-C, Carreel F, et al. (2012) The banana (Musa acuminata) genome and the evolution of monocotyledonous plants. Nature 488: 213–217. doi:10.1038/nature11241.
12. Fauré S, Noyer J-L, Carreel F, Horry J-P, Bakry F, et al. (1994) Maternal inheritance of chloroplast genome and paternal inheritance of mitochondrial genome in bananas (Musa acuminata). Curr Genet 25: 265–269. doi:10.1007/BF00357172.
13. Simmonds NW (1962) The evolution of the bananas. London: Longmans.
14. De Langhe E, Hřibová E, Carpentier S, Doležel J, Swennen R (2010) Did backcrossing contribute to the origin of hybrid edible bananas? Annals of Botany 106: 849–857. doi:10.1093/aob/mcq187.
15. Perrier X, De Langhe E, Donohue M, Lentfer C, Vrydaghs L, et al. (2011) Multidisciplinary perspectives on banana (Musa spp.) domestication. Proceedings of the National Academy of Sciences 108: 11311–11318. doi:10.1073/pnas.1102001108.
16. Carreel F, de Leon DG, Lagoda P, Lanaud C, Jenny C, et al. (2002) Ascertaining maternal and paternal lineage within Musa by chloroplast and mitochondrial DNA RFLP analyses. Genome 45: 679–692.
17. Boonruangrod R, Desai D, Fluch S, Berenyi M, Burg K (2008) Identification of cytoplasmic ancestor gene-pools of Musa acuminata Colla and Musa balbisiana Colla and their hybrids by chloroplast and mitochondrial haplotyping. Theor Appl Genet 118: 43–55. doi:10.1007/s00122-008-0875-3.
18. Lescot T (2011) The genetic diversity of the banana in figures. FruiTrop 189: 58–62.
19. Krzywinski M, Schein J, Birol İ, Connors J, Gascoyne R, et al. (2009) Circos: An information aesthetic for comparative genomics. Genome Research 19: 1639–1645. doi:10.1101/gr.092759.109.
20. Wyman SK, Jansen RK, Boore JL (2004) Automatic annotation of organellar genomes with DOGMA. Bioinformatics 20: 3252–3255. doi:10.1093/bioinformatics/bth352.
21. Yang M, Zhang X, Liu G, Yin Y, Chen K, et al. (2010) The Complete Chloroplast Genome Sequence of Date Palm (Phoenix dactylifera L.). PLoS ONE 5: e12762. doi:10.1371/journal.pone.0012762.
22. Uthaipaisanwong P, Chanprasert J, Shearman JR, Sangsrakru D, Yoocha T, et al. (2012) Characterization of the chloroplast genome sequence of oil palm (Elaeis guineensis Jacq.). Gene 500: 172–180. doi:10.1016/j.gene.2012.03.061.
23. Lowe TM, Eddy SR (1997) tRNAscan-SE: A Program for Improved Detection of Transfer RNA Genes in Genomic Sequence. Nucleic Acids Research 25: 0955–0964. doi:10.1093/nar/25.5.0955.
24. Charif D, Lobry J (2007) SeqinR 1.0–2: A Contributed Package to the R Project for Statistical Computing Devoted to Biological Sequences Retrieval and Analysis. In: Bastolla U, Porto M, Roman HE, Vendruscolo M, editors. Structural Approaches to Sequence Evolution. Biological and Medical Physics, Biomedical Engineering. Springer Berlin Heidelberg. 207–232. Available: http://dx.doi.org/10.1007/978-3-540-35306-5_10.
25. The Tomato Genome Consortium (2012) The tomato genome sequence provides insights into fleshy fruit evolution. Nature 485: 635–641. doi:10.1038/nature11119.
26. Katoh K, Misawa K, Kuma K, Miyata T (2002) MAFFT: a novel method for rapid multiple sequence alignment based on fast Fourier transform. Nucleic Acids Research 30: 3059–3066. doi:10.1093/nar/gkf436.
27. Posada D (2008) jModelTest: Phylogenetic Model Averaging. Molecular Biology and Evolution 25: 1253–1256. doi:10.1093/molbev/msn083.
28. Guindon S, Dufayard J-F, Lefort V, Anisimova M, Hordijk W, et al. (2010) New Algorithms and Methods to Estimate Maximum-Likelihood Phylogenies: Assessing the Performance of PhyML 3.0. Systematic Biology 59: 307–321. doi:10.1093/sysbio/syq010.
29. Goremykin VV, Hirsch-Ernst KI, Wölfl S, Hellwig FH (2003) Analysis of the Amborella trichopoda Chloroplast Genome Sequence Suggests That Amborella Is Not a Basal Angiosperm. Molecular Biology and Evolution 20: 1499–1505. doi:10.1093/molbev/msg159.
30. Thiel T, Michalek W, Varshney R, Graner A (2003) Exploiting EST databases for the development and characterization of gene-derived SSR-markers in barley (Hordeum vulgare L.). Theor Appl Genet 106: 411–422. doi:10.1007/s00122-002-1031-0.
31. Krumsiek J, Arnold R, Rattei T (2007) Gepard: a rapid and sensitive tool for creating dotplots on genome scale. Bioinformatics 23: 1026–1028. doi:10.1093/bioinformatics/btm039.

32. Rozen S, Skaletsky H (2000) Primer3 on the WWW for general users and for biologist programmers. In: Krawetz S, Misener S, editors. Bioinformatics Methods and Protocols in the series Methods in Molecular Biology. Totowa: Humana Press. 365–386.

33. Clegg MT, Gaut BS, Learn GH, Morton BR (1994) Rates and patterns of chloroplast DNA evolution. Proceedings of the National Academy of Sciences 91: 6795–6801.

34. Palmer JD (1983) Chloroplast DNA exists in two orientations. Nature 301: 92–93. doi:10.1038/301092a0.

35. Oldenburg DJ, Bendich AJ (2004) Most Chloroplast DNA of Maize Seedlings in Linear Molecules with Defined Ends and Branched Forms. Journal of Molecular Biology 335: 953–970. doi:10.1016/j.jmb.2003.11.020.

36. Shahmuradov I, Akbarova Y, Solovyev V, Aliyev J (2003) Abundance of plastid DNA insertions in nuclear genomes of rice and Arabidopsis. Plant Mol Biol 52: 923–934. doi:10.1023/A:1025472709537.

37. Cullis CA, Vorster BJ, Van Der Vyver C, Kunert KJ (2009) Transfer of genetic material between the chloroplast and nucleus: how is it related to stress in plants? Annals of Botany 103: 625–633. doi:10.1093/aob/mcn173.

38. Matsuo M, Ito Y, Yamauchi R, Obokata J (2005) The Rice Nuclear Genome Continuously Integrates, Shuffles, and Eliminates the Chloroplast Genome to Cause Chloroplast–Nuclear DNA Flux. The Plant Cell Online 17: 665–675. doi:10.1105/tpc.104.027706.

39. Shulaev V, Sargent DJ, Crowhurst RN, Mockler TC, Folkerts O, et al. (2011) The genome of woodland strawberry (Fragaria vesca). Nat Genet 43: 109–116. doi:10.1038/ng.740.

40. Degnan JH, Rosenberg NA (2009) Gene tree discordance, phylogenetic inference and the multispecies coalescent. Trends in Ecology & Evolution 24: 332–340. doi:10.1016/j.tree.2009.01.009.

41. Hendy MD, Penny D (1989) A Framework for the Quantitative Study of Evolutionary Trees. Systematic Biology 38: 297–309. doi:10.2307/2992396.

42. Bergsten J (2005) A review of long-branch attraction. Cladistics 21: 163–193. doi:10.1111/j.1096-0031.2005.00059.x.

43. Renoult J, Kjellberg F, Grout C, Santoni S, Khadari B (2009) Cyto-nuclear discordance in the phylogeny of Ficus section Galoglychia and host shifts in plant-pollinator associations. BMC Evolutionary Biology 9: 248.

44. Mardanov A, Ravin N, Kuznetsov B, Samigullin T, Antonov A, et al. (2008) Complete Sequence of the Duckweed (Lemna minor) Chloroplast Genome: Structural Organization and Phylogenetic Relationships to Other Angiosperms. J Mol Evol 66: 555–564. doi:10.1007/s00239-008-9091-7.

45. Morris LM, Duvall MR (2010) The chloroplast genome of Anomochloa marantoidea (Anomochlooideae; Poaceae) comprises a mixture of grass-like and unique features. American Journal of Botany 97: 620–627. doi:10.3732/ajb.0900226.

46. Chang C-C, Lin H-C, Lin I-P, Chow T-Y, Chen H-H, et al. (2006) The Chloroplast Genome of Phalaenopsis aphrodite (Orchidaceae): Comparative Analysis of Evolutionary Rate with that of Grasses and Its Phylogenetic Implications. Molecular Biology and Evolution 23: 279–291. doi:10.1093/molbev/msj029.

47. Wu F-H, Chan M-T, Liao D-C, Hsu C-T, Lee Y-W, et al. (2010) Complete chloroplast genome of Oncidium Gower Ramsey and evaluation of molecular markers for identification and breeding in Oncidiinae. BMC Plant Biol 10: 1–12. doi:10.1186/1471-2229-10-68.

48. Jheng C-F, Chen T-C, Lin J-Y, Chen T-C, Wu W-L, et al. (2012) The comparative chloroplast genomic analysis of photosynthetic orchids and developing DNA markers to distinguish Phalaenopsis orchids. Plant Science 190: 62–73. doi:10.1016/j.plantsci.2012.04.001.

49. Hansen DR, Dastidar SG, Cai Z, Penaflor C, Kuehl JV, et al. (2007) Phylogenetic and evolutionary implications of complete chloroplast genome sequences of four early-diverging angiosperms: Buxus (Buxaceae), Chloranthus (Chloranthaceae), Dioscorea (Dioscoreaceae), and Illicium (Schisandraceae). Molecular Phylogenetics and Evolution 45: 547–563. doi:10.1016/j.ympev.2007.06.004.

50. Doyle JJ, Davis JI, Soreng RJ, Garvin D, Anderson MJ (1992) Chloroplast DNA inversions and the origin of the grass family (Poaceae). Proceedings of the National Academy of Sciences 89: 7722–7726. doi:10.1073/pnas.89.16.7722.

51. Hiratsuka J, Shimada H, Whittier R, Ishibashi T, Sakamoto M, et al. (1989) The complete sequence of the rice (Oryza sativa) chloroplast genome: Intermolecular recombination between distinct tRNA genes accounts for a major plastid DNA inversion during the evolution of the cereals. Molec Gen Genet 217: 185–194. doi:10.1007/BF02464880.

52. Howe C, Barker R, Bowman C, Dyer T (1988) Common features of three inversions in wheat chloroplast DNA. Curr Genet 13: 343–349. doi:10.1007/BF00424430.

53. Katayama H, Ogihara Y (1993) Structural alterations of the chloroplast genome found in grasses are not common in monocots. Curr Genet 23: 160–165. doi:10.1007/BF00352016.

54. Wang R-J, Cheng C-L, Chang C-C, Wu C-L, Su T-M, et al. (2008) Dynamics and evolution of the inverted repeat-large single copy junctions in the chloroplast genomes of monocots. BMC Evol Biol 8: 1–14. doi:10.1186/1471-2148-8-36.

55. Terrab A, Paun O, Talavera S, Tremetsberger K, Arista M, et al. (2006) Genetic diversity and population structure in natural populations of Moroccan Atlas cedar (Cedrus atlantica; Pinaceae) determined with cpSSR markers. American Journal of Botany 93: 1274–1280. doi:10.3732/ajb.93.9.1274.

56. Grassi F, Labra M, Scienza A, Imazio S (2002) Chloroplast SSR markers to assess DNA diversity in wild and cultivated grapevines. Vitis.

57. Melotto-Passarin D, Tambarussi E, Dressano K, De Martin V, Carrer H (2011) Characterization of chloroplast DNA microsatellites from Saccharum spp and related species. Genet Mol Res 10: 2024–2033.

58. Kim K-J, Lee H-L (2004) Complete Chloroplast Genome Sequences from Korean Ginseng (Panax schinseng Nees) and Comparative Analysis of Sequence Evolution among 17 Vascular Plants. DNA Research 11: 247–261. doi:10.1093/dnares/11.4.247.

59. Kim J-S, Jung J, Lee J-A, Park H-W, Oh K-H, et al. (2006) Complete sequence and organization of the cucumber (Cucumis sativus L. cv. Baekmibaekdadagi) chloroplast genome. Plant Cell Rep 25: 334–340. doi:10.1007/s00299-005-0097-y.

60. Kuang D-Y, Wu H, Wang Y-L, Gao L-M, Zhang S-Z, et al. (2011) Complete chloroplast genome sequence of Magnolia kwangsiensis (Magnoliaceae): implication for DNA barcoding and population genetics. Genome 54: 663–673.

61. Kim Y-K, Park C, Kim K-J (2009) Complete chloroplast DNA sequence from a Korean endemic genus, Megaleranthis saniculifolia, and its evolutionary implications. Mol Cells 27: 365–381. doi:10.1007/s10059-009-0047-6.

62. Yi D-K, Kim K-J (2012) Complete Chloroplast Genome Sequences of Important Oilseed Crop Sesamum indicum L. PLoS ONE 7: e35872. doi:10.1371/journal.pone.0035872.

63. Raboin LM, Carreel F, Noyer J-L, Baurens F-C, Horry JP, et al. (2005) Diploid ancestors of triploid export banana cultivars: molecular identification of 2n restitution gamete donors and n gamete donors. Molecular Breeding 16: 333–341.

2

The Complete Genome Sequence of *Escherichia coli* EC958: A High Quality Reference Sequence for the Globally Disseminated Multidrug Resistant *E. coli* O25b:H4-ST131 Clone

Brian M. Forde[1], **Nouri L. Ben Zakour**[1], **Mitchell Stanton-Cook**[1], **Minh-Duy Phan**[1], **Makrina Totsika**[1], **Kate M. Peters**[1], **Kok Gan Chan**[2], **Mark A. Schembri**[1], **Mathew Upton**[3], **Scott A. Beatson**[1]*

1 Australian Infectious Diseases Research Centre, School of Chemistry & Molecular Biosciences, The University of Queensland, Queensland, Australia, 2 Division of Genetics and Molecular Biology, Institute of Biological Sciences, Faculty of Science, University of Malaya, Kuala Lumpur, Malaysia, 3 Plymouth University Peninsula Schools of Medicine and Dentistry, Plymouth, United Kingdom

Abstract

Escherichia coli ST131 is now recognised as a leading contributor to urinary tract and bloodstream infections in both community and clinical settings. Here we present the complete, annotated genome of *E. coli* EC958, which was isolated from the urine of a patient presenting with a urinary tract infection in the Northwest region of England and represents the most well characterised ST131 strain. Sequencing was carried out using the Pacific Biosciences platform, which provided sufficient depth and read-length to produce a complete genome without the need for other technologies. The discovery of spurious contigs within the assembly that correspond to site-specific inversions in the tail fibre regions of prophages demonstrates the potential for this technology to reveal dynamic evolutionary mechanisms. *E. coli* EC958 belongs to the major subgroup of ST131 strains that produce the CTX-M-15 extended spectrum β-lactamase, are fluoroquinolone resistant and encode the *fimH30* type 1 fimbrial adhesin. This subgroup includes the Indian strain NA114 and the North American strain JJ1886. A comparison of the genomes of EC958, JJ1886 and NA114 revealed that differences in the arrangement of genomic islands, prophages and other repetitive elements in the NA114 genome are not biologically relevant and are due to misassembly. The availability of a high quality uropathogenic *E. coli* ST131 genome provides a reference for understanding this multidrug resistant pathogen and will facilitate novel functional, comparative and clinical studies of the *E. coli* ST131 clonal lineage.

Editor: Ulrich Dobrindt, University of Münster, Germany

Funding: This work was supported by grants from the Australian National Health and Medical Research Council to MAS and SAB (APP1012076 and APP1067455) and a University of Malaya HIR Grant to KGC (UM-MOHE HIR Grant UM.C/625/1/HIR/MOHE/CHAN/14/1). MAS is supported by an Australian Research Council (ARC) Future Fellowship (FT100100662). MT is supported by an ARC Discovery Early Career Researcher Award (DE130101169). The funders had no role in study design, data collection and analysis, decision to publish, or preparation of the manuscript.

Competing Interests: The authors have declared that no competing interests exist.

* Email: s.beatson@uq.edu.au

Introduction

Many multidrug resistant (MDR) *Escherichia coli* strains belong to specific clones that are frequently isolated from urinary tract and bloodstream infections. These clones may originate in a specific locale, country or may be distributed globally without a clear place of origin. A major contributor to this phenomenon is *E. coli* ST131, a group of *E. coli* strains of multi-locus sequence type 131 (ST131) that have emerged rapidly and disseminated globally in hospitals and the community, causing MDR infections typically associated with frequent recurrences and limited treatment options [1–4]. *E. coli* ST131 strains are commonly identified among *E. coli* producing the CTX-M-15 type extended-spectrum β-lactamase (ESBL), currently the most widespread CTX-M ESBL enzyme worldwide [1,4,5]. The largest sub-clonal lineage of *E. coli* ST131 is resistant to fluoroquinolones and belongs to the *fimH*-based *H*30 group [6].

E. coli EC958 represents one of the most well characterised *E. coli* ST131 strains in the literature. *E. coli* EC958 is a phylogenetic group B2, CTX-M-15 positive, fluoroquinolone resistant, *H*30 *E. coli* ST131 strain isolated from the urine of an 8-year old girl presenting in the community in March 2005 in the United Kingdom (UK) [7]. The strain belongs to the pulse field gel electrophoresis defined UK epidemic strain A and has a O25b:H4 serotype [8]. *E. coli* EC958 contains multiple genes associated with the virulence of extra-intestinal *E. coli*, including those encoding adhesins, autotransporter proteins and siderophore receptors. *E. coli* EC958 expresses type 1 fimbriae and this is required for adherence to and invasion of human bladder cells, as well as colonization of the mouse bladder [7]. In mice, *E. coli* EC958 causes acute and chronic urinary tract infection (UTI) [9], as well as impairment of ureter contractility [10]. *E. coli* EC958 bladder infection follows a well-defined pathogenic pathway that involves the formation of intracellular bacterial communities (IBCs) in superficial epithelial cells and the subsequent release of rod-shaped and filamentous bacteria into the bladder lumen [9]. *E. coli* EC958 also causes impairment of uterine contractility [10], and is resistant to the bactericidal action of human serum [11]. The complement of genes that define the serum resistome of *E. coli* EC958 have been comprehensively defined [11].

Second generation sequencing (SGS) technologies have revolutionised genome research through the provision of a rapid, cost-effective method for generating sequence data. However, obtaining complete bacterial genomes using these technologies has been challenging. Short read lengths are a characteristic feature of SGS technologies and highly repetitive stretches of DNA, often present in multiple copies, are difficult to correctly resolve using these platforms. Typically, these assemblies are highly fragmented, prone to misassembly and require costly and time consuming finishing procedures [12–14]. Consequently, most genomes are not completely resolved; they are submitted as draft genomes, often containing hundreds of contigs that are generally unannotated or poorly annotated [15]. As a result, many of these genomes are of limited use for comparative, functional, clinical and epidemiological studies [16]. In contrast to other methods, the Pacific Biosciences (PacBio) single molecule real time (SMRT) sequencing platform [17] can produce read lengths of up to 30,000 bp that are capable of spanning large repeat regions (such as rRNA operons), thereby facilitating the generation of complete genome assemblies without the need for additional sequencing.

In order to enhance our knowledge of *E. coli* ST131 and its capacity to cause disease, a greater understanding of this clone is required at the genomic level. Four complete or draft *E. coli* ST131 genome sequences are currently available, namely EC958 (draft) [7], SE15 [18], NA114 [19] and most recently JJ1886 [20]. EC958, NA114 and JJ1886 are all phylogroup B2, CTX-M-15 positive, fluoroquinolone resistant, *H*30 strains which have recently been shown in two independent phylogenomic studies to belong to single clade (ST131 clade C) distinct from SE15 (ST131 clade A) [6,21]. A pair-wise comparison between SE15 and NA114 demonstrated that SE15 contains a number of differences in genome content despite being closely related at the core genome level [22]. Furthermore, we have shown that many of the genomic islands and prophage regions previously identified in the draft EC958 genome [7] are well conserved in most other fluoroquinolone resistant, clade C/*fimH*30 strains [21]. Here we used PacBio SMRT sequencing to determine the complete genome sequence of *E. coli* EC958. The *E. coli* EC958 genome represents as an accurate reference for future functional, comparative, phylogenetic and clinical studies of *E. coli* ST131.

Methods

Genome sequencing and assembly

Genomic DNA for *E. coli* EC958 was prepared using the Qiagen DNeasy Blood and Tissue kit, as per manufacturer's instructions. The genome of *E. coli* EC958 was sequenced by generating a total of 601,224 pre-filtered reads with an average length of 1,600 bp, from six SMRT cells on a PacBio RS I sequencing instrument, using an 8–12 kilobase (kb) insert library, generating approximately 200-fold coverage (GATC Biotech AG, Germany).

De novo genome assemblies were produced using PacBio's SMRT Portal (v2.0.0) and the hierarchical genome assembly process (HGAP) [23], with default settings and a seed read cut-off length of 5,000 bp to ensure accurate assembly across *E. coli* rRNA operons. Assemblies were performed multiple times using different combinations of between one and six SMRT cells of read data. The best assembly results were obtained with six SMRT cells which yielded approximately 547 Mb of sequence from 190,145 post-filtered reads (Table 1). The average read length was found to be 2,875 bp with an average single pass accuracy of 86.5%. During the preassembly stage 190,145 long reads were converted into 23,772 high quality, preassembled reads with an average length of 4,573 bp. Assembly of these reads returned seven contigs, three were greater than 500 kb. Furthermore, the largest contig (~3.8 Mb) was estimated to contain 74.5% of the chromosome of EC958. For all other assemblies total contig numbers exceeded 10 (Table 1). However, for assemblies using two or three SMRT cells, assembly metrics could be improved > 2-fold by reducing the seed read length (Table 1).

To determine their correct order and orientation, contigs from our six SMRT cell assembly were aligned to the complete genome of *E. coli* SE15 using Mauve v. 2.3.1 [24]. Contig ordering was confirmed by PCR. Overlapping but un-joined contigs, a characterised artefact of the HGAP assembly process [23], were manually trimmed based on sequence similarity and joined. All joins were manually inspected using ACT [25] and Contiguity (http://mjsull.github.io/Contiguity/).

A single contig representing the EC958 large plasmid pEC958 was identified and isolated by BLASTn comparison against the previous draft assembly of EC958 (NZ_CAFL00000000.1) [7]. Overlapping sequences on the 5′ and 3′ ends of the plasmid contig were then manually trimmed based on sequence similarity. Although the EC958 small plasmid (pEC958B) was too small to be assembled as part of the main assembly, 25 unassembled PacBio reads, with an average length of 2,031 bp, were found to align to the small 4,080 bp plasmid contig that had previously been assembled from 454 GS-FLX reads (emb|CAFL01000138).

To determine if reads containing unremoved adapter sequence have had an impact on the assembly of EC958 we first screened the filtered subreads for adapter sequence using BBMap version 31.40 (http://sourceforge.net/projects/bbmap/). A high level of adapter contamination would likely pose some risk of misassembly. Additionally, to eliminate the possibility that aberrant reads have resulted in the inclusion of assembly artefacts in the EC958 genome assembly, contig-ends were screened for hairpin artefacts using MUMmer version 3.23 [26].

Genome annotation and comparison

Initial annotation of the genome of EC958 was done by annotation transfer from the draft genome of EC958 (NZ_CAFL00000000.1) using the rapid annotation transfer tool (RATT) [27]. In addition, the genome of EC958 was subject to additional automatic annotation using Prokka (Prokka: Prokaryotic Genome Annotation System - http://vicbioinformatics.com/). All predicted protein coding sequences were searched (BLASTp) against the reannotated genome of *E. coli* UTI89 [28,29] with the aim of correcting CDS start sites and assigning correct gene names and an appropriate functional annotation. Whole genome nucleotide alignments for *E. coli* EC958, SE15 and NA114 were generated using BLASTn and visualised using Easyfig version 2.1 [30], Artemis Comparison Tool [25] and BRIG [31]. To compare the original 454 draft genome and the complete PacBio genome, 454 sequencing reads used for the draft assembly of *E. coli* EC958 [7] were mapped to the complete *E. coli* EC958 genome using SHRiMP v 2.0 [32]. SNP calling and insertion/deletion (indel) prediction were performed using the Nesoni package with default parameters (http://www.vicbioinformatics.com/software.nesoni.shtml). Additional platform-specific SNPs and indels were identified by comparison of the 454 draft genome contigs and the PacBio complete genome using MUMmer 3.23 [26]. The complete

Table 1. PacBio assembly statistics.

SMRT cells	Raw read data				Pre-assembly		Final assembly		
	Seed length[1]	Total reads	Total bases[2]	Average length[3]	Total bases[2]	Total reads	Assembly size[3]	Total contigs	N50
1	5	33736	89	2649	7	2381	1372346	162	8748
2	5	63802	177	2913	29	6244	5163106	154	56927
2	1.5	63802	177	2777	91	37720	5262395	44	225550
3	5	97231	286	2945	47	10407	5298899	40	216859
3	2.7	96187	268	2793	105	31531	5317490	20	594137
4	5	130044	383	2946	65	13934	5311243	18	1061190
4	3.5	125866	357	2844	108	27592	5314416	17	769937
5	5	159723	472	2958	81	17175	5320054	14	1100290
5	4.1	157332	449	2859	108	25345	5339571	16	710956
6	5	190145	546	2875	108	23772	5298989	7	3866706

[1]Kilobase-pairs;
[2]Megabase-pairs;
[3]Base-pairs.

annotated chromosome of EC958, large plasmid (pEC958A) and small plasmid (pEC958B) are available at the European Nucleotide Archive (ENA; http://www.ebi.ac.uk/ena) under the accession numbers HG941718, HG941719 and HG941720 respectively.

Phylogenetic analysis

To determine the phylogenetic relatedness of the four complete ST131 genomes, a single-nucleotide polymorphism (SNP) based phylogenetic tree was constructed. The pan-genome SNPs in EC958, 3 complete ST131 genomes (*E. coli* SE15, NA114 and JJ886), an additional 16 representative complete *E. coli* genomes: *E. coli* ED1A, CFT073, UTI89, 536, S88, APEC-01, IAI39, UMN026, HS, W3110, MG1655, BW2952, IAI1, SE11, Sakai, EDL933 [20,28,33–42] and the out-group species *E. fergusonii* ATCC35469 were identified using kSNP2 2.1.1 [43] (using default setting and a k-mer size of 21). In total, 261,214 SNPs were found to be common to all 21 *E. coli* genomes, including EC958. SNPs in each genome were concatenated into single contiguous sequences and aligned. The resulting SNP-based alignment was used for phylogenetic analysis. A maximum likelihood (ML) phylogenetic tree was constructed with PhyML 3.0 [44], using the GTR nucleotide substitution model and 1000 bootstrap replicates. The phylogenetic tree was plotted using FigTree 1.4.0 (http://tree.bio.ed.ac.uk/software/figtree/).

Genome assembly of EC958 using simulated Illumina paired-end reads

In an attempt to replicate the assembly protocol of *E. coli* NA114, simulated Illumina sequencing and assembly of *E. coli* EC958 was performed as described for *E. coli* NA114 in Avasthi et al [19]. The chromosome of EC958 was used as a reference to generate 500-fold coverage of simulated 54 bp, error free, Illumina paired-end reads with an average insert size of 300 bp. These simulated Illumina paired-end reads were then assembled using Velvet 1.2.7 [45]. Assembled contigs were ordered and orientated by aligning them to the genome of *E. coli* SE15 using Mauve and concatenated to produce a ~5 Mb pseudo-molecule.

Results

The complete PacBio genome assembly of *E. coli* EC958 reveals dynamic phage rearrangements

To determine the complete genome sequence of *E. coli* EC958 we carried out sequencing of genomic DNA using the PacBio RS I platform. An initial assembly of seven contigs representing the *E. coli* EC958 genome was produced by HGAP [21] using 190,145 post-filtered reads from 6 SMRT cells (Table 1). A circular chromosome was unambiguously assembled by trimming and joining the overlapping 3′ and 5′ ends from three large contigs of 3,866,718 bp, 715,826 bp and 541,428 bp, respectively. Contig joins were confirmed by PCR. Previously, we showed that a 14 scaffold draft 454 genome assembly of *E. coli* EC958 contained two additional replicons: a large antibiotic resistance plasmid (pEC958) and a small high-copy cryptic plasmid (pEC958B) [7]. In the PacBio assembly we found that pEC958 was represented as single circular contig of 135,602 bp that was consistent with the pEC958 scaffold in the original draft assembly (scaffold HG328349). In contrast, pEC958B was too small to be assembled using the HGAP parameters employed for rest of the chromosome, but it could be assembled from PacBio reads using a read-mapping approach.

The contig order and orientation in the original draft 454 assembly was contiguous with the complete PacBio assembly determined in this study. We also found a high degree of consensus concordance between the two technologies with only fifteen single nucleotide indels and a single substitution between the two assemblies, most of which could be accounted for by homopolymeric tract errors in the 454 assembly according to comparisons with independent *E. coli* genomes and manual read inspection (Table 2). We also noted two discrepant regions that exhibited a cluster of substitutions and indels in the GI-*leuX* genomic island and in the tail fibre region of prophage Phi1 that initially appeared to be PacBio assembly errors. Further investigation revealed that the GI-*leuX* discrepancies were within a 3727 bp repeat region also found within GI-*selC*, thus the differences were due to a collapsed repeat in the 454 assembly (Table 2). In contrast, the Phi1 prophage discrepancy corresponded to a 2773 bp segment in the tail fibre region that was also present in an inverted orientation within a separate 12.2 kb contig (Fig. 1A). This spurious contig resulted from the assembly of PacBio reads (approximately 50% of all reads in this region) that contained the 2.8 kb segment in an alternative orientation, suggesting that high-frequency allele switching had occurred during propagation of *E. coli* EC958 prior to DNA extraction. Prophage tail fibre allele switching mediated by a site-specific DNA invertase has long been recognised as a phenomenon for altering host specificity of phage by alternating in-frame C-terminal phage tail fibre protein fragments (for review see Sandmeier, 1994 [46]). Interestingly, we also identified PacBio contigs corresponding to alternative alleles of prophage tail fibre regions from prophage Phi2 and Phi4 that were separately assembled into 8.7 kb and 12.7 kb contigs, respectively, due to 2–3 kb inversions (Fig. 1B). SMRTbell adapter sequences were found to be present in only 620 of 217,502 subreads (0.29%). This low level of adapter contamination combined with the absence of any hairpin artefacts at contig break points make it highly unlikely that aberrant reads are responsible for the three small phage-associated contigs, and suggest these contigs represent real biological variation of tail fibre genes in the chromosome of EC958. All three invertible segments exhibited the 5′ and 3′ 26 bp crossover sites characteristic of DNA invertase mediated phage tail switching mechanisms [46] (Table 3).

E. coli EC958 general genome features

The genome of *E. coli* EC958 consists of a single circular chromosome of 5,109,767 bp with an average GC content of 50.7%. The chromosome encodes 4982 putative protein-coding genes, including 358 that were not previously annotated on the draft chromosome due their presence in repetitive regions that were not assembled as scaffolds. Seven rRNA loci, consisting of 16S, 23S and 5S rRNA genes, and 89 tRNA genes, representing all 20 amino acids, were identified on the chromosome. As described elsewhere [7], the virulence-associated gene complement of EC958 includes adhesins (e.g. *fimA-H*, *afa* and *curli*), autotransporters (e.g. *agn43*, *upaG*, *upaH*, *sat* and *picU*), iron receptors (e.g. *fepA*, *iutA*, *iha*, *chuA*, *hma* and *fyuA*) and a number of other virulence associated genes (e.g. *kpsM*, *usp*, *ompT*, *malX*). Four genes that were not annotated in the draft genome may be virulence related: *sitB* (EC958_5193), which encodes a component of an iron transport system that is up-regulated during *Shigella* intracellular growth [47]; and three hypothetical genes (EC958_4894, EC958_4977, EC958_4981) orthologous to genes previously identified as uropathogenic *E. coli* specific [48]. The EC958 large plasmid, pEC958, is predicted to contain 151 protein-coding genes, including a 22 kb locus encoding conjugal transfer (*tra*) genes and antibiotic resistance genes including *bla*$_{CTX-M-15}$ [7].

Whole genome comparison of *E. coli* EC958, NA114 and SE15

Phylogenetic analyses indicated that *E. coli* strains EC958, NA114 and JJ1886 cluster together in a clade discrete from *E. coli* SE15 within an ST131 specific lineage within the B2 phylogroup (Fig. 2). Whole-genome BLASTn comparisons showed that the major structural differences between the genomes of SE15 and the three *fimH30* ST131strains relate to the seven prophage loci (Phi1-Phi7) and four genomic islands (GI-*thrW*, GI-*pheV*, GI-*selC*, and GI-*leuX*) that were previously defined in the draft genome of *E. coli* EC958 [7] (Fig. 3A). The complete PacBio genome confirmed the position and size of these elements and was able to fill numerous gaps caused by insertion elements or other repetitive elements. These prophage and GI regions are absent in whole or in part from *E. coli* SE15, and from most of the 16 other *E. coli* representative strains surveyed (Fig. 3B). Additionally, GI-*selC* is largely absent from all ST131 strains except EC958, whereas GI-*thrW* and Phi7 are well conserved in all four ST131 strains (Fig. 3B). Genomic surveys with a greater number of ST131 strains from diverse origins will be necessary to determine the prevalence of prophage, genomic islands and other mobile genetic elements.

Large discrepancies between ST131 genomes are likely due to misassembly of *E. coli* NA114

At the core genome level EC958, NA114, JJ1886 and SE15 all display a high level of genome synteny, with major differences due to the number, content and location of integrated mobile elements giving rise to variation in chromosome length (Fig. 4). Whereas *E. coli* EC958 and *E. coli* JJ1886 chromosomes are 5.10 Mb and 5.12 Mb, respectively, *E. coli* NA114 is almost 200 kb smaller at 4.9 Mb, and *E. coli* SE15 has a 4.7 Mb chromosome. In addition to all seven defined EC958 prophages, the JJ1886 chromosome possess an additional prophage (Phi8) not present in the genomes of the other ST131 strains, but otherwise exhibits a high degree of synteny with the EC958 chromosome (Fig. 4). In contrast, the chromosome of *E. coli* NA114 shows multiple gaps relative to EC958, exhibits significant variation in both the number and content of prophages, and appears to lack the three largest defined EC958 genomic islands (GI-*pheV*, GI-*selC* and GI-*leuX*) (Fig. 4). Instead, *E. coli* NA114 has a ~160 kb region immediately upstream of *dnaJ* that consists of an assortment of GI and prophage sequence fragments that are found in several different locations in the EC958 and JJ1886 genomes. The *dnaJ* locus is not a known genomic island integration site and is well conserved in *E. coli* genomes from all phylogroups (Fig. 5). Together, these observations suggested to us that the *E. coli* NA114 genome has been misassembled.

To determine how a misassembly might have occurred, we replicated the NA114 assembly strategy and reassembled the genome of *E. coli* EC958 using simulated, error free, Illumina reads ordered against the *E. coli* SE15 chromosome (EC958-sim). We found that GI-*pheV*, GI-*selC*, GI-*leuX* and several of the prophage loci were placed incorrectly in EC958-sim relative to the complete *E. coli* EC958 genome (Fig. 6A). As expected, contigs associated with the EC958 genomic islands and prophages, which represent novel regions in the genome of EC958 compared to SE15, could not be correctly placed/ordered by alignment to SE15. Instead, these contigs have been randomly placed at the "end" of the chromosome in what might be mistaken for a large genomic island. Interestingly, the pattern of variation observed in the structure and location of EC958-sim mobile elements is similar to that observed in linear alignments of EC958 and NA114

Table 2. Comparison of complete PacBio EC958 genome with draft 454 EC958 genome.

Position[1]	Variant			454 contig information[2]			Genomic context	Comment[3]
	PacBio	454	SE15	Name	Position	Length		
786693	.	A	.	00007	131223	131239	Intergenic (EC958_0852 and EC958_0854)	Homopolymeric tract
955837–957967	N/A	N/A	N/A	00011	62258	64195	Phi1 phage tail region	2.8 Kb invertable region in Phi1 phage tail region (2 substitutions/20 indels)
985718	T	.	T	00014	19202	34445	Intergenic (EC958_1049 and EC958_1050)	Homopolymeric tract
1493217	.	T	T	00033	1143	74077	Tryptophan biosynthesis protein TrpCF (EC958_4924)	454 variant consistent with SE15 genome
2027391	A	.	A	00040	28390	75386	Chemotaxis protein CheA (EC958_2110)	Homopolymeric tract
2598139	A	.	A	00085	32434	57547	Intergenic (EC958_2623 and EC958_4988)	Homopolymeric tract
3098765	A	.	A	00073	3745	4934	Hypothetical protein (EC958_5038)	Homopolymeric tract
3377055	.	C	.	00057	28348	88502	Hypothetical protein (EC958_5205)	Homopolymeric tract
4149057	A	.	A	00104	7234	41331	Type II restriction enzyme (EC958_4083)	Homopolymeric tract
4308872	T	.	T	00148	3658	51318	Intergenic (EC958_4231 and EC958_4232)	Homopolymeric tract
4380208	T	.	T	00146	7992	42111	Hypothetical protein (EC958_4927)	Homopolymeric tract
4756303	A	.	A	00117	4118	15191	Intergenic (EC958_4610 and EC958_4611)	Homopolymeric tract
4762264	T	.	T	00117	10078	15191	Hypothetical protein (EC958_5122)	Homopolymeric tract
4762871	A	.	.	00117	10684	15191	Hypothetical protein (EC958_5123)	Homopolymeric tract; 454 variant consistent with SE15 genome
4776778	A	.	A	00119	7363	155673	Transcriptional activator CadC (EC958_4623)	Homopolymeric tract
4938457	G	T	G	00158	1028	1512	Transposase DDE domain protein (EC958_5125)	1.5 Kb repeat region duplicated in pEC958
4963207–4965562	N/A	N/A	N/A	00105	1144	3727	Repeat region in GI-SelC	3.7 Kb repeat region duplicated in GI-SelC (11 substitutions/4 indels)

[1]Nucleotide position (or range) in complete PacBio EC958 genome.
[2]Name, position of variant and length of 454 contig from draft 454 assembly.
[3]"Homopolymeric tract" indicates that variant falls within tract of 5 or more nucleotides of same type.

Table 3. Sites of DNA inversion within EC958 prophage genomes as determined by PacBio assembly of alternate alleles.

Crossover site	Sequence[1]	Location[2]	Comments[3]
Phi1_5prime	[gccgTTATCGAATACCTC^GGTTTACGAGAA – 478 bp]	c961070..961095	Part of 508 bp imperfect inverted repeat (77% nt identity); 2773 bp invertible segment
Phi1_3prime	[gccaTTATTTAAAACCTC^GGTTTACGAGAA – 478 bp]	958322..958347	-
Phi2_5prime	[TCCTCAATTACCTT^GGTTTAGGAGAA – 197 bp]	c1007582..1007607	Part of 227 bp imperfect inverted repeat (96% nt identity); 2067 bp invertible segment
Phi2_3prime	[GAGAGATAAACGTT^GGTTTGGGGGAA – 197 bp]	1005540..1005565	-
Phi4_5prime	[ccgccgTTATCGAATACCTC^GGTTTACAGGAA]	1484784..1484809	Part of 36 bp imperfect inverted repeat (3 mismatches); 3106 bp invertible segment
Phi4_3prime	[ccgccaTTATCTAAAACCTC^GGTTTACGAGAA]	c1487865..1487890	-
Consensus	TTCCC.TAAACGTT^CGTTTA.AAGAA	n/a	Based on consensus of crossover sites from.
	TT.A C C G T.GG		Mu, P1, e14, p15B and *S. boydii* DNA inversion systems, as previously determined by Sandmeier et al. 1994 [42]

[1]Predicted binding site for DNA invertase shown in capital letters; site of strand exchange is indicated by underlined central dinucleotide with ^ indicating downstream staggered cut; nucleotides in bold are consistent with the previously determined consensus DNA invertase crossover site [42]; square brackets indicate boundaries of larger imperfect inverted repeats that encode the crossover sites.
[2]Coordinates refer to start and end of 26 bp crossover site in EC958 complete genome; 5prime/3prime orientation is relative to the complete prophage tail fibre gene and prophage genome; c = complement.
[3]Phi1 and Phi4 5prime and 3prime 26 bp crossover sites differ by only 2 and 1 mismatches, respectively.

(Fig. 6B and Fig. 6C). Of the 77 gaps observed when EC958-sim contigs (>200 bp) were aligned with the complete *E. coli* EC958 chromosome, the majority corresponded with deletions or rearrangements at corresponding positions in the *E. coli* NA114 chromosome (Fig. 6C and Dataset S1).

Discussion

Here we report the complete genome sequence of the *E. coli* ST131 strain EC958. Sequencing the genome of *E. coli* EC958 with six SMRT cells of data followed by *de novo* assembly using the HGAP method and minimal post-processing produced a high quality finished genome comparable in terms of contiguity and error rate with a 454 GS-FLX mate-pair derived assembly. Since the sequence data for this genome was generated, the PacBio SMRT platform has transitioned from the RS I to the RS II instrument and improved chemistry, with average read lengths increasing to ~8 kb. Consequently, we expect that sequencing strategies utilising fewer than six SMRT cells on the PacBio RS II platform should be capable of producing fully assembled bacterial genomes with minimal intervention.

The sensitivity of PacBio for detecting dynamic prophage rearrangements is due to the length of PacBio reads, which allows them to span inverted regions and thus force the assembler to generate two alternative versions of regions that have undergone inversion in a subset of the bacterial population. In contrast, such mixed inversions are more difficult to detect in shorter read assemblies, which would normally require separate mapping and detection of discordant read-pairs to identify. Although there have been no other reports of phage tail inversion in PacBio assemblies

to date, others have noted that a ~7.5 kb "spurious contig" was produced in the assembly of the *E. coli* K-12 MG1655 genome [23]. PacBio thus offers a novel solution for studying the mechanism of phage tail fibre switching, and more generally, for the function of DNA invertase and other site-specific recombinases. For example, the DNA invertase gene has been severely truncated in the Phi4 prophage, suggesting that the inversion observed in this study must have been mediated by another enzyme *in trans*, as has been previously reported [49–51]. Notably, the Phi1 and Phi4 prophages encode near-identical 26 bp crossover sites at either end of their respective invertible segments (Table 3), suggesting that the Phi1 DNA invertase may be capable of mediating inversion at heterologous sites within the Phi4 prophage.

On a practical level, users should ensure that alternative allele contigs in PacBio assemblies are not integrated into the assembly of the main chromosome, which would lead to artefactual duplications in phage regions. Instead, we have annotated the EC958 chromosome to highlight the DNA invertase binding sites and invertible regions with misc_feature keys according to INSDC guidelines. We have also simplified the annotation of these regions to help avoid propagating genome-rot in *E. coli* genomes; for example, alternate phage tail gene 3′ fragments that contain the Phage Tail Collar domain but lack the Phage Tail Repeat domains are often auto-annotated as "Phage tail repeat domain proteins" due to their similarity to their full-length homologs. For *E. coli* assemblies, it is relatively straight-forward to determine which contigs are alternate versions of inverted loci as opposed to truly independent contigs, by first aligning all contigs to each other during post-assembly using tools such as ACT [25] or Contiguity

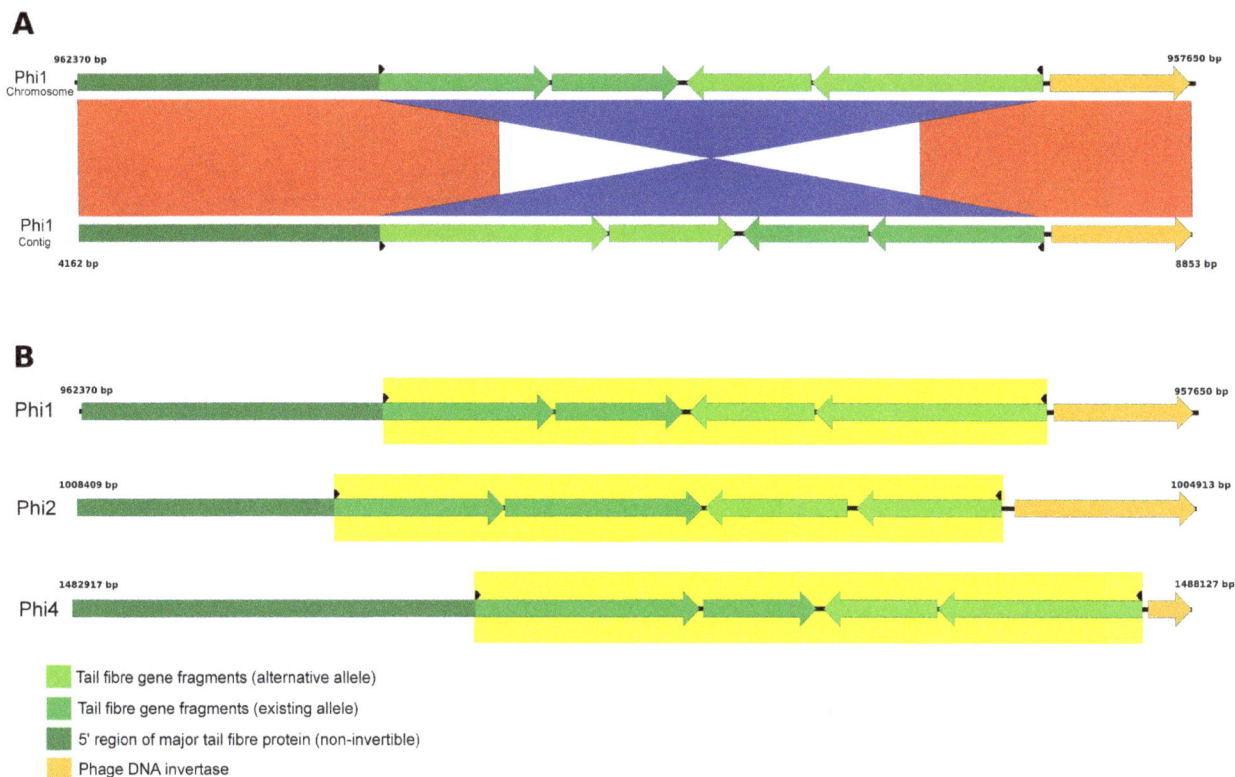

Figure 1. Prophage tail fibre allele switching in EC958. A. Alignment of the Phi1 alternative contig that contains the inversion of the tail fibre region to the genome of EC958. Phage tail fibre genes are coloured from dark green to light green. Phage DNA invertase genes are coloured orange. 26 bp crossover sites are indicated by black arrows. Red shading indicates nucleotide identity in the same orientation. Blue shading indicates nucleotide identity in the opposite orientation, highlighting the inversion in the phage tail fibre region. **B.** Genetic loci map of the tail fibre gene region of EC958 phages (Phi1, Phi2 and Phi4) and the location of recombination sites for DNA invertase. The major tail fibre gene is formed by a fusion of the stable 5′ region (dark green), encoding a series of Phage_fibre_2 tandem repeats (Pfam03406), with the invertible 3′ region (green) that encodes a Phage Tail Collar domain (Pfam07484). Downstream and presumably co-transcribed with the major tail fibre gene is a minor tail fibre gene (green). The alternate alleles form a mirror image of this arrangement, immediately downstream of the functional phage tail genes (lime green), enabling a new major tail fibre gene (and cognate minor tail fibre gene) to be formed by inversion of a 2–3 kb DNA segment. DNA invertase genes are coloured orange. The Phi4 prophage encodes a truncated DNA invertase (EC958_1582) that lacks the characteristic helix-turn-helix resolvase domain (PF02796). Invertible regions are highlighted in yellow. Figure prepared using Easyfig [27].

(http://mjsull.github.io/Contiguity/). However, care must be taken to ensure that "recombination" is not due to adapter sequences. Due to the high error rates associated with raw PacBio reads, occasionally adapters on the ends of the SMRTbell construct are not correctly identified and removed [52]. Failure to remove adapter sequences can result in chimeric subreads which consist of the insert sequence in the forward orientation followed by the adapter sequence and the insert sequence in the reverse orientation. Adapter sequences occur randomly within the reads and are removed during read correction but aberrant reads can be produced. Retaining these reads can result in false hairpins in assemblies and the generation of small spurious contigs. Users should also be aware that small plasmids are not necessarily assembled from PacBio reads using seed read length cut-offs in excess of the total plasmid size, as illustrated in this study with the 4.1 kb pEC958B plasmid. In this case we assembled pEC958B by utilising prior knowledge of the plasmid from the original 454 assembly, however, *de novo* assembly of the entire genome would be possible by iteratively reducing the seed read length cut-off within HGAP (data not shown).

We previously generated a high-quality draft sequence of *E. coli* EC958 [7], however, using only PacBio reads we were able to assemble a high-quality complete genome sequence. A comparison of the complete PacBio and draft 454 assemblies revealed a small number of discrepancies, the majority of which were due to homopolymeric tracts in the 454 assembly or collapsed repeats that were resolved in favour of the PacBio consensus after closer inspection. Although contig order and orientation in the original draft assembly was contiguous with the PacBio assembly, only the latter was able to resolve repetitive regions of the genome such as rRNA operons, extended tracts of tRNAs, prophage loci and insertion sequences (IS) within the GI-*pheV*, GI-*selC* and GI-*leuX* genomic islands. The long, multi-kilobase reads produced in SMRT sequencing can be unambiguously anchored with unique sequences flanking these repeats, allowing for their accurate and uninterrupted assembly. Given the rapid improvements in PacBio technology, and the HGAP assembly software [23], this technology may become the platform of choice for generating high-quality reference sequences for bacterial genomes.

Comparisons of the complete *E. coli* EC958 genome against other published ST131 genomes revealed the extensive nucleotide identity that exists between the core genomes of *E. coli* ST131 clade C strains EC958, NA114 and JJ1886. Although *E. coli* NA114 possesses many of the genes associated with genomic islands and prophages of EC958 and JJ1886, it lacks insertions at recognised *E. coli* integration hotspots, including the *pheV* tRNA

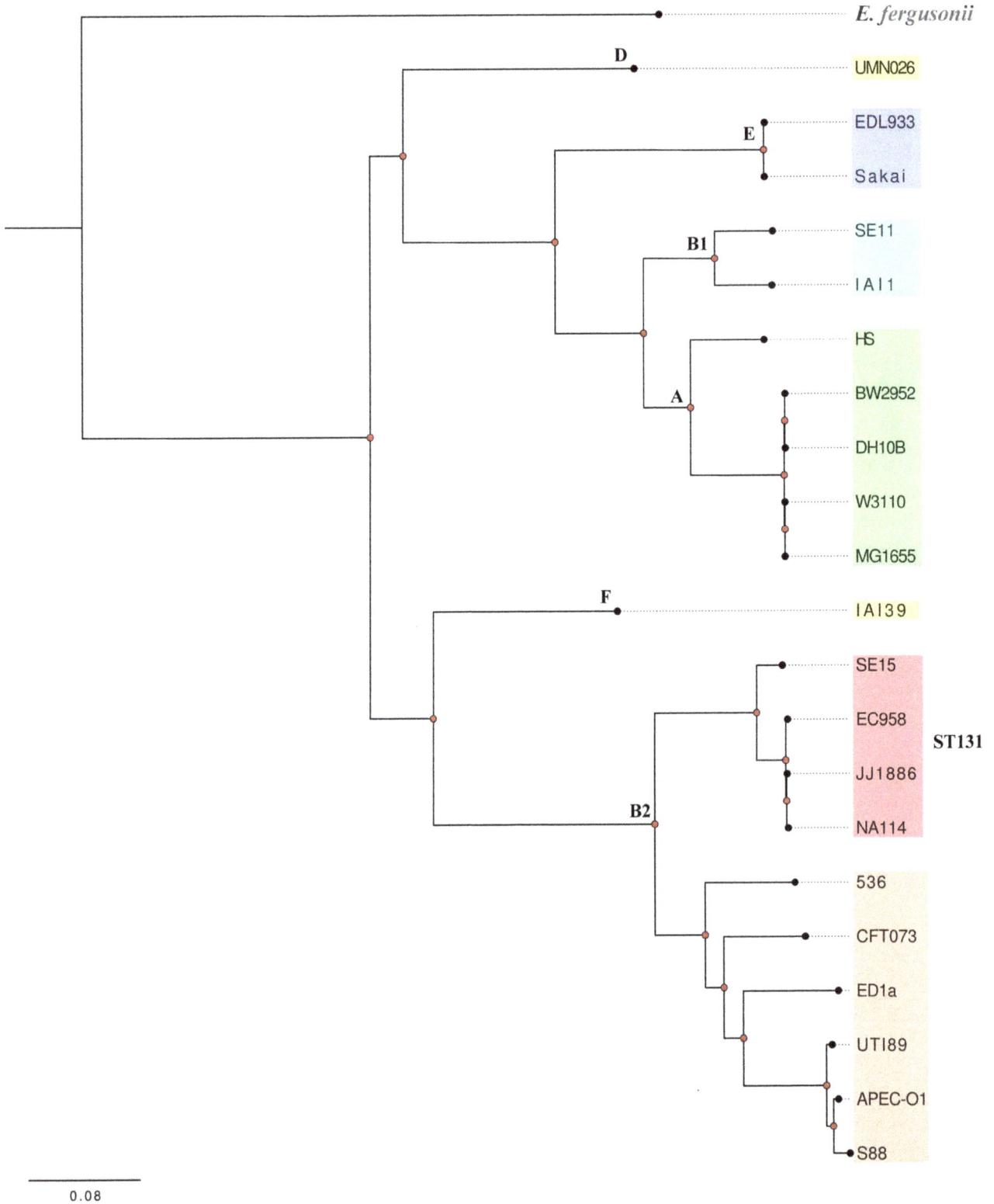

Figure 2. Maximum likelihood phylogenetic comparison of 4 ST131 and 17 representative *E. coli* isolates. The tree is rooted using the out-group species *E. fergusonii* ATCC35469. The phylogenetic relationships were inferred with the use of 261,214 SNPs identified between the genomes of the 22 *Escherichia* strains and 1000 bootstrap replicates. The major *E. coli* phylogroups are coloured as follows; phylogroup B2-ST131: SE15, NA114, JJ1886, EC958 (red); other phylogroup B2: APEC-01, S88, 536, UTI89, CFT073, ED1A (orange); phylogroup D: UMN026 (yellow); phylogroup F: IAI39 (yellow); phylogroup A: BW2952, MG1655, W3110, HS (green); phylogroup B1: SE11, IAI1 (aquamarine); phylogroup E: O157 EDL933, O157 Sakai (blue). Red nodes have 100% bootstrap support from 1000 replicates.

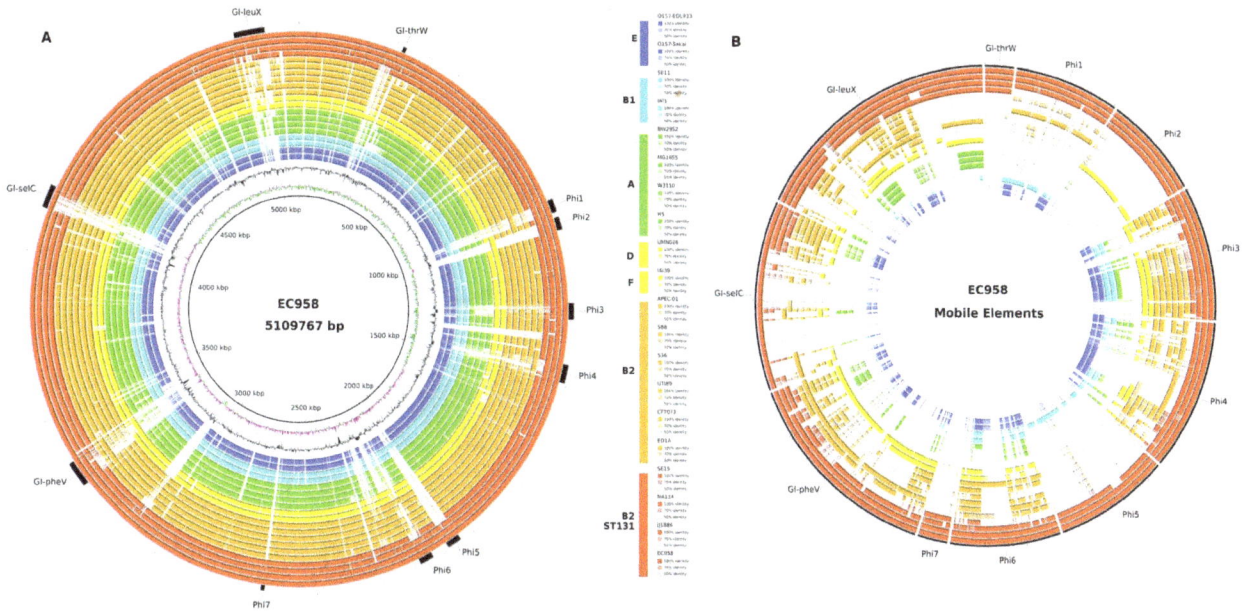

Figure 3. Distribution of EC958 mobile genetic elements in *E. coli*. A. Visualisation of the EC958 genome compared with three *E. coli* ST131 genomes and 16 other *E. coli* genomes using BLASTn. EC958 prophage (Phi1 – Phi7) and genomic islands (GI-*thrW*, GI-*pheV*, GI-*selC*, GI-*leuX*) are represented by black boxes in the outermost circle. The innermost circles represent the GC content (black) and GC skew (green/purple) of EC958. The remaining circles display BLASTn searches against the genome of EC958. **B.** A BRIG visualisation of the EC958 mobile elements compared with the 19 *E. coli* genomes. BLASTn searches of the 19 genomes against the EC958 prophage and genomic islands show that the EC958 GIs and prophage are well conserved in the ST131 clade C genomes but largely absent from the genomes of SE15 and the other 16 *E. coli* genomes, which are arranged inner to outer as follows: Group E strains O157 EDL933, O157 Sakai (blue); group B1 strains SE11, IAI1 (aquamarine); group A strains BW2952, MG1655, W3110, HS (green); group D strains UMN026, IAI39 (yellow); group B2 strains APEC-01, S88, 536, UTI89, CFT073, ED1A (orange); group B2 ST131 strains SE15, NA114, JJ1886, EC958 (red). Figure prepared using BRIG [28].

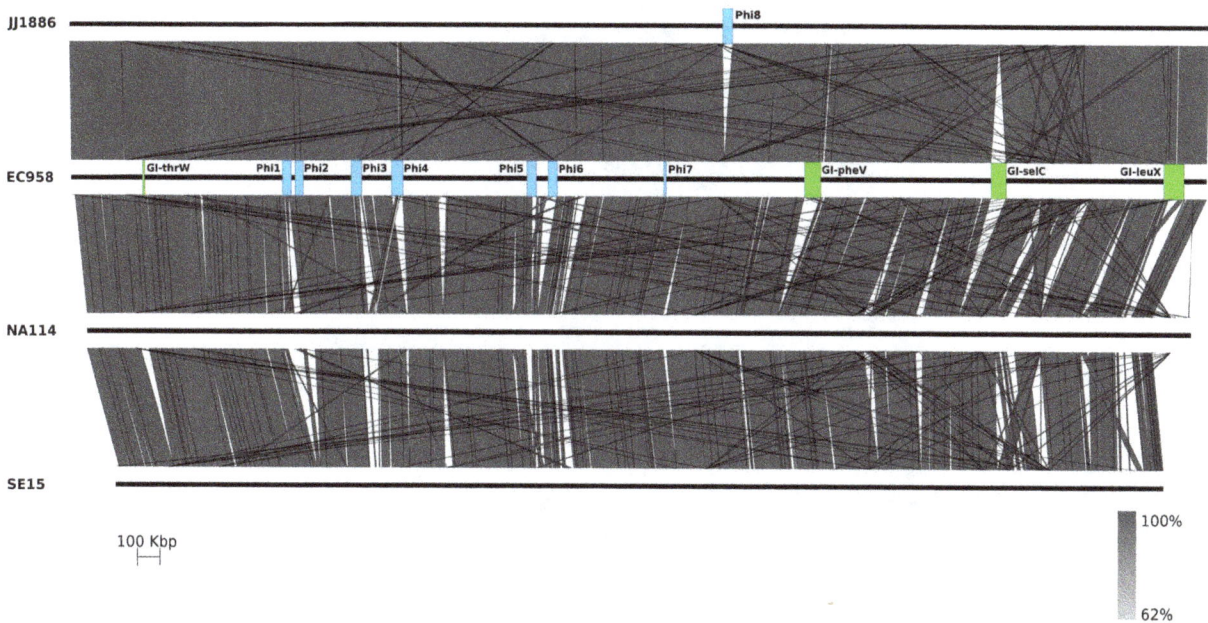

Figure 4. Nucleotide pairwise comparison of four *E. coli* ST131 chromosomes showing extensive variation in the structure and location of EC958 prophage elements (blue) and genomic islands (green). An additional prophage element present in JJ1886 has also been annotated here as Phi8 for clarity. ST131 genomes are arranged from top to bottom as follows: JJ1886, EC958, NA114, SE15. Grey shading indicates nucleotide identity between sequences according to BLASTn (62%–100%). Figure prepared using Easyfig [27].

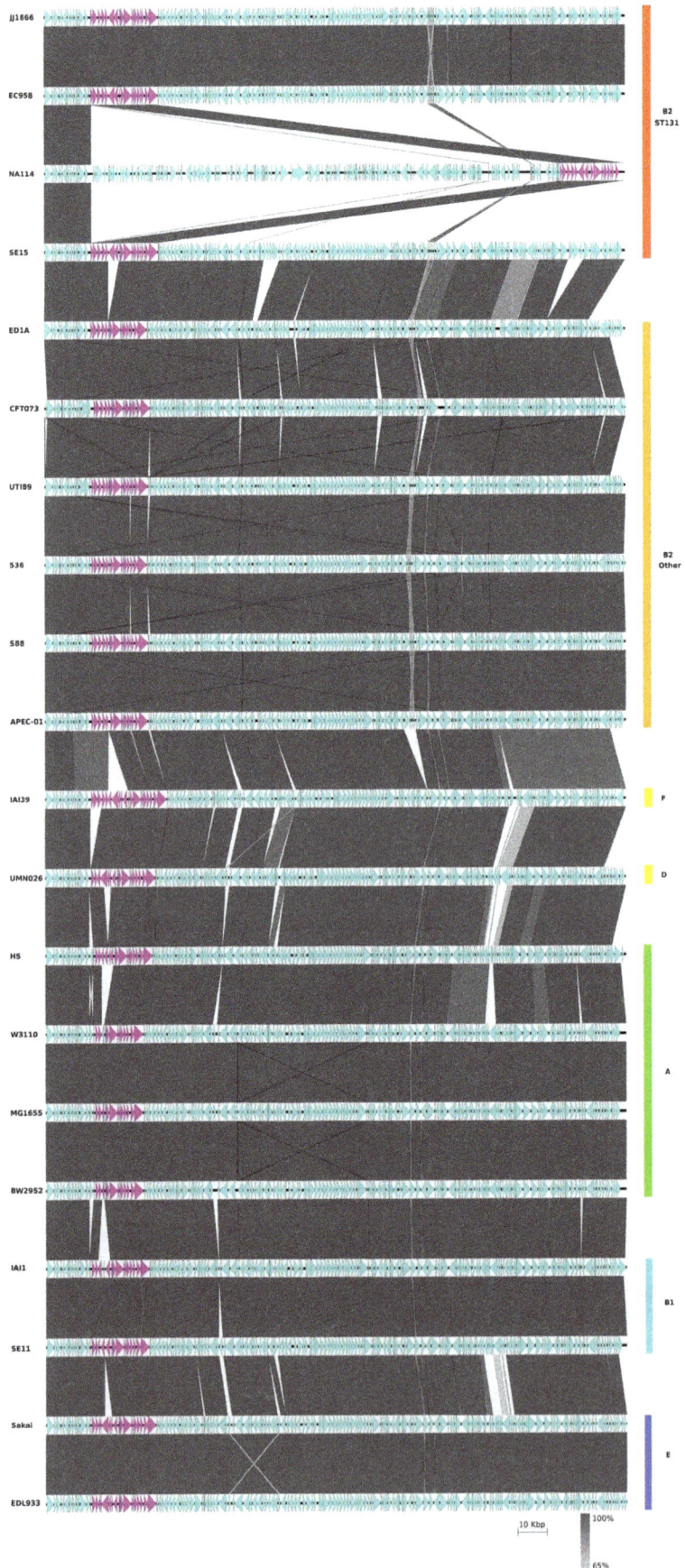

Figure 5. Nucleotide pairwise comparison of a 200 kb region (*thrA* to *degP*) from the genomes of the four ST131 and 16 other representative *E. coli* strains. Grey shading indicates nucleotide identity between sequences according to BLASTn (62%–100%). Coding regions immediately upstream of *dnaJ* are highlighted in purple. This region is well conserved in 19 of 20 *E. coli* genomes examined. However, a large insertion in the genome of NA114 located immediately upstream of *dnaJ* is clearly evident (white). *E. coli* genomes are arranged from top to bottom as follows: group B2 ST131 strains JJ1886, EC958, NA114, SE15 (red); group B2 strains ED1A, CFT073, UTI89, 536, S88, APEC-01 (orange); group F strain: IAI39 (yellow); group D strain UMN026 (yellow); group A strains HS, W3110, MG1655, BW2952 (green); group B1 strains IAI1, SE11 (aquamarine); group E strains O157 Sakai, O157 EDL933 (blue). Figure prepared using Easyfig [27].

gene [28]. Furthermore, it contains a highly atypical insertion of ~160 kb within a location that is consistent with the artefactual concatenation of contigs, "junked" at the end of the assembly, that could not be ordered against the SE15 reference genome. Our recent comparative genomic analysis has shown that, with the exception of GI-*selC* and Phi6, the genomic islands and prophages previously defined in EC958 are prevalent in nearly all other ST131 clade C strains [21]. Based on our whole genome comparisons of EC958, NA114, JJ1886 and SE15, and our simulated draft Illumina assembly (EC958-sim), we suggest that

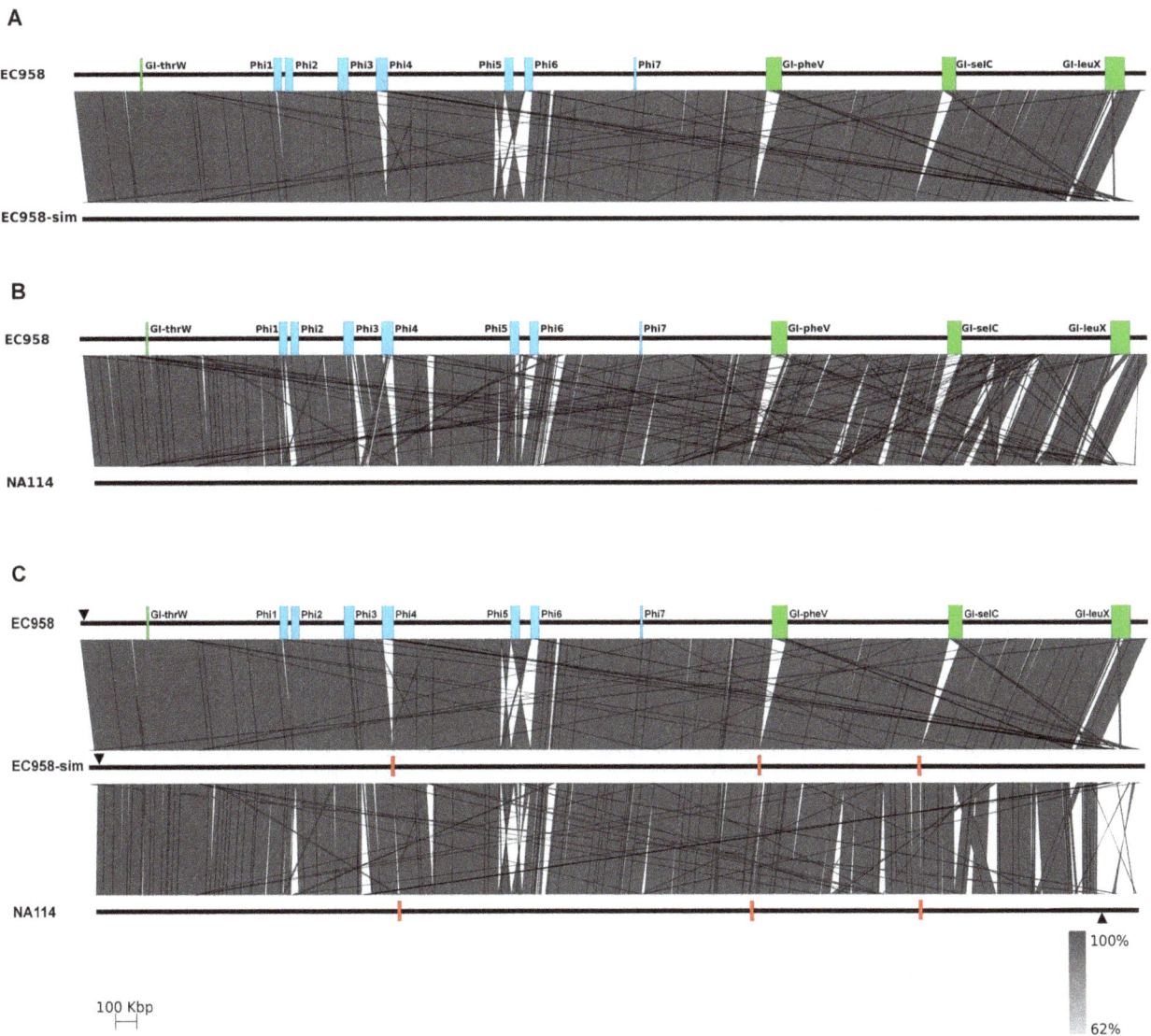

Figure 6. Nucleotide pairwise comparison between EC958, a simulated EC958 Illumina assembly and NA114. A. Nucleotide pairwise comparison of the EC958 chromosome (top) and a simulated EC958 chromosome assembly (EC958-sim, bottom). Linear alignments revealed extensive variations in the location and structure of mobile elements in EC958-sim when compared to EC958. Grey shading indicates nucleotide identity between sequences according to BLASTn (62%–100%). Prophage regions are annotated as blue boxes and genomic islands as green boxes. **B.** Nucleotide pairwise comparison of EC958 chromosome (top) and NA114 chromosome (bottom). **C.** Nucleotide pairwise comparison of EC958 (top), EC958-sim (centre) and NA114 (bottom) chromosomes. EC958 prophage and genomic islands misassembled in EC958-sim are similarly misassembled in the genome of NA114 (red boxes). Red boxes indicate positions in EC958-sim and NA114 where mobile genetic elements are present in EC958. The *dnaJ* gene is shown as a black triangle on each chromosome. Figure prepared using Easyfig [27].

much of the variation in mobile elements observed between NA114, EC958 and JJ1886 is not biologically relevant but rather the result of systematic errors introduced during the assembly of the *E. coli* NA114 genome.

Genome misassemblies are not only confined to draft genomes and have previously been identified in finished genomes [15]. Furthermore, in recent years a number of draft genomes have been erroneously deposited into the complete genome division of GenBank/EMBL/DDBJ, with reversal of sequence deposition very difficult due to the structure of these databases. Due to the clinical importance of uropathogenic *E. coli* we believe it is important to bring the misassembly of the *E. coli* NA114 genome to the attention of the community, particularly as it has been used recently in genome comparisons as if it was complete [22], and was used as the reference genome in a larger study of 100 *E. coli* ST131 isolates [6]. It should be more broadly recognised that it is not possible to generate an accurate representation of a complete *E. coli* genome by *de novo* assembly of Illumina, 454 or Ion Torrent reads alone. Ideally, a combination of paired-end and mate-pair libraries of varying insert length, often combined with PCR/Sanger sequencing, is necessary to correctly place contigs generated by SGS technologies and accurately close the gaps between them. In contrast, we show here that PacBio is able to act as a stand-alone platform for the generation of high-quality complete bacterial genome sequences. The availability of a complete, annotated genome of *E. coli* EC958 will provide an important resource for future comparative studies and reference guided assemblies of *E. coli* ST131 clade C/*fimH30* genomes.

Acknowledgments

We acknowledge Dr John Cheesbrough and staff at Preston Royal Infirmary bacteriology laboratories for original provision of the EC958 isolate and related clinical data.

Author Contributions

Conceived and designed the experiments: BMF MAS MU SAB. Performed the experiments: BMF SAB. Analyzed the data: BMF NLB MDP MT KGC MAS MU SAB. Contributed reagents/materials/analysis tools: KMP MSC. Wrote the paper: BMF MAS MU SAB.

References

1. Nicolas-Chanoine M-H, Blanco J, Leflon-Guibout V, Demarty R, Alonso MP, et al. (2008) Intercontinental emergence of Escherichia coli clone O25:H4-ST131 producing CTX-M-15. The Journal of antimicrobial chemotherapy 61: 273–281.
2. Lau SH, Reddy S, Cheesbrough J, Bolton FJ, Willshaw G, et al. (2008) Major uropathogenic Escherichia coli strain isolated in the northwest of England identified by multilocus sequence typing. J Clin Microbiol 46: 1076–1080.
3. Johnson JR, Johnston B, Clabots C, Kuskowski MA, Castanheira M (2010) Escherichia coli sequence type ST131 as the major cause of serious multidrug-resistant E. coli infections in the United States. Clinical infectious diseases: an official publication of the Infectious Diseases Society of America 51: 286–294.
4. Coque TM, Novais A, Carattoli A, Poirel L, Pitout J, et al. (2008) Dissemination of clonally related Escherichia coli strains expressing extended-spectrum beta-lactamase CTX-M-15. Emerging infectious diseases 14: 195–200.
5. Peirano G, Pitout JD (2010) Molecular epidemiology of Escherichia coli producing CTX-M beta-lactamases: the worldwide emergence of clone ST131 O25:H4. Int J Antimicrob Agents 35: 316–321.
6. Price LB, Johnson JR, Aziz M, Clabots C, Johnston B, et al. (2013) The epidemic of extended-spectrum-beta-lactamase-producing Escherichia coli ST131 is driven by a single highly pathogenic subclone, H30-Rx. MBio 4: e00377–00313.
7. Totsika M, Beatson SA, Sarkar S, Phan M-D, Petty NK, et al. (2011) Insights into a multidrug resistant Escherichia coli pathogen of the globally disseminated ST131 lineage: genome analysis and virulence mechanisms. PloS one 6: e26578.
8. Lau SH, Kaufmann ME, Livermore DM, Woodford N, Willshaw GA, et al. (2008) UK epidemic Escherichia coli strains A-E, with CTX-M-15 beta-lactamase, all belong to the international O25:H4-ST131 clone. J Antimicrob Chemother 62: 1241–1244.
9. Totsika M, Kostakioti M, Hannan TJ, Upton M, Beatson SA, et al. (2013) A FimH inhibitor prevents acute bladder infection and treats chronic cystitis caused by multidrug-resistant uropathogenic Escherichia coli ST131. The Journal of infectious diseases 208: 921–928.
10. Floyd RV, Upton M, Hultgren SJ, Wray S, Burdyga TV, et al. (2012) Escherichia coli-mediated impairment of ureteric contractility is uropathogenic E. coli specific. J Infect Dis 206: 1589–1596.
11. Phan M-D, Peters KM, Sarkar S, Lukowski SW, Allsopp LP, et al. (2013) The Serum Resistome of a Globally Disseminated Multidrug Resistant Uropathogenic Escherichia coli Clone. PLoS genetics 9: e1003834.
12. Salzberg SL, Phillippy AM, Zimin A, Puiu D, Magoc T, et al. (2012) GAGE: A critical evaluation of genome assemblies and assembly algorithms. Genome Res 22: 557–567.
13. Nagarajan N, Cook C, Di Bonaventura M, Ge H, Richards A, et al. (2010) Finishing genomes with limited resources: lessons from an ensemble of microbial genomes. BMC genomics 11: 242.
14. Kingsford C, Schatz MC, Pop M (2010) Assembly complexity of prokaryotic genomes using short reads. BMC bioinformatics 11: 21.
15. Phillippy AM, Schatz MC, Pop M (2008) Genome assembly forensics: finding the elusive mis-assembly. Genome biology 9: R55.
16. Ricker N, Qian H, Fulthorpe RR (2012) The limitations of draft assemblies for understanding prokaryotic adaptation and evolution. Genomics 100: 167–175.
17. Korlach J, Bjornson KP, Chaudhuri BP, Cicero RL, Flusberg BA, et al. (2010) Real-time DNA sequencing from single polymerase molecules. Methods in enzymology 472: 431–455.
18. Toh H, Oshima K, Toyoda A, Ogura Y, Ooka T, et al. (2010) Complete genome sequence of the wild-type commensal Escherichia coli strain SE15, belonging to phylogenetic group B2. Journal of bacteriology 192: 1165–1166.
19. Avasthi TS, Kumar N, Baddam R, Hussain A, Nandanwar N, et al. (2011) Genome of multidrug-resistant uropathogenic Escherichia coli strain NA114 from India. Journal of bacteriology 193: 4272–4273.
20. Andersen PS, Stegger M, Aziz M, Contente-Cuomo T, Gibbons HS, et al. (2013) Complete Genome Sequence of the Epidemic and Highly Virulent CTX-M-15-Producing H30-Rx Subclone of Escherichia coli ST131. Genome announcements 1.
21. Petty NK, Ben Zakour NL, Stanton-Cook M, Skippington E, Totsika M, et al. (2014) Global dissemination of a multidrug resistant Escherichia coli clone. Proc Natl Acad Sci U S A 111: 5694–5699.
22. Paul S, Linardopoulou EV, Billig M, Tchesnokova V, Price LB, et al. (2013) Role of homologous recombination in adaptive diversification of extraintestinal Escherichia coli. J Bacteriol 195: 231–242.
23. Chin C-S, Alexander DH, Marks P, Klammer AA, Drake J, et al. (2013) Nonhybrid, finished microbial genome assemblies from long-read SMRT sequencing data. Nature methods 10: 563–569.
24. Darling AE, Mau B, Perna NT (2010) progressiveMauve: multiple genome alignment with gene gain, loss and rearrangement. PloS one 5: e11147.
25. Carver T, Berriman M, Tivey A, Patel C, Bohme U, et al. (2008) Artemis and ACT: viewing, annotating and comparing sequences stored in a relational database. Bioinformatics 24: 2672–2676.
26. Kurtz S, Phillippy A, Delcher AL, Smoot M, Shumway M, et al. (2004) Versatile and open software for comparing large genomes. Genome Biol 5: R12.
27. Otto TD, Dillon GP, Degrave WS, Berriman M (2011) RATT: Rapid Annotation Transfer Tool. Nucleic acids research 39: e57.
28. Touchon M, Hoede C, Tenaillon O, Barbe V, Baeriswyl S, et al. (2009) Organised genome dynamics in the Escherichia coli species results in highly diverse adaptive paths. PLoS genetics 5: e1000344.
29. Chen SL, Hung C-S, Xu J, Reigstad CS, Magrini V, et al. (2006) Identification of genes subject to positive selection in uropathogenic strains of Escherichia coli: a comparative genomics approach. Proceedings of the National Academy of Sciences of the United States of America 103: 5977–5982.
30. Sullivan MJ, Petty NK, Beatson SA (2011) Easyfig: a genome comparison visualizer. Bioinformatics 27: 1009–1010.
31. Alikhan NF, Petty NK, Ben Zakour NL, Beatson SA (2011) BLAST Ring Image Generator (BRIG): simple prokaryote genome comparisons. BMC Genomics 12: 402.
32. David M, Dzamba M, Lister D, Ilie L, Brudno M (2011) SHRiMP2: sensitive yet practical SHort Read Mapping. Bioinformatics (Oxford, England) 27: 1011–1012.

33. Welch RA, Burland V, Plunkett G, Redford P, Roesch P, et al. (2002) Extensive mosaic structure revealed by the complete genome sequence of uropathogenic Escherichia coli. Proceedings of the National Academy of Sciences of the United States of America 99: 17020–17024.

34. Rasko DA, Rosovitz MJ, Myers GSA, Mongodin EF, Fricke WF, et al. (2008) The pangenome structure of Escherichia coli: comparative genomic analysis of E. coli commensal and pathogenic isolates. Journal of bacteriology 190: 6881–6893.

35. Perna NT, Plunkett G, Burland V, Mau B, Glasner JD, et al. (2001) Genome sequence of enterohaemorrhagic Escherichia coli O157:H7. Nature 409: 529–533.

36. Oshima K, Toh H, Ogura Y, Sasamoto H, Morita H, et al. (2008) Complete genome sequence and comparative analysis of the wild-type commensal Escherichia coli strain SE11 isolated from a healthy adult. DNA research: an international journal for rapid publication of reports on genes and genomes 15: 375–386.

37. Johnson TJ, Kariyawasam S, Wannemuehler Y, Mangiamele P, Johnson SJ, et al. (2007) The genome sequence of avian pathogenic Escherichia coli strain O1:K1:H7 shares strong similarities with human extraintestinal pathogenic E. coli genomes. Journal of bacteriology 189: 3228–3236.

38. Hayashi T, Makino K, Ohnishi M, Kurokawa K, Ishii K, et al. (2001) Complete genome sequence of enterohemorrhagic Escherichia coli O157:H7 and genomic comparison with a laboratory strain K-12. DNA research: an international journal for rapid publication of reports on genes and genomes 8: 11–22.

39. Hayashi K, Morooka N, Yamamoto Y, Fujita K, Isono K, et al. (2006) Highly accurate genome sequences of Escherichia coli K-12 strains MG1655 and W3110. Molecular systems biology 2: 2006.0007.

40. Ferenci T, Zhou Z, Betteridge T, Ren Y, Liu Y, et al. (2009) Genomic sequencing reveals regulatory mutations and recombinational events in the widely used MC4100 lineage of Escherichia coli K-12. Journal of bacteriology 191: 4025–4029.

41. Dobrindt U, Blum-Oehler G, Nagy G, Schneider G, Johann A, et al. (2002) Genetic structure and distribution of four pathogenicity islands (PAI I(536) to PAI IV(536)) of uropathogenic Escherichia coli strain 536. Infection and immunity 70: 6365–6372.

42. Blattner FR, Plunkett G, Bloch CA, Perna NT, Burland V, et al. (1997) The complete genome sequence of Escherichia coli K-12. Science (New York, NY) 277: 1453–1462.

43. Gardner SN, Hall BG (2013) When whole-genome alignments just won't work: kSNP v2 software for alignment-free SNP discovery and phylogenetics of hundreds of microbial genomes. PLoS One 8: e81760.

44. Guindon S, Dufayard JF, Lefort V, Anisimova M, Hordijk W, et al. (2010) New algorithms and methods to estimate maximum-likelihood phylogenies: assessing the performance of PhyML 3.0. Syst Biol 59: 307–321.

45. Zerbino DR, Birney E (2008) Velvet: algorithms for de novo short read assembly using de Bruijn graphs. Genome research 18: 821–829.

46. Sandmeier H (1994) Acquisition and rearrangement of sequence motifs in the evolution of bacteriophage tail fibres. Molecular microbiology 12: 343–350.

47. Fisher CR, Davies NMLL, Wyckoff EE, Feng Z, Oaks EV, et al. (2009) Genetics and virulence association of the Shigella flexneri sit iron transport system. Infection and immunity 77: 1992–1999.

48. Lloyd AL, Rasko DA, Mobley HLT (2007) Defining genomic islands and uropathogen-specific genes in uropathogenic Escherichia coli. Journal of bacteriology 189: 3532–3546.

49. Zhang L, Zhu B, Dai R, Zhao G, Ding X (2013) Control of directionality in Streptomyces phage phiBT1 integrase-mediated site-specific recombination. PLoS One 8: e80434.

50. Iida S, Hiestand-Nauer R (1987) Role of the central dinucleotide at the crossover sites for the selection of quasi sites in DNA inversion mediated by the site-specific Cin recombinase of phage P1. Mol Gen Genet 208: 464–468.

51. Iida S, Hiestand-Nauer R (1986) Localized conversion at the crossover sequences in the site-specific DNA inversion system of bacteriophage P1. Cell 45: 71–79.

52. English AC, Richards S, Han Y, Wang M, Vee V, et al. (2012) Mind the gap: upgrading genomes with Pacific Biosciences RS long-read sequencing technology. PLoS One 7: e47768.

3

Sequence and Ionomic Analysis of Divergent Strains of Maize Inbred Line B73 with an Altered Growth Phenotype

Martin Mascher[1,9], Nina Gerlach[2,9], Manfred Gahrtz[3], Marcel Bucher[2], Uwe Scholz[1], Thomas Dresselhaus[3]*

1 Department of Cytogenetics and Genome Analysis, Leibniz Institute of Plant Genetics and Crop Plant Research (IPK), Corrensstraße 3, Stadt Seeland, Germany, 2 Botanical Institute, Cologne Biocenter, Cluster of Excellence on Plant Sciences (CEPLAS), University of Cologne, Zülpicherstrasse 47b, Cologne, Germany, 3 Cell Biology and Plant Biochemistry, Biochemie-Zentrum Regensburg, University of Regensburg, Universitätsstraße 31, Regensburg, Germany

Abstract

Maize (*Zea mays*) is the most widely grown crop species in the world and a classical model organism for plant research. The completion of a high-quality reference genome sequence and the advent of high-throughput sequencing have greatly empowered re-sequencing studies in maize. In this study, plants of maize inbred line B73 descended from two different sets of seed material grown for several generations either in the field or in the greenhouse were found to show a different growth phenotype and ionome under phosphate starvation conditions and moreover a different responsiveness towards mycorrhizal fungi of the species *Glomus intraradices* (syn: *Rhizophagus irregularis*). Whole genome re-sequencing of individuals from both sets and comparison to the B73 reference sequence revealed three cryptic introgressions on chromosomes 1, 5 and 10 in the line grown in the greenhouse summing up to a total of 5,257 single-nucleotide polymorphisms (SNPs). Transcriptome sequencing of three individuals from each set lent further support to the location of the introgression intervals and confirmed them to be fixed in all sequenced individuals. Moreover, we identified >120 genes differentially expressed between the two B73 lines. We thus have found a nearly-isogenic line (NIL) of maize inbred line B73 that is characterized by an altered growth phenotype under phosphate starvation conditions and an improved responsiveness towards symbiosis with mycorrhizal fungi. Through next-generation sequencing of the genomes and transcriptomes we were able to delineate exact introgression intervals. Putative *de novo* mutations appeared approximately uniformly distributed along the ten maize chromosomes mainly representing G:C -> A:T transitions. The plant material described in this study will be a valuable tool both for functional studies of genes differentially expressed in both B73 lines and for research on growth behavior especially in response to symbiosis between maize and mycorrhizal fungi.

Editor: Hector Candela, Universidad Miguel Hernández de Elche, Spain

Funding: This work was supported by the German Federal Ministry of Education and Research (BMBF) in the frame of OPTIMAS [FKZ 0315430]. The funders had no role in study design, data collection and analysis, decision to publish, or preparation of the manuscript.

Competing Interests: The authors have declared that no competing interests exist.

* E-mail: thomas.dresselhaus@ur.de

9 These authors contributed equally to this work.

Introduction

Maize (Z. *mays*) is an important cereal crop and has been a major plant model species for genetic research since the first half of the 20th century. It is an extremely diverse species whose genome abounds with single-nucleotide polymorphisms (SNPs) [1,2] as well as with copy number and presence-absence variation between inbred lines [2,3]. Dedicated stock centers maintain phenotypically characterized mutant lines as well as cultivars of former and present agricultural importance. A community-driven online database [4] provides a searchable catalogue of morphological and cytological variation captured in elite cultivars, landraces as well as wild accessions and facilitates the distribution of seed material to researchers around the world.

Like its wild progenitor teosinte, maize is a predominantly outcrossing crop. However, artificial self-pollination is possible and commonly used in breeding programs. Inbred lines, i.e. nearly

homozygous individuals, can be easily generated and maintained by repeated self-fertilization. One of the most widely used inbred lines of maize is named as B73. Developed at Iowa State University, B73 is among the founder lines of the so-called stiff-stalk germplasm group. Consequently, a recent re-sequencing study found more than 50 inbred lines from different breeding programs to be closely related to B73 [5]. In a scientific context, B73 is one parent of the IBM intermated recombinant inbred line (RIL) mapping population [6] and is the common parent shared by all lines of the maize nested association mapping population [7]. The genome of B73 had been sequenced to high quality using a map-based clone-by-clone strategy [8]. This sequence resource now represents an invaluable resource for re-sequencing studies. Haplotype maps (HapMaps) that include hundred of thousands to millions of variant positions genotyped in hundreds to tens of thousands of individuals have been constructed by whole-genome or reduced representation sequencing [1,2,5]. The B73 reference sequence thus provides the backbone for genome-wide association studies [9] and map-based cloning projects [10] for the crop and model plant maize.

Maize like most other terrestrial plants can form symbiotic relationships with arbuscular mycorrhizal (AM) fungi. These soil-born fungi are obligate symbionts that colonize plant roots and assist their hosts in the uptake of water and nutrients, in particular phosphate (Pi), while obtaining plant carbohydrates in exchange (see [11] for a review). Transporters specific for nutrients like P, S or Zn or plant sugars functioning at the plant/fungus interface mediate mycorrhiza-specific exchange processes [12–14]. The physiological response of plants to AM symbiosis is variable and strongly depends on environmental conditions and the genotypic background [15–17]. Even between replicate experiments, mycorrhizal responsiveness has been described to vary depending on growth seasons [18]. In maize, Kaeppler et al. [19] evaluated the responsiveness of different maize lines including B73 towards mycorrhizal colonization. In this study, line B73 robustly showed a strong increase in biomass generation under low Pi condition in the presence of AM fungi. Moreover, one QTL controlling mycorrhiza responsiveness was found on chromosome #2 whereas three QTLs for Pi starvation response in absence of mycorrhiza were located on other chromosomal regions [19].

Here, we describe an inbred line of maize that is nearly isogenic to B73 used to generate the reference genome. This line was propagated for more than 20 years exclusively in the greenhouse without intended exposure to the mycorrhizal fungus *G. intraradices*. It phenotypically differs significantly from B73 in its responsiveness towards symbiosis with AM under limited availability of Pi. Additionally, it shows a different growth phenotype and ion composition under low Pi conditions in absence of mycorrhizal fungi. Through whole genome and transcriptome comparison with the B73 reference genome and a set of B73 plants that were obtained more recently from the stock center and exclusively grown in the field, we show that the genome of this nearly isogenic line (NIL) harbors three well-delineated segments from a different maize genotype in a B73 background summing up to a total of 5,257 SNPs. *De novo* mutation were found throughout the genome mostly representing G:C -> A:T transitions. Finally, we report and discuss >120 genes differently expressed between the two B73 strains potentially associated with the above described growth phenotypes.

Results

In the course of a project that involved the collection of transcriptome, ionome and metabolome data from a large number of maize inbred lines [20], we grew, amongst others, plants of maize inbred line B73 under various growth conditions. B73 seeds had been obtained from two different sources. One set of seeds (set A) is descended from seed material obtained from the USDA stock center in 2007 (Acc.-No.: PI 550473) and was further propagated after self-pollination for three generations in the field. The second set of seeds (set B) had been obtained as B73 seeds from the USDA stock center in the early 1990s. Plants descended from this seed material were propagated after self-pollination for more than 20 generations exclusively in green houses at the universities of Hamburg and Regensburg, respectively.

Growth Response towards Phosphate Starvation and Arbuscular Mycorrhizal Colonisation

In the course of our experiments, plants from both sets together with other inbred lines were grown in a bicompartmented system including or lacking arbuscular mycorrhizal fungi (+AM, −AM).

One compartment was supplemented with Pi and closed by a hyphae-permeable membrane (hyphal compartment, HC) as already described [21]. By using this method, fungal hyphae link the Pi source with the maize root system and therefore enable Pi uptake via the mycorrhizal Pi uptake pathway. Under these conditions, −AM plants highly suffer from Pi deficiency, while +AM were supplied by Pi out of the HC [21].

As shown in Figure 1A we observed obvious differences in plant growth behavior between the two sets of B73 plants with and without colonization by AM fungi. Especially under −AM conditions where plants were highly Pi-starved, plants of set A grow taller with erected leaves while set B plants stay smaller with overhanging leaves. Differences in the growth phenotype are also present under +AM condition. Here plants of set B seem to form an increase in overall leaf health, which is visible by greener, thicker and even more overhanging leaves. These growth differences were apparent in two independent experiments. To quantify these observations, dry weight, leaf number and size of the plants have been determined in autumn 2010. Here, plants from set B showed a significant difference in dry weight and number of green, non-senescent leaves in −AM versus +AM conditions (Figure 1B). In other words, under these conditions set B plants but not set A plants exhibited a strong positive mycorrhizal growth response with *G. intraradices* when access to Pi was limited to the mycorrhizal uptake pathway. Under high Pi conditions (plants have been grown at the universities of Hamburg and Regensburg in semi-sterilized soil supplemented with fertilizer) a significant difference in plant growth behavior was not observed between sets A and B (data not shown).

Analysis of Elemental Composition

The measurement of the total elemental composition (ionomics) by inductively-coupled plasma mass spectrometry (ICP-MS) in source leaves of maize plants underlined the observed growth differences between plants of set A and B under +AM and −AM conditions (Figure 2A). A principal component analysis (PCA), which reduces all analyzed traits per treatment into few components showed a strong discrimination between −AM and +AM plants of set B (PC1 = 36.6%) while plants of set A reveal just a weak effect in PC1. In particular, P concentration is significantly increased in mycorrhizal set B plants (Figure S1). In plants of set A, a slight discrimination of +AM and −AM plants by a third PC is visible (PC3 = 11.9%). Ionomic data from the seeds used for all studies of this work showed significant differences between both sets of B73 plants. PCA analysis separated seeds of set A from seeds of set B by a PC1 of 54.3% (Figure 2B). This includes a higher accumulation of diverse elements (e.g. K, Fe, Mn, Zn, Cu, Ni, Co) in seeds of set A accompanied by a reduction of Na, Se, Cs and Mo concentrations (Figure 2C).

Whole Genome Sequencing

As indicated above the observed differences between both sets of B73 plants may have been caused by genetic or epigenetic effects occurred during propagation of seed material for many generations either in the green house or in the field. As the genome of inbred line B73 has been sequenced and assembled to high-quality, we therefore sequenced the genome of plants from both sets. We performed whole genome shotgun (WGS) sequencing of a single plant from both set A and set B (plant A and plant B), respectively. We sequenced both plants to ~15x whole genome coverage using the Illumina sequencing-by-synthesis platform (Table 1). Sequence reads were mapped against the maize reference sequence (AGPv2). Around 94% of all reads could be mapped. *In silico* detection of single nucleotide polymorphisms

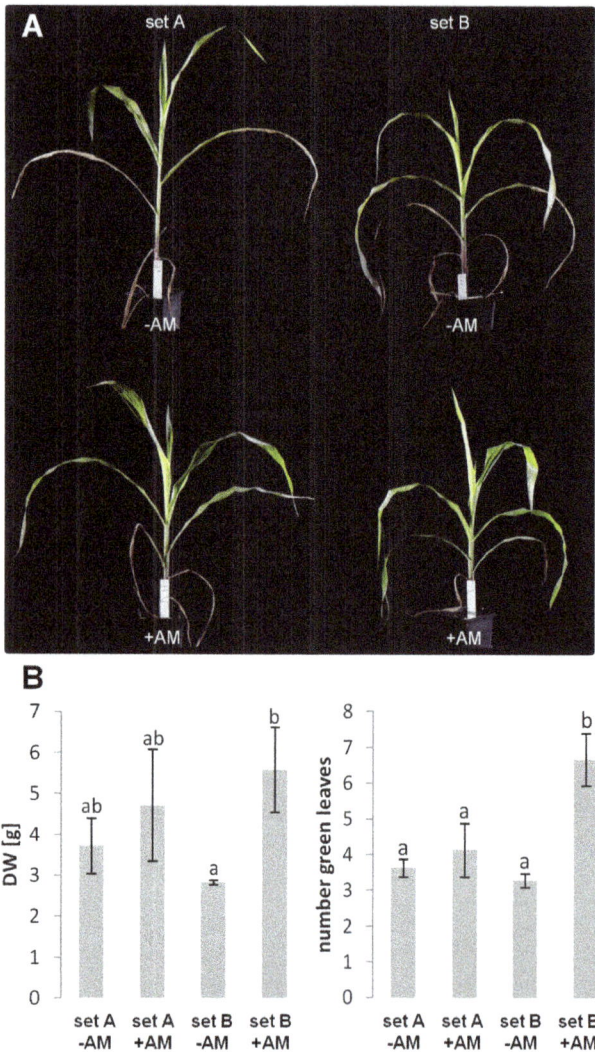

Figure 1. Phenotypic comparison of progeny plants from two maize B73 inbred lines grown for generations either exclusively in the field (set A) or in the greenhouse (set B). Plants were grown in compartmented pots with (+AM) or without (−AM) arbuscular mycorrhiza with phosphate addition to the hyphal compartment. **(A)** Comparison of plant growth phenotype. Pictures have been taken before harvest, approx. 7 weeks after sowing. **(B)** Comparison of dry weight (DW) and number of green leaves from 3–4 pooled plants of both inbred lines (set A and set B) analyzed in autumn 2010. Significant differences between the treatments are indicated by different letters (n = 3–4, $p \leq 0.05$, one-way ANOVA).

(SNPs) was performed using the SAMtools pipeline. The maize reference is characterized by a high content of repetitive elements and highly similar copies of genes as a consequence of recent allopolyploidy. Likewise, the maize reference sequence is not to be considered a finished reference sequence as the sequence has only been placed to BAC-level resolution (100–200 kb), but sequence contigs have not been ordered within single BACs. These factors caused some obstacles to accurate read mapping and variant calling from short read NGS data. We therefore focused our attention on SNPs that had sufficient ($\geq$10x) read coverage in both samples and which were called as "homozygous alternative" in one sample and "homozygous reference" in the other. A total of 5,615 SNPs met our criteria. Of these, 358 had the non-B73 allele

in plant A and 5,257 had the non-B73 allele in plant B. We calculated the number of SNPs in non-overlapping 100 kb bins. SNPs were distributed uniformly across the genome in set B (Figure 3A). While there were at most three SNPs per bin in plant A, we found 19 bins with more than 100 SNPs in plant B. The maximal SNP count per bin was 285. All bins with more than two SNPs were located in three genomic regions: Chr. 1, 292.1–293.2 Mb; Chr. 5, 208.1–208.6 Mb; Chr. 10, 142.3–145.3 Mb (Figure 3A, Figure 4A, Table 2). Of the 5,615 SNPs, 1,097 (19.5%) were located in annotated exons. Among these, 506 were predicted to have an effect on the protein sequence with four of them located on Chr. 10 introducing premature stop codons and one generating a start-codon loss (Table S1). While the latter gene (GRMZM5G874366_T01) encodes a precusor of a receptor-like protein kinase 1, two of the other candidate genes encode proteins with homology to a serine/threonine-protein phosphatase (GRMZM2G096107_P01) and a N-lysine methyltransferase-like protein (AC204437.3_FGP004).

To validate our SNP calls, we compared them against the maize HapMap version 2, which includes the genotypes of 103 lines of domesticated maize and teosinte at 55 million SNP positions. Only 24 (6.7%) of the 358 SNPs between the B73 reference and plant A were present in HapMap2. By contrast, 3,962 (70.5%) of 5,257 SNP of plant B were polymorphic in the HapMap panel. This led us to the conclusion, that plant B most likely carries an introgression from a different maize genotype in a B73 background, whereas the genome of plant A is almost identical to the B73 reference, with most SNPs attributable to errors in read mapping or SNP calling. We tried to find a possible donor genotype of the introgressed segment among the 103 HapMap lines. However, a HapMap genotype that is highly similar to plant B could not be identified.

Transcriptome Sequencing

As we only sequenced one individual plant grown from each set A and B, respectively, we could not rule out the possibility that further introgressions are present in other plants from set B. We therefore performed transcriptome sequencing of plants grown under the same conditions from three randomly selected seeds of both sets (Table 1). RNA-seq data was used to compare SNPs called in the transcriptome data against the SNPs discovered in the genomic data and to detect consistent differences in transcript abundance between both sets. RNA was extracted from 16 d old seedlings to (i) avoid maternal effects from differentially propagated seeds and (ii) to avoid long-term environmental influences, which may alter gene expression in older partially stressed plants. Sequencing was performed using an Illumina instrument. Resulting sequence reads were mapped against the maize reference sequence. Approximately 82–86% of all reads could be mapped against the B73 reference sequence (Table 1). We called genotypes from the mapping files of each sample at the 5,257 putative variant positions discovered by whole genome sequencing. We required that at least 10 RNA-seq reads be present to make a genotype call. A total of 567 SNPs were successfully typed in the three samples from set A. Genotype calls were in complete agreement with the calls made from the WGS shotgun data. In five cases, plants from set A had the non-B73 allele. In the plant from set B, genotypes for 309 SNPs were called in all three replicates. Genotypes of all but one SNP agreed with the WGS data. The single discordant SNP was called heterozygous in the RNA-seq data, while it was called homozygous non-B73 in the WGS data. Among the other SNPs, 303 (98%) had the non-B73 allele and five had the B73 allele (Table 2).

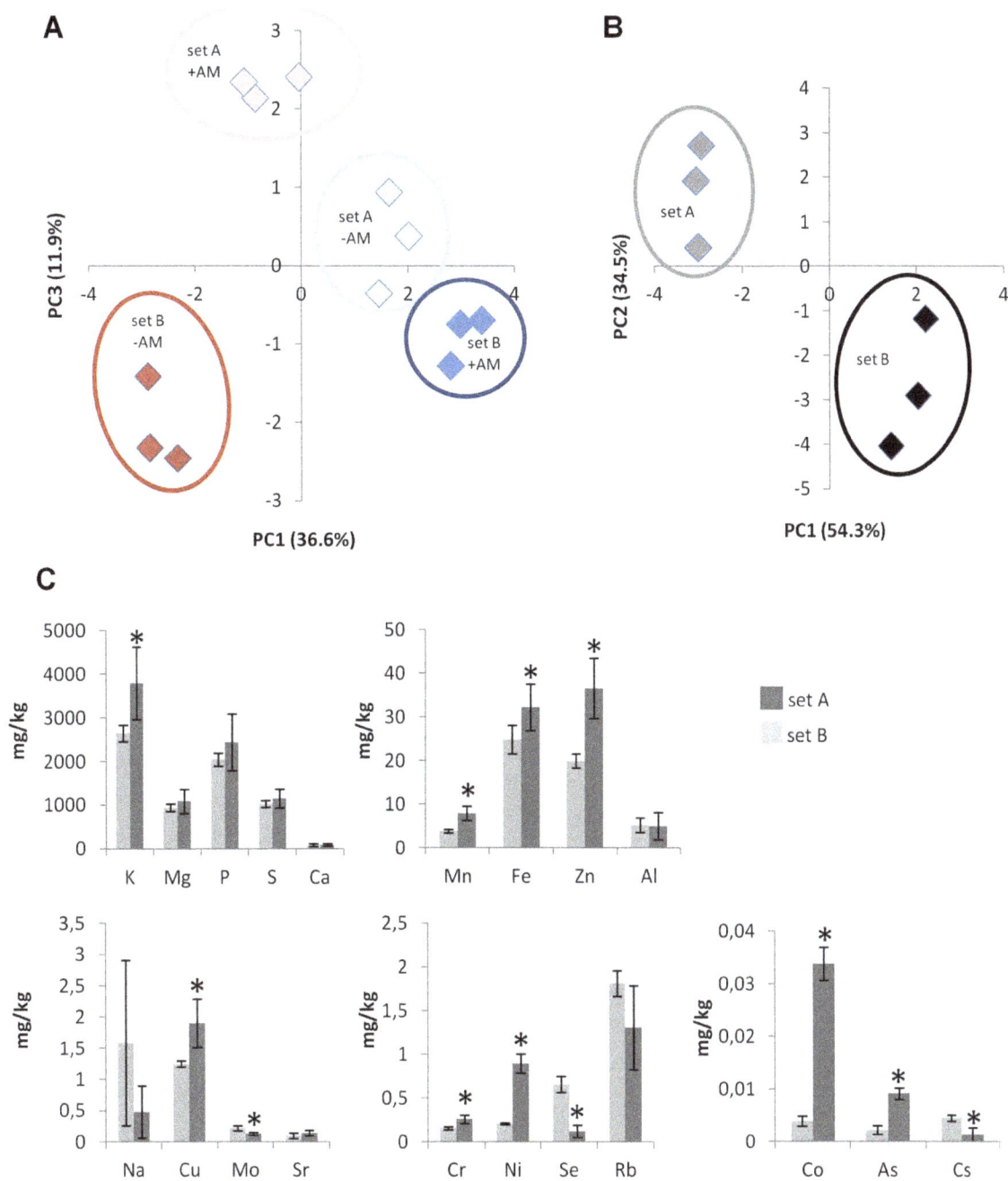

Figure 2. Comparison of elemental composition in source leaves and dry seeds of progeny plants from both B73 inbred lines (set A and B). (**A**) PCA-analysis of ionomics data from source leaves. Three plants were each grown in compartmented pots with (+AM) or without (−AM) arbuscular mycorrhiza with phosphate addition to the hyphal compartment. (**B**) PCA-analysis of ionomic data from dry seeds (n = 3). (**C**) Depiction of 20 elements separately as bar charts. Significant differences between plants of set A and set B are labeled by an asterisk (n = 4–5, student's t-test, $p \leq$ 0.05).

We also determined the number of SNPs called across all six samples and differentiating between set A and B (i.e. all plants from set A are homozygous for one allele and all plants from set B are homozygous for the other allele) without making use of SNP positions discovered in the WGS data. Visualization of the number of these SNPs in 100 kb bins identified the same intervals of increased SNP density (Figure 3B). The agreement between SNPs detected in the WGS and in the RNA-seq data supports the hypothesis that all set B plants harbor the same non-B73 segment.

To exclude the possibility of other unnoticed non-B73 segments in either set, we plotted the counts of heterozygous and homozygous SNPs (relative to the B73 reference) in the eight sequenced individuals (Figures S2 and S3). No other intervals with an increased number of homozygous SNPs were discovered. The number of heterozygous calls was higher in all samples compared to the number of homozygous calls. Several isolated 100 kb bins showed an elevated number of heterozygous SNPs. Three bins (Chr3: 186.6 Mb, Chr7: 162.2 Mb and Chr9: 7.5 Mb) had more

Table 1. Whole genome sequence (WGS; genomic) and transcriptome (RNA) data generated in this study from B73 inbred lines of set A and B plants, respectively.

Seed set	Sample	Read type[a]	All reads	Mapped reads
A	Genomic	PE	196 M	185 M (94.6%)
	Genomic	SE	132 M	124 M (94.0%)
	RNA replicate 1	PE	145 M	123 M (85.7%)
	RNA replicate 2	PE	166 M	139 M (84.6%)
	RNA replicate 3	PE	168 M	140 M (83.3%)
B	Genomic	PE	212 M	200 M (94.0%)
	Genomic	SE	140 M	132 M (93.9%)
	RNA replicate 1	PE	130 M	108 M (82.7%)
	RNA replicate 2	PE	118 M	101 M (85.9%)
	RNA replicate 3	PE	126 M	105 M (83.1%)

[a]PE - paired end reads (2×100 bp), SE – single end reads (100 bp).

than 20 heterozygous SNPs in at least one RNA-seq sample and at least 14 SNPs in all six RNA-seq samples. These intervals are most likely paralogous regions that are collapsed in the B73 genome assembly. Two other bins (Chr1: 293.0 Mb and Chr10: 142.4 Mb) within the putative introgression intervals (Table 2) had at least 17 SNPs in all three replicates from set B and less than two SNPs in the samples from set A. These intervals could be the results of small regions of residual heterozygosity or paralogous SNPs in tandem duplicated regions that occur in single copy number in B73, but in higher copy number in the unknown donor of the introgression.

We used the RNA-seq data collected from the six samples to analyze differential transcript abundance between plants from set A and set B. Overall 121 genes (Table 3, Table S2) were found to be differentially expressed (DE) at a liberal threshold (corrected p-value ≤0.05 and log fold change ≥2). DE genes were also present outside of the introgression intervals on chromosome 1, 5, and 10 (Figure 3B). Only nine out of 121 DE genes were located in the introgression intervals (Table S2). This is not unexpected as sequence polymorphisms within the introgressed segments may alter the abundance of transcripts or their encoded proteins and these may function subsequently as *trans*-acting factors influencing the expression of transcripts from loci in other genomic regions. Transcripts >10 times up-regulated in set A compared with set B include three genes involved in the anthocyanin biosynthesis pathway: the Anthocyanin regulatory C1 protein (GRMZM2G005066), a Dihydroflavonol-4-reductase (GRMZM2G026930) and an Anthocyanidin 3-O-glucosyltrans-

ferase (GRMZM2G165390). Activation of the anthocyanin pathway is a well-known reaction of plants to Pi stress [22] and may be related to the different growth phenotype observed between set A and set B plants. Other strongly up-regulated genes in set A are involved in plant hormone metabolism such as auxin conjugation (Indole-3-acetate beta-glucosyltransferase; GRMZM2G078465) and generation of active cytokinin (Cis-zeatin O-glucosyltransferase; GRMZM2G110511) as well as hormonal responses (SAUR14 - auxin-responsive SAUR family member; GRMZM2G447151) that may explain the AM independent differences in growth behavior. Genes strongly up-regulated in set B are involved, for example, in pyruvate metabolism (Isochorismatase family protein rutB), gene regulation (Homeobox-leucine zipper protein with homology to ATHB-4; GRMZM2G126239) and RNA metabolism (Splicing factor U2af 38 kDa subunit; GRMZM2G031827). The NIL identified in this study thus additionally provides an elegant tool to study the genetic and biochemical effects of genes that are differentially expressed between both sets of B73 inbred lines.

De Novo Mutations

Resequencing studies in several model organisms, such as *Arabidopsis thaliana* [23], yeast [24] and *Caenorhabditis elegans* [25], have revealed that genomes accumulate spontaneous mutations at a rate of the order of 10^{-9} to 10^{-8} per site per generation. We mined our genomic variants calls for SNPs to the B73 reference sequence that occur either in set A or set B plants (but not in both) and are not part of HapMap2, and considered these "*de novo*

Table 2. Introgression intervals in the genome of set B plants of maize inbred line B73.

Region	No. of SNPs between sets A and B in WGS data[a]	No. of SNPs between sets A and B in transcriptome data[a]
Chr1: 292.1–293.2 Mb	1,103 (20.9%)	46 (15.2%)
Chr5: 208.1–208.6 Mb	450 (8.5%)	28 (9.2%)
Chr10: 142.3–145.3 Mb	2,809 (53.4%)	224 (73.9%)
Complete genome	5,257 (100%)	303 (100%)

[a]only SNPs within the non-B73 allele in set B plants were counted.

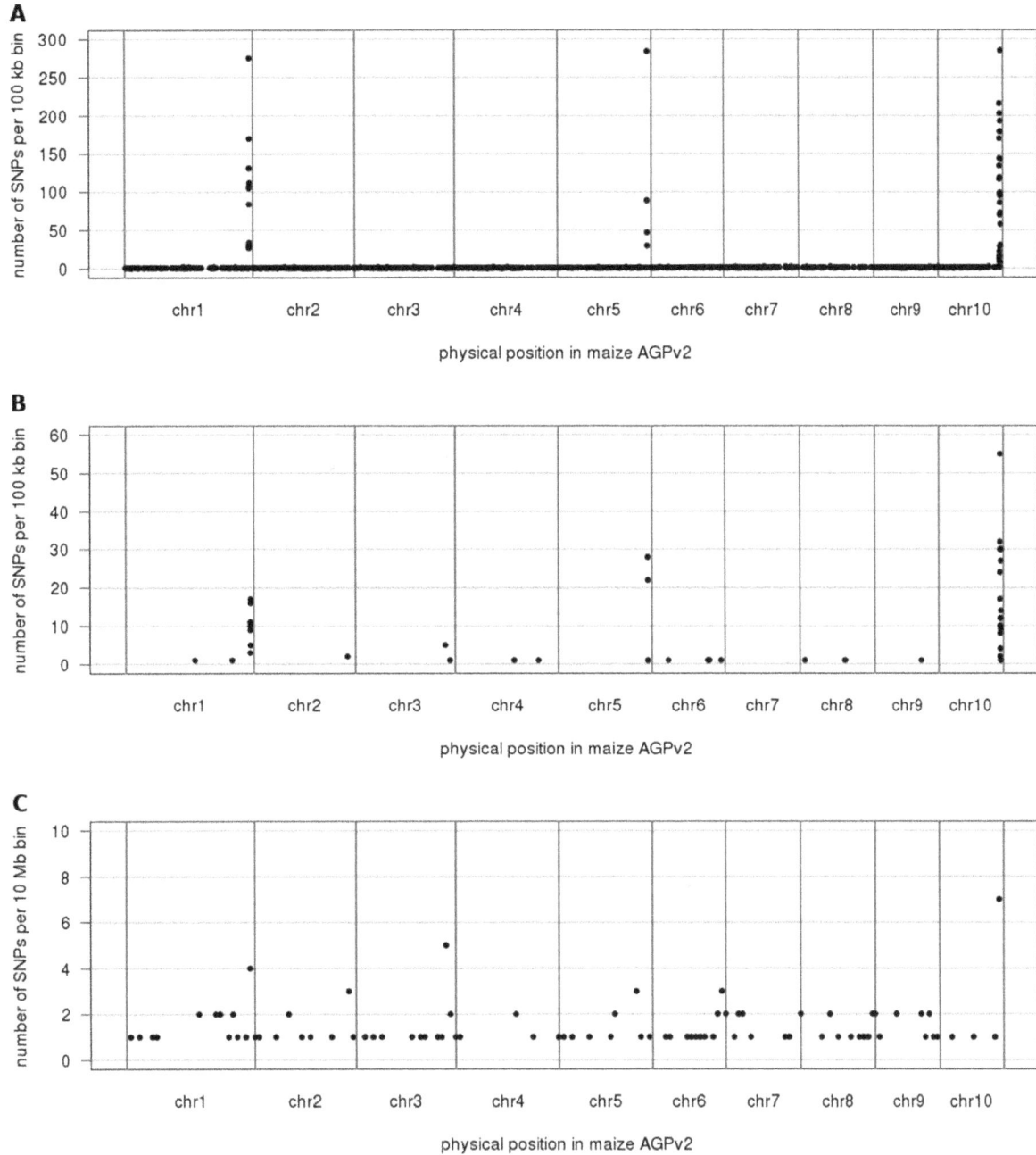

Figure 3. Distribution of SNPs and differentially expressed genes along the length of the 10 maize chromosomes (separated by blue lines). (A) Number in non-overlapping 100 kb windows of genomic SNPs differentiating between plants of set A and B, respectively. (B) Number in non-overlapping 100 kb windows of transcriptome SNPs differentiating between the three replicates from set A and the three replicates from set B. (C) Number in non-overlapping 10 Mb windows of differentially expressed genes (q value ≤0.05, log fold change ≥2).

SNPs" to be the result of spontaneous mutations that occurred during propagation of the seed stock. A total of 358 and 883 *de novo* SNPs were found in the genome of set A and set B plants, respectively. These putative *de novo* mutations were located approximately uniformly along the ten maize chromosomes (Figure 4A). Taking the mutation rate in *A. thaliana* (7×10^{-9} base substitutions per site per generation) [23] as a proxy for the mutation rate in maize, ~16 mutations occur in the 2.3 Gbp genome of maize per generation, indicating that the B73 reference is ~22 generations removed from the genomes of set A plants and

~55 generations from the genome of set B plants. Note that these numbers count propagation cycles starting from a putative common ancestor of both the reference genome and set A/B pkants. The spectrum of *de novo* mutations was largely similar between set A and B plants (Figure 4B). The majority of mutations were G:C -> A:T transitions, similar to findings in *A. thaliana* [26]. Mutations were distributed more or less uniformly along the chromosomes. Spontaneous mutations in protein-coding genes may result in differential gene expression and could be a possible cause for the phenotypic differences of set A and B plants. Seven of

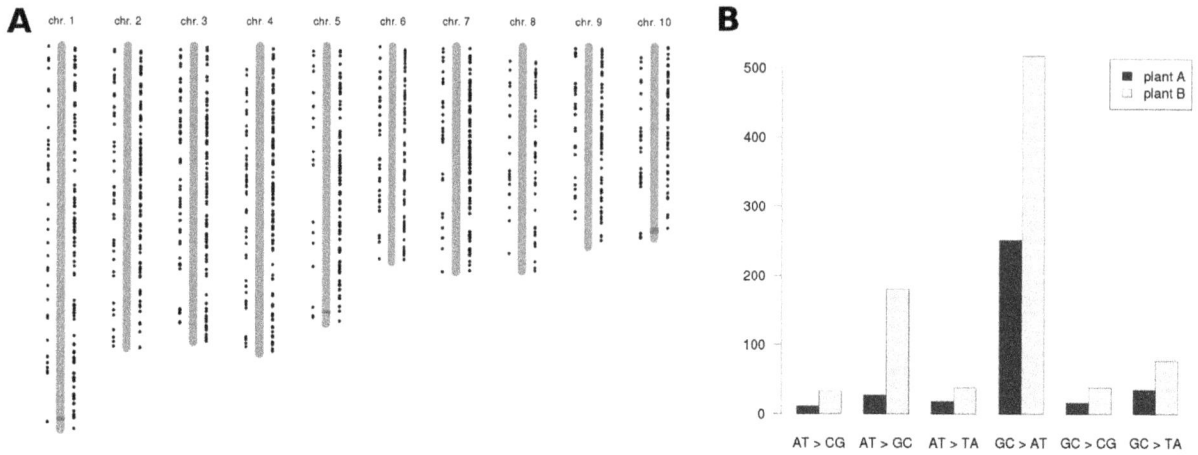

Figure 4. Analysis of putative *de novo* **SNPs and introgression loci in both B73 inbred lines.** (**A**) Distribution of putative *de novo* SNPs along the maize genome. *De novo* SNPs of set A plants are shown to the left of the chromosome ideograms, SNPs of set B plants are shown to the right. The locations of the putative introgressions on chromosomes 1, 5, and 10 in plant B are highlighted in red and drawn to scale. (**B**) Spectrum of *de novo* mutations in set A and set B plants, respectively.

Table 3. Top20[a] differentially expressed genes between B73 inbred lines of set A and B with functional annotation.

Gene	Locus	FPKM_A	FPKM_B	Functional annotation[b]
GRMZM2G049021	5:133618125–133626411	0.12	54.09	Isochorismatase family protein rutB
GRMZM2G078465	7:140007358–140009147	51.34	1.32	Indole-3-acetate beta-glucosyltransferase
GRMZM2G026930	3:216304733–216306568	2.21	0.1	Dihydroflavonol-4-reductase
GRMZM2G031827	8:70136196–70144793	0.77	15.21	Splicing factor U2af 38 kDa subunit
GRMZM2G110511	8:167407655–167409403	1.06	0.08	Cis-zeatin O-glucosyltransferase 2 (cisZOG2)
GRMZM2G126239	3:214857326–214861719	0.47	5.52	Homeobox-leucine zipper protein ATHB-4
GRMZM2G016890	10:34232716–34238135	0.41	4.25	Beta-glucosidase, chloroplastic
GRMZM2G000620	10:144143810–144147098	11.18	1.07	Receptor-like kinase
GRMZM2G165390	9:11774732–11776491	1.62	0.16	Anthocyanidin 3-O-glucosyltransferase
GRMZM2G161905	9:155596297–155597233	11.15	1.13	Glutathione S-transferase GST 25 Fragment
GRMZM2G005066	9:9740802–9741876	6.13	0.64	Anthocyanin regulatory C1 protein
GRMZM2G073916	8:91681549–91689395	0.12	1.13	Protein aq_1857
GRMZM2G047368	7:41539129–41540625	147.58	17.41	Aquaporin PIP2-6
GRMZM2G447151	1:258800054–258800675	25.56	3.29	SAUR14 - auxin-responsive SAUR family member
GRMZM2G179294	5:180731982–180733210	0.45	3.12	High affinity nitrate transporter
GRMZM2G035444	1:172897702–172899238	0.21	1.26	UPF0497 membrane protein 3
GRMZM2G151227	2:223888705–223892691	15.35	2.58	Chalcone synthase WHP1
GRMZM2G063244	8:1141429–1145642	1.57	0.27	Peptidyl-prolyl cis-trans isomerase
GRMZM2G046952	10:132098076–132099489	0.29	1.54	FIP1
GRMZM2G025459	8:120217908–120221032	0.52	2.79	SNF1-related protein kinase regulatory subunit beta-1

[a] a full list of all differentially expressed genes is available as Table S2.
[b] only genes with functional annotation are shown.

the putative *de novo* SNPs of set A plants and 14 of those of set B plants are located in annotated exons. Out of these exonic SNPs, only one was found within a differentially expressed gene (GRMZM2G132956). The difference in transcript abundance of this gene was statistically significant (p = 0.003), but the log2 fold change was only 0.77. Given the absence of a functional annotation, we did not consider this gene as a good candidate for a causal gene underlying the described growth and ionome phenotypes.

Discussion

We have described here a nearly isogenic line (NIL) of maize inbred line B73. The NIL was found by chance when we compared the response of field and green house grown plants to Pi starvation and AM fungi in a wider panel of maize inbred lines. Phenotypic changes were accompanied by alterations in transcript levels and the ionomic composition of leaves and seeds. Through whole-genome and transcriptome sequencing, we could ascertain that the NIL is pure and harbors three small introgression intervals on chromosome 1, 5 and 10 in a B73 genomic background. The analysis of differential gene expression revealed consistent differences in transcript abundance between both sets of plants.

We were not able to trace back the origin of this NIL. It could be speculated that the introgression was introduced by unintentional and unnoticed cross-fertilization with an unknown parent during seed propagation at the maize stock center. Several rounds of backcrossing to "genuine" B73 may have decreased the size of the genomic regions inherited by the non-B73 parent. Subsequent rounds of self-fertilization may subsequently have led to fixation of the non-B73 allele in a homozygous state in three introgression intervals. This theory requires that in the "development" of the NIL backcrossing by unsupervised pollination in a plot of supposedly homogeneous B73 plants was supplanted with artificial self-pollination. Another possible explanation would be that B73 is not an entirely homogeneous inbred line, but to some extent a "fluid concept": B73 lines maintained at different locations (or at the same location at different time points) exhibit slight differences in sequence composition that result only in minor phenotypic alterations, which in most cases escape notice. It is thus likely that more B73 NILs are around and are used by various labs and plant breeders.

Whatever the exact crossing scheme may be, the NIL seems now to be maintained as an (almost) pure (i.e. (nearly) homozygous) line at the University of Regensburg as plants grown from different B73 seed stocks showed an altered growth phenotype under low Pi conditions and between +AM and –AM treatment. Likewise, sequencing of RNA samples of three B73 plants randomly selected at Regensburg (set B plants) revealed the same three genomic intervals in all three samples and in the plant sequenced by whole-genome shotgun sequencing. As only controlled self-pollination has been carried out in the greenhouse and the number of SNPs is very small, it is unlikely that cross-pollination occurred at the Universities of Hamburg and Regensburg after seeds had been obtained from the stock center. The combined size of all three introgression intervals is about 4.6 Mb, i.e. ~0.2% of the entire maize genome. The introgressed segments are located in highly recombinogenic subtelomeric regions of the long arms of the respective chromosomes. The recombination rate at these regions is approximately 3 cM per Mb [27]. The genetic length of the introgression interval is approximately 14 cM (0.6–0.8% of the genetic length of the maize genome [27]). An initial hybrid has to be backcrossed to B73 for seven generations and subsequently selfed for several generations

to decrease the introgressions size to this tiny proportion of the genome and to obtain homozygous lines. If this process is carried out unintentionally, it would most likely not result in homogeneous material. Thus, it is likely that seeds carrying the undesired introgression had already been shipped by the stock center.

This finding raises some concerns about the homogeneity of material from different sources that is supposed to have the same genetic background. Cryptic introgressions in the genomes of presumably identical individuals may result in phenotypic differences that become evident only under very special growth conditions. If slightly divergent material is inadvertently used in line development (e.g. in map-based cloning projects), mutant phenotypes may be obfuscated or, in the worst case, even emulated by variation in diverging chromosomal regions. In view of this worrying perspective, we advise that the identity of plant material, in particular when originating from different laboratories, should always be double-checked. Other reports corroborate our concerns about the purity of seed material. A small segment (1.5 Mb) that differs between different B73 lines has been reported previously [1]. Moreover, the maize haplotype map has revealed that many inbred lines are not completely homozygous, but retain considerable heterozygosity in rarely recombining pericentromeric regions [1]. Genotyping-by-sequencing (GBS) of the US maize seed bank [5] revealed residual heterozygoity as a common source of intra-accession variability. Next generation sequencing in conjunction with cost-effective means of genomic complexity reduction [28,29] has enabled the simultaneous interrogation of thousands of molecular markers in a large number of individuals. In the future, it may become possible to monitor the genomic identity of each individuals involved in a specific experiment as well as to validate the identity and homozygosity of samples distributed by genebanks [5].

We are fully aware that the data reported in this study does not constitute a definite proof that a gene or sequence variant underlying the different growth phenotype is located in the three introgressions intervals delineated by high-throughput sequencing. We have found several hundred putative *de novo* SNPs of the two lines relative to the B73 reference genome. Only one of these was located in the coding exon of a differentially expressed gene and this gene was not an obvious causal candidate underlying the growth phenotype. It is possible that, for example, a complex structural rearrangement such as a transposon insertion into the regulatory sequence of a gene residing outside the putative introgressions has affected the expression of one or several genes, which subsequently brought about the differential growth response to Pi starvation in the presence or absence of AM fungi. We can also not rule out epigenetic differences between both lines, though these have to be stably inherited as we saw consistent differences between all plants of the two B73 lines.

We moreover cannot exclude that overall growth responses could be influenced by the different initial nutrient content in the two seed batches (Figure 2 B, C). This may affect seedling's growth in the first weeks after germination on low nutrient substrate as shown for dicotyledonous species [30]. Under our experimental conditions non Pi-fertilized plants suffer from Pi-limitation approximately two weeks after germination. Latest at this time point internal nutritional reserves should be completely depleted. This was also one reason why we have chosen seedlings at 16 days after sowing for transcriptomic studies. Interestingly higher nutrient concentration of set A seeds for many growth-relevant elements like K, Mn, Zn and Fe disagrees with a finally higher biomass accumulation of six week old set B plants. Moreover no additional differences in P concentration of both seed batches (Figure 2 C) but significantly higher P-concentrations in mycorrhizal source leaves of set B (Figure S1) were found. This altogether points against maternal effects by the different nutrient content of the seeds and to a direct influence of the altered genetic background of set B plants towards improved mycorrhizal responsiveness including altered nutrient uptake. Particularly the mycorrhizal uptake pathway of set B plants could be induced by alterations in the mycorrhizal signal transduction cascade, which may result into an increased uptake of essential nutrients (e. g. P) and therefore improved growth. This hypothesis is supported by the finding that genes encoding a novel leucine-rich repeat receptor-like protein kinase family protein as well as a SNF1 protein kinase regulatory subunit are strongly induced in set B plants (see also Table 3 and Suppl. Table S2). Further analysis of gene expression in roots of both B73 maize lines may highlight putative differential responses of nutrient transporter genes like the mycorrhiza-specific phosphate transporter Pht1; 6 [12] or ammonium and nitrate transporters [31]. The improved ability of nutrient transport within set B plants like nitrogen-transport is reflected, for example, by the increased expression of a high affinity nitrate transporter (GRMZM2G179294) (Table 3).

As gene expression analysis has been conducted with young seedlings under fully fertilized conditions without AM fungi, thus a direct link between increased mycorrhizal responsiveness and the differential gene expression results requires further experimentation. Nevertheless it is conspicuous that differential gene regulation between set A and B plants points to an overall elevated stress response level of set A plants. In particular the strongly increased expression of genes involved in the flavonoid and especially anthocyanin biosynthesis pathway (such as Dihydroflavonol-4-reductase, Anthocyanidin 3-O-glucosyltransferase, Chalcone synthase WHP1, Phenylalanine ammonia-lyase and the major transcriptional regulator of the pathway, C1 [32], see also Table 3 and Suppl. Table S2) might not only mirror higher Pi-stress of set A plants even under full fertilizer conditions, but additionally indicate higher vitality of these plants to general stress conditions. Under Pi-limited conditions these plants also visually look more Pi-starved with erected leaves and an overall non-compact growth phenotype compared with set B plants (Figure 1). Moreover expression of the heat shock protein genes HSP26 and HSP101 is increased in set A seedlings, which is typically expressed at high levels under stress reactions [33] as well as the strongly increased expression of a Glutathione S-transferase gene, which is required to detoxify endogenous compounds and xenobiotics such as herbicides [34].

In summary the strong up-regulation of flavonoid biosynthesis and stress related genes in set A plants suggest a higher fitness of these plants, while an increase of kinase genes might be associated with signaling for mycorrhizal uptake pathways in set B plants. These hypothesis require further experimentation, but the NIL described in the present study can now form the basis for further research to fine-map, for example, the loci responsible for the altered growth phenotype of set B plants. Moreover, genetic mapping could be employed to confirm or reject the association of genomic introgression intervals and the phenotype. For example, plants of an F2 population obtained by crossing the AM line with B73 could be scored for their growth phenotype and genotyped with cost-efficient GBS technology [29]. As all three introgression intervals are located in distal, highly recombinogenic regions, mapping in even a small F2 population of 50–100 plants may delimit a reasonably small target region, which can then be mined for candidate genes with the assistance of the genomic and transcriptomic data reported in this study. Finally, the NIL reported in this study provides a valuable resource for functional studies of above described genes differentially expressed between

both sets of plants such as the genes involved in flavonoid biosynthesis.

Materials and Methods

Plant Material and Growth Conditions Prior to DNA/RNA Extraction for Sequencing

B73 maize lines were obtained from the USDA stock center in the early 1990s (named as set B in this study) and in 2007 (Acc.-No.: PI 550473; named as set A). Both sets were propagated exclusively after self-pollination either in the field (set A) or in the greenhouse (set B). Seed material of set B plants is available by request to the corresponding author. Plants were grown in the greenhouse with supplemental light under long-day conditions (16 h light/8 h dark) at 27°C during the day and 18.5°C during the night.

Genomic DNA Preparation for Illumina Sequencing

Leaf material from 12 day old plants at stage V3 were sampled and immediately frozen in liquid nitrogen. One gram of leaf tissue each was ground in a mortar under liquid nitrogen and genomic DNA was prepared using the DNeasy Plant Maxi Kit (Qiagen) according to the instructions of the manufacturer.

RNA Preparation for Illumina Sequencing

Material of the third leaf from sixteen day old plants at stage V4 were sampled and immediately frozen in liquid nitrogen. Hundred mg frozen leaf tissue each was ground in a swing mill and RNA was prepared using the RNeasy Mini Kit (Qiagen) according to the instructions of the manufacturer.

Sample Preparation and Illumina Sequencing

Genomic DNA was fragmented to a target size of 200 bp using the Covaris S2 and the DNA microTUBE protocol (KBiosciences). Indexed libraries were prepared with the TruSeq DNA Sample Preparation Kit (Illumina) and were subsequently quantified with the KAPA Library Quantification Kit for Illumina (KAPA Biosystems) and the DNA 1000 assay on the 2100 Bioanalyzer (Agilent).

Single end reads: Cluster generation on the cBot was performed using the TruSeq SR Cluster Kit v2, followed by a 101 cycles standard single-read sequencing run on the HiScanSQ using the TruSeq SBS Kit v2. *Paired end reads*: Cluster generation on the cBot was performed using the TruSeq PE Cluster Kit v3, followed by a $2 \times 101 + 7$ cycles multiplexed paired-end run on the HiSeq 2000 instrument using the TruSeq SBS Kit v3.

RNAseq and preparation of the respective libraries were carried out as described in the Illumina TruSeq RNA Sample Preparation Guide, the Illumina HiScan 1000 System User Guide (Illumina), and the KAPA Library Quantification Kit - Illumina/ABI Prism User Guide (Kapa Biosystems). In brief, 1 μg of total RNA was used for purifying the poly-A containing mRNA molecules using poly-T oligo-attached magnetic beads. Following purification, mRNA was fragmented to an average insert size of 250–450 bases using divalent cations under elevated temperature (94°C for 4 minutes). Cleaved RNA fragments were copied into first strand cDNA using reverse transcriptase and random primers followed by second strand cDNA synthesis using DNA Polymerase I and RNase H. The resulting cDNA fragments subsequently went through an end repair process, the addition of a single 'A' base, the ligation of the adapters, and a purification step. Finally cDNA libraries were created by PCR enrichment. Libraries were quantified using the KAPA SYBR FAST ABI Prism Library Quantification Kit. Equimolar amounts of each library were used for cluster generation using the cBot (TruSeq PE Cluster Kit v3). Sequencing runs were performed on a HiSeq 1000 instrument using the indexed, 2×100 cycles paired end (PE) protocol and the TruSeq SBS v3 Kit. Image analysis and base calling resulted in.bcl files, which were converted into.fastq files by the CASAVA1.8.2 software.

Genomic DNA and RNA fragmentation, library generation and sequencing were performed at the local genomics core facility "KFB - Center of Excellence for Fluorescent Bioanalytics" at the University of Regensburg. Sequence data have been deposited under the SRA accession numbers PRJEB4837 (WGS data) and PRJEB4838 (RNA-seq data).

Bicompartmented System

Maize plants were grown in a bicompartmented system consisting of a root hyphal compartment (RHC) and a hyphal compartment (HC) containing 80 mg KH_2PO_4. By this experimental approach, Pi could be taken up by the mycorrhizal phosphate uptake pathway. Individual maize plants were grown in 1.1 kg of a sand:soil (0.71–1.25 mm; Quarzwerke GmbH)/soil (Stender Vermehrungssubstrat A210, Stender) mixture (9:1) in a rectangular container (root and hyphal compartment, RHC). The substrate was supplemented with roots of *Plantago lanceolata* colonized by *Glomus intraradices* Schenck and Smith (BEG75) (syn: *Rhizophagus irregulare*) (72 g/kg substrate) for the +AM treatment. As a hyphal compartment a small plastic vial was added upside down to the container including 40 g of a sand/soil mixture without inoculum and 80 mg of KH_2PO_4 as a Pi source. The plastic vial separates maize plant and Pi source via a semipermeable membrane, which is only permeable by fungal hyphae but not by roots of the plant. Three times a week plants were fertilized with a modified single strength Hoagland's nutrient solution without Pi (NH_4PO_4 replaced by NH_4Cl). Water-holding capacity was maintained at 75%. Plants were grown in the greenhouse at 24°C/20°C day/night temperature for 6–8 weeks. The first completely developed leaf (first source leaf) was harvested for ionomic analysis. For each condition/genotype three to four replicates were analyzed. For determination of the colonization degree by trypan blue staining a representative root sample was harvested per plant [35]. Above- and belowground plant material was dried to stable weight for biomass determination. Leaf number and the number of green leaves was determined.

Ionomics Analysis

A microwave system (Multiwave 3000, Anton Paar) was employed for complete digestion of plant material for ICP-MS analysis. Approximately 0.1 g of dried and homogenized plant material (leaves, seeds) was digested using 4 ml of concentrated HNO_3 (66%) and 2 ml of H_2O_2 (33%). The microwave run started with a 10 min power ramp followed by 30 min at 1400 W and finished with 15 min of cool down. A certified reference material "hay powder" (Community Bureau of Reference, No 129) was used for control of the digestion quality. A 7700 ICP-MS instrument (Agilent) was employed for determination of approx. 20 elements following the manufacturer's instructions.

Read Mapping and Variant Calling

WGS reads were mapped against the maize reference sequence (AGPv2) with BWA [36] version 0.5.9. The BWA command "aln" was called with the parameters "-I –q 20" for trimming off bad-quality at the ends of reads. Default BWA parameters were used otherwise. Variant and genotype calling were performed with SAMtools [37] version 0.1.19. The command "samtools mpileup" was called with the parameter "-D" to record per-sample read depth. The resulting VCF file was filtered with an AWK script (available as Text S3 of [38]). SNP position were retained if they were both samples had at least 10-fold read coverage, the SAMtools SNP quality score was at least 40, both samples were called homozygous (reference or alternative allele) with minimum genotype score of 10 and both samples had different genotype calls. Maize HapMap2 [2] genotype calls were downloaded from Panzea [39]. HapMap position were intersected with our SNP set using Tabix [40]. Aggregation of variants in 100 kb bins and visualization were performed with R scripts (http://www.r-project.org). R source code is provided as Text S1. Functional annotation of variant sites was performed with snpEff (version 3.2a) [41] using the filtered gene set of maize (version 5b.60, http://www.maizesequence.org).

RNA-seq reads from six samples were mapped against the maize B73 reference sequence with the Tophat spliced-alignment program [42] version 2.0.4 (parameters "-mate-std-dev 90 -r 0"). Fragment size statistics were determined by mapping reads with BWA against maize cDNA sequences downloaded from Phytozome [8,43]. Genotypes in all samples were called at the SNP positions determined from genomic reads. A read pileup at variant positions was generated with "samtools mpileup" for each sample. Variant positions were supplied with the parameter "-l". The resulting VCF files were imported into the R statistical environment. A genotype call was set to missing if the read depth was less than 10.

A multi-sample calling was performed with the SAMtools pipeline from RNA-seq reads using the same parameters as described for the WGS data. For the visualization of all homozygous and heterozygous SNPs called in the genomic and RNA-seq samples, SNPs were filtered for minimal coverage ($\geq$ 10x), SNP quality score ($\geq$40) and genotype score ($\geq$10). Filtered variant call files are available as Datasets S1 (WGS data) and Dataset S2 (RNA-seq data).

Analysis of Differential Gene Expression

Cufflinks [44] version 2.1.1 was used for the analysis of differential gene expression. The command "cuffdiff" was supplied with the maize reference annotation [8] (filtered gene set 5b60) and the Tophat mapping file of all six RNA-seq samples divided into two contrasting groups (set A and set B). In addition, the parameters "-u" and "-b" were used for multi-read correction and fragment bias correction, respectively. Tables containing the expression values and test results for each reference gene were imported into the R environment. The number of genes that were differentially expressed at a significance threshold of 0.05 after Benjamini-Hochberg correction with a minimal log-fold change $\geq$ 2 were counted in 10 Mb windows and visualized along the length of the maize genome. Predictions of molecular functions were downloaded from Gramene [45] and merged with the list of DE genes.

Acknowledgments

We thank all members of the OPTIMAS consortium for their encouragement and helpful discussions. We are grateful to Doreen Stengel for help with sequence data submission and acknowledge Caroline Gutjahr for valuable suggestions regarding the interpretation of the transcriptomic data.

Author Contributions

Conceived and designed the experiments: TD MB US. Performed the experiments: NG MG. Analyzed the data: MM NG. Contributed reagents/materials/analysis tools: TD MM NG MG MB US. Wrote the paper: MM NG MG TD. Analyzed sequence data: MM. Analyzed growth behavior and ionome data: NG. Provided seed material, NGS data and greenhouse space: TD MG. Provided growth chambers, a bicompartmented system and ICP-MS data: NG MB. Provided software analysis tools: MM US. Reviewed and approved the final manuscript: TD MM NG MG MB US.

References

1. Gore MA, Chia JM, Elshire RJ, Sun Q, Ersoz ES, et al. (2009) A first-generation haplotype map of maize. Science 326: 1115–1117.
2. Chia JM, Song C, Bradbury PJ, Costich D, de Leon N, et al. (2012) Maize HapMap2 identifies extant variation from a genome in flux. Nat Genet 44: 803–807.
3. Springer NM, Ying K, Fu Y, Ji T, Yeh CT, et al. (2009) Maize inbreds exhibit high levels of copy number variation (CNV) and presence/absence variation (PAV) in genome content. PLoS Genet 5: e1000734.
4. Lawrence CJ, Dong Q, Polacco ML, Seigfried TE, Brendel V (2004) MaizeGDB, the community database for maize genetics and genomics. Nucleic Acids Res 32: D393–397.
5. Romay MC, Millard MJ, Glaubitz JC, Peiffer JA, Swarts KL, et al. (2013) Comprehensive genotyping of the USA national maize inbred seed bank. Genome Biol 14: R55.
6. Lee M, Sharopova N, Beavis WD, Grant D, Katt M, et al. (2002) Expanding the genetic map of maize with the intermated B73 x Mo17 (IBM) population. Plant Mol Biol 48: 453–461.
7. Yu J, Holland JB, McMullen MD, Buckler ES (2008) Genetic design and statistical power of nested association mapping in maize. Genetics 178: 539–551.
8. Schnable PS, Ware D, Fulton RS, Stein JC, Wei F, et al. (2009) The B73 maize genome: complexity, diversity, and dynamics. Science 326: 1112–1115.
9. Tian F, Bradbury PJ, Brown PJ, Hung H, Sun Q, et al. (2011) Genome-wide association study of leaf architecture in the maize nested association mapping population. Nat Genet 43: 159–162.
10. Miclaus M, Wu Y, Xu JH, Dooner HK, Messing J (2011) The maize high-lysine mutant opaque7 is defective in an acyl-CoA synthetase-like protein. Genetics 189: 1271–1280.
11. Strack D, Fester T, Hause B, Schliemann W, Walter MH (2003) Arbuscular mycorrhiza: biological, chemical, and molecular aspects. J Chem Ecol 29: 1955–1979.
12. Willmann M, Gerlach N, Buer B, Polatajko A, Nagy R, et al. (2013) Mycorrhizal phosphate uptake pathway in maize: vital for growth and cob development on nutrient poor agricultural and greenhouse soils. Front Plant Sci 4: 533.
13. Casieri L, Gallardo K, Wipf D (2012) Transcriptional response of Medicago truncatula sulphate transporters to arbuscular mycorrhizal symbiosis with and without sulphur stress. Planta 235: 1431–1447.
14. Jansa J, Mozafar A, Frossard E (2003) Long-distance transport of P and Zn through the hyphae of an arbuscular mycorrhizal fungus in symbiosis with maize. Agronomie 23: 481–488.
15. Subramanian KS, Charest C (1997) Nutritional, growth, and reproductive responses of maize (Zea mays L) to arbuscular mycorrhizal inoculation during and after drought stress at tasselling. Mycorrhiza 7: 25–32.
16. Lu S, Braunberger PG, Miller MH (1994) Response of Vesicular-Arbuscular Mycorrhizas of Maize to Various Rates of P-Addition to Different Rooting Zones. Plant and Soil 158: 119–128.
17. Sawers RJ, Gutjahr C, Paszkowski U (2008) Cereal mycorrhiza: an ancient symbiosis in modern agriculture. Trends in Plant Science 13: 93–97.
18. Fester T, Fetzer I, Buchert S, Lucas R, Rillig MC, et al. (2011) Towards a systemic metabolic signature of the arbuscular mycorrhizal interaction. Oecologia 167: 913–924.
19. Kaeppler SM, Parke JL, Mueller SM, Senior L, Stuber C, et al. (2000) Variation among maize inbred lines and detection of quantitative trait loci for growth at low phosphorus and responsiveness to arbuscular mycorrhizal fungi. Crop Science 40: 358–364.
20. Colmsee C, Mascher M, Czauderna T, Hartmann A, Schluter U, et al. (2012) OPTIMAS-DW: a comprehensive transcriptomics, metabolomics, ionomics, proteomics and phenomics data resource for maize. BMC Plant Biol 12: 245.
21. Smith SE, Smith FA, Jakobsen I (2003) Mycorrhizal fungi can dominate phosphate supply to plants irrespective of growth responses. Plant Physiology 133: 16–20.
22. Ballachanda N, Devaiah, Ramaiah M, Athikkattuvalasu S, Karthikeyan, et al. (2009) Phosphate starvation responses and gibberellic acid biosynthesis are regulated by the MYB62 transcription factor in Arabidopsis. Molecular Plant 2: 43–58.
23. Ossowski S, Schneeberger K, Lucas-Lledo JI, Warthmann N, Clark RM, et al. (2010) The Rate and Molecular Spectrum of Spontaneous Mutations in Arabidopsis thaliana. Science 327: 92–94.
24. Lynch M, Sung W, Morris K, Coffey N, Landry CR, et al. (2008) A genome-wide view of the spectrum of spontaneous mutations in yeast. Proceedings of the National Academy of Sciences of the United States of America 105: 9272–9277.
25. Denver DR, Morris K, Lynch M, Thomas WK (2004) High mutation rate and predominance of insertions in the Caenorhabditis elegans nuclear genome. Nature 430: 679–682.
26. Schneeberger K, Ossowski S, Lanz C, Juul T, Petersen AH, et al. (2009) SHOREmap: simultaneous mapping and mutation identification by deep sequencing. Nat Methods 6: 550–551.
27. Ganal MW, Durstewitz G, Polley A, Berard A, Buckler ES, et al. (2011) A large maize (Zea mays L.) SNP genotyping array: development and germplasm genotyping, and genetic mapping to compare with the B73 reference genome. PLoS One 6: e28334.
28. Poland JA, Brown PJ, Sorrells ME, Jannink JL (2012) Development of high-density genetic maps for barley and wheat using a novel two-enzyme genotyping-by-sequencing approach. PLoS One 7: e32253.
29. Elshire RJ, Glaubitz JC, Sun Q, Poland JA, Kawamoto K, et al. (2011) A robust, simple genotyping-by-sequencing (GBS) approach for high diversity species. PLoS One 6: e19379.
30. Milberg P, Lamont BB (1997) Seed/cotyledon size and nutrient content play a major role in early performance of species on nutrient-poor soils. New Phytologist 137: 665–672.
31. McFarland JW, Ruess RW, Kielland K, Pregitzer K, Hendrick R, et al. (2010) Cross-Ecosystem Comparisons of In Situ Plant Uptake of Amino Acid-N and NH4 (+). Ecosystems 13: 177–193.
32. Dooner HK, Robbins TP, Jorgensen RA (1991) Genetic and Developmental Control of Anthocyanin Biosynthesis. Annual Review of Genetics 25: 173–199.
33. Pegoraro C, Mertz LM, da Maia LC, Rombaldi CV, de Oliveira AC (2011) Importance of heat shock proteins in Maize. Journal of crop science and Biotechnology 14: 85–95.
34. Dixon DP, Skipsey M, Edwards R (2010) Roles for glutathione transferases in plant secondary metabolism. Phytochemistry 71: 338–350.
35. Brundrett M, Bougher N, Dell B, Grove T, Malajczuk N (1996) Working with mycorrhizas in forestry and agriculture. Australian Centre for International Agricultural Research, Canberra.
36. Li H, Durbin R (2009) Fast and accurate short read alignment with Burrows-Wheeler transform. Bioinformatics 25: 1754–1760.
37. Li H (2011) A statistical framework for SNP calling, mutation discovery, association mapping and population genetical parameter estimation from sequencing data. Bioinformatics 27: 2987–2993.
38. Mascher M, Wu S, Amand PS, Stein N, Poland J (2013) Application of genotyping-by-sequencing on semiconductor sequencing platforms: a comparison of genetic and reference-based marker ordering in barley. PLoS One 8: e76925.
39. Zhao W, Canaran P, Jurkuta R, Fulton T, Glaubitz J, et al. (2006) Panzea: a database and resource for molecular and functional diversity in the maize genome. Nucleic Acids Res 34: D752–757.
40. Li H (2011) Tabix: fast retrieval of sequence features from generic TAB-delimited files. Bioinformatics 27: 718–719.

41. Cingolani P, Platts A, Wang le L, Coon M, Nguyen T, et al. (2012) A program for annotating and predicting the effects of single nucleotide polymorphisms, SnpEff: SNPs in the genome of Drosophila melanogaster strain w1118; iso-2; iso-3. Fly (Austin) 6: 80–92.

42. Trapnell C, Pachter L, Salzberg SL (2009) TopHat: discovering splice junctions with RNA-Seq. Bioinformatics 25: 1105–1111.

43. Goodstein DM, Shu S, Howson R, Neupane R, Hayes RD, et al. (2012) Phytozome: a comparative platform for green plant genomics. Nucleic Acids Res 40: D1178–1186.

44. Trapnell C, Roberts A, Goff L, Pertea G, Kim D, et al. (2012) Differential gene and transcript expression analysis of RNA-seq experiments with TopHat and Cufflinks. Nat Protoc 7: 562–578.

45. Ware D, Jaiswal P, Ni JJ, Pan XK, Chang K, et al. (2002) Gramene: a resource for comparative grass genomics. Nucleic Acids Research 30: 103–105.

4

A SUPER Powerful Method for Genome Wide Association Study

Qishan Wang[1,9], Feng Tian[2*,9], Yuchun Pan[1*], Edward S. Buckler[3,4], Zhiwu Zhang[4,5,6*]

1 School of Agriculture and Biology, Shanghai Jiaotong University, Shanghai, China, 2 National Maize Improvement Center of China, China Agricultural University, Beijing, China, 3 United States Department of Agriculture (USDA) – Agricultural Research Service (ARS), Ithaca, New York, United States of America, 4 Institute for Genomic Diversity, Cornell University, Ithaca, New York, United States of America, 5 Department of Animal Science, Northeast Agricultural University, Harbin, China, 6 Department of Crop and Soil Science, Washington State University, Pullman, Washington, United States of America

Abstract

Genome-Wide Association Studies shed light on the identification of genes underlying human diseases and agriculturally important traits. This potential has been shadowed by false positive findings. The Mixed Linear Model (MLM) method is flexible enough to simultaneously incorporate population structure and cryptic relationships to reduce false positives. However, its intensive computational burden is prohibitive in practice, especially for large samples. The newly developed algorithm, FaST-LMM, solved the computational problem, but requires that the number of SNPs be less than the number of individuals to derive a rank-reduced relationship. This restriction potentially leads to less statistical power when compared to using all SNPs. We developed a method to extract a small subset of SNPs and use them in FaST-LMM. This method not only retains the computational advantage of FaST-LMM, but also remarkably increases statistical power even when compared to using the entire set of SNPs. We named the method SUPER (Settlement of MLM Under Progressively Exclusive Relationship) and made it available within an implementation of the GAPIT software package.

Editor: Yun Li, University of North Carolina, United States of America

Funding: This study was supported by NSF-Plant Genome Program (DBI- 0820619), National Natural Science Foundation of China (grant no 31370043, 31272414), National 948 Project of China (2011-G2A,2012-Z26), National High Technology Research and Development Program of China (2012AA101104 and 2012AA10A307), and the United States Department of Agriculture's Agricultural Research Service. The funders had no role in study design, data collection and analysis, decision to publish, or preparation of the manuscript.

Competing Interests: The authors have declared that no competing interests exist.

* Email: ft55@cau.edu.cn (FT); panyc@sjtu.edu.cn (YP); zhiwu.zhang@wsu.edu (ZZ)

9 These authors contributed equally to this work.

Introduction

Genome-Wide Association Study (GWAS) has become the leading method to identify genes underlying human diseases and agriculturally important traits. However, the genetic variants identified so far only explain a small portion of phenotypic variation [1]. Rare genes and genes without large effect still remain unidentified due to lack of statistical power [2]. Statistical power is determined by many factors such as gene effect, allele frequency, sample size, marker density, and null distribution of type I error [3]. Inflation of type I error (false positives) leads to more false discoveries than expected [4,5].

Population stratification and cryptic relationships are two common reasons for the inflation of false positives [6,7]. Compared to the general linear model (GLM), the Mixed Linear Model (MLM) method effectively eliminates false positives by incorporating these two factors simultaneously [8]. The population stratification is fit as a fixed effect through population structure [6] or principal components [9]. The cryptic relationship among individuals is joined with variance components to collectively define variance and covariance of the random genetic effects from individuals.

The number of individuals in the population largely determines the size of a MLM equation [10]. The computing complexity of solving a MLM is a cubic function of the number of individuals. It is prohibitive to solve a MLM with large number of individuals, especially with iterations to estimate unknown variance components [11]. Several advances have partially solved the computational problem. The Efficient Mixed-Model Association (EMMA) algorithm turns the two-dimensional optimization of genetic and residual variance components into one dimensional optimization by deriving the likelihood as a function of their ratio [12].

Efforts have been made to change the computational function from cubic to quadratic, especially for marker screening, which dominates the entire computation for data with high marker density. The Population Parameter Previously Determined (P3D), or Efficient Mixed-Model Association eXpedited (EMMAX), estimates variance components (or their ratio) only once and then fixes them to test genetic markers [13,14]. Furthermore, an exact method, Genome-wide Efficient Mixed-Model Association (GEMMA), was developed to estimate the population parameters for each testing marker with the similar computational efficiency of P3D or EMMAX [15].

The method of compressed MLM [13] clusters individuals into groups and fits the groups as the random effect. The computing complexity function is thus reduced from the cubic of the number of individuals to the cubic of a smaller number of groups. However, the cubic property still remains. In practice, the maximum compression (i.e., the average number of individuals per group) observed is only about twenty-fold [16]. Consequently, solving a MLM is still prohibitive with extremely large numbers of individuals.

The Factored Spectrally Transformed Linear Mixed Model (FaST-LMM) partitions the cubic function of computing complexity as the product of two parts: 1) the number of individuals and 2) the square of the rank of the relationship among individuals [17]. When all the genetic markers (usually much larger than the number of individuals) are used to define the relationship among individuals, the kinship among individuals has full rank (i.e., is the same as the number of individuals). The computing complexity is still cubic to the number of individuals. Using a small subset of randomly selected markers to define a rank-reduced relationship has been suggested [17]. When the small subset has a constant number of Single Nucleotide Polymorphisms (SNPs) relative to the number of individuals, the computing complexity becomes linear to the number of individuals. The authors of FaST-LMM show a few examples using a small subset of randomly selected markers to define kinship that have similar results to those using all genetic markers [17]. Further the study demonstrated that a small set of associated genetic markers has better statistical power than a small set of genetic markers selected randomly. The small set of associated genetic markers are used in such way that some of these markers are removed for defining individual relationship if they are from the same region of the testing markers (e.g., within 2 Mb) [18]. The size and content of the set of markers selected becomes critical for computing speed and statistical power.

In this study, we developed a method that dramatically reduces the number of genetic markers used to define individual relationships and remarkably increases statistical power. First, we divide the whole genome into small bins. Each bin is represented by the most significant marker. Second, we select only the influential bins. Third, we use a maximum likelihood method to optimize the size and number of bins selected as the pseudo Quantitative Trait Nucleotides (QTNs) underlying the phenotypes. Fourth, in the final test of each marker, the small set of markers is used to define the relationship among the individuals by excluding the markers that are in Linkage Disequilibrium (LD) to the testing marker, regardless local distance. We call the algorithm the Settlement of MLM Under Progressively Exclusive Relationship (SUPER).

Materials and Methods

SUPER method

We developed the SUPER method in the framework of a standard MLM approach, which decomposes the observation ($\mathbf{Y}$) into fixed effect ($\boldsymbol{\beta}$), random genetic effect ($\mathbf{u}$) and residual ($\mathbf{e}$) as follows.

$$\mathbf{y} = \mathbf{X}\boldsymbol{\beta} + \mathbf{Z}\mathbf{u} + \mathbf{e} \tag{1}$$

where $\mathbf{u}$ is a vector of size n (number of individuals) for unknown random polygenic effects having a distribution with mean of zero and covariance matrix of $\mathbf{G} = 2\mathbf{K}\sigma_a^2$, where $\mathbf{K}$ is the kinship (co-ancestry) matrix with element $\mathbf{K}_{ij}$ (i, j = 1, 2, ..., n) calculated from genetic markers, and is an unknown additive genetic variance. $\mathbf{X}$ and $\mathbf{Z}$ are the incidence matrices for $\boldsymbol{\beta}$ and $\mathbf{u}$, respectively, and random residual effects $\mathbf{e}$ are normally distributed with zero mean and covariance $\mathbf{R} = \mathbf{I}\sigma_e^2$, where $\mathbf{I}$ is the identity matrix and is the

unknown residual variance. Solving equation (1) involves determining all the unknown parameters under which the observations ($\mathbf{y}$) have the maximum likelihood, defined as the following:

$$L(\mathbf{y}|\beta,\sigma_a^2,\sigma_e^2) \tag{2}$$

To perform a GWAS, marker effect ($\mathbf{v}$) is added to equation (1), one at a time:

$$\mathbf{y} = \mathbf{W}\mathbf{v} + \mathbf{X}\boldsymbol{\beta} + \mathbf{Z}\mathbf{u} + \mathbf{e} \tag{3}$$

where $\mathbf{W}$ is the incidence matrix for $\mathbf{v}$. Solving equation (3) by using P3D [13] or EMMAX [14] only involves optimization of $\mathbf{v}$ and $\boldsymbol{\beta}$ to optimize following likelihood:

$$L(y|v,\beta,\hat{\sigma}_a^2,\hat{\sigma}_e^2) \tag{4}$$

where, $\hat{\sigma}_e^2$ are estimates to maximize equation (2).

Kinship (K) is a known parameter, which is derived from genetic markers. Consequently, different sets of genetic markers create different kinships. This is the only difference among all the methods compared in this study. We used the efficient algorithm [19] of Van Raden et.al. (implemented in GAPIT [20]) to calculate the kinship matrix. The first method is to use the QTNs only. The second method is to use all the SNPs including QTNs. The third method is to use all SNPs except QTNs. The second and third methods are barely different when the number of SNPs is large. The fourth method is similar to the first method in respect of using QTNs. The difference is that a QTN is excluded for deriving the kinship when the testing SNP is the same as the QTN. The kinship is called complementary trait specific kinship. The fifth method is similar to the fourth method except that the QTNs are masked and have to be identified by estimation. Therefore, the method can be used in practice where the true QTNs are unknown. We developed a procedure to find QTN-like SNPs, called pseudo QTNs.

Our procedure consists of three steps. The first two steps perform the inclusion of pseudo QTNs. The last step performs GWAS with exclusion of the pseudo QTNs that are in LD with the tested SNP.

Step 1: To sort SNPs on their p values or effects through a preliminary GWAS or genomic prediction for a specific trait.

Step 2: For each bin (segment) on a chromosome, choose the most influential SNP (e.g., with the lowest P value) as the representative for the bin. Then select s most influential bins to build kinship. The size of bins and number of bins chosen are treated as parameters to maximize the restricted maximum likelihood for a trait. The s selected SNPs (each represent a bin) are then used as a base of a SNP pool to define individual relationships for the later association test. More precisely, we optimize the following likelihood:

$$L(y|\beta,\sigma_a^2,\sigma_e^2,s,b) \tag{5}$$

where s and b are the number and size of bins.

Step 3: When testing a SNP in equation (3), we exclude the SNPs in the SNP pool that are in LD with the testing SNP to derive a complementary trait specific kinship. We call this method as the Settlement Under Progressively Exclusive Relationship (**SUPER**).

Solving equation (3) only involves the optimization of **v** and **β** to optimize following likelihood:

$$L(y|v,\beta,\sigma_a^2,\sigma_e^2,\hat{s},\hat{b}) \tag{6}$$

Where $\hat{\sigma}_a^2$, $\hat{\sigma}_e^2$ and $\hat{b}$ are estimates to maximize equation (5).

Real Data

Six published datasets from dog, maize, rice, *Arabidopsis*, mouse, and human were examined. The datasets from dog, maize, and rice were the same datasets used in our previous study [13,16]. The dog dataset was sampled from a dataset used for mapping Quantitative Trait Loci (QTLs) underlying canine hip dysplasia [21] and a dataset used to estimate heritability of canine hip dysplasia [22]. The data contained 292 dogs from two breeds (Labrador Retriever and Greyhound) and their crosses (F_1, F_2, and two backcrosses). All dogs were genotyped with 23,500 SNPs at genome-wide coverage.

The maize data contained 282 inbred lines. The genotypes (2,911 SNPs) were released as a tutorial dataset of the TASSEL and GAPIT software packages [23].

The rice data contained 374 inbred lines, 50,000 SNPs randomly sampled from the one million SNPs from genotyping by sequencing technology [16].

The *Arabidopsis* dataset included 199 landraces genotyped by 216,130 SNPs [24]. We randomly sampled 50,000 SNPs for this study.

The mouse data contained 688 34th generation advanced intercross lines (AIL) derived from two inbred strains (SM/J and LG/J). The genotype data contained 3,117 SNPs [25]. The methamphetamine-induced locomotor activity on day 3 was used to compare SUPER with other methods.

The Human Framingham Heart Study (FHS) data were downloaded from the database of Genotypes And Phenotypes (dbGAP) databases (phg000005.v5). The total Cholesterol (Off-spring exams 7) was used as the phenotype for the association study. The present study sample comprised 806 FHS offspring participants who were genotyped using the 100K Affymetrix GeneChip and have fasting blood lipid traits for exams 7. We imputed the missing values using mean values by the program GCTA [26]. The genotype data consist of 57,581 SNPs on 22 autosomes after exclusion of rare SNPs with Minor Allele Frequency(MAF) less than 0.1 and SNPs with missing genotypes more than 5%. We adjusted the test to control for age, gender, and body mass index to perform GWAS.

Phenotype simulations

A set of SNPs was randomly sampled as causal QTNs for the simulated traits (27, 20, 24, and 20 QTNs for maize, *Arabidopsis*, rice, and dog, respectively). The location of QTNs were restricted under two scenarios. One scenario was implemented for all the species without any restriction, e.g., a QTN could be any SNP. The other scenario was implemented on the maize dataset only where the last chromosome was excluded to sample QTNs. The last chromosome in the second scenario was used to investigate the effect of a clear null distribution, i.e., no genetic correlation existed between QTNs and non-QTN SNPs.

The distribution of these QTN effects followed a normal distribution with a mean of 0 and variance of 1. Phenotypes were simulated as the following equation: y = additive + residual. For each individual, the total additive effect is calculated as the sum of additive effects across all QTNs. The residual variance was calculated as Ve = Va(1-h2)/h2, where Va is the additive genetic variance and h2 is the heritability. A residual error following a normal distribution with mean of 0 and variance of Ve was added to the total additive effect to form the simulated phenotype for each individual. Heritability was set to 0.75 for examination of statistical power in all datasets. Another five levels of heritability (h2 = 0, 0.25, 0.4, 0.5 and 1) were set to further compare the statistical power of SUPER with other methods by using the maize dataset.

Null distribution and power examination

The association tests on the markers were performed by conducting F tests. In the scenario that sampled QTNs without any restriction, the empirical distribution of the non-QTN markers was used as the null distribution of type I error. For the second scenario—last chromosome was excluded for sampling QTNs—the empirical distribution of the markers on the last chromosome was used as the null distribution of type I error. The power is examined as the proportion of QTNs that pass a testing threshold for a given type I error (5%). A total of 100 replications were conducted for each method and the average over the 100 replicates was reported.

Ethics statement

All the datasets analyzed herein have been previously published. This study did not obtain actual samples from human or animals.

Results

Through simulations, we demonstrated that the effective components in the small set of selected genetic markers are the QTNs underlying a trait. To remove the confounding between the QTNs and testing markers, the exclusion of QTNs is more effective when LD is used instead of local distance. We examined our proposed method for the practical situations where QTNs are unknown.

We compared SUPER and other popular mixed model methods through a series of simulations. The difference among these methods is how to build kinship. We showed that a small subset of randomly selected genetic markers will not always produce the equivalent statistical power compared to using all genetic markers (**Figure 1a**). The average statistical power of the small subset of randomly selected genetic markers was significantly less than the power by using all genetic markers (p<0.01). The statistical power was about 50% when using all the markers in a maize dataset with 282 individuals. It does not make difference to include or exclude QTNs as the number of markers is usually much larger than the number of QTNs underlying a trait. Exclusion of all the markers in LD with QTNs, does not make difference compared to using all the markers to build kinship (**Figure 1c, Table 1**).

In the above simulation study, 35% of the time the small set of randomly selected SNPs had higher power than using all SNP kinship. This finding indicates that the gold-standard kinship of using all SNPs is definitely not the best choice. So, the interesting question is: what type of small subset of SNPs produces higher power than using all the SNPs? We were motivated by the fact that a trait specific kinship derived from weighted SNPs has better prediction accuracy than the kinship derived from all the SNPs in genomic prediction [27].

However, when we applied kinship from all the QTNs for GWAS, we found that statistical power decreased to about 30%, which was much lower than using kinship derived from all SNPs. This result is not surprising because the kinship derived from all

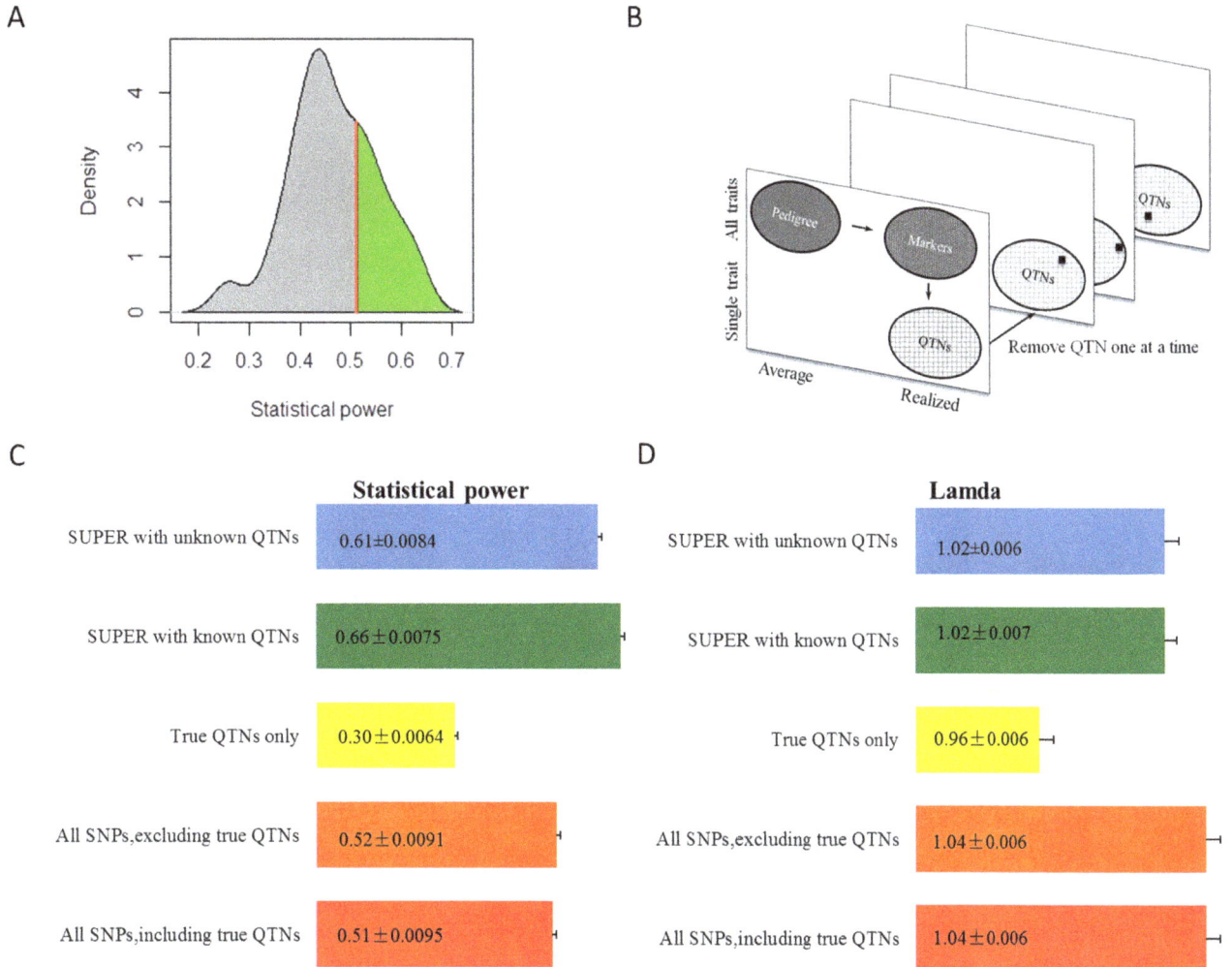

Figure 1. Conception and performances of different methods. A) Distribution of statistical power by using a kinship derived from a set of SNPs selected randomly. The dataset contained ~3,000 SNPs genotyped on 282 maize inbred lines. The number of selected SNPs was the same as the number of individuals used to derive kinship. Power was examined on a trait simulated from 27 causative mutations, i.e. Quantitative Trait Nucleotide (QTNs), sampled from the ~3,000 SNPs except the ones on the last chromosome. The SNPs on the last chromosome were used to derive the null distribution of Type I error. The heritability of the trait was set to 0.75. A total of 100 replications were conducted. The average and the median power are 0.476 and 0.444. The power of using kinship derived from all SNPs is 0.511 (red line). **B)** Conception of kinship for association study. Pedigree is the first available information used to calculate kinship. It is the expectation for a pair of individuals to be identical by descent at any locus, (e.g., full siblings have a kinship of 50% in cases of no inbreeding). Pedigree kinship can be used across traits. A realized kinship derived from genetic markers covering entire genome is more precise than pedigree based (e.g., full siblings could have a kinship of 60% - or 40% - instead of 50%). However, it is still general and can be used for all traits. A complete trait specific realized kinship is using all the QTNs underlying the trait. This complete trait specific kinship is ideal for genome prediction, but not for GWAS. The ideal kinship for GWAS is its complement (using all QTNs except the one being tested) to remove the confounding between the kinship and the tested SNPs. **C)** and **D)** display the performance of statistical power and effectiveness of genomic control of inflation factor by using different kinship. The statistical power is about 50% when using all the SNPs. Inclusion or exclusion of the 27 QTNs did not have a significant impact. When only the 27 QTNs were used to derive a complete trait specific kinship, the statistical power was dramatically reduced to 30%. When each of the 27 QTNs was tested by using the complementary trait specific kinship derived from the other 26 QTNs (SUPER with known QTNs), the statistical power was boosted to 66%. A statistical power of 61% was retained by using SUPER with masked QTNs. The genomic control of SUPER was similar with known QTNs and with masked QTNs, closer to expectation (1.00) than other methods.

QTNs is confounded with the effect of the tested SNP if this SNP is one of the QTNs.

This finding confirmed the strategy for selecting the kinship method for GWAS. When testing a SNP, we remove the SNP from the QTN list if the SNP is a QTN. We then use the remaining QTNs to derive a complementary trait specific relationship for the SNP (**Figure 1b**). When the complementary trait specific relationship is applied to GWAS, statistical power is boosted to 66% for the 282 maize dataset, which is much higher than using all SNPs.

For the real situation, where QTNs are unknown, we developed an algorithm to derive a set of pseudo-QTNs for the SUPER method. The algorithm involves three steps. The first step is to perform a preliminary GWAS to sort SNPs. The second step determines the size and number of bins that give the maximum likelihood for a specific trait. Then, for each bin, the most associated SNP is used as the pseudo-QTN to represent that bin.

Table 1. Statistical power of using different kinship for four species (*Arabidopsis*, Rice, Dog and Maize).

Method to build kinship	Arabidopsis	Rice	Dog	Maize
All SNPs, including true QTNs	0.63±0.0070	0.52±0.0063	0.59±0.0079	0.51±0.0095
All SNPs, excluding true QTNs	0.63±0.0072	0.52±0.0061	0.59±0.0083	0.52±0.0091
True QTNs only	0.42±0.0066	0.29±0.0064	0.40±0.0083	0.30±0.0064
SUPER with known QTNs	0.75±0.0065	0.65±0.0057	0.72±0.0076	0.66±0.0075
SUPER with unknown QTNs	0.72±0.0063	0.60±0.0059	0.68±0.0078	0.61±0.0084

A set of SNPs was randomly sampled as causal QTNs for the simulated traits (0.04%, 0.05%, 0.085% and 1%, of the total number of SNPs for *Arabidopsis*, Rice, Dog, and Maize, respectively). The statistical power was estimated with heritability of 0.75. Power is defined as the proportion of QTNs detected under type I error of 5%. A total of 100 replications was conducted for each method. The statistical power shown here is the average of 100 replications.

The size and number of bins are the two parameters chosen for optimization. The third step is to perform the complementary process in GWAS by excluding the pseudo-QTNs that are in LD with the tested SNP. The remaining pseudo-QTNs are used to define the complementary relationship among individuals. In the simulation study where the QTNs were masked, we obtained a statistical power of 61%, lower than the situation with known QTNs, but still much higher than using all SNPs (**Figure 1c, 1d**).

We extended our examination of statistical power against the type I error for four methods: SUPER, FaST-LMM-Select, EMMAX, and GLM (**Figure 2a**). The SUPER method is consistently better than the others over the entire range of type I errors. The GLM is consistently the worst. The FaST-LMM-Select and EMMAX performed better than GLM. We also compared the statistical power under different levels of heritability. When a trait is more heritable, (e.g., heritability 0.25), the four methods perform differently from each other. FaST-LMM-Select performs better than EMMAX and GLM. SUPER performs better than FaST-LMM-Select (**Figure 2b**).

We explored several ways to reduce the computing time of SUPER. First, we examined the effect using P3D/EMMAX [13,28]. We found SUPER works well with P3D/EMMAX to reduce computing time and retain similar statistical power. No significant difference (p>0.05) in power was found whether we used P3D/EMMAX or not. Thus, re-estimating population parameters (e.g., genetic variance, residual variance, or their ratio) for testing each SNP is unnecessary. This completely eliminates the iteration time to optimize these population parameters for screening SNPs (**Figure S1**).

Second, we explored speeding up computation by using a fast method to derive the P values at the first stage of SUPER. Three methods were compared: GLM, MLM [8], and Compressed MLM (CMLM) [13]. The GLM method is much faster than the other two methods. Although using GLM in the first step tends to have less power than using the other two methods, the difference is not significant. Thus, even when using GLM to keep computing cost low, the statistical power of the SUPER method is not affected significantly (**Figure S2**).

Third, we provided a procedure to determine the threshold of LD between tested SNPs and QTNs. When the threshold is too high (e.g., $r^2 = 100\%$), QTNs are barely removed. The result should be similar to the complete trait specific kinship. In the opposite case, where the threshold is too low (e.g., $r^2 = 0.01\%$), QTNs are hardly survived in the exclusion process. The kinship matrix does not retain much information and the results would be similar to the GLM. We observed that a threshold of $r^2 = 10\%$ was best for both maize and rice. This threshold also worked well for the other species (dog and *Arabidopsis*) we examined (**Figure S3**).

Nevertheless, this finding only provides guidance for the optimizations, which might be necessary for other populations or species.

We examined our findings for a variety of circumstances. We verified the effect of the correlation between QTNs and the non-QTN SNPs. The non-QTN SNPs were used to derive the empirical null distribution of type I error. Two scenarios were examined. In the first scenario, no correlation was found because QTNs and non-QTN SNPs were sampled from different chromosomes. In the second scenario, correlation was possible because random sampling might place QTNs and non-QTN SNPs next to each other. We observed that, in either case, our findings still held. That is, 1) SUPER with known QTNs had the highest statistical power, 2) complete trait specific kinship had the lowest power, 3) kinship from all SNPs was in the middle, and 4) SUPER with unknown QTNs fell between SUPER with known QTNs and the kinship from all SNPs (**Figure S4**).

We then examined the impact of the magnitude of QTN effect (**Figure S5**) and heritability (**Figure S6**). We observed the same trend in statistical power as above. SUPER with known QTNs is the best and SUPER with unknown QTNs is the second best.

We expanded the comparisons of SUPER with EMMAX and FaST-LMM-Select to real traits. The first is from the Advanced Intercross Line (AIL) mouse data [25]. Manhattan plots of all mouse data for the three different methods are shown in **Figure 3** (A to C). The SNPs identified using SUPER at a Bonferroni correction threshold of 0.05 and a False Discovery Rate (FDR) less than 0.1 are listed in **Table 2**. Using the SUPER method, we identified all the associations previously detected by the original paper. Two of these significant SNPs were located in known genes (Rsrc2 and Pitpnm2) [29,30]. EMMAX and FaST-LMM-Select did not identify significant SNPs that reached the same threshold.

The second real data is from the Human Framingham Heart Study (FHS) project. Missing genotypes were imputed. As FaST-LMM-Select does not accept dosage genotypes, the comparison was performed between SUPER and EMMAx. Manhattan plots of total cholesterol for SUPER and EMMAX are shown in **Figure 3** (D and E). Neither method identified significant SNPs that reached the Bonferroni correction threshold of 0.05. However, using the SUPER method, we identified two significant SNPs (rs1599231 and rs898408) at a FDR less than 0.1. The identified SNPs are located in known gene (*CACNA1D*) associated with cholesterol [31]. EMMAX did not identify significant SNPs at this FDR threshold.

With the SUPER method, the restriction of the computationally efficient FaST-LMM method is no longer a problem. Their joint usage retains the similar computing speed while remarkably improving the statistical power.

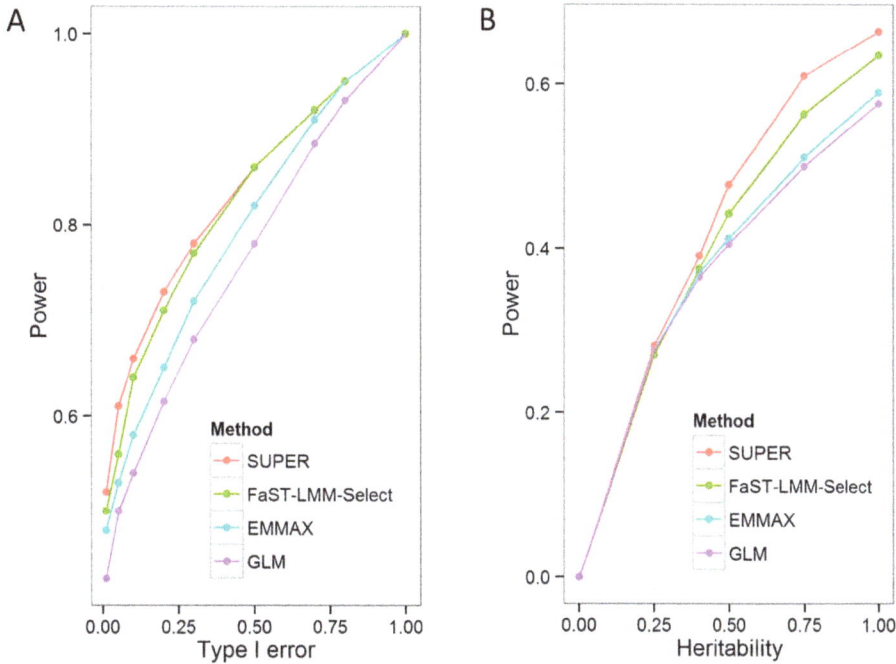

Figure 2. Statistical power under different ranges of type 1 error and heritability. A) Statistical power was examined on a trait simulated from 27 causative mutations (QTNs) sampled from SNPs on chromosomes 1 to 9 in maize data. The SNPs on the last chromosome (10) were used to derive null distribution. Power was defined as the proportion of detected QTNs under type I error of 5%. A total of 100 replications was conducted for each method. The heritability of the trait was set to 75%. Four methods were examined: 1) SUPER; 2) EMMAX; 3) FaST-LMM-Select; and 4) General linear model (GLM). **B**) Statistical power of four methods under different heritability levels. The four methods are SUPER, LMM-Selected, EMMAX and GLM.

Discussion

The concept of complementary trait specific kinship reflects a landmark in the development of kinship. As essential information in population and quantitative genetics, kinship is traditionally derived from pedigree as an expected chance that two individuals share the same allele by descent [32]. The pedigree-kinship relationship has been widely used to study human diseases and predict breeding values for animals and plants [33,34].

An alternative way to derive kinship is to rely on genetic markers [35,36]. This marker-based kinship more precisely specifies the actual difference between individuals. Some of these differences are not distinguishable using the kinship derived from pedigree [37]. For example, all full siblings have the same relationship with each other based on pedigree. These relationships become distinguishable with genetic markers. The realized kinship revealed by markers could be quite different from the kinship derived from pedigree due to factors like allele sampling and segregation distortion [37]. The realized kinship is superior to the pedigree kinship for ranking individuals for their genetic merit [38]. When the realized kinship is used jointly with population structure in a MLM for GWAS, it performs well in controlling false positives [8]. Furthermore, the realized kinship can be derived for a specific trait using the markers that are influential to the trait. This trait specific kinship produces higher prediction accuracy than the universal realized kinship [27].

Obviously, the best kinship to define individual genetic relationship on a complex trait is the one derived from all the QTNs underlying the trait as they define it [39]. Adding additional SNPs (non QTNs) would dilute the actual relationship. Complete trait specific kinship works the best for genomic prediction [27]. But, when used for GWAS, the markers defining the kinship are confounded with the tested markers, consequently decreasing the statistical power of GWAS.

However, less obvious, is that a small proportion of randomly sampled SNPs would have higher statistical power than using all SNPs. The increased power might result from the combination of the following factors: 1) sampled SNPs contain QTNs or SNPs in LD with QTNs, 2) fewer non-QTN SNPs result in less dilution, and 3) a portion of QTNs, or SNPs in LD with QTNs are excluded and become more detectable.

There is a random chance that a small subset of SNPs selected randomly could have higher power than using all SNPs. In general, the randomly selected subsets of SNPs have less power. Therefore, randomly selecting a small set of SNPs is unsafe. The goal of this study was to find a better method to find subsets. Ideally, the subset contains fewer SNPs than number of individuals and has the same or higher power than using all the SNPs.

FaST-LMM-Select has been undertaken to find small subsets of SNPs [18]. Similar to SUPER, the strategy works best for a scenario in which a complex trait is controlled by genes with large effect, small effect, and anything between. For an extreme case having only a few (e.g., 1 to 3) genes with major effects and the rest (e.g., 500) with very small effects, the power will be saturated to 100% for the major genes even with a small sample and a simple method. However, the rest of the genes will have no power regardless of method, including FaST-LMM-Select or SUPER proposed in this study if the sample is not large enough.

Our study was unique in a several ways. Overall, our study gives the biological, inside-view for the statistical phenomena observed in the FaST-LMM-Select study. Through a series of simulations, we proved that their finding — that using a small set of randomly selected SNPs generates the equivalent statistical power as using all

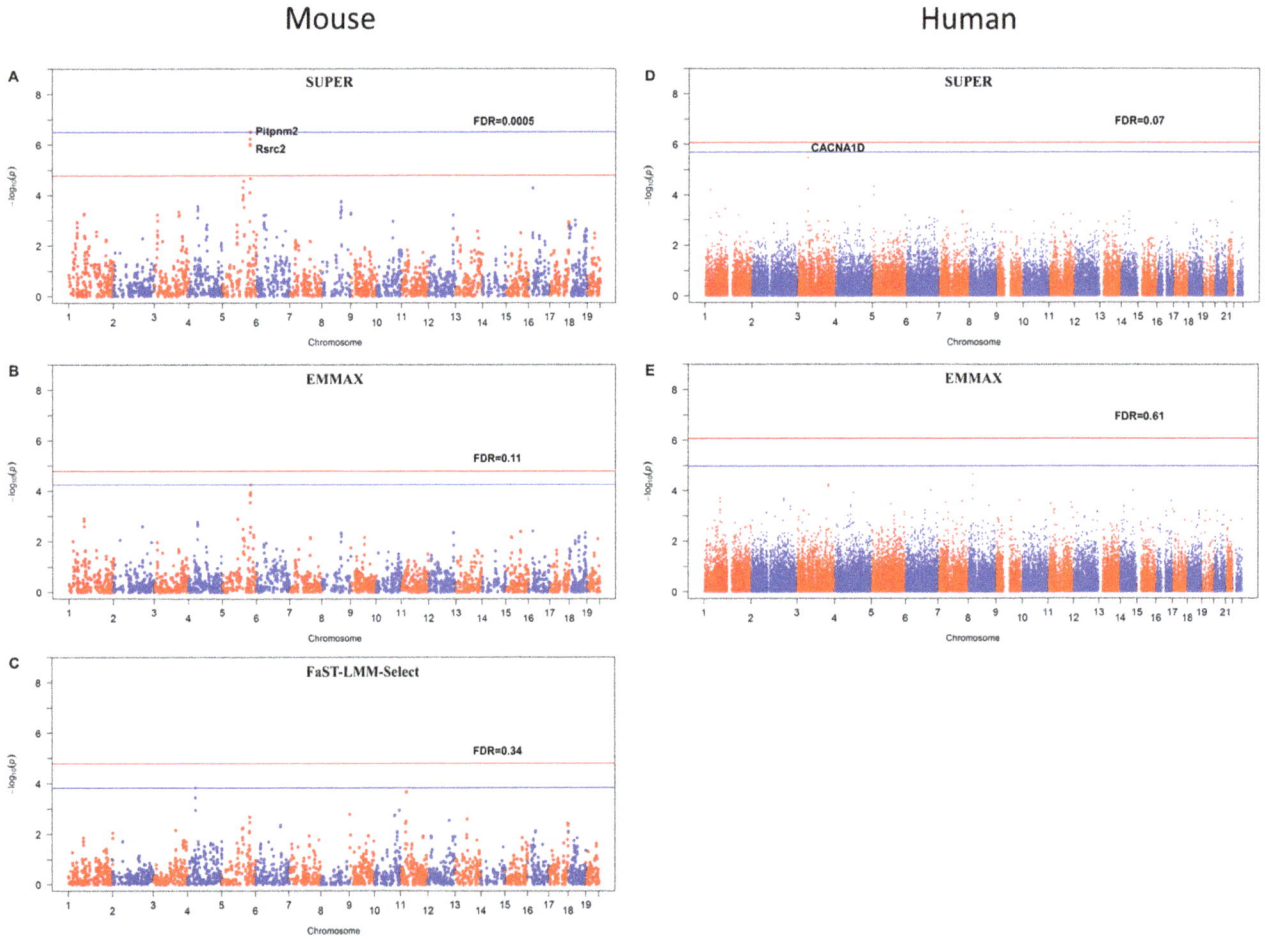

Figure 3. Results of association studies on real mouse and human phenotypes. The mouse phenotype is methamphetamine-induced locomotor activity on day 3 measured on 688 Advanced Intercross Lines (AIL). The human phenotype is cholesterol collected by the Framingham Heart Study (FHS) Project. Each dataset was analyzed with three different methods (SUPER, EMMAX, and FaST-LMM-Select) except the combination between FaST-LMM-Select and human data. The missing genotypes in the human data were imputed in format of dosage, which is not accepted by FaST-LMM-Select. The most significant SNP is highlighted by a horizontal blue line and labeled by its corresponding False Discover Rate (FDR). The p value threshold of 0.05 (after bonferroni multiple test correction) is indicated by a horizontal red line.

the SNPs — is not always true. In fact, statistical power can be reduced significantly. This result is not surprising as a small random sample of SNPs is less informative than using all the SNPs [40].

Furthermore, we explained why the kinship for GWAS should be specific for a trait and complementary to a testing SNP. We started with known QTNs and showed how different scenarios

impact statistical power, such as using all QTNs or using QTNs excluding the one being tested. These studies demonstrate how the inclusion of all QTNs confounds with the effects of testing SNPs when compared to all SNPs and how the exclusion of QTNs eliminates the confounding.

We applied the method derived from situations with known QTNs to real-life situations with unknown QTNs. We developed

Table 2. SNPs found to be significant by SUPER and other three methods for AIL mouse data.

SNP	Chromosome	Position	EMMAX	Fast_LMM_Select	SUPER	Gene
5-122651666	5	125405148	5.60E-05	2.15E-03	**3.22E-07**	
mUC-rs13478501	5	124051672	1.47E-04	2.81E-02	**6.09E-07**	
NES14715162	5	124119050	1.28E-04	8.81E-03	**9.82E-07**	Pitpnm2
5-122053167	5	124768242	1.15E-04	7.75E-03	**9.98E-07**	
5-121026072	5	123740172	2.80E-04	4.46E-02	**1.04E-06**	Rsrc2

P values that reached the Bonferroni correction threshold (1.6E-5) are shown in bold.

the algorithm to find their representatives (pseudo-QTNs) and demonstrated that the SUPER approach has statistical power close to that achieved with known QTNs. We determined the set of pseudo-QTNs by optimizing bin size and bin number to define the trait through a method of maximum likelihood. This set of pseudo-QTNs is the best combination among all SNPs compared with the FaST-LMM-Select study, which selects only the top significant SNPs. That we demonstrated a higher power by using SUPER, compared to FaST-LMM-Select, is not surprising.

The top significant SNPs selected in the FaST-LMM-Select study likely include multiple SNPs from each association peak in GWAS. These SNPs are in strong LD among themselves. One obvious disadvantage is that this SNP selection method causes severe dilution. The other disadvantage is that computational time increases by including more SNPs than necessary. The SUPER method avoids this problem by using the pseudo-QTNs. Only one SNP is selected from many SNPs on each peak. Consequently, the optimum number of SNPs used to derive kinship is much smaller.

Moreover, LD is not only caused by local genetic linkage. Many other factors can cause LD between SNPs (e.g., population structure), even when SNPS are on different chromosomes. Therefore, our complementary process is performed genome-wide, and is not limited to the nearby SNPs (FaST-LMM-Select uses a 2 cM interval).

FaST-LMM-Select uses an arbitrary interval (2 cM) as the threshold of exclusion for LD. We use a precise LD parameter (R^2). We demonstrated that R^2 of 10% was robust enough to give the highest statistical power in all species we examined.

Last, but certainly not least, FaST-LMM-Select complements our method. FaST-LMM-Select provides an elegant algorithm to reduce computation time by conducting single value decomposition only once. Thus, the joint usage of these two methods will provide powerful and flexible tools.

We anticipate that the SUPER method could be used jointly with the CMLM to further improve statistical power. Each individual would still have its group assignment. However, the kinship of groups would be replaced by the assignment of individual QTN to groups. The effects of different assignments remain an open research question.

SUPER has been implemented in the publicly available software package, GAPIT. This method makes it possible to detect a gene with smaller samples, or alternatively, to detect a smaller effect gene with the same sample size.

URLs: Computer programs (R source code) are available at http://www.zzlab.net/GAPIT/.

Supporting Information

Figure S1 P3D (Population Parameter Previously Determined) can be used in SUPER. Similar to kinship derived from other methods, the statistical power of SUPER with unknown QTNs was the same for using or not using P3D. The other methods include the kinship derived from all the SNPs including true QTNs, the kinship derived from all SNPs excluding true QTNs, SUPER with known QTNs, and the complete trait specific kinship (True QTNs only).

Figure S2 Effect of the methods to derive the P values at the first stage of SUPER. Three methods were compared: General Linear Model (GLM), Mixed Linear Model (MLM) and Compressed Mixed Linear Model (CMLM).

Figure S3 The effect of linkage disequilibrium threshold to exclude QTNs for testing SNPs. The scenarios were implemented on the Maize, *Arabidopsis*, Rice, and Dog datasets, respectively. When the threshold is large, e.g. $r^2 = 100\%$, QTNs are barely removed. The result should be similar to the complete trait specific kinship. In the opposite case, when the threshold is too small, e.g. $r^2 = 0.01\%$, QTNs hardly survived the exclusion process. The kinship does not retain much information and the result would be similar to GLM. Interestingly, we observed that the threshold of $r^2 = 10\%$ work well for all species.

Figure S4 Effect from the relation between the QTNs and the other SNPs to derive the null distribution of test statistics. The power was examined on a trait simulated from 27 causative mutations (QTNs) sampled from the Maize dataset under a type I error of 0.05. A total of 100 replications were conducted for each method. No linkage was found between QTNs and the null SNPs in the ideal situation, when the QTNs and the null SNPs were sampled from different chromosomes. In the opposite situation (regular SNPs), when QTNs and the null SNPs were randomly sampled from the entire SNPs, a potential linkage was found between QTNs and the null SNPs. The statistical power was the same between these two scenarios for all methods. These methods include SUPER with known QTNs, SUPER with unknown QTNs, the complete trait specific kinship (true QTNs only), kinship from all SNPs including true QTNs, and kinship from all SNPs except QTNs.

Figure S5 Statistical power of five methods under different magnitudes of QTN effect. The power was examined on a trait underlying causative mutations (QTNs) sampled from ~3000 SNPs in maize. A total of 100 replications was conducted for each method. The statistical power shown here is the average of 100 replications. The heritability of the trait was 50%. The five methods are: 1) complete trait specific kinship (true QTNs only), 2) complementary trait specific kinship with known QTNs (SUPER with known QTNs), 3) complementary trait specific kinship with unknown QTNs (SUPER with unknown QTNs), 4) all SNPs including QTNs, and 5) all SNPs except QTNs.

Figure S6 Statistical power of five methods under different heritability levels. The power was examined on a trait simulated from 27 causative mutations (QTNs) sampled from ~3000 SNPs in maize. A total of 100 replications was conducted for each method. The statistical power shown here is the average of 100 replications. The heritability of the trait varied from 0 to 1. The differences between complete trait specific kinship (true QTNs only) and complementary trait specific kinship (SUPER with known QTNs) were greater when heritability was between 0 and 1. The difference between SUPER with known QTNs and kinship derived from all SNPs increases with heritability. No significant difference was found between kinship derived from all SNPs and all SNPs except QTNs.

Acknowledgments

The authors thank Sara J. Miller and Linda R. Klein for editing the manuscript.

Author Contributions

Conceived and designed the experiments: ZZ YP ESB. Performed the experiments: QW FT. Analyzed the data: QW ZZ. Contributed reagents/materials/analysis tools: QW ZZ FT. Wrote the paper: QW ZZ.

References

1. Yang J, Benyamin B, McEvoy BP, Gordon S, Henders AK, et al. (2010) Common SNPs explain a large proportion of the heritability for human height. Nat Genet 42: 565–569.
2. Buckler ES, Holland JB, Bradbury PJ, Acharya CB, Brown PJ, et al. (2009) The genetic architecture of maize flowering time. Science 325: 714–718.
3. Pe'er I, de Bakker PI, Maller J, Yelensky R, Altshuler D, et al. (2006) Evaluating and improving power in whole-genome association studies using fixed marker sets. Nat Genet 38: 663–667.
4. Moonesinghe R, Khoury MJ, Janssens AC (2007) Most published research findings are false-but a little replication goes a long way. PLoS Med 4: e28.
5. Ioannidis JPA (2005) Why most published research findings are false. Plos Medicine 2: 696–701.
6. Pritchard JK, Stephens M, Rosenberg NA, Donnelly P (2000) Association mapping in structured populations. American Journal of Human Genetics 67: 170–181.
7. Zhang Z, Buckler ES, Casstevens TM, Bradbury PJ (2009) Software engineering the mixed model for genome-wide association studies on large samples. Brief Bioinform 10: 664–675.
8. Yu JM, Pressoir G, Briggs WH, Bi IV, Yamasaki M, et al. (2006) A unified mixed-model method for association mapping that accounts for multiple levels of relatedness. Nature Genetics 38: 203–208.
9. Zhao K, Aranzana MJ, Kim S, Lister C, Shindo C, et al. (2007) An Arabidopsis example of association mapping in structured samples. PLoS Genet 3: e4.
10. Henderson CR (1984) Applications of Linear Models in Animal Breeding. University of Guelph, Guelph, Ontario, Canada.
11. Gilmour AR, Thompson R, Cullis BR (1995) Average Information REML: An Efficient Algorithm for Variance Parameter Estimation in Linear Mixed ModelsAverage Information REML: An Efficient Algorithm for Variance Parameter Estimation in Linear Mixed Models Biometrics 51: 1440–1450.
12. Kang HM, Zaitlen NA, Wade CM, Kirby A, Heckerman D, et al. (2008) Efficient Control of Population Structure in Model Organism Association Mapping. Genetics 178: 1709–1723.
13. Zhang Z, Ersoz E, Lai CQ, Todhunter RJ, Tiwari HK, et al. (2010) Mixed linear model approach adapted for genome-wide association studies. Nat Genet 42: 355–360.
14. Kang HM, Sul JH, Service SK, Zaitlen NA, Kong SY, et al. (2010) Variance component model to account for sample structure in genome-wide association studies. Nat Genet 42: 348–354.
15. Zhou X, Stephens M (2012) Genome-wide efficient mixed-model analysis for association studies. Nat Genet 44: 821–824.
16. Huang X, Wei X, Sang T, Zhao Q, Feng Q, et al. (2010) Genome-wide association studies of 14 agronomic traits in rice landraces. Nat Genet 42: 961–967.
17. Lippert C, Listgarten J, Liu Y, Kadie CM, Davidson RI, et al. (2011) FaST linear mixed models for genome-wide association studies. Nat Methods 8: 833–835.
18. Listgarten J, Lippert C, Kadie CM, Davidson RI, Eskin E, et al. (2012) Improved linear mixed models for genome-wide association studies. Nat Methods 9: 525–526.
19. VanRaden PM (2008) Efficient methods to compute genomic predictions. J Dairy Sci 91: 4414–4423.
20. Lipka AE, Tian F, Wang Q, Peiffer J, Li M, et al. (2012) GAPIT: genome association and prediction integrated tool. Bioinformatics 28: 2397–2399.
21. Zhou Z, Sheng X, Zhang Z, Zhao K, Zhu L, et al. (2010) Differential Genetic Regulation of Canine Hip Dysplasia and Osteoarthritis. PLoS ONE 5: e13219.

22. Zhang Z, Zhu L, Sandler J, Friedenberg SS, Egelhoff J, et al. (2009) Estimation of heritabilities, genetic correlations, and breeding values of four traits that collectively define hip dysplasia in dogs. American Journal of Veterinary Research 70: 483–492.
23. Bradbury PJ, Zhang Z, Kroon DE, Casstevens TM, Ramdoss Y, et al. (2007) TASSEL: software for association mapping of complex traits in diverse samples. Bioinformatics 23: 2633–2635.
24. Atwell S, Huang YS, Vilhjalmsson BJ, Willems G, Horton M, et al. (2010) Genome-wide association study of 107 phenotypes in Arabidopsis thaliana inbred lines. Nature 465: 627–631.
25. Cheng R, Lim JE, Samocha KE, Sokoloff G, Abney M, et al. (2010) Genome-wide association studies and the problem of relatedness among advanced intercross lines and other highly recombinant populations. Genetics 185: 1033–1044.
26. Yang J, Lee SH, Goddard ME, Visscher PM (2011) GCTA: a tool for genome-wide complex trait analysis. Am J Hum Genet 88: 76–82.
27. Zhang Z, Liu JF, Ding XD, Bijma P, de Koning DJ, et al. (2010) Best Linear Unbiased Prediction of Genomic Breeding Values Using a Trait-Specific Marker-Derived Relationship Matrix. PLoS ONE 5.
28. Kang HM, Sul JH, Service SK, Zaitlen NA, Kong S-y, et al. (2010) Variance component model to account for sample structure in genome-wide association studies. Nature genetics 42: 348–354.
29. Kawai J, Shinagawa A, Shibata K, Yoshino M, Itoh M, et al. (2001) Functional annotation of a full-length mouse cDNA collection. Nature 409: 685–690.
30. Diez-Roux G, Banfi S, Sultan M, Geffers L, Anand S, et al. (2011) A high-resolution anatomical atlas of the transcriptome in the mouse embryo. PLoS Biol 9: e1000582.
31. Kathiresan S, Manning AK, Demissie S, D'Agostino RB, Surti A, et al. (2007) A genome-wide association study for blood lipid phenotypes in the Framingham Heart Study. BMC Med Genet 8 Suppl 1: S17.
32. Wright SI (1922) Coefficient of inbreeding and relationship. The American Naturalist 56: 330–338.
33. Henderson CR (1953) Estimation of Variance and Covariance Components. Biometrics 9: 226–252.
34. Bernardo R (2003) Parental selection, number of breeding populations, and size of each population in inbred development. Theor Appl Genet 107: 1252–1256.
35. Hardy OJ, Vekemans X (2002) spagedi: a versatile computer program to analyse spatial genetic structure at the individual or population levels. Molecular Ecology Notes 2: 618–620.
36. Zhang Z, Todhunter RJ, Buckler ES, Van Vleck LD (2007) Technical note: Use of marker-based relationships with multiple-trait derivative-free restricted maximal likelihood. J Anim Sci 85: 881–885.
37. Myles S, Peiffer J, Brown PJ, Ersoz ES, Zhang Z, et al. (2009) Association Mapping: Critical Considerations Shift from Genotyping to Experimental Design. Plant Cell 21: 2194–2202.
38. Hayes BJ, Visscher PM, Goddard ME (2009) Increased accuracy of artificial selection by using the realized relationship matrix. Genet Res 91: 47–60.
39. Pirinen M, Donnelly P, Spencer CC (2012) Including known covariates can reduce power to detect genetic effects in case-control studies. Nat Genet 44: 848–851.
40. Yu J, Zhang Z, Zhu C, Tabanao DA, Pressoir G, et al. (2009) Simulation Appraisal of the Adequacy of Number of Background Markers for Relationship Estimation in Association Mapping. Plant Genome 2: 63–77.

5

Orthology Detection Combining Clustering and Synteny for Very Large Datasets

Marcus Lechner[1]*, **Maribel Hernandez-Rosales**[2,3,4,5], **Daniel Doerr**[6,7], **Nicolas Wieseke**[8], **Annelyse Thévenin**[6,7], **Jens Stoye**[6,7], **Roland K. Hartmann**[1], **Sonja J. Prohaska**[9], **Peter F. Stadler**[2,3,4,10,11,12,13]

1 Institut für Pharmazeutische Chemie, Philipps-Universität Marburg, Marburg, Germany, 2 Bioinformatics Group, Department of Computer Science, Universität Leipzig, Leipzig, Germany, 3 Interdisciplinary Center for Bioinformatics, Universität Leipzig, Leipzig, Germany, 4 Max Planck Institute for Mathematics in the Sciences, Leipzig, Germany, 5 Departamento de Ciência da Computação, Instituto de Ciências Exatas, Universidade de Brasília, Brasília, Brasil, 6 Genome Informatics, Faculty of Technology, Bielefeld University, Bielefeld, Germany, 7 Institute for Bioinformatics, Center for Biotechnology, Bielefeld University, Bielefeld, Germany, 8 Faculty of Mathematics and Computer Science University of Leipzig, Leipzig, Germany, 9 Computational EvoDevo Group, Department of Computer Science, Universität Leipzig, Leipzig, Germany, 10 Institute for Theoretical Chemistry, University of Vienna, Vienna, Austria, 11 Center for non-coding RNA in Technology and Health, University of Copenhagen, Frederiksberg, Denmark, 12 The Santa Fe Institute, Santa Fe, New Mexico, United States of America, 13 RNomics Group, Fraunhofer Institut for Cell Therapy and Immunology, Leipzig, Germany

Abstract

The elucidation of orthology relationships is an important step both in gene function prediction as well as towards understanding patterns of sequence evolution. Orthology assignments are usually derived directly from sequence similarities for large data because more exact approaches exhibit too high computational costs. Here we present PoFF, an extension for the standalone tool Proteinortho, which enhances orthology detection by combining clustering, sequence similarity, and synteny. In the course of this work, FFAdj-MCS, a heuristic that assesses pairwise gene order using adjacencies (a similarity measure related to the breakpoint distance) was adapted to support multiple linear chromosomes and extended to detect duplicated regions. PoFF largely reduces the number of false positives and enables more fine-grained predictions than purely similarity-based approaches. The extension maintains the low memory requirements and the efficient concurrency options of its basis Proteinortho, making the software applicable to very large datasets.

Editor: Christos A. Ouzounis, Hellas, Greece

Funding: We acknowledge support for the Article Processing Charge by the German Research Foundation and the Open Access Publication Fund of Bielefeld University Library. This work was supported in part by the Deutsche Forschungsgemeinschaft grants no. GRK-1384, MA5082/1- 1, MI439/14-1. DD receives a scholarship from the CLIB Graduate Cluster Industrial Biotechnology. AT is a research fellow of the Alexander von Humboldt Foundation. The funders had no role in study design, data collection and analysis, decision to publish, or preparation of the manuscript.

Competing Interests: The authors have declared that no competing interests exist.

* Email: lechner@staff.uni-marburg.de

Introduction

Detailed knowledge on the history of large gene families is crucial to the understanding of their patterns of sequence evolution and their functional interpretation. Throughout this contribution we use the term "gene" to denote any genomic feature that can be represented as a sequence interval. No further functional or structural properties are implied. An important step towards this goal is the elucidation of orthology relationships. Two genes are orthologs if they arose via a speciation event from their last common ancestor in the gene tree. In contrast, paralogs originate from a gene duplication event [1,2]. The definition of orthology implies that an event-annotated gene tree is available, and thus a gene tree and its reconciliation with the underlying species tree must be known to determine with certainty which pairs of genes are orthologs. Since ancestral states are in general experimentally inaccessible, the orthology relation, just like the gene phylogeny, has to be inferred from extant sequence data.

A large class of orthology detection tools therefore attempts to explicitly infer gene phylogenies and their reconciliation with species trees, e.g. Orthology analysis using MCMC [3], Multi-MSOAR [4], LOFT [5], Ensembl Compara [6], and Synergy [7]. Although this tree-based approach is often considered the most accurate, it suffers from high computational costs and is hence limited in practice to a moderate number of species and genes. Moreover, all practical issues that hamper phylogenetic inference (e.g. variability of evolutionary rate, mistaken homology, homoplasy, and horizontal gene transfer) limit the accuracy of both the gene and the species trees.

The second class of algorithms bypasses the construction of gene and species trees by directly deriving orthology assignments from similarity data. Approaches of this type are COG [8], OrthoMCL [9,10], OMA [11,12], InParanoid [13], eggNOG [14], Homolo-Gene [15], Roundup 2.0 [16], or EGM2 [17]. Since orthology is not a transitive relation, the problem of orthology detection is fundamentally different from clustering or partitioning of the input gene set. In particular, a set A of genes can be orthologous to another gene $x \notin A$ but the genes within A are not necessarily orthologous to each other. In this case, the genes in A are called co-orthologs to gene x [18]. A common feature of most of the methods mentioned above is that they do not produce an estimate for the pairwise orthology relations but return orthologous groups containing genes which are mutually orthologous to the greatest extent but also comprise co-orthologous genes. We refer to these groups as orthologous groups in the following. In addition to OMA and Proteinortho [19], only Synergy, EGM2, and InParanoid attempt to resolve the orthology relation at the level of gene pairs. The latter two tools can only be used for the analysis of two species at a time, while Synergy is not available as standalone tool and therefore cannot be applied to arbitrary user-defined datasets. The use of these tools is limited to the species offered through the databases published by their authors.

The orthology relation can be represented as a graph on the set of genes. It forms a cograph rather than a partition [20]. Clustering approaches identify dense subgraphs of these cographs and hence introduce false-positive edges corresponding to recent paralogs. On the other hand, ancient paralogs are often separated into different groups of co-orthologs. Despite this theoretical shortcoming, cluster-based methods have consistently been reported to yield very good results [21–23]. Since they are much faster than tree-based algorithms, they can be applied to very large datasets.

The clustering method and, in many cases, user-defined parameters determine the granularity of the orthologous groups and thus the tolerance to false positive orthology assignments. Some methods are very inclusive [5], but the aim typically is to remove as many paralogs as possible to approach a one-to-one orthology relation. These simple relationships are especially useful for phylogenetic analysis and for exact functional predictions. Phylogenomic studies typically employ pipelines such as HaMStR [24] to restrict the data to one-to-one orthologs. When the phylogenetic range of interest includes duplication events however, such approaches are bound to fail [25].

Here we focus on an intermediate balance. Our main aim is to avoid false positive orthology assignments within the phylogenetic range of the reported orthologous groups, while we tolerate recent in-paralogs (speciation preceding duplication) as unavoidable contamination. Clustering approaches for orthology detection are usually based on the "best match method", which attempts to find orthologs as the sequence in another genome that is most similar to the query. It often fails in the presence of paralogs with comparable similarity to the query. Best match approaches are nevertheless routinely used to gain insight into relationships of genes among phylogenetically very diverse organisms. These approaches are used in particular for gene annotation in newly sequenced genomes for which a well studied close relative is lacking. However, the large number of sequencing projects of the last decade have largely reduced the gaping holes in phylogenetic coverage and most large-scale comparative studies nowadays focus on closely related species or even strains [26,27]. As a result, the evolutionary distances within a phylogeny of interest are often rather small, hence additional information to resolve evolutionary relationships between genes can be obtained from genomic context. Furthermore synteny, i.e., the conservation of gene order (also referred to as gene context) provides information independent of sequence similarity, which can help to sort paralogs. Both Synergy and EGM2 incorporate synteny information to compute orthology relations. The Synergy algorithm achieves high accuracy due to the fact that it reconstructs gene family trees

[28]. EGM2 considers synteny by identifying similar genomic regions to detect orthologs. However, this tool is not suitable for large datasets due to its restriction to only two genomes at a time. Genes with a common ancestry that are functionally linked with each other frequently show a conservation in local gene order over long evolutionary distances [29,30]. Thus, synteny is frequently used to disentangle complex duplication histories, see e.g. [31] and references therein. The intricacies of conserved synteny and positional orthology have been reviewed recently [32].

The computational prediction of syntenic regions usually relies on the detection of genomic neighborhoods that are conserved between genomes of related species. Proximity relations among genes, such as adjacencies [33] (two genes encoded adjacent to each other in several genomes), generalized adjacencies [34] or conserved intervals [35], are used to assess genomic neighborhoods. Typical methods for the detection of syntenic regions utilize gene family information, similarity scores or conserved distances to establish putative homologies and then apply chaining or clustering algorithms. When paralogous genes are considered, the underlying computational problems become prohibitive because many alternative synteny assignments are possible. Exact algorithms are therefore slow and limited to small datasets. In fact, the problem of computing the syntenic distance between two genomes is NP-hard [36,37]. Efficient heuristics are therefore employed to deal with large datasets.

If gene family information is available, popular synteny tools such as i-ADHoRe 3.0 [38] and MCScanX [39] can efficiently detect homologous regions even in large-scale analyses. Otherwise, using local alignments of sequences, tools such as CYNTENATOR [40] and DAGchainer [41] allow for detection of syntenic regions based on pairwise similarity scores of sequence intervals. The heuristic method FFAdj-MCS [42] has proven to be a good compromise in terms of both, speed and accuracy, as it takes a different approach by calculating a matching whose objective function maximizes towards a balance between adjacencies and similarity scores of genes.

In this contribution we describe PoFF, an extension of Proteinortho [19], to include synteny information in a systematic way. More precisely, a pair of genes ($A1$, $A2$) in genome A is considered syntenous with another pair of genes ($B1$, $B2$) in genome B, if both $A1$, $B1$ and $A2$, $B2$ are potential orthologs (as determined by sequence similarity), and both ($A1$, $A2$) as well as ($B1$, $B2$) are adjacent gene pairs on their corresponding chromosomal locations. In case other genes are located between ($A1$, $A2$) or ($B1$, $B2$), these must not be orthologous to any other genes in genomes A or B. Proteinortho applies an adaptive best match method together with spectral clustering to define (co-)orthologs. Its performance in terms of accuracy has been shown to be comparable to other clustering-based methods. At the same time it has modest requirements in terms of memory and computation time and is thus suitable for very large datasets. Complementing the evaluation of pairwise sequence similarities, we incorporate here the efficient heuristic algorithm FFAdj-MCS that computes ortholog assignments by maximizing the above synteny measure between pairs of genomes. Following a recent suggestion [43], true orthologs among multiple candidates were defined as those that retained their original genomic context. In the course of this work, we adapted FFAdj-MCS to include multiple linear chromosomes within single organisms and extended it for the detection of duplicated genes and large duplicated genomic regions. We note that the algorithm may also be applied to circular chromosomes at the expense of losing synteny information for at most two pairs of genes at the very ends of the linearized representation. This minor shortcoming should have no or only a vanishingly low effect in the overall process of orthology assignment.

Figure 1 illustrates the idea of the synteny-enhanced version of `Proteinortho`. In this example, four genes ($A1$, $A2$, $B1$, $B2$) in two species (A and B) are considered. The gene tree in Figure 1a shows a duplication preceding a speciation event. $A1$ and $B1$ as well as $A2$ and $B2$ are orthologous to each other as they derived from a common ancestor by speciation. Given sufficient similarity, however, all four genes would be reported as an orthologous group using regular sequence similarity-based approaches. The gene order depicted in Figure 1b allows one to distinguish the genes 1 and 2 from each other. The combined approach therefore predicts the two distinct orthologous groups $\{A1, B1\}$ and $\{A2, B2\}$ and thus avoids false positive orthology assignments.

We argue that the level of granularity achieved in this way is more useful in most cases than an arbitrary separation of groups solely based on sequence similarity scores which tend to lack significance when sequences are closely related. The same holds compared to inclusive strategies which hardly discriminate subgroups. Assuming that numerous extant genes have derived from a limited set of common ancestors by a series of duplication events, inclusive strategies will include entire gene families, and hence lead to very large groups with a significant amount of actually non-orthologous genes. An emphasis on including all pairwise orthology relations when reporting orthologous groups thus seems to be of little use.

We evaluated `PoFF` using several sets of simulated protein-coding genes. Each set was derived from event-annotated gene trees. Thus, for each pair of genes, the true relationship regarding orthology is unambiguously defined and used to validate the predictions. Our results reveal a significant improvement with respect to true negative and false positive predictions at the expense of only a marginal decrease of the true positive rate.

Materials and Methods

Conceptual Outline

Our starting point for orthology detection is a directed graph Γ whose vertices are all the genes of all input genomes. A directed edge $x \rightarrow y$ is introduced if (i) x and y are taken from two different genomes (A and B) and (ii) the similarity $s(x, y)$ is not much smaller than the gene z in B that is most similar to x, i.e., if

$s(x,y) \geq f \times \max_{z \in B} s(x,z)$ for some stringency parameter $f \leq 1$. Since any true ortholog of $x \in A$ in genome B should be among the most similar sequences that can be found in B, Γ should have few false negatives (i.e. missing true edges) as long as the stringency is not set to a value that is too restrictive. The idea is, therefore, to remove edges from the graph Γ that are likely false positives. Since orthology is a symmetric relation, we only retain edges $x \rightarrow y$ if $y \rightarrow x$ is also contained in Γ.

Synteny information determined by `FFAdj-MCS` provides an additional filter for the edge set of Γ. By construction, the subgraph $\Gamma[A \cup B]$ induced by the genes in A and B is bipartite. Synteny is modeled as the relative order of edges along both genomes. Synteny as a filter reduces the edge set of $\Gamma[A \cup B]$ to a matching that maximizes a trade-off between the total number of edges and the number of conserved adjacencies. Among similar paralogs, this strategy favors the one with the best-conserved local gene order as representative of the orthologous group. In the final step, a clustering algorithm [19] is employed to extract groups of co-orthologs from Γ, which contains all subgraphs $\Gamma[A \cup B]$ for all pairs of genomes.

Implementation

`Proteinortho` uses the `blast` bit score to determine potential homologs in another species and to measure sequence similarity. The definition of the edge set above makes it possible to construct Γ directly from pairwise comparisons. Thus, this initial state can be trivially parallelized and does not require the storage of genome-wide `blast` comparison data in memory. As the `FFAdj-MCS` algorithm applies to pairs of genomes A and B, it can be added to the workflow without breaking these advantageous properties. The algorithm requires information on gene order and pairwise gene similarity for two genomes and determines a matching that maximizes a weighted sum of edge weights and weights of conserved adjacencies. To this end `FFAdj-MCS` matches genes in regions with conserved gene order that locally maximize the objective of FF-Adjacencies [42]. These regions are called maximum common substrings (MCSs). Since the `blast` scores $s(x, y)$ are not symmetric, they are symmetrized (taking the average of both scores) for use in `FFAdj-MCS`. The combination `PoFF` of `Proteinortho` and `FFAdj-MCS` yields, for each pair of genomes, a pruned set of edges that is highly enriched in true orthologous pairs. The workflow of our extension is illustrated in Figure 2.

Figure 1. Synteny-enhanced orthology prediction. Four genes ($A1$, $A2$, $B1$, $B2$) in two species (A and B). a) The gene tree with a duplication (filled double circle) and a speciation event (empty circle). b) Gene order in the genomic context of both genes. Genes $A'x$ and $B'x$ are orthologous to each other. Lines depict suggested partners based on sequence similarity of which the dashed were neglected by the gene order algorithm.

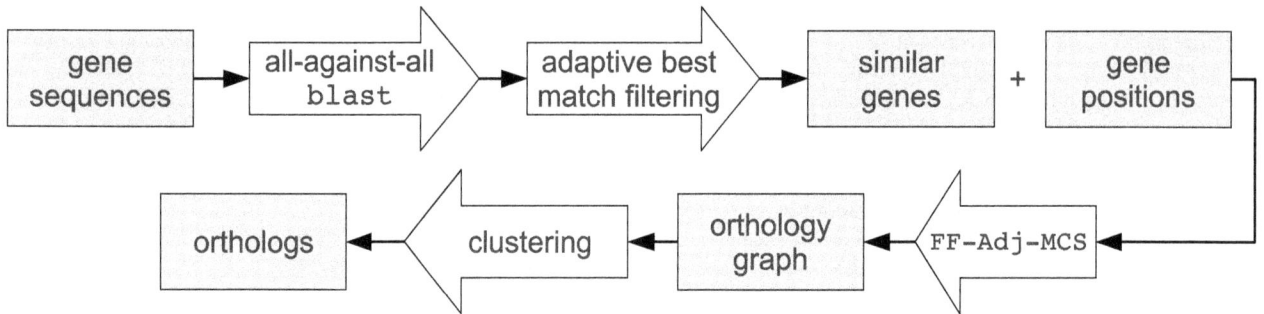

Figure 2. Workflow of PoFF. Similar gene sequences are determined by an all-against-all blast search. Top reciprocal matches are ordered by their positions in the respective genomes. The FFAdj-MCS algorithm is applied to determine the maximum matching with respect to sequence similarity and gene order. As a result the orthology graph Γ only contains the remaining edges from pairwise comparisons. Finally, orthologous groups are extracted by clustering.

We added three extensions to the FFAdj-MCS program as presented in [42]. Firstly, it was adapted to allow for more than one chromosome per genome. Secondly, the detection of duplicated genes and large duplicated regions was implemented: The heuristic was adapted to repeat a user-defined number of complete matchings, where edges selected by preceding matchings are removed before each subsequent matching. Thirdly, FFAdj-MCS allows to filter the size of MCSs obtained from subsequent matchings by means of a user-defined minimal size $\beta \in [2 \ldots n]$ that defines the minimum number of gene pairs in each MCS.

Finally, we relaxed the criteria for very similar neighboring genes: If two adjacent genes x and y in A both have their best alignment to the same gene $z \in B$, we include both edges $\{x, y\}$ and $\{y, z\}$ in Γ, since x and y are likely in-paralogs and a decision for one of the two edges based on a small score difference is not reasonable from a biological perspective. Even though this makes PoFF more inclusive, we argue that this behavior is more reasonable because such in-paralogs can be quite easily detected in a post-processing step if required for a particular application.

PoFF has several parameters that can be set by the user – in particular score thresholds and coverage requirements of the blast searches. We used the default settings throughout. The stringency parameter f defines the fraction of the bit score of the best blast hit that must be reached by an alternative candidate ortholog. Proteinortho's default value, $f = 0.95$, has been shown to work also in conjunction with the synteny filter. The FFAdj-MCS algorithm provides an adjustable parameter $\alpha \in [0 \ldots 1]$ that controls the relative importance of edge weights and the weights of adjacencies. Benchmarking PoFF did not reveal a strong dependency of the results on this parameter, likely because nucleotide sequences and order of genes evolve in parallel and with comparable speeds. We therefore used the default value $\alpha = 0.5$ throughout. By default, we perform one matching iteration with $\beta = 3$ to cover the detection of large duplicated regions. If multiple copies of a region are expected in a dataset, this number of iterative matchings can be increased further. Practical experience with Proteinortho also led to the decision to increase the default E-value threshold from 10^{-10} to 10^{-5} in order to improve coverage of less conserved orthologs.

Benchmarking

Since implementations of competing tools that generate fully resolved orthology relations are not publicly available, we cannot employ the usual evaluation strategy of comparing all tools on series of benchmarking datasets of our choice. Instead we apply both Proteinortho and PoFF to several reference datasets that either comprise simulated data for which the underlying gene trees are known, or real data which defines orthologous groups and/or pairwise orthologous relationships by extensive analysis, often including manual curation. Results are then compared to the published performance of alternative tools.

For Proteinortho and PoFF we used standard parameters, including an E-value threshold of 10^{-5}. However, the more recent blastp+ software [44] instead of the original blastp implementation [45] was applied to find the initial matches. This can be set by a parameter in Proteinortho.

The generation of simulated data is described below. As some of these sets were sufficiently small, we also applied OrthoMCL, OMA and InParanoid in order to evaluate the results. Again, standard parameters were used, including an E-value threshold of 10^{-5} for OrthoMCL. Real life data was taken from various sources also described hereafter. The YGOB dataset [46], was used in a previous study to evaluate the Synergy approach [7]. Hence, we took the opportunity to include the available results to the benchmark.

Simulated data. In the absence of extensive gold standard datasets comprising sequence and synteny data as well as the underlying gene trees that could be used for benchmarking our orthology prediction method, we simulated sequence evolution and genomic rearrangements on a single chromosome for three example datasets comprising 50, 80 and 100 gene families (proteins) in 20 species (named hereafter $F50$, $F80d$ and $F100$, respectively). All test sets feature duplications of both individual genes and gene clusters. The set $F80d$ in addition includes whole genome duplications. Table 1 gives a closer look to the composition of all three datasets as well as to their average breakpoint distances determined by PoFF. The simulation pipeline is available in the Online Supplemental Material.

Species trees were simulated according to the *Age Model* [47]. These trees are balanced and edge lengths are normalized so that the total length of the path from the root to each leaf is 1. For each species tree S, we then simulated gene trees using the following rules:

1. The root of S contains an ordered list of ancestral genes one for each gene family. The number of families is a user-defined parameter.

2. S is traversed in a depth first order. All changes to the genome are simulated independently for each edge of S with constant rates.

Table 1. Composition of simulated datasets.

Dataset	Families	Proteins	ø Family size	ø Breakpoint distance
F50	50	8,363	167 proteins	13
F80d	80	15,296	191 proteins	19
F100	100	27,258	273 proteins	14

The simulated datasets differ by the number of gene families present in the species as well as by the size of these families. The larger the families the more diversity among the set of species can be considered. Set F80d additionally comprises whole genome duplications.

3. At each internal node of S, the ordered gene list received from its parental edge is copied without change to both offspring edges.

4. Along each edge of S a number of events is sampled from a Poisson Process $P_{\lambda,l}$, where the parameter $\lambda \in [0,1]$ is the probability of the event to happen and l is the branch length. The process may generate none, one, or several events of the following types: gene duplication, cluster duplication, genome duplication, and gene loss. Here we used the parameters $\lambda=0.9$ for gene duplication, $\lambda=0.5$ for cluster duplication, $\lambda=0.5$ for gene loss. For the dataset $F80d$ we consider genome duplications with $\lambda=0.03$ instead.

5. A special rule applies to recently duplicated genes to account for the deletion of redundant gene copies before they can be stabilized by sufficient functional divergence or subfunctionalization [48,49]. We model this by a probability of 0.3.

6. To obtain an order of the generated genes, rearrangements are carried out for each edge of S using translocation and inversion operations on the ordered list of genes that "survived" until the next speciation. Rearrangements are picked randomly and the number of inversion operations is chosen uniformly proportional to the branch length [50].

The result of this simulation is a gene tree G_i for each family i together with a true reconciliation map to the species tree S. All gene lineages terminating in a deletion event are pruned from the gene tree so that we retain a gene tree G_i in which only extant genes appear as its leaves. The known reconciliation furthermore provides us with a labeling of the internal nodes of G_i with *duplication* or *speciation* events, see Figure 3 for an example. This in turn determines the true orthology relation for all genes received in the leaves of S. In addition, the gene orders within their respective genomes are obtained. The simulations were performed using a simulation environment for large gene families [51].

Since large-scale orthology analysis are usually performed for protein sequences, we use indel-Seq-Gen [52] to generate simulated amino acid (aa) sequences for the gene trees G_i. For each gene family a random seed sequence is initiated with a length between 100 and 1,000 aa. Then, to define the offspring genes, indel-Seq-Gen introduces substitutions according to PAM substitution matrix and insertions and deletions with a Zipfian probability distribution [53] with maximal length between 1% to 10% of the sequence length. For the gene trees a branch scale factor of 0.5 was used. This is the frequency of a single amino acid to be substituted. Hence, approximately half of the amino acids are changed during the simulation on the path from the root to the leaf.

We remark that the Artificial Life Framework (ALF) [54] for simulating sequence evolution could in principle have been used for simulating test data. However, in its current version, this tool does not support genome-wide duplications and selective loss of recently duplicated genes. We therefore opted to construct our own simulation framework.

Real life data. COG: We used proteome data from the COG-database, which provides manually curated orthology relations (ftp://ftp.ncbi.nih.gov/pub/COG/COG/, 2009/10/15), for the following set of 16 species covering three bacterial groups: *Bacillus halodurans, Bacillus subtilis, Lactococcus lactis, Listeria innocua, Streptococcus pneumoniae* TIGR4, *Streptococcus pyogenes* M1 GAS from the Gram-positive bacilli class, *Buchnera sp.* APS, *Escherichia coli* K12, *Pasteurella multocida, Salmonella typhimurium* LT2, *Vibrio cholerae, Yersinia pestis* from the gamma proteobacteria class and *Brucella melitensis, Caulobacter vibrioides, Mesorhizobium loti, Rickettsia prowazekii* from the alpha proteobacteria class. According to PoFF, the average breakpoint distance of this set is 642.

To obtain the gene orders we retrieved the genomes from the NCBI-database (ftp://ftp.ncbi.nih.gov/genomes/Bacteria/, 2012/12/13, see supplement). When several strains were available, we picked the one with the smallest uid as they represent the older genomes preferentially included in secondary databases. All genes were then located in the genomes using tblastn+ with an E-value threshold of 10^{-8}. The best match was considered to be the gene of interest. A small minority of genes (98 out of 53,264) could not be located unambiguously and was thus removed from the dataset.

As we used an extract of the COG-database (16 out of 66 species), only COG-groups covering at least eight proteins within the set of the chosen 16 species were considered to estimate the orthology matrix as described below (see Evaluation). Otherwise, their classification might have been based on species not in the dataset used here, which would make a comparison of approaches unreasonable.

OrthoBench: We also used the reference annotation Ortho-Bench [23]. Manually curated orthologous groups were downloaded from http://eggnog.embl.de/orthobench/ at 2013/01/05. The set comprises 12 metazoan proteomes and is based on the Ensembl $v60$ genome annotation [55] which was downloaded from ftp://ftp.ensembl.org/pub/release-60/ at 2013/01/11. According to PoFF, the average breakpoint distance of this set is 5,433. 124 out of 1,692 proteins stated in OrthoBench could not be located in the $v60$ set and were excluded from the analysis.

YGOB: From this dataset we obtained orthology assignments of five ascomycete fungi *Ashbya gossypii, Saccharomyces cerevisiae, Candida glabrata, Kluyveromyces lactis,* and *Kluyveromyces waltii* that have been used in the evaluation of Synergy in the original study [7]. According to PoFF, the average breakpoint distance of this set is 2, 697. The data provided by the authors included pairwise blast results with an E-value threshold of 10^{-5}, which we directly used in our analysis, omitting the blast step. In this way, the initial blast data on which Synergy, Proteinortho, and PoFF operated was assured to be identical. We then compared

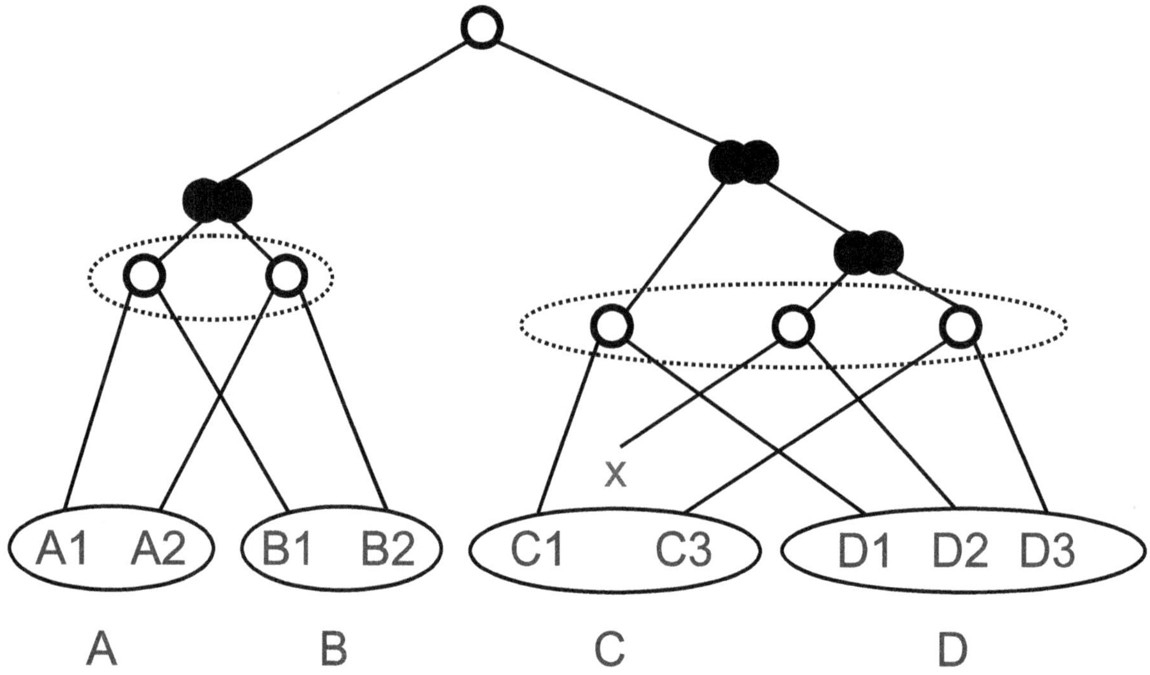

Figure 3. A reconciled tree for gene families. The gene tree is embedded in the species tree. Internal nodes represent either gene duplication (filled double circle) or speciation events (empty circles). Gene loss is depicted by ×.

orthologs predicted by the three approaches to orthologs from the YGOB dataset (v1, 2005) [46]. We excluded genes from the predictions that were not contained in the YGOB dataset (6, 218, 6, 076 and 6, 817 out of 23, 134 for Proteinortho, PoFF and Synergy, respectively). In this way, we avoid to bias our evaluation with data that is not present in the reference dataset.

Evaluation. For each gene family/orthologous group in the reference sets, we compared the pairwise orthologous relationships between its members to the predictions, counting true positives (*tp*), false positives (*fp*), true negatives (*tn*) and false negatives (*fn*) as well as the number of orthology relations between reference groups. These data were then used for statistics as follows:

$$\text{Precision} = \frac{tp}{tp+fp}, \text{ recall} = \frac{tp}{tp+fn},$$

$$\text{accuracy} = \frac{tp+tn}{tp+tn+fp+fn}$$

$$\text{and } tn \text{ rate} = \frac{tn}{tn+fp}.$$

For evaluation of PoFF and Proteinortho, we used the orthology graph returned in addition to orthologous groups which contains information on pairwise orthology relations. OMA returns this graph equivalently in the PairwiseOrthologs output. InParanoid was applied to all pairs of species successively. After merging the results, this resulted in pairwise orthology relations for the whole dataset as well. OrthoMCL on the other hand, does not return the orthology graph directly. We extracted the information on pairwise orthology relations from the MCL clustering output file. Connected components in there are used by OrthoMCL to determine orthologous groups, making this output file similar to the orthology graph returned by PoFF/Proteinortho. We note

however, that the file was not meant to be used as orthology graph. Given the mode of calculation applied by OrthoMCL, it contains numerous orthology relations for paralogs of the same species which cannot occur using PoFF/Proteinortho. For Synergy and YGOB, pairwise orthology relations were present. For COG and OrthoBench, however, only data on orthologous groups was provided. The pairwise orthology relations had to be estimated. We did this by assuming each protein of an orthologous group to be orthologous to each other protein in the same group, except when both proteins belong to the same species. We emphasize that this strategy strictly overestimates the number of orthologous relationships in the dataset. Nonetheless, this method makes it possible to compare the results on a pairwise level.

The simulated data also provides gene trees. These were used to acquire pairwise orthology relations. Two genes of a simulated gene family are orthologous to each other, if and only if their most recent common ancestor event was a speciation.

Results and Discussion

In order to estimate how PoFF performs with respect to closely related species and compared to the original Proteinortho implementation, we simulated and subsequently evaluated three datasets (F50, F80d, F100), for which the gene histories and hence the true orthology relations are defined. The datasets differ in number and size of gene families, thus representing increasing levels of diversity among closely related species. The results are summarized in Table 2. Proteinortho already performs very efficiently. However, as the number of paralogs with similar sequences increases, the basic algorithm becomes less effective in precisely predicting the correct orthology relations within these gene families, a trend that exacerbates with increasing size of gene families. The use of the synteny information provided by the FFAdj-MCS algorithm efficiently counteracts this tendency and substantially improves the precision. Other performance statistics

Table 2. Comparison using simulated data.

Dataset	Method	Precision	Recall	Accuracy	*tn* rate	Runtime
F50	OrthoMCL	3.06%	7.26%	86.18%	89.71%	7 h, 22 min
	OMA	38.64%	9.62%	95.49%	99.32%	1 day, 14 h
	InParanoid	98.01%	5.02%	95.94%	99.99%	2 days, 2 h
	Proteinortho	80.63%	23.11%	97.62%	99.83%	0 h, 36 min
	PoFF	96.15%	24.18%	97.53%	99.96%	0 h, 36 min
F80d	OrthoMCL	0.92%	0.88%	87.44%	93.43%	15 h, 46 min
	OMA	43.97%	5.25%	93.51%	99.54%	3 days, 23 h
	InParanoid	97.67%	0.89%	93.65%	99.99%	8 days, 23 h
	Proteinortho	79.36%	16.64%	97.68%	99.88%	1 h, 29 min
	PoFF	93.98%	15.52%	97.30%	99.96%	1 h, 30 min
F100	OrthoMCL/OMA/InParanoid	-	-	-	-	>31 days
	Proteinortho	23.99%	20.48%	99.37%	99.71%	6 h, 39 min
	PoFF	90.16%	18.17%	99.62%	99.99%	6 h, 44 min

Comparison of computational results with orthology relations derived from simulated datasets with different gene family sizes. Statistical values are explained in Materials and Methods. *tn* rate refers to true negative rate. Running time was measured on a quad core CPU (Intel core i7 at 2.9 GHz) with eight threads.

as well as the runtime remain nearly unchanged, which indicates a significant advantage of PoFF over the original Proteinortho tool.

It would be desirable to include several other orthology detection tools to directly compare the results achieved using the simulated datasets. To our knowledge, only OrthoMCL, OMA, InParanoid and Roundup 2.0 are available as standalone tools that can be used for large input datasets. Since Roundup 2.0 largely relies on a commercial implementation of blast, we were only able to include the first two tools in the benchmark. We observed that OrthoMCL is very inclusive. It returns huge orthologous groups comprising whole gene families but, according to the results, does not reflect pairwise orthology to a reasonable extent. This results in a large number of false positive predictions. It also requires extensive computational resources: We terminated the analysis of the biggest dataset ($F100$) after 31 days of runtime without obtaining a result (using an Intel core i7 quad core CPU at 2.9 GHz). OMA and InParanoid required even more computational resources. We also had to terminate the analysis of the biggest dataset without obtaining a result from these tools. The results obtained for the two other datasets, however, were superior to those obtained from OrthoMCL. InParanoid reports the smallest amount of orthology relations (only ~1–5% recall) and exhibits the longest runtime. The results however hardly include any false positives.

Since FFAdj-MCS acts as an efficient filter against false orthology predictions, we tested whether we could rely entirely on the synteny information. After all, this information is also derived from the alignment scores determined by blast, hence low-scoring edges are unlikely to enter the final matching and would thus be dismissed either way. We therefore removed Proteinortho's filter for near-optimal alignment scores by setting $f = 0$, which includes all reciprocal alignments above the given E-value threshold. We observed that this did not improve the quality of the predictions but increased the CPU time by a factor of 20 to 40 on the simulated datasets. A cutoff value of f close to 1 thus not only saves computational resources but also contributes to the identification of the correct edges in Γ

independent of FFAdj-MCS. This observation justifies the design decision to run the gene order filter only on the nearly optimal orthology candidates.

In addition to simulated data, we performed benchmarks using estimated orthology relationships from several real life datasets. The COG-database [8] was used as complete reference annotation for a set of 16 prokaryotes. All proteins present in this set are assigned to some group. OrthoBench [23] and YGOB [46] provided a partial annotation for a number of reference proteins in twelve metazoan and five fungal species, respectively. The YGOB dataset was used in a previous study to evaluate the tool Synergy [7]. While the latter is not publicly available, the results of its application to YGOB have been published, which allowed us to compare Synergy and PoFF on this dataset (see Table 3 and discussion below).

For real life datasets, PoFF predicts 4 to 57% fewer pairwise orthology relations than Proteinortho. This tendency is even more pronounced for the very similar simulated datasets (23 to 77%, data not shown). The reduced number of pairwise orthology relations allows separating the orthologous groups in a more fine-grained way and reduces the number of false positive assignments. In turn, however, the number of true positive assignments is reduced as well. For the real life datasets, which comprise far more distant species than the simulated data, this results in reduced recall and sometimes also reduced accuracy (Table 3).

We emphasize that neither the COG nor the OrthoBench data are ideal benchmarking sets for fine-grained orthology predictions. Both provide orthologous groups rather than pairwise orthology relations which, in turn, had to be estimated for evaluation (see Materials and Methods). Moreover, many of these groups are rather large as they contain numerous paralogs, which were – as we would argue – correctly clustered into subgroups by PoFF and/ or Proteinortho. The COG-database was originally constructed using 13 Archaea, three Eukarya and 50 Bacteria. For evaluation, we used a bacterial subset of 16 species. This in turn makes duplications specific to the chosen subset harder to detect. The combination of these issues leads to artifacts in the reference datasets that might have a negative impact on recall and accuracy.

Table 3. Comparison using real data.

Dataset	Method	Precision	Recall	Accuracy	*tn* rate
COG	Proteinortho	99.50%	23.80%	29.12%	98.45%
	PoFF	99.52%	22.50%	27.93%	98.47%
YGOB	Synergy	61.36%	42.82%	99.64%	99.89%
	Proteinortho	59.10%	38.35%	99.62%	99.89%
	PoFF	59.07%	36.97%	99.62%	99.89%
OrthoBench	Proteinortho	100%	17.68%	24.71%	100%
	PoFF	100%	9.72%	17.44%	90.27%

Comparison of tools on the basis of estimated orthology relations from real data sets. Statistical values are explained in Materials and Methods. *tn* rate refers to true negative rate.

Both, PoFF and Proteinortho tend to split the groups annotated in the reference sets into smaller subgroups. This effect of subdividing is more pronounced for PoFF. OrthoBench groups contain on average 23.5 genes while comprising only up to 12 species. On average these groups are divided into 3.8 subgroups by Proteinortho and 5.4 groups using PoFF. COG groups contain 18.4 genes on average. These groups are divided into 3.0 and 3.1 subgroups, respectively (see File S1).

Only the YGOB dataset offers pairwise orthology data and can thus be regarded as more exact than the other two sets. Here, the results of Proteinortho and PoFF are quite similar. Again we find the slight decrease in recall observed already for the simulated dataset. Increased phylogenetic distance decreases the positive impact on precision, which was found for the more closely related simulated datasets. The predictions for this dataset achieved by Synergy are slightly better than those of Proteinortho and PoFF. However, the algorithm relies on genome-wide reconstruction of phylogenetic gene trees and is thus far more time-consuming. Moreover, a standalone tool that applies the algorithm is currently not available.

The strategy pursued by PoFF is particularly useful to separate large orthologous groups with many co-orthologs into smaller subgroups. Typically, there is one major group for each gene family in each simulated dataset that spans all species of the original group but includes only one or a small number of genes from each species. In addition, we observe one or more "minor" groups of duplicates that contain diverged and/or largely rearranged paralogs. Using the real life dataset OrthoBench we see this trend in particular for Otoferlin, Dilute myosin heavy chain, GPS domain-containing GPCRs and S-adenosylmethionine synthetase isoform families. This type of subdivision appears useful and desirable in most practical applications of automatic orthology detectors.

The increase in runtime introduced by FFAdj-MCS is marginal for small genomes (e.g. Bacteria). For simulated data as well as the COG set we observed an increase by 1–3%. For large genomes as present in the OrthoBench set the increase was 5–10% and thus more notable. For example, the analysis of *Rattus norvegicus* and *Pan troglodytes* took 12.5h using Proteinortho and 13.5 h using PoFF applying a single thread. The memory requirements remained unchanged.

Conclusions

Dissecting large gene families from many genomes into clusters of orthologs is not a well-posed problem. Orthology, as defined by Fitch [1,2], is a binary relation of the set of genes. Gene duplication events typically appear in many different locations of the underlying phylogenetic tree and give rise to a complex structure of co-orthologs and paralogs at different levels. The resulting cograph nevertheless contains dense clusters that can be meaningfully associated with orthologous groups. Clustering-based orthology detection is therefore a useful pragmatic way to easily and correctly identify orthologous groups, provided duplications are absent within the phylogenetic range of the input data. It is a common feature of orthology methods, in particular those geared towards large datasets, that the orthology is approximated by a partition of the genes into groups of co-orthologs. The tool PoFF described here also follows this paradigm but provides pairwise orthology predictions in addition.

Several orthology prediction methods that avoid the explicit use of gene and species trees have been described in the literature. Most of them can be applied to large datasets only at high performance computing facilities. Their pre-computed results are usually available in databases, whereas the software itself is not available for public use or restricted in practice to small datasets. This limits their usefulness since poorly studied or newly sequenced organisms that are not (yet) available in the pre-computed results cannot easily be included in large-scale studies. PoFF is specifically designed to overcome these limitations and provides users a tool for compiling large-scale orthology datasets with moderate computational resources. Here we have shown that the combination of the fast, clustering-based orthology heuristic, Proteinortho, with the equally efficient heuristic for large-scale synteny assessment, FFAdj-MCS, leads to a substantial improvement of the data quality for related species without loss of performance. Synteny information proves to be a highly efficient filter against false-positive orthology assignments without a huge increase of the false negative rate. The extended approach, PoFF, is capable of boosting large-scale comparative studies which focus on closely related species or even strains.

Orthologous groups can provide a convenient starting point for more detailed analyses of the history of entire gene families. To this end, it is necessary to reduce in particular false positive orthology assignments. Figure 4 illustrates that the filtering and clustering strategy can have a strong influence on both the false positive and false negative rates of orthology assignments. Orthology is only defined as a pairwise relationship which is not transitive. Hence, reducing the false positive rate within orthologous groups will ultimately lead to a reduction of true positive rates when the pairwise definition is applied, as we did here (see Figure 4, e.g., separating the paralogs $B1$ and $B2$ into two distinct

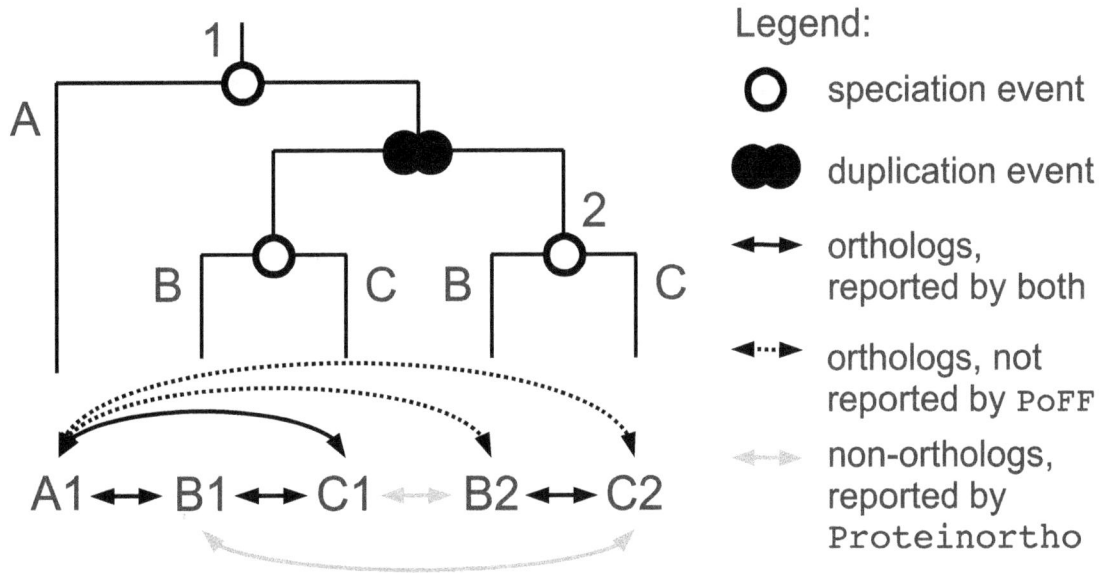

Figure 4. The false negative issue. The genes {A1, B1, B2, C1, C2} form an orthologous group. {B1, B2} as well as {C1, C2} are not orthologous to each other but co-orthologous with respect to A1 (A1 and {B1, B2, C1, C2} are separated by a speciation event). Pairwise true orthology relationships are marked by black arcs, false ones are grey. Proteinortho is more inclusive, it would report all five genes as one group, yielding six true and two false positives (grey). Assuming that the gene copies 1 and 2 exhibit distinct genomic neighborhoods in all three species A to C, PoFF would report two separate groups, namely {A1, B1, C1} and {B2, C2}. This more fine-grained method avoids false positive orthology assignments. However, it introduces false negative assignments. Two in this example, depicted by dashed arcs.

orthologous groups requires to discard the true orthology relation to A1 for one of them, otherwise both genes would be connected via A1). Given this, we had expected PoFF to perform much worse regarding true positives, which was, however, not the case.

While conserved synteny is a powerful feature to support the confidence in orthology predictions [56], gene orders evolve faster than protein sequences [57]. This fact is reflected by the benchmark results of the closely related simulated datasets compared to the real-life sets including more distantly related species, where the advantage of PoFF regarding pairwise orthology prediction was clearly reduced (see Tables 2 and 3). However, PoFF yields orthologous groups that are more fine-grained and contain fewer paralogs. We argue that this is a practical improvement for subsequent analyses, such as gene function prediction, genome annotation, marker development and phylogenetics. There, the presence of many-to-many relations in orthologous groups due to co-orthologs may lead to inconclusive results. In turn, these groups are often omitted and single-copy orthologs (a single gene per species) are used only [58–60]. This fact could make an application of PoFF desirable, even for more distant species.

The extension of Proteinortho by FFAdj-MCS leads to a very moderate increase in runtime and does not increase the hardware requirements, making this combined method applicable to very large datasets further on. The current approach of combining sequence similarity, conserved synteny and clustering entails a significant improvement when comparing closely related species. As gene orders generally evolve faster than protein sequences [57],

the improvement decreases with growing phylogenetic distance of species in the set, which may even compromise precision. Future extensions of the approach could thus aim at deciding on a case-by-case basis if the FFAdj-MCS algorithm should be used as additional filter for the comparison of two species, e.g., based on the respective breakpoint distance. Alternatively, a less restrictive synteny measure (e.g. common intervals instead of adjacencies) could be applied.

Supporting Information

File S1 Table S1: Accuracy of separation of Proteinortho and PoFF evaluated in reference dataset Orthobench. Table S2: Accuracy of separation of Proteinortho and PoFF evaluated in reference dataset COG.

Acknowledgments

We are thankful with Chunfang Zheng for providing us her scripts and guidance for genome rearrangements.

Author Contributions

Conceived and designed the experiments: ML MHR JS SJP PFS. Analyzed the data: ML. Contributed reagents/materials/analysis tools: ML MHR DD AT NW. Contributed to the writing of the manuscript: ML RKH PFS. Performed the simulations: ML DD. Read and approved the final manuscript: ML MHR DD NW AT JS RKH SJP PFS.

References

1. Fitch WM (1970) Distinguishing homologous from analogous proteins. Syst Zool 19: 99–113.
2. Fitch WM (2000) Homology a personal view on some of the problems. Trends Genet 16: 227–31.
3. Arvestad L, Berglund AC, Lagergren J, Sennblad B (2003) Bayesian gene/species tree reconciliation and orthology analysis using MCMC. Bioinformatics 19 (Suppl.1): 7–15.
4. Shi G, Peng MC, Jiang T (2011) MultiMSOAR 2.0: an accurate tool to identify ortholog groups among multiple genomes. PLoS One 6.

5. van der Heijden RT, Snel B, van Noort V, Huynen MA (2007) Orthology prediction at scalable resolution by phylogenetic tree analysis. BMC Bioinformatics 8: 83.
6. Hubbard TJ, Aken BL, Beal K, Ballester B, Caccamo M, et al. (2007) Ensembl 2007. Nucleic Acids Res 35: D610–617.
7. Wapinski I, Pfeffer A, Friedman N, Regev A (2007) Automatic genome-wide reconstruction of phylogenetic gene trees. Bioinformatics 23: 549–558.
8. Tatusov RL, Galperin MY, Natale DA, Koonin EV (2000) The COG database: a tool for genomescale analysis of protein functions and evolution. Nucleic Acids Res 28: 33–36.
9. Li L, Stoeckert CJ Jr, Roos DS (2003) OrthoMCL: identification of ortholog groups for eukaryotic genomes. Genome Res 13: 2178–2189.
10. Chen F, Mackey AJ, Stoeckert CJ Jr, Roos DS (2006) Orthomcl-db: querying a comprehensive multi-species collection of ortholog groups. Nucleic Acids Res 34: D363–368.
11. Schneider A, Dessimoz C, Gonnet GH (2007) OMA browser–exploring orthologous relations across 352 complete genomes. Bioinformatics 23: 2180–2182.
12. Altenhoff AM, Schneider A, Gonnet GH, Dessimoz C (2011) OMA 2011: orthology inference among 1000 complete genomes. Nucleic Acids Res 39: D289–D294.
13. Berglund AC, Sjölund E, Ostlund G, Sonnhammer EL (2008) InParanoid 6: eukaryotic ortholog clusters with inparalogs. Nucleic Acids Res 36: D263–266.
14. Jensen LJ, Julien P, Kuhn M, von Mering C, Muller J, et al. (2008) eggNOG: automated construction and annotation of orthologous groups of genes. Nucleic Acids Res 36: D250–254.
15. Wheeler DL, Barrett T, Benson DA, Bryant SH, Canese K, et al. (2008) Database resources of the National Center for Biotechnology Information. Nucleic Acids Res 36: D13–21.
16. DeLuca TF, Cui J, Jung JY, St Gabriel KC, Wall DP (2012) Roundup 2.0: enabling comparative genomics for over 1800 genomes. Bioinformatics 28: 715–716.
17. Mahmood K, Webb GI, Song J, Whisstock JC, Konagurthu AS (2012) Efficient large-scale protein sequence comparison and gene matching to identify orthologs and co-orthologs. Nucleic Acids Res 40.
18. Koonin EV (2005) Orthologs, paralogs, and evolutionary genomics. Annu Rev Genet 39: 309–38.
19. Lechner M, Findeiss S, Steiner L, Marz M, Stadler PF, et al. (2011) Proteinortho: detection of (co-)orthologs in large-scale analysis. BMC Bioinformatics 12: 124.
20. Hellmuth M, Hernandez-Rosales M, Huber KT, Moulton V, Stadler PF, et al. (2013) Orthology relations, symbolic ultrametrics, and cographs. J Math Biol 66: 399–420.
21. Altenhoff AM, Dessimoz C (2009) Phylogenetic and functional assessment of orthologs inference projects and methods. PLoS Comput Biol 5: e1000262.
22. Chen F, Mackey AJ, Vermunt JK, Roos DS (2007) Assessing performance of orthology detection strategies applied to eukaryotic genomes. PLoS One 2: e383.
23. Trachana K, Larsson TA, Powell S, Chen WH, Doerks T, et al. (2011) Orthology prediction methods: a quality assessment using curated protein families. Bioessays 33: 769–780.
24. Ebersberger I, Strauss S, von Haeseler A (2009) HaMStR: Profile hidden markov model based search for orthologs in ESTs. BMC Evol Biol 9: 157.
25. Shi G, Zhang L, Jiang T (2010) Msoar 2.0: Incorporating tandem duplications into ortholog assignment based on genome rearrangement. BMC Bioinformatics 11: 10–10.
26. Grigoriev IV, Cullen D, Goodwin SB, Hibbett D, Jeffries TW, et al. (2011) Fueling the future with fungal genomics. Mycology 2: 192–209.
27. Cao J, Schneeberger K, Ossowski S, Günther T, Bender S, et al. (2011) Whole-genome sequencing of multiple *Arabidopsis thaliana* populations. Nat Genet 43: 956–963.
28. Wapinski I, Pfeffer A, Friedman N, Regev A (2007) Natural history and evolutionary principles of gene duplication in fungi. Nature 449: 54–61.
29. Dandekar T, Snel B, Huynen M, Bork P (1998) Conservation of gene order: a fingerprint of proteins that physically interact. Trends Biochem Sci 23: 324–328.
30. Suyama M, Bork P (2001) Evolution of prokaryotic gene order: genome rearrangements in closely related species. Trends Genet 17: 10–13.
31. Lehmann J, Stadler PF, Prohaska SJ (2008) SynBlast: Assisting the analysis of conserved synteny information. BMC Bioinformatics 9: 351.
32. Dewey CN (2011) Positional orthology: putting genomic evolutionary relationships into context. Briefings Bioinf 12: 401–412.
33. Waterson G, Ewens W, Hall T, Morgan A (1982) The chromosome inversion problem. J Theor Biol 99: 1–7.
34. Bergeron A, Corteel S, Raffinot M (2002) The algorithmic of gene teams. In: Guigó R, Gusfield D, editors, WABI 2002. Heidelberg: Springer, volume 2452, pp. 464–476.
35. Bergeron A, Stoye J (2006) On the similarity of sets of permutations and its applications to genome comparison. J Comp Biol 13: 1340–1354.
36. Liben-Nowell D (2002) Gossip is synteny: Incomplete gossip and the syntenic distance between genomes. J Algorithms 43: 264–283.
37. Ting C, Yong HE (2006) Optimal algorithms for uncovering synteny problem. J Comb Optimization 12: 421–432.
38. Proost S, Fostier J, De Witte D, Dhoedt B, Demeester P, et al. (2012) i-ADHoRe 3.0–fast and sensitive detection of genomic homology in extremely large data sets. Nucleic Acids Res 40: e11.
39. Wang Y, Tang H, Debarry JD, Tan X, Li J, et al. (2012) MCScanX: a toolkit for detection and evolutionary analysis of gene synteny and collinearity. Nucleic Acids Res 40: e49.
40. Roedelsperger C, Dieterich C (2010) CYNTENATOR: Progressive gene order alignment of 17 vertebrate genomes. PLoS ONE 5: e8861.
41. Haas BJ, Delcher AL, Wortman JR, Salzberg SL (2004) DAGchainer: a tool for mining segmental genome duplications and synteny. Bioinformatics 20(18): 3643–3646.
42. Doerr D, Thévenin A, Stoye J (2012) Gene family assignment-free comparative genomics. BMC Bioinformatics 13: S3 19.
43. Braga MD, Machado R, Ribeiro LC, Stoye J (2011) Genomic distance under gene substitutions. BMC Bioinformatics 12 (Suppl. 9): S8.
44. Camacho C, Coulouris G, Avagyan V, Ma N, Papadopoulos J, et al. (2009) BLAST+: architecture and applications. BMC Bioinformatics 10: 421.
45. Altschul SF, Gish W, Miller W, Myers EW, Lipman DJ (1990) Basic local alignment search tool. J Mol Biol 215: 403–10.
46. Byrne KP, Wolfe KH (2005) The Yeast Gene Order Browser: combining curated homology and syntenic context reveals gene fate in polyploid species. Genome Res 15: 1456–1461.
47. Keller-Schmidt S, Tuğrul M, Eguíluz VM, Hernández-García E, Klemm K (2010) An age dependent branching model for macroevolution. Technical Report 1012.3298v1, arXiv.
48. Ohno S (1999) Gene duplication and the uniqueness of vertebrate genomes circa 1970–1999. Seminars in Cell and Developmental Biology 10: 517–522.
49. Lynch M, Conery JS (2000) The evolutionary fate and consequences of duplicate genes. Science 290: 1151–1155.
50. Xu W, Zheng C, Sankoff D (2007) Paths and cycles in breakpoint graph of random multichromosomal genomes. J Comput Biol 14: 423–435.
51. Hernandez-Rosales M, Wieseke N, Hellmuth M, Stadler PF (2014) Simulation of gene family histories. BMC Bioinformatics 15: S3–A8.
52. Strope CL, Abel K, Scott SD, Moriyama EN (2009) Biological sequence simulation for testing complex evolutionary hypotheses: indel-Seq-Gen version 2.0. Mol Biol Evol 26: 2581–2593.
53. Chang MSS, Brenner SA (2004) Empirical analysis of protein insertions and deletions determining parameters for the correct placement of gaps in protein sequence alignments. J Mol Biol 341: 617–631.
54. Dalquen DA, Anisimova M, Gonnet GH, Dessimoz C (2012) ALF–a simulation framework for genome evolution. Mol Biol Evol 29: 1115–1123.
55. Flicek P, Amode MR, Barrell D, Beal K, Brent S, et al. (2011) Ensembl 2011. Nucleic Acids Res 39: 800–806.
56. Rolland T, Neuvéglise C, Sacerdot C, Dujon B (2009) Insertion of horizontally transferred genes within conserved syntenic regions of yeast genomes. PLoS One 4.
57. Kristensen DM, Wolf YI, Mushegian AR, Koonin EV (2011) Computational methods for Gene Orthology inference. Brief Bioinform 12: 379–391.
58. Creevey CJ, Muller J, Doerks T, Thompson JD, Arendt D, et al. (2011) Identifying single copy orthologs in Metazoa. PLoS Comput Biol 7.
59. Franzén O, Jerlström-Hultqvist J, Einarsson E, Ankarklev J, Ferella M, et al. (2013) Transcriptome profiling of *Giardia intestinalis* using strand-specific RNA-seq. PLoS Comput Biol 9.
60. Liu H, Guo X, Wu J, Chen GB, Ying Y (2013) Development of universal genetic markers based on single-copy orthologous (COSII) genes in Poaceae. Plant Cell Rep 32: 379–388.

6

Genome Analysis of *Bacillus amyloliquefaciens* Subsp. *plantarum* UCMB5113: A Rhizobacterium That Improves Plant Growth and Stress Management

Adnan Niazi[1]*, **Shahid Manzoor**[1,3], **Shashidar Asari**[2], **Sarosh Bejai**[2], **Johan Meijer**[2], **Erik Bongcam-Rudloff**[1]

1 Department of Animal Breeding and Genetics, SLU Global Bioinformatics Centre, Swedish University of Agricultural Sciences, Uppsala, Sweden, 2 Department of Plant Biology, Linnéan Center for Plant Biology, Uppsala Biocenter, Swedish University of Agricultural Sciences, Uppsala, Sweden, 3 University of the Punjab, Lahore, Pakistan

Abstract

The *Bacillus amyloliquefaciens* subsp. *plantarum* strain UCMB5113 is a Gram-positive rhizobacterium that can colonize plant roots and stimulate plant growth and defense based on unknown mechanisms. This reinforcement of plants may provide protection to various forms of biotic and abiotic stress. To determine the genetic traits involved in the mechanism of plant-bacteria association, the genome sequence of UCMB5113 was obtained by assembling paired-end Illumina reads. The assembled chromosome of 3,889,532 bp was predicted to encode 3,656 proteins. Genes that potentially contribute to plant growth promotion such as indole-3-acetic acid (IAA) biosynthesis, acetoin synthesis and siderophore production were identified. Moreover, annotation identified putative genes responsible for non-ribosomal synthesis of secondary metabolites and genes supporting environment fitness of UCMB5113 including drug and metal resistance. A large number of genes encoding a diverse set of secretory proteins, enzymes of primary and secondary metabolism and carbohydrate active enzymes were found which reflect a high capacity to degrade various rhizosphere macromolecules. Additionally, many predicted membrane transporters provides the bacterium with efficient uptake capabilities of several nutrients. Although, UCMB5113 has the possibility to produce antibiotics and biosurfactants, the protective effect of plants to pathogens seems to be indirect and due to priming of plant induced systemic resistance. The availability of the genome enables identification of genes and their function underpinning beneficial interactions of UCMB5113 with plants.

Editor: Mark Gijzen, Agriculture and Agri-Food Canada, Canada

Funding: This work was supported by the Swedish Research Council for Environment, Agricultural Sciences and Spatial Planning (FORMAS), Carl Tryggers Stiftelse, Nilsson-Ehle Stiftelsen, Helge-Ax:son Johnsons Stiftelse, the Swedish University of Agricultural Sciences (SLU) and the Higher Education Commission of Pakistan (HEC). Funding for plant growth facilities were provided in part by KFI-VR. The funders had no role in study design, data collection and analysis, decision to publish, or preparation of the manuscript.

Competing Interests: The authors have declared that no competing interests exist.

* Email: Adnan.Niazi@slu.se

Introduction

Plant growth-promoting rhizobacteria (PGPR) colonize the plant rhizosphere and develop close physical and biochemical contacts with plants and enhance growth of the plant [1]. Many PGPR can also help plants to tolerate abiotic and or biotic stress better [2,3]. PGPR can mediate plant growth and protection in several ways. For instance, competition for growth space and essential nutrients, and production of a wide range of antibiotics and enzymes (like proteases and chitinases) counteract harmful microorganisms. Moreover production of siderophores also protect plants by solubilizing and scavenging iron from the environment, hence making it unavailable for other and more deleterious microorganisms [4]. Phytohormones such as auxins, gibberellins, and cytokinins produced by certain bacteria stimulate growth of the plants [5].

Many *Bacillus* ssp. have been found to provide beneficial effects to different plant species [6,7]. The Gram-positive *Bacillus* group represents a large genetic and habitat diversity and has several properties of interest for applied use. Many *Bacillus* that can serve as biofertilizers and biopesticides are regarded non-pathogenic which makes handling easier. These bacteria are mobile and show good rhizosphere competence and are also facultative anaerobes enabling survival in soil at different environmental conditions. *Bacillus* sporulate under unfavorable conditions and the spores are very resistant to harsh conditions providing long shelf-life useful for commercial applications. Several *Bacillus* ssp. have high secretory capacity and certain strains are used as "cell factories" for industrial production of enzymes. *Bacillus subtilis* is one of the best studied bacteria out of many aspects impoving understanding of many processes and features. *B. amyloliquefaciens* subsp. *plantarum* strains are capable to enhance plant growth and confer protection by producing phytohormones and antimicrobial compounds. The type strain of the *B. amyloliquefaciens* subsp. *plantarum* group, FZB42, is known to have a great capacity for non-ribosomal synthesis of secondary metabolites including lipopeptides and polyketides by some gene clusters with antimicrobial and antifungal activity [8,9].

The plant root-colonizing strain *B. amyloliquefaciens* subsp. *plantarum* UCMB5113 was isolated from soil in the Karpaty mountains of Ukraine and was identified initially as a member of the *B. amyloliquefaciens* group on the basis of phenotypic properties and partial gene sequence analysis [10]. The UCMB5113 strain can promote growth of both underground and aboveground tissues of different plants. This is observed as increased root branching, increased total root surface area, and increased leaf area providing increased photosynthesis capacity, and nutrient and water uptake. The UCMB5113 strain can restrict the growth of several fungal pathogens on oilseed rape such as *Alternaria brassicae*, *Botrytis cinerea*, *Leptosphaeria maculans* and *Verticillium longisporum* [11,12]. Disease suppression by UCMB5113 has also been observed using *Arabidopsis thaliana* infected with fungal or bacterial pathogens (unpublished observations). Further, improved tolerance of UCMB5113 treated wheat seedlings to abiotic stress factors like drought, cold and heat has been demonstrated [13,14,15]. Thus, the UCMB5113 strain seems to have the capacity to operate on different plants, improve different kinds of stress management and stimulate plant growth making it an interesting candidate for use in agriculture to support more sustainable crop production.

In this study, we describe the analysis of the *B. amyloliquefaciens* subsp. *plantarum* UCMB5113 genome sequence, and through comparison with the model species of the *B. subtilis* group, make an attempt to target genes that contribute to the beneficial interaction between bacteria and plants that ultimately results in growth promotion and improved stress management of the host plant.

Materials and Methods

Genome sequencing and assembly

The genomic DNA isolated from *B. amyloliquefaciens* subsp. *plantarum* UCMB5113 was sequenced through multiplexed sequencing process using Illumina technology and read data from two sequencing lanes were used. A total of 16,399,248 paired-end reads of length 75 bp with the average insert size of 230 bp were generated. MIRA ver_3.4 [16] and Velvet ver_1.1.04 [17] was used for the assembly of reads-data with the similar assembly approach as previously applied [18]. The published genome of *B. amyloliquefaciens* FZB42 (accession no. NC_009725) [19] was used as template to identify genomic rearrangements using Mauve genome alignment software [20]. All gap-filled regions and indels were verified through genomic PCR and subsequent Sanger sequencing of the amplicons.

Genome annotation and comparative analysis

The genome annotation was accomplished via the Magnifying Genome (MaGe) Annotation Platform [21] which is embedded with tools for structural annotation and functional annotation. Putative functions to encoding genes were automatically assigned via BlastP similiarity searches against the UniProt database. PRIAM was used to predict enzyme-coding genes, and transmemberane domains were identified by THMM. All of these tools are stitched up in MaGe annotation pipeline. Lastly, the predictions were reviewed and curated manually.

Comparative analysis of the genome was performed using the EDGAR software framework for the prokaryotic genomes [22] to identify homologs and non-homologs in all of the selected genomes of closely related *Bacillus* species. For phylogenetic analysis, complete sequences of nine housekeeping genes from *B. amyloliquefaciens* and other *Bacillus* species were obtained and concatenated after performing multiple alignment using MUS-

CLE [23]. Phylogenetic tree was constructed by the Maximum-Likelihood method with bootstrap analysis (1000 replications) using MEGA5 [24]. The genetic content comparison was conducted at protein level using BLASTP program (min length ≥ 80 and identity ≥ 80) between the completely available genomes of '*plantarum*' and '*amyloliquefaciens*' species. Similarity between complete genomes and local genome regions were visualized with GenoPlotR.

Regions of genome plasticity (RGP) were identified using RGP finder, SIGI, and IVOM programs embedded in the MAGE platform, by taking into account some genomic properties which include synteny breakage, sequence composition (GC-content and codon bias), tRNA, phage, and mobile elements spots such as recombinases, IS elements, integrases and transposases. ISFinder (http://www.is.biotoul.fr/) was used for the identification of IS elements. The prediction of secretory proteins containing signal peptides was done using SignalP v4.0 [25], lipoproteins containing cleavage sites for signal peptidase II (SPII) were predicted using LipoP v1.0 [26], and proteins with Twin-arginine cleavage sites were identified by TatP1.0 [27].

The complete nucleotide genome sequence of *B. amyloliquefaciens* subsp. *plantarum* UCMB5113 has been deposited in European Nucleotide Archive (ENA) database under accession number HG328254.

Carbohydrate fermentation analysis

Fermentation analysis of several sugars was performed using the API 50 CH system (BioMerieux). The fermentation test strips were inoculated with UCMB5113 resuspended in API 50 CHB/E Medium and was incubated for 24–48 hours at 28°C under aerobic conditions.

Enzyme activity assays

A single bacterial colony was streaked and/or 50 µl of 1×10^7 CFU ml^{-1} bacterial culture was inoculated at the center of agar plates containing different carbon source media and incubated at 28°C for 2–4 days. The assays were conducted as described using blood agar assay [28]; drop collapse test [29]; siderophore assay [30]; chitinase assay – crab-shell chitin [31] dissolved in M9 minimal medium without glucose; phosphate solubilization assay [32]; starch hydrolysis [33] and urease activity [34]. All the experiments were repeated at least twice with similar results and a representative experimental result is shown.

Swarming motility

In three independent experiments, aliquots of 50 µl of 1×10^7 bacterial culture were inoculated at the center of PDA (Potato dextrose agar) plates and incubated at 28°C for 4 days.

UCMB5113 mediated growth promotion of *Brassica napus* and *Arabidopsis thaliana*

Brassica napus cv. Westar and *Arabidopsis thaliana* ecotype Col-0 seeds were surface sterilized (1 min in 70% ethanol, 5 min in 10% bleach containing chlorine and Tween 20, and 5 min in water repeated 3 times). The seeds were dipped in a spore suspension of 1×10^7 CFU/ml^{-1} of *Bacillus* UCMB5113 and incubated at 28°C for 2 h. The seeds were then sown on sterile soil (s-soil, Hasselfors Garden), one seed per pot and grown at 22°C, 16/8 h photo period, for 18 to 21 days.

In vitro growth promotion studies were carried out using *Arabidopsis* seeds sterilized as above but germinated on Murashige & Skoog (MS) nutrient medium containing 0.6% bacto agar for 12 days at 22°C, 16/8 h photo period. Later the seedlings were dipped in a solution with 1×10^7 CFU ml^{-1} *Bacillus* strain UCMB 5113 and transplanted on 22×22 cm square plates containing MS with 0.8% bacto agar. The plants were grown at 22°C, 16/8 h photo period for two weeks further and growth promotion parameters were recorded.

Results and Discussion

Genomic structure and features of *B. amyloliquefaciens* UCMB5113

The general genomic structure of the *B. amyloliquefaciens* subsp. *plantarum* UCMB5113 chromosome of 3,889,530 bp is depicted in Figure 1. The genome was predicted to encode 3,656 coding sequences (CDSs) including 4 fragmented coding sequences (fCDSs). Putative biological functions were assigned to 3,106 CDS (85%) after manual curation. There were 506 (13.7%) conserved hypothetical proteins, and 44 (1.2%) hypothetical proteins having no homology to any previously reported sequence. The genome contains 10 copies of 5S, 16S, and 23S rRNA genes. Analysis of tRNAs showed 89 tRNA genes with specificities for all 20 standard amino acids on the chromosome.

The functional classification assigned 3,437 CDS in different COG (cluster of orthologs groups) functional groups (Figure 2). This analysis revealed that, apart from essential housekeeping genes, the genome seems biased towards three major functional gene classes: amino acid (E), carbohydrate (G), inorganic ion (P) transport and metabolism, secondary metabolite biosynthesis (Q), and defense mechanisms (V) representing 26% of all CDS. The high proportion of these genes in the genome indicates the inherent potential of UCMB5113 to compete in the rhizosphere with other microorganisms by having an efficient uptake of nutrients, along with the set of genes needed for defense mechanisms and secondary metabolite biosynthesis to antagonize other microorganisms.

Comparative analysis and phylogeny

A comparative analysis of the *B. amyloliquefaciens* subsp. *plantarum* UCMB5113 genome with other *Bacillus* species is shown in Table 1. The over all core genome of these species is comprised of 2391 orthologs (Figure 3). In total, 3,077 coding sequences are shared by the genomes of *B. subtilis* 168, *B. amyloliquefaciens* FZB42 and *B. amyloliquefaciens* UCMB5113. The genomes of UCMB5113 and FZB42 appear to be most similar by sharing 3,421 orthologs, whereas 3,345 orthologs were found between UCMB5113 and the non plant-associated type strain *B. amyloliquefaciens* subsp. *amyloliquefaciens* DSM7^T. A total of 112 unique coding sequences in the genome of *B. amyloliquefaciens* subsp. *plantarum* UCMB5113 were identified in the comparison. Moreover, the seventh allele of the 16S rRNA gene was also found to be truncated similar to FZB42 indicating a close relationship and may explain the evolutionary aspects of the two strains. Phylogenetic analysis performed (Figure 4) using sequences of nine housekeeping genes (*gyrA, dnaX, glyA, cysS, glpF, gmk, gyrB, ligA* and *recN*) from *B. amyloliquefaciens* strains and other *Bacillus* species further illustrated the close relationship between these organisms. The tree also show the clear distinction that can be made between the subsp. *plantarum* and subsp. *amyloliquefaciens* [35].

The genome of *B. amyloliquefaciens* subsp. *plantarum* UCMB5113 encodes several transcription regulators e.g. 17 sigma factors, 13 anti-sigma factors, and 2 anti-anti-sigma factors. The sigma factor encoding genes: *sigM, sigV, sigW, and sigX* were found while the genome is devoid of *sigY, sigZ* and *ylaC* and their antagonists *yxlC, sdss,* and *ylaD,* respectively. In addition, a putative sigma factor encoding gene *BASU_0627* was found similar to *sigV* (33% amino acid sequence identity).

Regions of genome plasticity

The multiple genome alignment highlighted that the genome of *B. amyloliquefaciens* subsp. *plantarum* UCMB5113 shared large blocks within which the gene order and synteny appears to be remarkably conserved between the closely related organisms analyzed (Figure S1). However, UCMB5113 contains defined inserted regions that seem to have originated with the transfer of mobile genetic elements, including phages, transposases and insertion sequences. Several regions of genome plasticity where horizontal gene transfer often occurs were identified using RGP Finder. In total 29 sites of genome plasticity were identified by comparing the UCMB5113 genome with the available genome sequences of *B. subtilis* 168, *B. amyloliquefaciens* DSM7 and *B. amyloliquefaciens* FZB42. These regions carry genes involved in sporulation, competence development, metal and drug resistance, biosynthesis of metabolites, phage derived proteins, sugar transport (PTS system) as well as other transporters and regulators. Three of those regions (RGP3a, RGP16, RGP22) containing: nitroreductases, oxidoreductases, hydrolases, and putative NRPS related genes were specific to UCMB5113 (Figure 1, Table S1 in File S1), whereas one region corresponding to *B. subtilis* prophage elements was identified. Indels identified in the UCMB5113 genome are summarized in Table S2 in File S1.

Prophages and transposable elements

The *B. amyloliquefaciens* subsp. *plantarum* UCMB5113 genome comprises 53 phage related genes including remnants and two putative phage integrases. A cluster of 16 phage-related genes (*xkdEFGHIJKMNOPQRSTU*) similar to *B. subtilis* was found on the chromosome. A significant portion of prophage derived endonuclease *yokF* was found truncated in the genome. Disruption of *yokF* in *B. subtilis* showed sensitivity to mitomycin C but increased transformation efficiency [36]. Two putative transposase encoding genes, *orfA* and *orfB,* identified with IS Finder (http://www-is.biotoul.fr/), as part of the only insertion sequence element, "ISBsu1" having a size of 1,289 bp from the IS3 family, exist in the UCMB5113 genome. Similar transposase genes seem conserved in most species of the *B. subtilis* group. Furthermore, three putative integrase (*BASU_1646, BASU_1806, BASU_1970*) and two resolvase encoding genes (*recU, yrrK*) involved in DNA recombination and decreasing susceptibility to DNA damaging agents were found on the chromosome.

Extracellular enzymes and metabolic activities

In the *B. amyloliquefaciens* subsp. *plantarum* UCMB5113 genome, 298 out of the 3,693 genes were predicted to encode proteins in the secretory pathway including 149 proteins that lack any transmembrane domain as predicted by THMM2 (Table S3 in File S1). Of these, 200 were Sec dependent having signal peptidase I as a site for cleavage, 87 were identified as lipoprotein consisting (SPII/Lsp), and 11 contained a TAT (twin-arginine translocation) motif for protein secretion. Out of all these, 78 were assigned putative functions, 26 were uncharacterized protein encoding genes, and 14 were hypothetical or conserved hypothetical genes. Of the remaining 180 genes, 111 were found to be enzyme encoding genes. Further examination revealed 11 unique putative CDSs in the genome predicted to be secreted in the extracellular space. In accordance with its symbiotic lifestyle, the secretome of UCMB5113 comprises various enzymes that probably hydrolyze polysaccharides, proteins and other com-

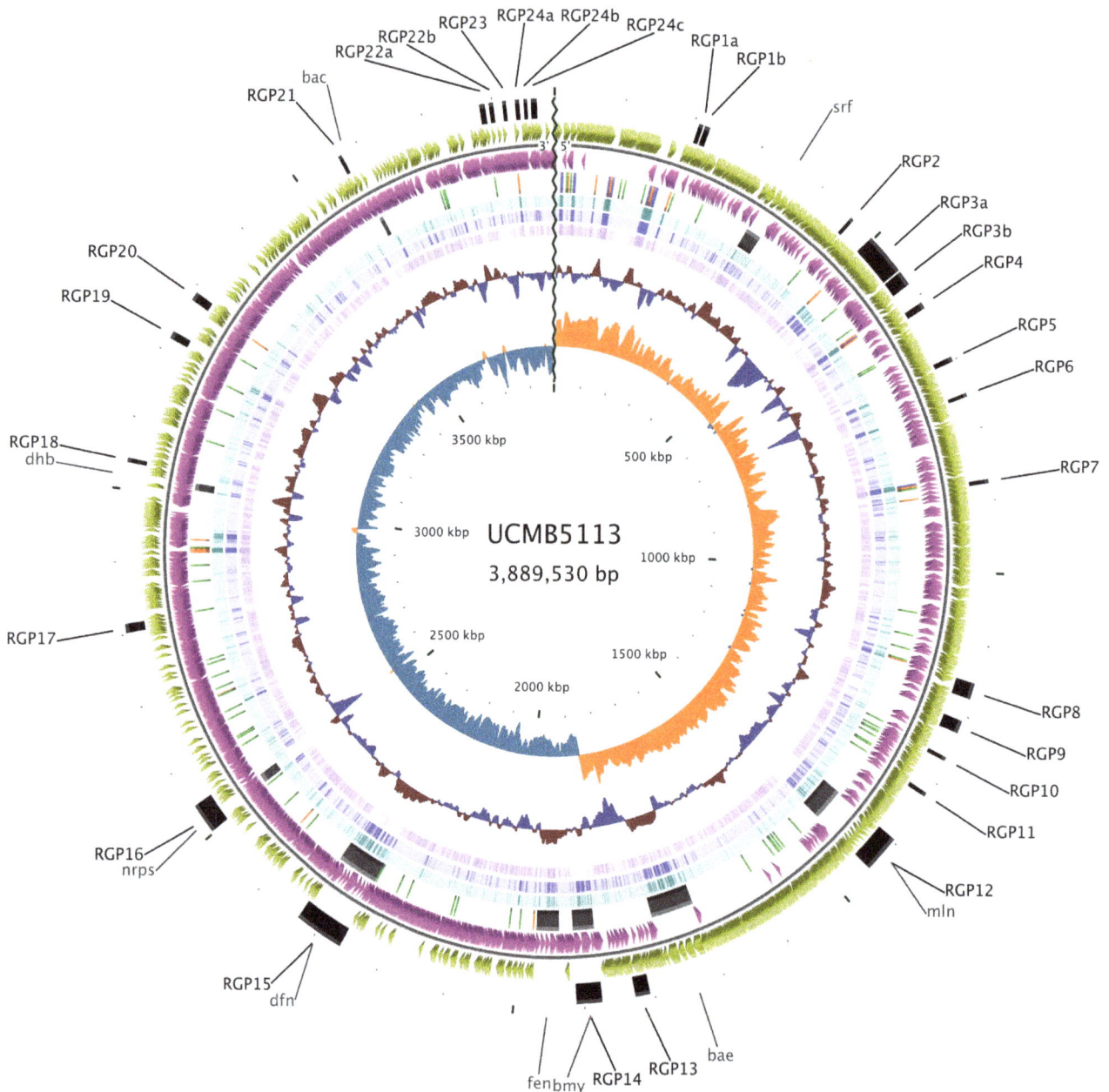

Figure 1. Graphical representation of genomic features of *B. amyloliquefaciens* **subsp.** *plantarum* **UCMB5113.** Circles display (from the outside): (1) Sites of genome plasticity. (2) Predicted CDSs transcribed in the clockwise direction. (3) Predicted CDSs transcribed in the counterclockwise direction. (4) rRNA (blue), tRNA (orange), non-coding RNA (green), and NRPS/PKS gene clusters (grey). (5,6,7) Blast comparison of UCMB5113 genome with type strain FZB42[T] and *B. subtilis* 168, respectively. (8) GC percent deviation (GC window - mean GC) in a 1000-bp window. (9) GC skew (G+C/G-C) in a 1000-bp window.

pounds available in the rhizosphere. Part of this enzyme repertoire may be used by UCMB5113 to process plant surfaces to allow colonization at the rhizoplane. For instance, members of different glycoside hydrolase families (GH5, GH9, GH11, GH16, GH44, and GH48) may allow bacteria to grow on a broad range of polysaccharides by degrading hemicellulosic components present in plant cell walls. However, no putative members of the GH9, GH44, GH48 families were found in the UCMB5113 genome. Further degradative genes in UCMB5113 includes: *abnA* encoding endo-arabinase (GH43), *xynA* encoding xylanase (GH11), *bglS* encoding lichenase (GH16) and *eglS* encoding endo-glucanase/

cellulase (GH5). Also members of the polysaccharide lyase family were identified which include pectate lyase (pel and BASU_3156) and pectin lyase (pelB) that may also be involved in degradation of plant tissues. All of the gene products encoded for polysaccharide degradation were predicted to have secretory signal peptides allowing degradation and assimilation of molecules present in the rhizosphere or rhizoplane. Tests to ensure the presence of certain prominent genes encoding hydrolytic enzymes revealed that UCMB5113 was able to utilize chitin, urea and starch (Figure 5).

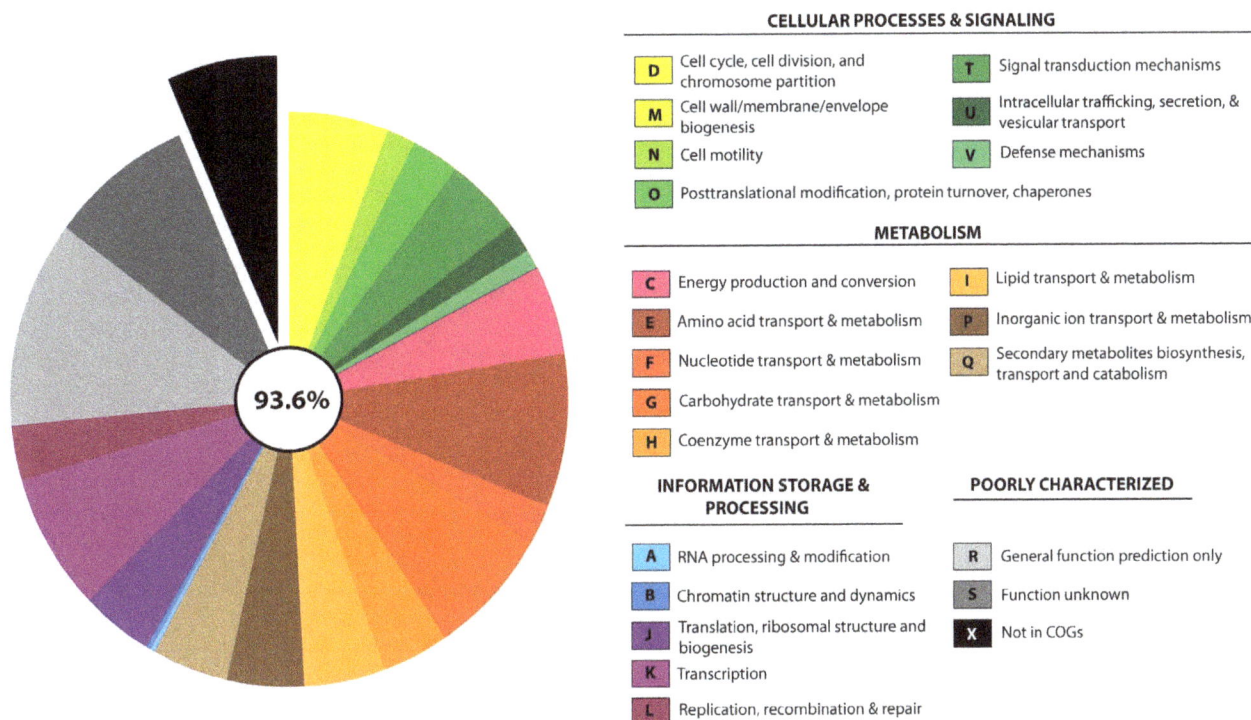

CELLULAR PROCESSES & SIGNALING

D — Cell cycle, cell division, and chromosome partition

M — Cell wall/membrane/envelope biogenesis

N — Cell motility

O — Posttranslational modification, protein turnover, chaperones

T — Signal transduction mechanisms

U — Intracellular trafficking, secretion, & vesicular transport

V — Defense mechanisms

METABOLISM

C — Energy production and conversion

E — Amino acid transport & metabolism

F — Nucleotide transport & metabolism

G — Carbohydrate transport & metabolism

H — Coenzyme transport & metabolism

I — Lipid transport & metabolism

P — Inorganic ion transport & metabolism

Q — Secondary metabolites biosynthesis, transport and catabolism

INFORMATION STORAGE & PROCESSING

A — RNA processing & modification

B — Chromatin structure and dynamics

J — Translation, ribosomal structure and biogenesis

K — Transcription

L — Replication, recombination & repair

POORLY CHARACTERIZED

R — General function prediction only

S — Function unknown

X — Not in COGs

Figure 2. Functional classification of protein-coding genes in UCMB5113. Distibution of UCMB5113 coding sequences (93.6%) in COG functional classes. Genes that did not have any inferred COG annotation were assigned to category X.

Secondary metabolites, siderophores and antibiotics

The genome of *B. amyloliquefaciens* subsp. *plantarum* UCMB5113 harbors gene clusters which are responsible for biosynthesis of several bioactive lipopeptides via nonribosomal peptide synthetases (NRPSs) including; surfactin (*srf*), bacillomycin D (*bmy*) and fengycin (*fen*), with known antagonistic activities (Table 2). A gene cluster for the dipeptide antibiotic bacilysin [37,38], synthesized in several strains of *B. subtilis*, *B. amyloliquefaciens*, *B. pumilus*, and *B. licheniformis* [39], was also found in the UCMB5113 genome. In addition to NRPS operons, a macrolactin synthesizing *mln* operon that has been reported to exhibit antibacterial activity in *B. subtilis* AT29 [40] and *B. amyloliquefaciens* CHO104 [41] was located in the UCMB5113 genome. Furthermore, gene clusters for synthesis of bacillaene (*bae*) and difficidin (*dfn*) were identified in the chromosome. Both *bae* and *dfn* have been characterized to exhibit antimicrobial activity in *B. subtilis* and *B. amyloliquefaciens* strains [9,42,43]. The organization of gene clusters in UCMB5113 was observed to be similar to the corresponding genomic segments in FZB42 (Figure 6) with high identity at the amino acid level. Remnants of the *fen* gene cluster was located in the genome of *B. amyloliquefaciens* DSM7, whereas no counterparts for *mln* and *dfn* operons was found in the chromosomes of DSM7, *B. subtilis* 168, or sequenced species of *B. licheniformis*, *B. cereus*, and *B. pumilus*. In addition, a putative NRPS cluster of five genes (*BASU_2336-BASU_2340*) probably encoding a novel antibiotic was found on the chromosome of *B. amyloliquefaciens* subsp. *plantarum* UCMB5113. Apparently the UCMB5113 strain has great capacity for antibiosis and an earlier investigation showed that this strain could antagonize several phytopathogens using *in vitro* assays [11].

Certain bacteria have developed diverse iron uptake mechanisms to compete for iron ions in the rhizosphere that often is a limiting factor for growth. Such mechanisms include iron uptake transporters, synthesis of siderophores and siderophore receptors, which contribute to confer protection of host plants from pathogens. The *B. amyloliquefaciens* subsp. *plantarum* UCMB5113 genome contains a gene cluster (*dhbABCDEF*) that is responsible for the synthesis of the iron-siderophore bacillibactin in *B. subtilis* [44]. The respective genes show identities of 95–99% for the *dhb* operon at the amino acid level. UCMB5113 grown on blood agar plates showed hemolytic activity within two days by the presence of a clear zone around the bacteria (Figure 5) indicating production of biosurfactants that can serve in antibiosis. Unlike *B. amyloliquefaciens* FZB42, no counterpart for the *nrs* operon, probably responsible for non-ribosomal synthesis of siderophores [17], was found in the genome of UCMB5113.

Sporulation and competence genes

Some bacteria undergo morphological changes, i.e. sporulation, induced by nutritional depletion in the environment. Similar to *B. subtilis*, the chromosome of UCMB5113 possesses genes implicated in sporulation, including the *spsABCDEFG* operon responsible for the synthesis of spore coat polysaccharides. The exception, in contrast to *B. subtilis*, is the absence of *sdpABC* genes, which are involved in sporulation-delay followed by cannibalism [45]. However, three genes, *BASU_2886-BASU_2888* similar to the *sdp* operon were identified which may be a substitution for cannibalism in UCMB5113. Internalization of exogenous DNA (competence development) depends on several genes, some being expressed early during growth while others are activated later. The *B. amyloliquefaciens* subsp. *plantarum* UCMB5113 genome encodes all such genes as *comC*, *comK*, and *comEFG* operons, except the *comS* gene that is involved in the establishment of genetic competence [46,47]. However, we uncovered three genes in the GI-06 region with substantially lower GC content (37%)

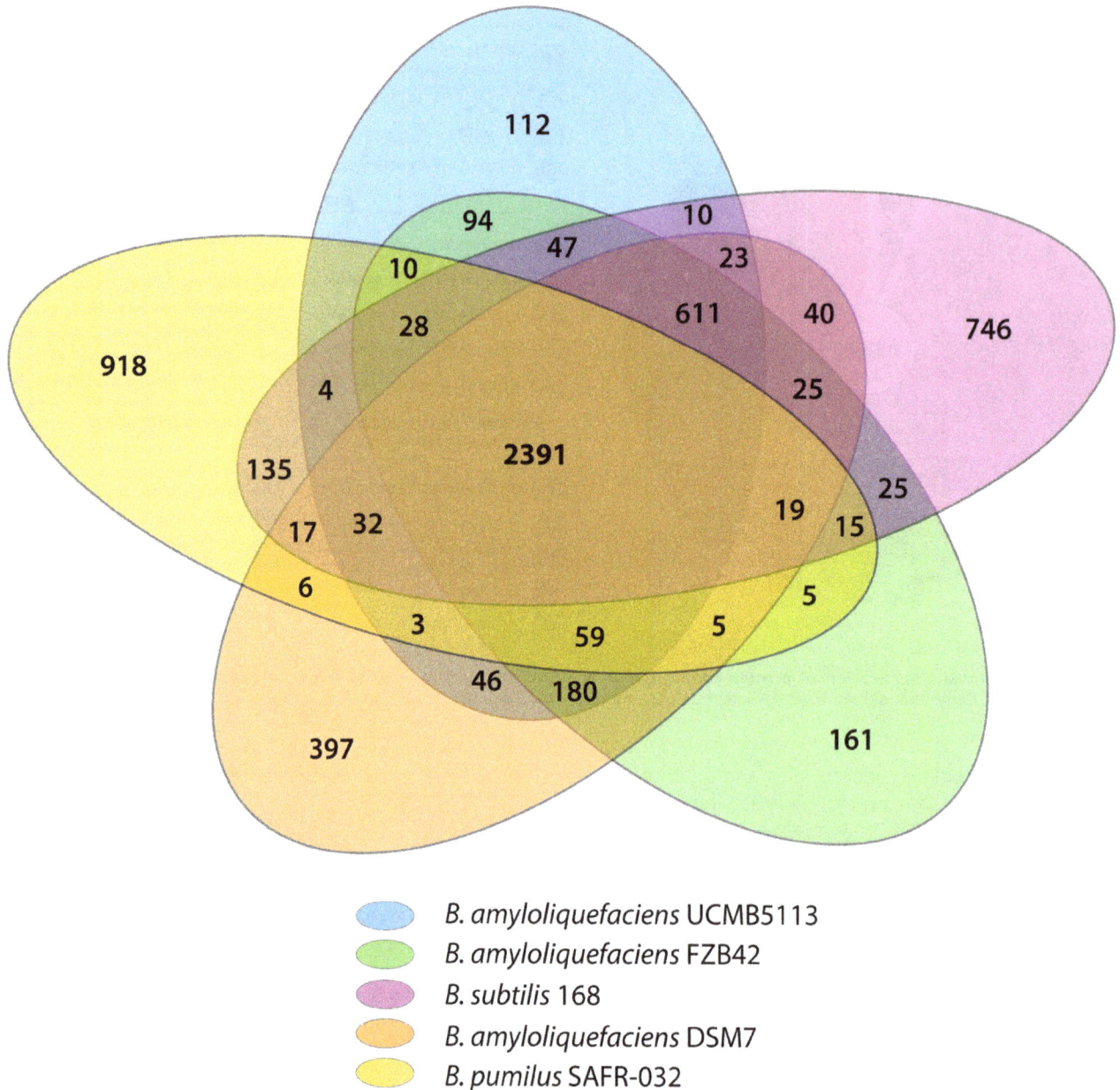

Figure 3. Numbers of shared and genome-specific genes. The Venn Diagram shows the number of shared and genome-specific genes in *B. amyloliquefaciens* subsp. *plantarum* UCMB5113, *B. amyloliquefaciens* subsp. *plantarum* FZB42[T], *B. subtilis* 168, *B. amyloliquefaciens* subsp. *amyloliquefaciens* DSM7[T] and *B. pumilus* SAFR-032.

that could putatively encode HsdS, HsdM, and HsdR subunits sharing significant identity to type-I restriction-modification systems.

Sugar Transporters

B. amyloliquefaciens subsp. *plantarum* UCMB5113 encodes a plethora of distinct carrier proteins to transport the organic and inorganic nutrients available in the rhizosphere. Similarity searches and conserved domain analysis of protein sequences found categorizes of 430 putative carrier proteins including efflux and permeases (Table S4 in File S1). Out of this collection, 136 were ABC-type transporters (including the phosphate and zinc-

specific operons *pst* and *znu*, respectively), 68 classified as MFS family transporters, 24 belonged to the phosphotransferase system (PTS) family of transporters, one putative membrane fusion protein (MFP), *BASU_1625*, belonging to the RND family of transporters. A member of the MMPL family with unknown function was also identified. It has been suggested that members of the MMPL family may be involved in lipid transport [48]. Presence of PTS transporters and other hydrolases/isomerases enable UCMB5113 to break down and utilize several plant-derived compounds as carbon sources. To investigate this, we tested the fermentation ability of UCMB5113 with several sugars. The sugars fermented by UCMB5113 are summarized in Table 3.

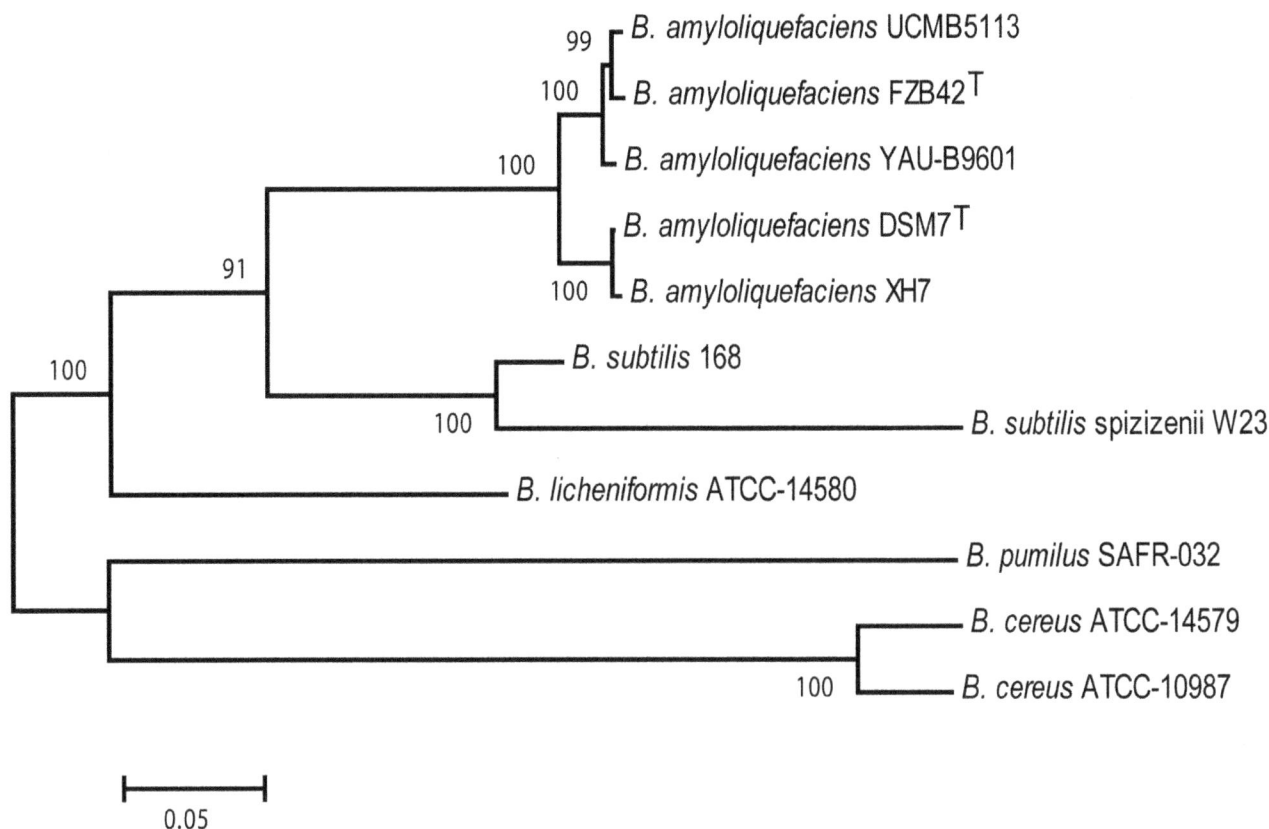

Figure 4. Neighbour-joining phylogenetic tree. The position of *B. amyloliquefaciens* strain UCMB5113 in relation to other species within the genus *Bacillus*. The numbers above the branches are support values obtained from 1,000 bootstrap replicates.

In addition to common sugars this strain has the ability to metabolize acyclic polyols (sugar alcohols) such as sorbitol and D-mannitol by converting them to fructose via sorbitol dehydrogenase and mannitol-1-phosphate dehydrogenase. The UCMB5113 strain thus has a dynamic usage of different sugars although the differences in inducibility of sugar degradation capacity observed between 24 and 48 h, suggest preferences for certain types of sugars.

Resistance to drugs and heavy metals

B. amyloliquefaciens subsp. *plantarum* UCMB5113 seems resistant to several antimicrobial compounds in the rhizosphere improving competitiveness in microbial antagonism and may also support establishment of a symbiotic relationship with the host plants. The genome encodes a putative tetB protein that contributes to tetracycline resistance by decreasing its accumulation in *B. subtilis* [49]. UCMB5113 growth on tetracycline containing LB plates confirmed this property with good growth

Table 1. Comparison of genomic features of *B. amyloliquefaciens* subsp. *plantarum* UCMB5113 with genomes of other *Bacillus* spp. belonging to the *B. subtilis* group.

Features	*B. amyloliquefaciens*				
	UCMB5113	**FZB42[T]**	**DSM7[T]**	**B. subtilis 168[T]**	**B. pumilus SAFR-032**
Genome size (bp)	3,889,530	3,918,591	3,980,199	4,214,630	3,704,465
G+C content (%)	46.71	46.49	46.1	43.51	41.3
Protein-coding sequences	3656	3693	3893	4106	3679
Number of CDS shared	n.a*	3421	3345	3153	2541
Unique CDS (not shared by UCMB5113)	n.a	235	311	503	1115
rRNA operons	10	10	10	10	7
tRNA genes	89	89	94	86	69

*n.a: not applicable.

Swarming motility

Urease activity

CAS agar with iron

Chitinase activity

Hemolytic activity

Phosphate solubilization

Starch hydrolysis

Droplet collapse

UCMB5113   Control

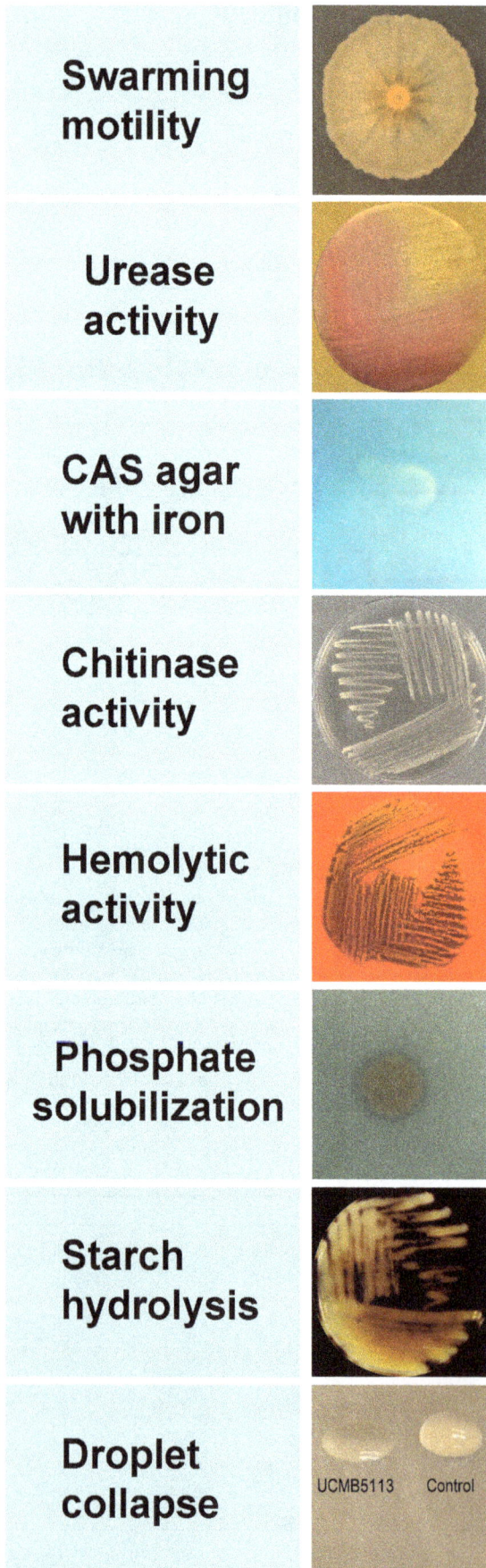

Figure 5. *B. amyloliquefaciens* **UCMB5113 related activity on different substrates.** A branch pattern with massive groups of bacteria observed after 4 days of incubation period indicated swarming motility; bright pink color indicated the hydrolysis of urea to carbon dioxide and ammonia; siderophore biosynthetic cluster produced a clear zone on CAS agar; chitin degradation and utilization as carbon source; expressed hemolytic activity on blood agar; phosphate solubilization around bacteria apparent as a transparent zone; amylase activity on starch medium; aqueous drop collapse to assess production of biosurfactants.

recorded at 5 μg tetracycline ml^{-1}. Another putative resistance gene, *yyaR*, was found displaying similarity to the *sat-4* gene that encodes a streptothricin acetyltransferase in *Campylobacter coli* BE/G4 [50]. Additionally, two novel genes (*BASU_2335* and *BASU_3689*) putatively encoding multidrug exporters were located that may also support resistance to several antibiotics.

The *B. amyloliquefaciens* subsp. *plantarum* UCMB5113 chromosome carries genes that also confer resistance to heavy metals such as zinc, copper, cobalt, and cadmium. UCMB5113 grown on LB plates containing up to 10 mM of cadmium, cobalt, copper or zinc exhibited good metal tolerance at all concentrations tested except for zinc (at 10 mM), and cadmium where some inhibition was noted at 1 and more at 10 mM concentration. The strain does not encode the genes *arsABC* involved in arsenic/arsenate resistance. However, a putative arsenic pump membrane protein YdfA share significant similarity to ArsB and may contribute to arsenic resistance [51,52]. Altogether this property makes UCMB5113 a promising biocontrol agent candidate to serve also in metal contaminated soils or soils that by agricultural practices get higher metal content e.g, through addition of mud from sewage plants.

Oxidative stress

B. amyloliquefaciens subsp. *plantarum* UCMB5113 seems capable to produce enzymes with proteolytic/hydrolytic like activities in response to oxidative stress in the rhizoplane, prior to colonization. Primarily, these enzymes include superoxide dismutases (SodA, SodC, SodF), three hydrogen peroxide decomposing catalases (KatA, KatE, KatX), manganese catalase (YdbD), three alkyl hydroperoxide reductases (AhpC and AhpF, BASU_0830), thiol peroxidase (tpx), glutathione peroxidase (gpo), bacillopeptidase F (bpr), gamma-glutamyl transpeptidase (ggt), and an operon (*ohrARB*) for resistance to organic peroxides [53]. Furthermore, the flavohemoprotein nitric oxide dioxygenase encoded by the genes *hmp* and *BASU_2738*, seem to protect the bacterium from nitrosative stress [54]. Accordingly, UCMB5113 has a battery of enzymes to prevent damage from reactive oxygen species that could be formed by non-enzymatic reactions in soil or the result of various biological processes including plant defense reactions at the colonization stage. The UCMB5113 battery of antioxidant genes could be of benefit also for the plant if such components are produced in the rhizosphere, e.g. as a result of abiotic stress, but quenched by the root colonizing bacteria.

Swarming motility and chemotaxis

B. amyloliquefaciens subsp. *plantarum* UCMB5113 exhibits swarming motility (Figure 5). The presence of gene clusters (*flg*, *flh*, *fli*) for the production and assembly of functional flagellar components together with the *swrA* gene that up-regulates the expression of flagellar genes and increases swarming motility [55,56] supports the high rhizosphere competence observed for UCMB5113. Production of lipopeptide biosurfactants is also an

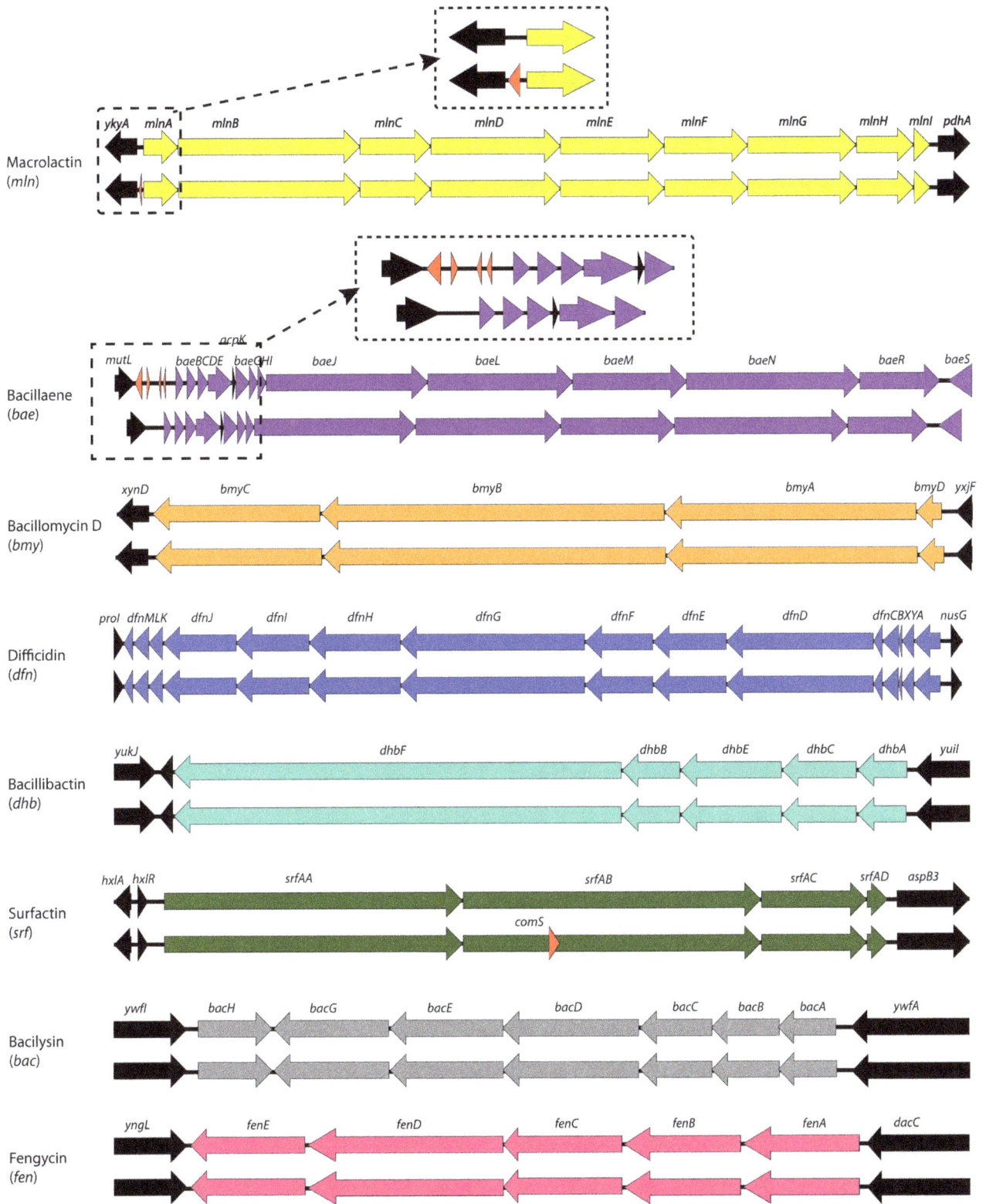

Figure 6. Blast comparison of NRPS/PKS clusters in UCMB5113 (above) and FZB42^T(below). Arrows indicate gene clusters; Macrolactin (light green), Bacilllaene (purple), BacillomycinD (orange), Difficidin (blue), Bacillibactin (turquoise), Surfactin (green), Bacilysin (grey), Fengycin (pink). Genes highlighted in red represent the differences where as black represent other genes flanking in each cluster.

Table 2. NRPS and PKS gene clusters involved in synthesis of secondary metabolites in *B. amyloliquefaciens*.

Compound	UCMB5113	FZB42	DSM7^T	Identity %
Surfactin	srfABCD	srfABCD	srfABCD	98–99
BacillomycinD	bmyCBAD	bmyCBAD	Not present	96–98
Fengycin	fenABCDE	fenABCDE	fenDE	98–99
Putative peptide	Not present	nrsABCDEF	Not present	0
Bacillibactin	dhbACEBF	dhbACEBF	dhbACEBF	95–99
Bacilysin	bacABCDEGH	bacABCDEGH	bacABCDEGH	99–100
Macrolactin	mlnABCDEFGHI	mlnABCDEFGHI	Not present	98–99
Bacillaene	baeBCDE, acpk, baeGHIJLMNRS	baeBCDE, acpk, baeGHIJLMNRS	baeBCDE, acpk, baeGHIJLMNRS	97–99
Difficidin	dfnAYXBCDEFGHIJKLM	dfnAYXBCDEFGHIJKLM	Not present	98–99
Putative peptide	nrpsGFEC	Not present	nrpsGFEC	0

essential feature that facilitates motility by lowering the surface tension of solid surfaces changing interfacial interaction. In addition, genetic determinants of chemotaxis (*che*) were also identified in the genome. All these genes will greatly favour host plant sensing and identification to assure proper root colonization and formation of a suitable environment that enables proliferation of UCMB5113. A drop collapse test to evaluate the biosurfactant activity of UCMB5113 indeed showed that bacterial exudates caused a rapid loss of surface tension of aqueous drops indicating the presence of bioactive compounds (Figure 5).

Biofilm formation and root colonization

Biofilm formation is essential for efficient surface colonization by bacteria, and this matrix is comprised of a variety of extracellular polymeric products. Exopolysaccharides are important components of the extracellular matrix making up the biofilm. Two operons, *epsA-O* and *yqxM-sipW-tasA*, that are required for the formation of robust biofilms in *B. subtilis* [57,58,59] were located in the UCMB5113 genome. Besides this, the protein YuaB is also probably encoded, which not only facilitates the assembly of a biofilm matrix [60], but also is responsible for biofilm surface repellency by forming a hydrophobic layer on the surface of the biofilm [61], which probably provides resistance to a broad spectrum of antimicrobial agents. In addition, *ypqP* probably encodes a capsular polysaccharide biosynthesis protein, with homology to CapD that has been implicated to epimerase UDP-galactose to UDP-glucose. This reaction contribute to the production of exopolysaccharide or lipopolysaccharide slime layers that surrounds the bacterium [62]. By contrast, an insertion of phage-related genes caused fragmentation of *ypqP* in the genome of *B. subtilis* 168 decreasing the biofilm production ability.

Root colonization by *B. amyloliquefacines* subsp. *plantarum* strains mainly determined by the chemotaxis and biofilm formation ability, is important for establishing associations with host plants, initiated after extensive and intricate cross talk. However, many steps in this process are not well understood thus far. Most probably factors like externalized polysaccharides, adhesins, and motility functions play significant roles in root colonization. Two of the *B. amyloliquefaciens* subsp. *plantarum* UCMB5113 genes, *BASU_0726* and *BASU_0727*, encode proteins with a collagen-like GXT structural motif expected to form extended fiber-like protein structures that may be involved in surface adhesion observed to occur for UCMB5113 on plant seeds

and seedlings [10]. In addition, inhibition of the host innate immunity system or avoidance of recognition by plant's pattern recognition receptors is needed in order for UCMB5113 to avoid rejection from the plant.

Plant growth promotion hormones and volatile compounds

The production or stimulation of plant formation of phytohormones such as auxin, gibberillin, ethylene, abscisic acid, and cytokinin, is a characteristic feature of many PGPR. Plant-associated bacteria having the capacity to contribute to the host plants hormone pool can manipulate plant physiology and bring outcomes that favor their own survival. Auxin or indole-3-acetic acid (IAA) that serves as a master regulator of root growth and development is synthesized by several *Bacillus* species. Tryptophan dependent IAA synthesis involving putative IAA acetyltransferase (YsnE) and putative nitrilase (YhcX) and their effect on plant growth promotion has been demonstrated in *B. amyloliquefaciens* [63]. The presence of these two genes in the UMCB5113 genome tempts us to hypothesize that they play a prominent role in the enhanced plant growth observed when UCMB5113 is added to *Brassica napus* and *Arabidopsis thaliana* (Figure 7). Additionally, the gene *ywkB* putatively encoding an auxin efflux carrier protein that may be involved in the transport and redistribution of auxin to the roots was found in the UCMB5113 genome. However, the existance of intermediate compounds involved in IAA biosynthesis is not well known and has not been demonstrated for any *B. amyloliquefaciens* subsp. *plantarum* strain.

Volatile organic compounds (VOCs) like 3-hydroxy-2-butanone (acetoin) and 2,3-butanediol emitted by the rhizobacteria *B. subtilis* GB03 and *B. amyloliquefaciens* IN937a may not only trigger plant growth but a role in igniting ISR has also been implicated in plant-bacteria systems [64,65]. Acetoin production in the two *Bacillus* strains involves two enzyme encoding genes, *alsS* and *alsD*, which encode acetolactate synthase and acetolactate decarboxylase, respectively. The biosynthesis of 2,3-butanediol is catalyzed by the enzyme (*R,R*)-butanediol dehydrogenase, which is encoded by *bdhA*. The UCMB5113 chromosome possesses genes that ferment pyruvate to acetoin and 2,3-butanediol. The growth promotion effect observed on plants by UCMB5113 can most likely be attributed to the action of these genes.

Table 3. Fermentation of sugars by UCMB5113 analyzed using API strips.

Substrate	Utilization	
	after 24 hrs	after 48 hrs
Glycerol	++++	++++
Erythritol	+	+
D-Arbinose	+	+
L-Arbinose	++++	++++
D-Ribose	++++	++++
D-Xylose	−	++
L-Xylose	−	+
D-Adonitol	−	+
Methyl-βD-Xylopyranoside	−	+
D-Galactose	+	+
D-Glucose	++++	++++
D-Fructose	++++	++++
D-Mannose	++++	++++
L-Sorbose	−	+
L-Rhamnose	−	+
Dulcitol	−	+
Inositol	++	++
D-Mannitol	++++	++++
D-Sorbitol	++++	++++
Methyl-αD-Mannopyranoside	−	+
Methyl-αD-Glucopyranoside	++	+++
N-Acetylglucosamine	−	−
Amygdalin	+++	++++
Arbutin	+++	v
Esculin, ferric citrate	++++	++++
Salicin	+++	+++
D-Cellobiose	++++	++++
D-Maltose	++++	++++
D-Lactose	++	+++
D-Melibiose	−	++
D-Saccharose	++++	+
D-Trehalose	++++	+
Inulin	−	−
D-Melizitose	−	−
D-Raffinose	−	+++
Amidon (starch)	−	+++
Glycogen	−	+++
Xylitol	−	+
Gentiobiose	+	+
D-Turanose	−	−
D-Lyxose	−	−
D-Tagatose	−	−
D-Fucose	−	−
L-Fucose	−	−
D-Arabitol	−	−
L-Arabitol	−	−
Potassium gluconate	−	−

Table 3. Cont.

Substrate	Utilization	
	after 24 hrs	after 48 hrs
Potassium 2-Ketogluconate	−	−
Potassium 5-Ketogluconate	−	−

++++: Highly positive reaction (Bright Yellow).
+++: Positive reaction (Dark Yellow).
++: Medium reaction (Bright Orange).
+: Weak reaction (Dark orange).
−: Negative (Red).
v: Variable reaction (positive to undetermined).

Conserved 'plantarum' species coding genes

Gene content comparison of the two groups of *B. amyloliquefaciens* species performed using the proteome of *B. amyloliquefaciens* subsp. *plantarum* UCMB5113 as reference, identified characteristic features of '*plantarum*' species (Figure 8). Interestingly, two large genomic regions of size 70 kb and 53 kb (occupied by the *dfn* and *mln* gene clusters, respectively) were found to be conserved only in the genomes of plant-associated species. Conservation of *dfn* and *mln* gene clusters strongly suggests their role is not confined only to antimicrobial activities but may be linked with plant-associated activities as to become accepted as a beneficial partner since certain polyketides have been indicated to serve both as virulence and avirulence factors [66]. In total, 80 protein-coding genes were identified with no counterparts in the genomes of the strains belonging to the *B. amyloliquefaciens* subsp. *amyloliquefaciens* group (Table S5 in File S1). The identified genetic elements may represent characteristics of the plantarum species i.e. host plant association and rhizosphere competence. Results from expression analysis of a selection of nine of these genes confirmed that the predicted genes were indeed expressed (Figure S2), and thus may have role for plant interaction.

Conclusions

The genome assembly of *B. amyloliquefaciens* subsp. *plantarum* UCMB5113 and its comparison with the genomes of model

species of *B. subtilis* group gave solid basis for gene annotation. The UCMB5113 strain seems to have high capacity to produce different kinds of antibiotics and also secrete a large number of enzymes to improve nutrient acquisition in the rhizosphere. This strain also seems to have potential for production of hormones and volatile compounds that support plant growth and thereby improve plant roots and biomass improving plant quality as a colonization partner and at the same time also increase surface area for colonization and the nutrient resource for the bacteria. UCMB5113 seems able to use a wide range of sugars and other organic compounds that could be present in root exudates. In return the bacteria antagonize detrimental soil microorganisms and strengthen plants improving their nutrient and stress handling capabilities. The ability of UCMB5113 to quench reactive oxygen species should be beneficial for the plant to decrease stress damage. The annotation of UCMB5113 from this study will pave way in elucidating the mechanism involved in plant-bacterial interation. Comparison with other related *B. amyloliquefaciens* genomes that differ in their capability of plant colonization, growth promotion and stress tolerance provides an excellent basis for *in silico* predictions of gene candidates and regulatory factors that are involved in these processes. We are currently searching for plant genotypes that vary in the interaction with *Bacillus* strains as a basis to pinpoint plant genes important for the interactions. Deciphering of the molecular determinants for these processes open up possibilities to identify even more efficient *Bacillus* strains,

Figure 7. Plant growth promotion by UCMB5113 on *Arabidopsis thaliana* Col-0. (A). Plants grown on MS agar and treated with UCMB5113 display bigger leaves and increased root branching. (B) Plants grown on soil and treated with UCMB5113 have bigger leaves compared to control plants. The experiment was performed at least three times, and similar results were obtained in each case.

(A)

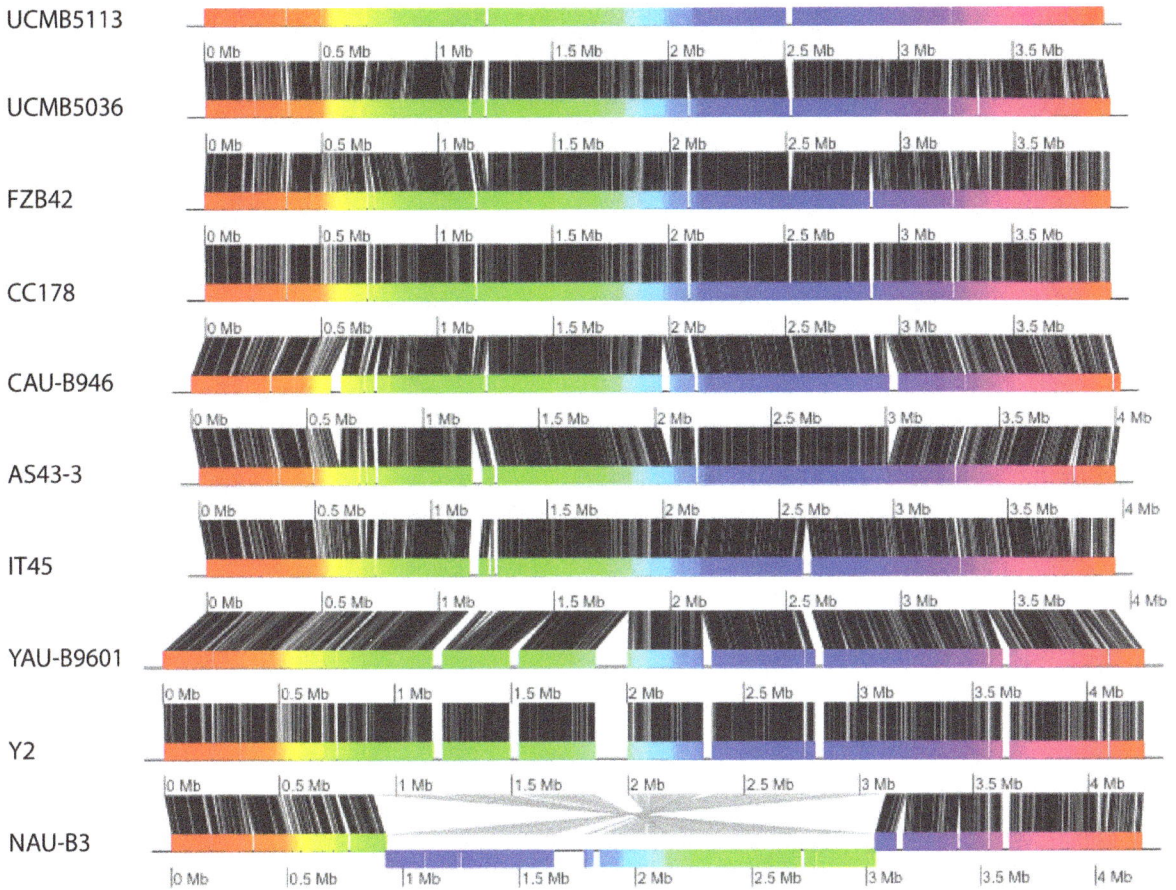

(B)

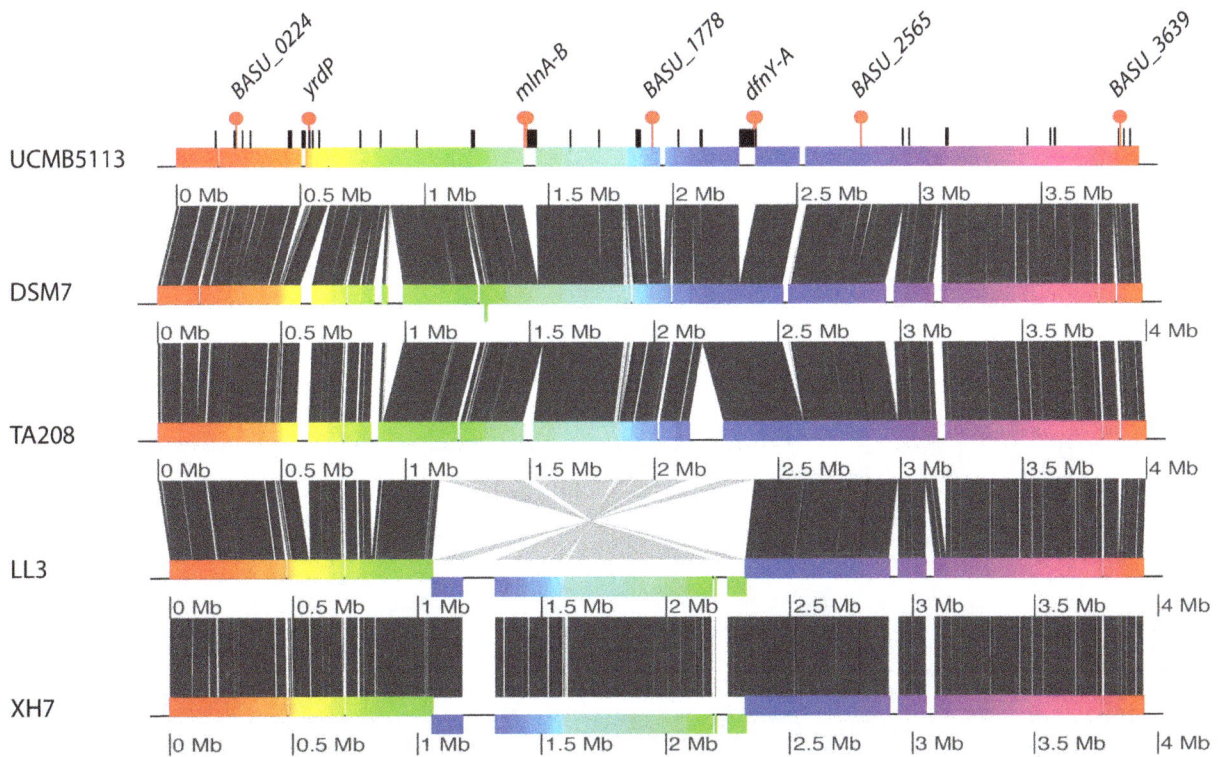

Figure 8. Genomic overview of the similarity between completely sequenced *B. amyloliquefaciens* strains. (A) Genomic comparison map of plant-associated *B. amyloliquefaciens* subsp. *plantarum* strains. The grey blocks indicate similarity and sequence conservation whereas gaps between the blocks show differences in genomic content between genomes. The rainbow color bar shows synteny between genomes. (B) Comparison of the plant growth promoting strain UCMB5113 (as representative of *plantarum* species) with non plant-associated subsp. *amyloliquefaciens* strains. Vertical bars (black and red) on the top show the location of *plantarum* group specific genes in UCMB5113, whereas red dots indicate nine of the selected *plantarum*-specific genes that were shown to be expressed (Figure S2).

engineer improved strains, and optimize conditions that favour interaction and colonization to support durable crop production.

Supporting Information

Figure S1   Global alignment of bacterial chromosomes. Shows highly conserved regions between the genomes of *B. amyloliquefaciens* subsp. *plantarum* UCMB5113, *B. amyloliquefaciens* DSM7, *b. subtilis* 168 and *B. pumilus* SARF-032

Figure S2   Expression analysis of UCMB5113 genes specific to *plantarum* species. The genes expressed during the exponential growth phase of UCMB5113. Each lane was loaded with 5ul of RT-PCR amplified product. The *tetB* gene was used as an expression control.

File S1   Contains the files: **Table S1**: Regions of genomic plasticity (RGP) in *Bacillus amyloliquefaciens* UCMB5113 genome. **Table S2.** Deletions occurring in the UCMB5113 genome in comparison to strain FZB42. In some cases deletions were partially substituted by RGPs or smaller insertions. **Table S3**: Putative CDS with secretory signal peptides in *B. amyloliquefaciens* UCMB5113 genome. **Table S4**: List of Transporter proteins encoded by *B. amyloliquefaciens* UCMB5113. **Table S5**. List of *plantarum* species-specific genes.

Acknowledgments

We thank the SNP&SEQ Technology Platform of the Genomics platform of Science for Life Laboratory in Uppsala for *Bacillus* DNA sequence analysis. The management and processing of the NGS data were supported by the SLU Global Bioinformatics Centre (http://sgbc.slu.se), supported by BILS (http://www.bils.se).

Author Contributions

Conceived and designed the experiments: JM EBR. Performed the experiments: AN SA SB. Analyzed the data: AN SM. Contributed reagents/materials/analysis tools: JM EBR. Contributed to the writing of the manuscript: AN SM SA.

References

1. Lugtenberg B, Kamilova F (2009) Plant-growth-promoting rhizobacteria. Annu Rev Microbiol 63: 541–556.
2. Compant S, Duffy B, Nowak J, Clément C, Barka EA (2005) Use of plant growth-promoting bacteria for biocontrol of plant diseases: principles, mechanisms of action, and future prospects. Appl Environ Microbiol 71: 4951–4959.
3. Yang J, Kloepper JW, Ryu CM (2009) Rhizosphere bacteria help plants tolerate abiotic stress. Trends Plant Sci 14: 1–4.
4. Whipps JM (2001) Microbial interactions and biocontrol in the rhizosphere. J Exp Bot 52: 487–511.
5. Hayat R, Ali S, Amara U, Khalid R, Ahmed I (2010) Soil beneficial bacteria and their role in plant growth promotion: a review. Ann Microbiol 60: 579–598.
6. Kloepper JW, Ryu C-M, Zhang S (2004) Induced systemic resistance and promotion of plant growth by *Bacillus* spp. Phytopathology 94: 1259–1266.
7. Choudhary DK, Johri BN (2009) Interactions of *Bacillus* spp. and plants—with special reference to induced systemic resistance (ISR). Microbiol Res 164: 493–513.
8. Koumoutsi A, Chen XH, Henne A, Liesegang H, Hitzeroth G, et al. (2004) Structural and functional charaterization of gene clusters directing nonribosomal synthesis of bioactive cyclic lipopeptides in *Bacillus amyloliquefaciens* FZB42. J Bacteriol 186: 1084–1096.
9. Chen XH, Vater J, Piel J, Franke P, Scholz R, et al. (2006) Structural and functional characterization of three polyketide synthase gene clusters in *Bacillus amyloliquefaciens* FZB42. J Bacteriol 188: 4024–4036.
10. Reva ON, Dixelius C, Meijer J, Priest FG (2004) Taxonomic characterization and plant colonizing abilities of some bacteria related to *Bacillus amyloliquefaciens* and *Bacillus subtilis*. FEMS Microbiol Ecol 48: 249–259.
11. Danielsson J, Reva O, Meijer J (2007) Protection of oilseed rape (*Brassica napus*) toward fungal pathogens by strain of plant-associated *Bacillus amyloliquefaciens*. Microb Ecol 54: 134–140.
12. Sarosh BR, Danielsson J, Meijer J (2009) Transcript profiling of oilseed rape (*Brassica napus*) primed for biocontrol differentiate genes involved in microbial interactions with beneficial *Bacillus amyloliquefaciens* from pathogenic *Botrytis cinerea*. Plant Mol Biol 70: 31–45.
13. Kasim WA, Osman ME, Omar MN, Abd El-Daim IA, Bejai S, et al. (2013) Control of drought stress in wheat using plant-growth-promoting bacteria. J Plant Growth Regul 32: 122–130.
14. Osman ME, Kasim WA, Omar MN, Abd El-Daim IA, Bejai S, et al. (2013) Impact of bacterial priming on some stress tolerance mechanisms and growth of cold stressed wheat seedlings. Internat J Plant Biol 4:e8.
15. Abd El-Daim IA, Bejai S, Meijer J (2014) Improved heat stress tolerance of wheat seedlings by bacterial seed treatment. Plant and Soil,379: 337–350.
16. Chevreux B, Wetter T, Suhai S (1999) Genomic sequence assembly using trace signals and additional sequence information. Computer Science and Biology: Proceedings of the German Conference on Bioinformatics (GCB) 99, pp. 45–56.
17. Zerbino DR, Birney E (2008) Velvet: algorithms for de novo short read assembly using de Bruijn graphs. Genome Res 18: 821–829.
18. Manzoor S, Niazi A, Bejai S, Meijer J, Bongcam-Rudloff E (2013) Genome sequence of a plant-associated bacterium, *Bacillus amyloliquefaciens* strain UCMB5036. Genome Announc 1:e0011113.
19. Chen XH, Koumoutsi A, Scholz R, Eisenreich A, Schneider K, et al. (2007) Comparative analysis of the complete genome sequence of the plant growth-promoting bacterium *Bacillus amyloliquefaciens* FZB42. Nat Biotechnol 25: 1007–1014.
20. Darling AE, Mau B, Perna NT (2010) progressiveMauve: multiple genome alignment with gene gain, loss and rearrangement. PLoS One 5:e11147.
21. Vallenet D, Labarre L, Rouy Z, Barbe V, Bocs S, et al. (2006) MaGe: a microbial genome annotation system supported by synteny results. Nucleic Acids Res 34: 53–65.
22. Blom J, Albaum S, Doppmeier D, Pühler A, Vorhölter F-J, et al. (2009) EDGAR: a software framework for the comparative analysis of prokaryotic genomes. BMC Bioinformatics 10: 154.
23. Edgar RC (2004) MUSCLE: multiple sequence alignment with high accuracy and high throughput. Nucleic Acids Res 32: 1792–1797.
24. Tamura K, Peterson D, Peterson N, Stecher G, Nei M, et al. (2011) MEGA5: Molecular Evolutionary Genetics Analysis using Maximum Likelihood, Evolutionary Distance, and Maximum Parsimony Methods Mol Biol Evol 28: 2731–2739.
25. Petersen TN, Brunak S, von Heijne G, Nielsen H (2011) SignalP 4.0: discriminating signal peptides from transmembrane regions. Nat Methods 8: 785–786.
26. Juncker AS, Willenbrock H, Von Heijne G, Brunak S, Nielsen H, et al. (2003) Prediction of lipoprotein signal peptides in Gram-negative bacteria. Protein Sci 12: 1652–1662.
27. Bendtsen JD, Nielsen H, Widdick D, Palmer T, Brunak S (2005) Prediction of twin-arginine signal peptides. BMC Bioinformatics 6: 167.
28. Beecher DJ, Wong AC (1994) Identification of hemolysin BL-producing *Bacillus cereus* isolates by a discontinuous hemolytic pattern in blood agar. Appl Environ Microbiol 60: 1646–1651.
29. Tugrul T, Cansunar E (2005) Detecting surfactant-producing microorganisms by the drop-collapse test. World J Microbiol Biotechnol 21: 851–853.
30. Schwyn B, Neilands JB (1987) Universal chemical assay for the detection and determination of siderophores. Anal Biochem 160: 47–56.
31. Walden KR, Claude PS (1988) Plant and bacterial chitinase differ in antifungal activity. J Gen Microbiol 134: 169–176.

32. Edi Premono M, Moawad AM, Vlek PLG (1996) Effect of phosphate-solubilizing *Pseudomonas putida* on the growth of maize and its survival in the rhizosphere. Indones. J Crop Sci 11: 13–23.

33. Zimbro MJ, Power DA, Millwer SM, Wilson GE, Johnson JA (2009) Difco and BBL manual - Manual of biological culture media. 10th ed. p. 879-880: Becton Dickinson and Co., Sparks, MD.

34. Christensen WB (1946) Urea decomposition as a means of differentiating *Proteus* and paracolon cultures from each other and from *Salmonella* and *Shigella* types. J Bacteriol 52: 461–466.

35. Borriss R, Chen XH, Rueckert C, Blom J, Becker A, et al. (2011) Relationship of *Bacillus amyloliquefaciens* clades associated with strains DSM 7T and FZB42T: a proposal for *Bacillus amyloliquefaciens* subsp. *amyloliquefaciens* subsp. nov. and *Bacillus amyloliquefaciens* subsp. *plantarum* subsp. nov. based on complete genome sequence comparisons.Int J Syst Evol Microbiol 61: 1786–1801.

36. Sakamoto JJ, Sasaki M, Tsuchido T (2001) Purification and characterization of a *Bacillus subtilis* 168 nuclease, YokF, involved in chromosomal DNA degradation and cell death caused by thermal shock treatments. J Biol Chem 276: 47046–47051.

37. Kenig M, Abraham EP (1976) Antimicrobial activities and antagonists of bacilysin and anticapsin. J Gen Microbiol 94: 37–45.

38. Tschen M (1990) Gegen Pilze wirksame antibiotika der *Bacillus subtilis*-gruppe. Forum Mikrobiol 3: 156–163.

39. Steinborn G, Hajirezaei M-R, Hofemeister J (2005) bac genes for recombinant bacilysin and anticapsin production in *Bacillus* host strains. Arch Microbiol 183: 71–79.

40. Yoo JS, Zheng CJ, Lee S, Kwak JH, Kim WG (2006) Macrolactin N, a new peptide deformylase inhibitor produced by *Bacillus subtilis*. Bioorg Med Chem Lett 16: 4889–4892.

41. Lee SJ, Cho JY, Cho JI, Moon JH, Park KD, et al. (2004) Isolation and characterization of antimicrobial substance macrolactin A produced from *Bacillus amyloliquefaciens* CHO104 isolated from soil. J Microbiol Biotechnol 14: 525–531.

42. Butcher RA, Schroeder FC, Fischbach MA, Straight PD, Kolter R, et al. (2007) The identification of bacillaene, the product of the PksX megacomplex in *Bacillus subtilis*. Proc Natl Acad Sci USA 104: 1506–1509.

43. Zimmerman SB, Schwartz CD, Monaghan RL, Pelak BA, Weissberger B, et al. (1987) Difficidin and oxydifficidin: novel broad spectrum antibacterial antibiotics produced by *Bacillus subtilis*. I. Production, taxonomy and antibacterial activity. J Antibiot 40: 1677–1681.

44. Miethke M, Klotz O, Linne U, May JJ, Beckering CL, et al. (2006) Ferri-bacillibactin uptake and hydrolysis in *Bacillus subtilis*. Mol Microbiol 61: 1413–1427.

45. González-Pastor JE, Hobbs EC, Losick R (2003) Cannibalism by sporulating bacteria. Science 301: 510–513.

46. Hamoen LW, Eshuis H, Jongbloed J, Venema G, Van Sinderen D (1995) A small gene, designated *comS*, located within the coding region of the fourth amino acid-activation domain of *srfA*, is required for competence development in *Bacillus subtilis*. Mol Microbiol 15: 55–63.

47. D'Souza C, Nakano MM, Zuber P (1994) Identification of *comS*, a gene of the *srfA* operon that regulates the establishment of genetic competence in *Bacillus subtilis*. Proc Natl Acad Sci USA 91: 9397–9401.

48. Tekaia F, Gordon SV, Garnier T, Brosch R, Barrell BG, et al. (1999) Analysis of the proteome of *Mycobacterium tuberculosis* in silico. Tuber Lung Dis 79: 329–342.

49. Sakaguchi R, Amano H, Shishido K (1988) Nucleotide sequence homology of the tetracycline-resistance determinant naturally maintained in *Bacillus subtilis* Marburg 168 chromosome and the tetracycline-resistance gene of *B. subtilis* plasmid pNS1981. Biochim Biophys Acta 950: 441–444.

50. Jacob J, Evers S, Bischoff K, Carlier C, Courvalin P (1994) Characterization of the *sat4* gene encoding a streptothricin acetyltransferase in *Campylobacter coli* BE/G4. FEMS Microbiol Lett 120: 13–17.

51. Silver S, Phung LT (1996) Bacterial heavy metal resistance: new surprises. Annu Rev Microbiol 50: 753–789.

52. Harvie DR, Andreini C, Cavallaro G, Meng W, Connolly BA, et al. (2006) Predicting metals sensed by ArsR-SmtB repressors: allosteric interference by a non-effector metal. Mol Microbiol 59: 1341–1356.

53. Zuber P (2009) Management of oxidative stress in *Bacillus*. Annu Rev Microbiol 63: 575–597.

54. Meilhoc E, Cam Y, Skapski A, Bruand C (2010) The response to nitric oxide of the nitrogen-fixing symbiont *Sinorhizobium meliloti*. Mol Plant-Microbe Interact 23: 748–759.

55. Ghelardi E, Salvetti S, Ceragioli M, Gueye SA, Celandroni F, et al. (2012) Contribution of surfactin and SwrA to flagellin expression, swimming, and surface motility in *Bacillus subtilis*. Appl Environ Microbiol 78: 6540–6544.

56. Kearns DB, Chu F, Rudner R, Losick R (2004) Genes governing swarming in *Bacillus subtilis* and evidence for a phase variation mechanism controlling surface motility. Mol Microbiol 52: 357–369.

57. Chu F, Kearns DB, Branda SS, Kolter R, Losick R (2006) Targets of the master regulator of biofilm formation in *Bacillus subtilis*. Mol Microbiol 59: 1216–1228.

58. Branda SS, Chu F, Kearns DB, Losick R, Kolter R (2006) A major protein component of the *Bacillus subtilis* biofilm matrix. Mol Microbiol 59: 1229–1238.

59. Romero D, Vlamakis H, Losick R, Kolter R (2011) An accessory protein required for anchoring and assembly of amyloid fibres in *B. subtilis* biofilms. Mol Microbiol 80: 1155–1168.

60. Ostrowski A, Mehert A, Prescott A, Kiley TB, Stanley-Wall NR (2011) YuaB functions synergistically with the exopolysaccharide and TasA amyloid fibers to allow biofilm formation by *Bacillus subtilis*. J Bacteriol 193: 4821–4831.

61. Kobayashi K, Iwano M (2012) BslA(YuaB) forms a hydrophobic layer on the surface of *Bacillus subtilis* biofilms. Mol Microbiol 85: 51–66.

62. Santhanagopalan V, Coker C, Radulovic S (2006) Characterization of RP 333, a gene encoding CapD of *Rickettsia prowazekii* with UDP-glucose 4-epimerase activity. Gene 369: 119–125.

63. Idris EE, Iglesias DJ, Talon M, Borriss R (2007) Tryptophan-dependent production of indole-3-acetic acid (IAA) affects level of plant growth promotion by *Bacillus amyloliquefaciens* FZB42. Mol Plant-Microbe Interact 20: 619–626.

64. Ryu CM, Farag MA, Hu CH, Reddy MS, Wei HX, et al. (2003) Bacterial volatiles promote growth in *Arabidopsis*. Proc Natl Acad Sci USA 100: 4927–4932.

65. Ryu CM, Farag MA, Hu CH, Reddy MS, Kloepper JW, et al. (2004) Bacterial volatiles induce systemic resistance in *Arabidopsis*. Plant Physiol 134: 1017–1026.

66. Collemare J, Lebrun MH (2012) Fungal secondary metabolites: ancient toxins and novel effectors in plant-microbe interactions. In: Martin F, Kamoun S, editors. Effectors in plant-microbe interactions. Chichester, UK: John Wiley and Sons, Inc. pp. 377–399.

7

Characterization of Small Interfering RNAs Derived from *Sugarcane Mosaic Virus* in Infected Maize Plants by Deep Sequencing

Zihao Xia[1], Jun Peng[1,2], Yongqiang Li[1], Ling Chen[1], Shuai Li[1], Tao Zhou[1], Zaifeng Fan[1]*

1 State Key Laboratory of Agro-biotechnology and Ministry of Agriculture Key Laboratory for Plant Pathology, China Agricultural University, Beijing, China, 2 Ministry of Agriculture Key Laboratory of Integrated Pest Management on Tropical Crops, Environmental and Plant Protection Institute, Chinese Academy of Tropical Agricultural Sciences, Haikou, Hainan, China

Abstract

RNA silencing is a conserved surveillance mechanism against viruses in plants. It is mediated by Dicer-like (DCL) proteins producing small interfering RNAs (siRNAs), which guide specific Argonaute (AGO)-containing complexes to inactivate viral genomes and may promote the silencing of host mRNAs. In this study, we obtained the profile of virus-derived siRNAs (vsiRNAs) from *Sugarcane mosaic virus* (SCMV) in infected maize (*Zea mays* L.) plants by deep sequencing. Our data showed that vsiRNAs which derived almost equally from sense and antisense SCMV RNA strands accumulated preferentially as 21- and 22-nucleotide (nt) species and had an adenosine bias at the 5′-terminus. The single-nucleotide resolution maps revealed that vsiRNAs were almost continuously but heterogeneously distributed throughout the SCMV genome and the hotspots of sense and antisense strands were mainly distributed in the HC-Pro coding region. Moreover, dozens of host transcripts targeted by vsiRNAs were predicted, several of which encode putative proteins involved in ribosome biogenesis and in biotic and abiotic stresses. We also found that *ZmDCL2* mRNAs were up-regulated in SCMV-infected maize plants, which may be the cause of abundant 22-nt vsiRNAs production. However, *ZmDCL4* mRNAs were down-regulated slightly regardless of the most abundant 21-nt vsiRNAs. Our results also showed that SCMV infection induced the accumulation of *AGO2* mRNAs, which may indicate a role for AGO2 in antiviral defense. To our knowledge, this is the first report on vsiRNAs in maize plants.

Editor: Neena Mitter, Department of Primary Industries and Fisheries, Australia

Funding: This research was supported by the National Basic Research Program of China (#2012CB114004), and grants from the Ministry of Education (IRT1042) and State Key Laboratory of Biotechnology (SKLAB). The funders had no role in study design, data collection and analysis, decision to publish, or preparation of the manuscript.

Competing Interests: The authors have declared that no competing interests exist.

* E-mail: fanzf@cau.edu.cn

Introduction

RNA silencing is a conserved antiviral defense mechanism in plants. The antiviral silencing can be triggered by viral double-stranded RNA (dsRNA) and highly structured single-stranded RNA (ssRNA), which can be recognized and cleaved by Dicer-like (DCL) proteins and processed into virus-derived small interfering RNAs (vsiRNAs) that vary in length from 21 to 24 nucleotides (nt) in virus-infected plants [1–4]. The vsiRNAs are then loaded into Argonaute (AGO)-containing complexes known as RNA-induced silencing complexes (RISCs), promoting the degradation of both genomic and subgenomic viral RNAs and the silencing of host mRNAs in a sequence-specific manner [5–9]. Two classes of vsiRNAs are generated during virus infections: primary siRNAs, which derived from DCL-mediated cleavage of an initial trigger RNA, and secondary siRNAs, whose biogenesis requires an RNA dependent RNA polymerase (RDR) [10–13].

DCL4 and DCL2 play key roles in the generation of vsiRNAs derived from positive-strand RNA viruses to produce 21- and 22-nt vsiRNAs, respectively [11,12,14]. Plants infected with positive-strand RNA viruses mainly accumulate 21-nt vsiRNAs processed by DCL4, but when the activity of DCL4 is reduced or inhibited by viruses, DCL2, as the substitute, is known to produce 22-nt vsiRNAs [12,14–17]. However, recent findings have suggested that there is a difference between 21- and 22-nt vsiRNAs in antiviral defense, and DCL2-dependent 22-nt vsiRNAs alone do not guide efficient silencing [18]. In addition, it is demonstrated that the production of viral secondary siRNAs mainly depends on host RDR1, RDR2, or RDR6 in *Arabidopsis* infected by distinct positive-strand RNA viruses [12,13,19–21]. Moreover, it was reported that RDR1 and RDR6 exhibited specificity in targeting the genome sequences of *Cucumber mosaic virus* (CMV) in amplifying viral secondary siRNAs [13]. vsiRNAs are associated with specific AGO complexes to function in RNA silencing [22–25]. In plants, the recruiting small RNAs of a particular AGO complex is preferentially, but not exclusively, dictated by their 5′-terminal nucleotides [18,25–27]. In *Arabidopsis*, there are higher levels of viral RNA accumulation in hypomorphic *ago1*, null *ago2* and *ago7* mutants, and AGO1, AGO2, and AGO5 proteins can bind vsiRNAs, suggesting an antiviral role for these AGOs [17,18,28–33]. Moreover, it was reported that vsiRNAs could be recruited into AGOs 1, 2, 3, 5, 7 and 10, which were demonstrated to exhibit *in vitro* slicer activity [25]. Recent studies also revealed that AGO2 plays an antiviral role in *Nicotiana benthamiana* [34]. Other components involved in RNA silencing also participate in antiviral defense in plants, including dsRNA-binding protein 4 (DRB4), suppressor of gene silencing 3 (SGS3) and HUA ENHANCER 1 (HEN1) [11,22,35–39].

It was predicted previously that vsiRNAs could target host transcripts at post-transcriptional level, as endogenous miRNAs or siRNAs. To date, only a few studies have provided experimental evidence to verify the targeting of host genes, although many host transcripts potentially targeted by vsiRNAs have been predicted using bioinformatics [6,7,40]. Early studies suggested that some of the vsiRNAs may target host transcripts for post-transcriptional regulation by BLAST search and 5′ RACE [3,41,42]. Recently, two research groups confirmed that vsiRNA derived from the Y-satellite of CMV could specifically and directly cleave *ChlI* mRNA in *N. benthamiana* and modulate the virus disease symptoms [6,7]. Moreover, it was demonstrated that siRNA containing the pathogenic determinant of a chloroplast-replicating viroid guided the degradation of the mRNA encoding the chloroplastic heat-shock protein 90 as predicted by RNA silencing [43]. It was also reported that vsiRNAs promoted the silencing of host mRNAs in a sequence-specific manner by degradome analysis and 5′ RACE [9].

Sugarcane mosaic virus (SCMV), a member of the genus *Potyvirus*, can infect various crops (e.g., sugarcane, sorghum, and maize) which leads to symptoms such as mosaic, chlorosis and dwarfing, and causes considerable losses in different field crops in the world [44,45]. Our previous studies showed that SCMV was the major causal agent of maize dwarf mosaic disease in China, and the Beijing isolate (SCMV-BJ) belonged to the prevalent strain [46]. It was reported that SCMV infection could elicit the accumulation of *RDR1* mRNA, and silenced *RDR1* maize plants were more susceptible to SCMV infection [47]. Co-expression assay demonstrated that the HC-Pro encoded by SCMV suppressed the RNA silencing induced by sense RNA and dsRNA, and down-regulated the accumulation of *RDR6* mRNA [48]. These results suggested that RDR1 and RDR6 may be involved in SCMV infection and plant antiviral defense. Other reports have investigated the interaction between SCMV and maize, including protein-protein interaction and the possible genes involved in the defense responses to SCMV infection [49–54]. However, the roles of the vsiRNAs played in the interaction between SCMV and maize were still unknown. In this study, the profile of vsiRNAs derived from SCMV in infected maize (*Zea mays* L.) plants was obtained by deep sequencing. We analyzed the characters of vsiRNAs and predicted the targets of some vsiRNAs. Moreover, the relative accumulation level of *ZmDCLs* and *ZmAGO2* mRNAs in SCMV-infected maize plants were detected.

Results

21- and 22-nt vsiRNAs accumulated at high levels in maize plants inoculated with SCMV

The profile of vsiRNAs can help to decipher the mechanisms and components involved in their biogenesis and function. To obtain the profile of vsiRNAs produced during SCMV infection, small RNAs obtained from maize plants inoculated with SCMV or with phosphate buffer (mock) were analyzed by deep sequencing using the Illumina Solexa platform. A total of 17,630,207 and 14,736,470 reads were obtained from small RNA library of either mock- or SCMV-inoculated maize plants, respectively (Figure 1A). Reads ranging from 18- to 28-nt were mapped to the viral genome in sense and antisense orientations. The sequences within 2 mismatches were regarded as vsiRNAs in the libraries (Figure 1A). In total, 6,220,433 vsiRNA reads were identified in SCMV-

inoculated maize plants, accounting for more than half of 18–28 nt reads. However, only 8,246 reads matched to the SCMV genome in the mock-inoculated library, which corresponded to approximately 0.08% of 18–28 nt reads (Figure 1A). In SCMV-infected maize plants, 21- and 22-nt vsiRNAs accumulated to high levels, representing 49.42% and 43.79% of total vsiRNAs, respectively (Figure 1B), which suggested that the maize homologue of DCL4 and DCL2 may be the predominant Dicer ribonucleases involved in vsiRNA biogenesis. We then compared the overall profile of small RNAs between mock- and SCMV-inoculated libraries. The results showed that 21- and 22-nt reads increased significantly in the SCMV-inoculated library, while 24-nt reads decreased (Figure 1C). Interestingly, the increase of the 21- and 22-nt small RNAs was mainly attributed to the accumulation of vsiRNAs (Figure 1D), suggesting that SCMV infection produced amounts of vsiRNAs and the high levels of vsiRNAs seemed to be a result of the antiviral RNA silencing mechanism or a specific SCMV-host interaction.

The characteristics of vsiRNAs

In *Arabidopsis*, it has been reported that the selective loading of small RNAs into specific AGOs is influenced by their 5′-terminal nucleotides [26]. To determine potential interactions between vsiRNAs with distinct AGO complexes, we analyzed the relative abundance of vsiRNAs according to their 5′-terminal nucleotides (Figure 2A). For the 21- and 22-nt vsiRNAs, A was the most abundant nucleotide at the 5′-end (32.99% and 35.50%, respectively), while U was the least abundant (19.25% and 21.07%, respectively). These results suggested that 21- and 22-nt vsiRNAs might be potentially loaded into diverse AGO-containing complexes with most of vsiRNAs preferentially loaded into AGO2 and/or AGO4, which showed a preference for A [26].

To explore the origin of the vsiRNAs, the polarity distribution of vsiRNAs was further characterized. Almost equivalent amounts of sense (51.59%) and antisense (48.41%) vsiRNAs suggested that vsiRNAs derived from both sense and antisense SCMV RNA strands to a similar extent (Figure 2B). To examine the genomic distribution of the vsiRNAs, 21- and 22-nt vsiRNA sequences were mapped along the SCMV genome (Figure 3A and 3B). The single-nucleotide resolution maps indicated that vsiRNAs from both polarities were almost continuously but heterogeneously distributed throughout the SCMV genome (Figure 3B and Table S1). To better understand the hotspots of vsiRNAs distribution, we counted and summed up the reads of single-nucleotide resolution maps of 21–24 nt vsiRNAs, and defined the region that the number of at least 21 consecutive single-nucleotide reads should be not less than 30,000 as a hotspot (Table S2). Further estimation of the vsiRNA-generating hotspots showed that the number of hotspots derived from the sense strand was more than that from antisense strand, and the region corresponding to HC-Pro contained more hotspots (Figure 3B and Table S2). Moreover, we calculated the GC content of each hotspot on both sense and antisense strand and found that the GC content of most hotspots were less than 50% (Table S2), not as high (GC content within hotspots) as reported [55,56]. The results we obtained also indicated that most prominent peaks of sequence abundance corresponding to 21-nt vsiRNAs usually localized to the same genomic regions as peaks corresponding to 22-nt vsiRNAs (Table S1). Nevertheless, the positions 4540-4561 on sense strand and positions 460-481 on antisense stand had a preference to 21-nt and 22-nt, respectively (Table S2). The results indicated that different DCLs have a similar but slightly different targeting preference toward the same regions along the viral genome.

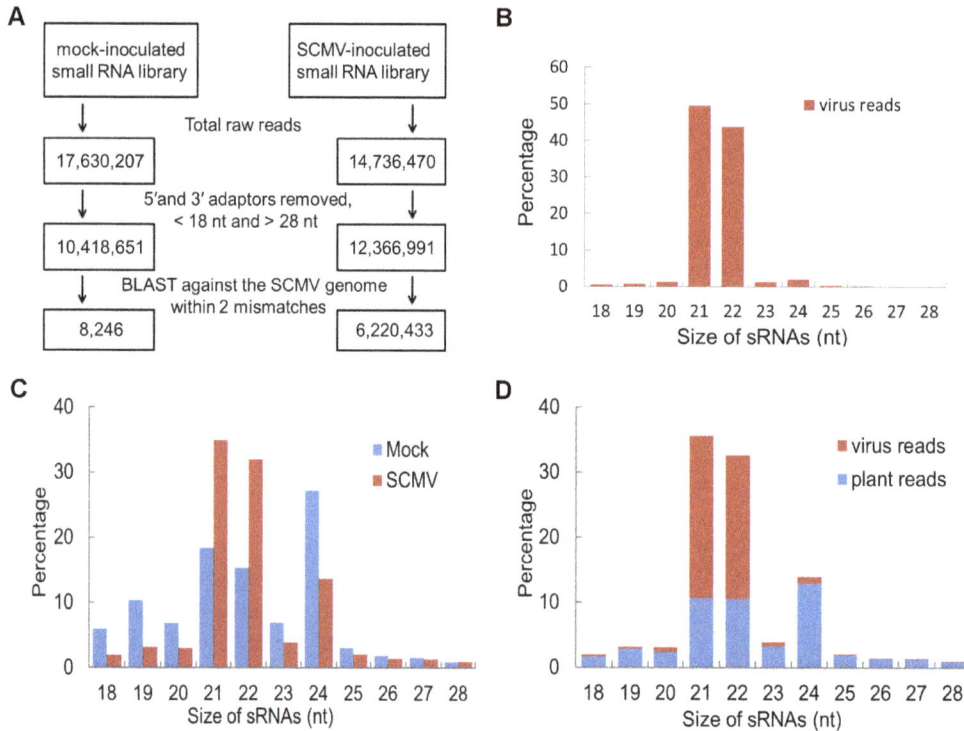

Figure 1. 21- and 22-nt vsiRNAs accumulated at high levels in SCMV-inoculated maize. A: Diagram showing the stepwise computational extraction of vsiRNA reads from small RNA libraries recovered from mock-inoculated and SCMV-inoculated systemic leaves. B: Histogram representation of total vsiRNA reads in each size class. C: Size distribution of total small RNAs in libraries prepared from either mock-inoculated or SCMV-inoculated maize plants. D: Size distribution of total small RNAs in the library from SCMV-inoculated maize plants.

There were large amounts of vsiRNAs accumulated in the host plants when virus infection triggered the RNA silencing mechanism. To confirm the existence of vsiRNAs, approximately 15 µg of total RNAs was used to analyze the accumulation of vsiRNAs derived from different SCMV genome positions by Northern blotting. The results showed that there were almost equivalent 21- and 22-nt vsiRNAs in the SCMV-infected maize plants, except that vsiR157 (+), vsiR109 (-) and vsiR460 (-) had a preference for 22-nt and vsiR4541 (+) had a preference for 21-nt (Figure 3C), which was consistent with the results of deep sequencing (Table S1). These results indicated that there were indeed large amounts of vsiRNAs accumulation in SCMV-infected maize plants and DCLs played different roles in processing different positions of viral RNAs. However, vsiR1065 (-) hardly had any detectable signal, which implied that little such vsiRNAs accumulated.

Plant transcripts targeted by vsiRNAs

MiRnada is an algorithm for finding genomic targets for miRNAs [57]. In this study, we used this method to identify maize mRNAs targeted by vsiRNAs derived from SCMV. Due to the vast variety of vsiRNAs, only some vsiRNAs with high abundant reads were selected (Table S3) and only the targets whose scores were not less than 180 were presented in Table S4. The results showed that most vsiRNAs derived from the sense strand had only one target in the given condition, while most vsiRNAs derived from antisense strand had more than one targets (Table S4), indicating that the vsiRNAs from different strands might play distinct roles in regulating the expression of host transcripts. Moreover, some vsiRNAs had multiple targets, for example, vsiR2304 (+), vsiR4318 (+), vsiR8469 (+), vsiR699 (-) and vsiR7454 (-), and in most cases, they could target different transcripts from

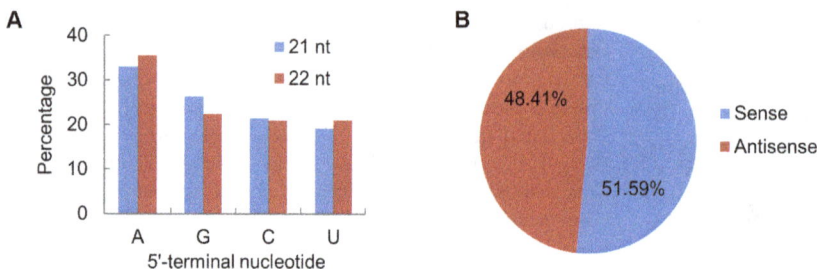

Figure 2. Relative frequency of 5'-terminal nucleotide of vsiRNAs and accumulation of sense and antisense vsiRNAs. A: Relative frequency of distinct 5'-terminal nucleotides in 21- and 22-nt vsiRNAs of SCMV-inoculated library. B: Accumulation of sense and antisense vsiRNAs. Percentage for each class of vsiRNAs from the SCMV-inoculated library is shown within the pie graph.

A

B

C

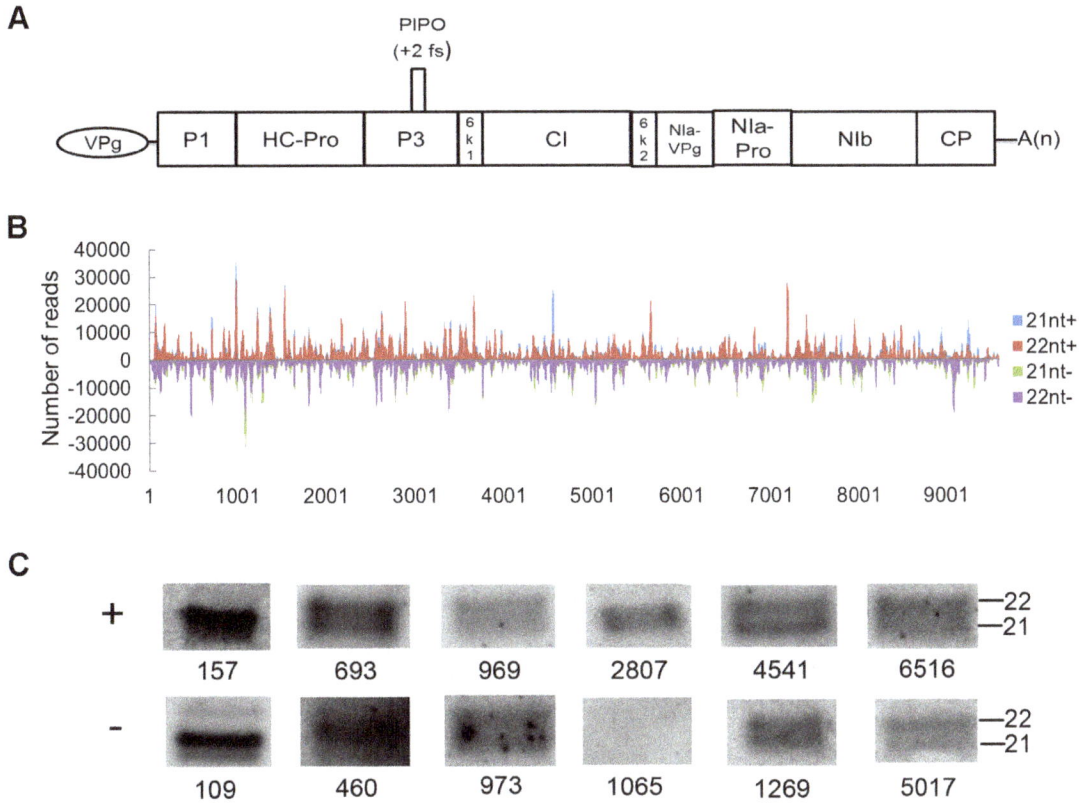

Figure 3. Profile of vsiRNAs derived from SCMV-inoculated library. A: Schematic diagram of SCMV genome. B: Maps of 21- and 22-nt vsiRNAs from SCMV-inoculated maize plants at single-nucleotide resolution. The graphs plot the number of 21- and 22-nt vsiRNA reads at each nucleotide position of the 9595-nt SCMV genome; Bars above the axis represent sense reads starting at each respective position; those below represent antisense reads ending at the respective position. C: Northern blotting of vsiRNAs from distinct regions. "+" indicates vsiRNAs derived from sense strand of SCMV genome; "-" indicates vsiRNAs derived from antisense strand of SCMV genome.

one gene (Table S4), which suggested that they might have versatile functions in different physiological pathways. The results also showed that predicted targets were involved in many different physiological pathways, including gene expression, energy metabolism, signal transduction, transcriptional regulation and cell defense (Table S4). The broad range of targets suggested that the identified vsiRNAs possibly played significant roles in SCMV-inoculated maize plants.

SiRNAs are known to down-regulate targets at the post-transcriptional level. To determine whether vsiRNAs from SCMV promoted the degradation of target transcripts, quantitative real-time reverse transcription-polymerase chain reaction (qRT-PCR) was carried out to examine the accumulation of target transcripts. Some predicted targets (whose corresponding vsiRNAs had higher number reads) that had high scores, were selected to perform qRT-PCR, except for T2807 (+) and T973 (-), whose score were 179 (Table S5). The accumulation of T973 (-) was significantly down-regulated in SCMV-infected maize plants, while T6516 (+) was up-regulated and there was no significant change in other predicted targets (Figure 4). The results indicated that these targets might be involved in several pathways rather than only be regulated by vsiRNAs at the post-transcriptional level.

To understand the roles of the predicted vsiRNA target genes in maize, the target gene sequences were used to query the Gene Ontology (GO) database [58]. Since the scores of the majority of the predicted targets were low, targets with scores not less than 180 were analyzed with GO annotations (Table S4). The vsiRNA

target genes were grouped into three root GO categories: molecular function (MF), biological process (BP) and cellular components (CC) (Figure 5). In addition to unknown genes (accounted for 49.13% vsiRNA target genes which showed no matches in the GO database), the most abundant target genes were classified as BP GO term (32/173), including reproduction,

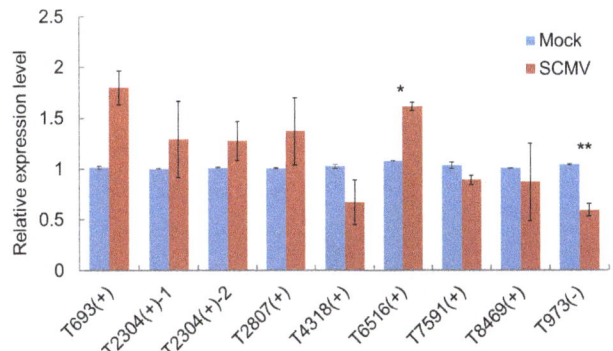

Figure 4. The expression level of the predicted target mRNAs of vsiRNAs in mock- (blue) and SCMV-inoculated (red) maize plants. For each target, the asterisk(s) indicates significant differences (*P<0.05; **P<0.01) of SCMV-inoculated versus mock-inoculated maize plants. The information and primer sequences of the predicted targets were listed in Table S5.

cellular process and metabolic process functions, followed by MF GO term (31/173), which consisted of binding and catalytic activity function (Figure 5 and Table S4). Other targets were classified as CC GO term (25/173) (Figure 5), and the secondary classification of these targets were overlapped (Table S4), suggesting that they may play different roles as cellular components.

Differential expression of *ZmDCLs* and *ZmAGO2* mRNAs after SCMV infection

Our results demonstrated that there was abundant 21- and 22-nt vsiRNAs accumulation after SCMV infection. To gain insights into the effects of SCMV infection on the RNA silencing pathways, we characterized the accumulation of *ZmDCLs* mRNAs involved in the biogenesis of vsiRNAs using qRT-PCR. The results indicated that the accumulation of *ZmDCL2* mRNA was significantly up-regulated, while *ZmDCL4* was down-regulated and there were no significant differences in the levels of *ZmDCL1*, *ZmDCL3a* and *ZmDCL3b* mRNAs between mock- and SCMV-inoculated maize plants (Figure 4A). Considering the 5′-terminal nucleotides of most of vsiRNAs were A, we explored the expression of *ZmAGO2* mRNA. The results showed that the accumulation of *ZmAGO2* mRNA was significantly induced after SCMV infection (Figure 4B), indicating a role for ZmAGO2 in antiviral defense. Taken together, these results might represent a distinct mechanism involved in the interaction between SCMV and maize plants.

Discussion

RNA silencing is a small RNA-mediated repression mechanism of gene regulation in eukaryotes and plays a critical role in the defense against viruses in plants. Virus infection triggers the production of vsiRNAs in infected plant cells. In this study, a Solexa-based deep-sequencing approach was used to profile vsiRNAs populations from SCMV-inoculated maize plants.

Sequence analysis of the deep-sequencing data revealed that SCMV infection triggered the production of large amounts of vsiRNAs, which accounted for 50.30% of the 18–28 nt reads. Our results also showed that there were more abundant small RNAs accumulation in SCMV-inoculated maize plants than that in mock-inoculated plants by ethidium bromide (EtBr) staining of size-separated RNAs (data not shown). In positive-strand RNA virus-infected plants, DCL4-dependent 21-nt vsiRNAs are the most abundant species [12,14,16,], whereas DCL2-dependent 22-nt vsiRNAs accumulated to higher levels in the absence of DCL4

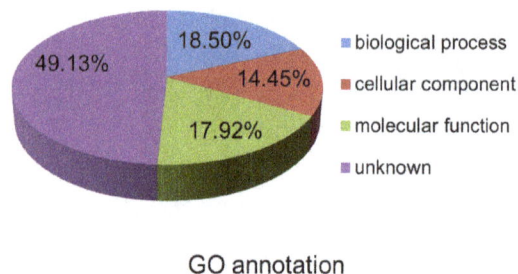

Figure 5. Functional classification of some predicted vsiRNAs target transcripts according to BLAST2GO. The GO classification includes biological process, molecular function and cellular component. The detailed GO annotation information of each target transcript was listed in Table S4.

[15]. However, 22-nt vsiRNAs accumulated predominately in *Tobacco rattle virus* (TRV)-infected *N. benthamiana* plants and *Cotton leafroll dwarf virus* (CLRDV)-infected cotton plants [10,59]. In TRV-infected *N. benthamiana* leaves, TRV-derived siRNAs of 22-nt (44.7%) were cloned to the same extent as 21-nt (42.5%), whereas 21-nt siRNA species were overrepresented (65.2%) in TRV-infected *Arabidopsis* [10]. Different size class distribution of vsiRNAs suggested the difference of the biosynthetic pathways of siRNAs in *N. benthamiana* and those in *Arabidopsis* [10]. In our study, 21- and 22-nt vsiRNAs accumulated at high levels (49.42% and 43.79%, respectively) in SCMV-infected maize plants, suggesting that DCL4 and DCL2 worked redundantly and, perhaps, synergistically in the production of vsiRNAs, which is consistent with the model that cooperative interaction between DCL4 and DCL2 was necessary during systemic antiviral silencing in TuMV-infected *Arabidopsis* [12], as all the experimental samples we used were maize systemic leaves. In SCMV-infected maize plants, *ZmDCL2* mRNA was up-regulated (Figure 6A), confirming the role ZmDCL2 played in the production of vsiRNAs. Nevertheless, the *GhDCL2* mRNA was down-regulated in CLRDV-infected cotton plants and the predominance of 22-nt vsiRNAs associated with CLRDV infection would be hypothesized to be the result of GhDCL2 activity [59]. Surprisingly, though *ZmDCL4* mRNA was down-regulated (Figure 6A), yet there existed the most abundant 21-nt vsiRNAs accumulation in SCMV-infected maize plants (Figure 1B), indicating that ZmDCL4 still played the major role in biosynthesis of vsiRNAs. In previous reports, TCV infection was associated with an abundance of 22-nt vsiRNAs, which seemed to be related to the activity of the suppressor protein P38 that could indirectly block AtDCL4 activity by suppressing AGO1 function [15,28]. Although HC-Pro had been proved to function as a viral suppressor of RNA silencing (VSR) and down-regulate the accumulation of 3′ secondary siRNA and *RDR6* mRNA [48], the possible correlation between HC-Pro and DCLs is still unknown.

Interestingly, as demonstrated by deep sequencing (Table S1) and Northern blotting results (Figure 3C), hotspots of each vsiRNA size class typically co-localized within the same regions of SCMV genome in SCMV-inoculated maize plants, especially 21- and 22-nt vsiRNAs, indicating similar, although hierarchical, targeting affinities among the DCL enzymes [37]. It has been reported that DCL activities could be favored by a higher GC content within hotspots rendering dsRNA structures more stable [20,55,56]. In this study, the GC content of each hotspot of sense and antisense strand had been obtained (Table S2), while there were no obvious correlations between higher GC content and hotspots of vsiRNAs. To date, it is not yet clear what structural features ultimately influence the accessibility, affinity or processing of DCLs [32,37].

In previous reports, AGO1 played a dominant role in defending against RNA viruses [31–33]. However, in SCMV-infected maize plants, the vsiRNAs with a 5′-terminal U, which would be loaded into AGO1, accounted for the smallest proportion (Figure 2A), suggesting that this may be a new mechanism of weakening RNA silencing against SCMV. Recently, more and more reports showed that AGO2 plays an antiviral role in different plant spices [29,30,34]. Moreover, the vsiRNAs loaded into different AGOs to form RISC had been demonstrated by AGO immunoprecipitates test [18,25]. In this study, the majority of the 21- and 22-nt vsiRNAs in SCMV-inoculated library showed a bias for sequences beginning with a 5′-terminal A, indicative of their association with AGO2 and/or AGO4. Interestingly, our data showed that SCMV infection induced the accumulation of *ZmAGO2* mRNA (Figure 6B), which further increases the possibility that ZmAGO2

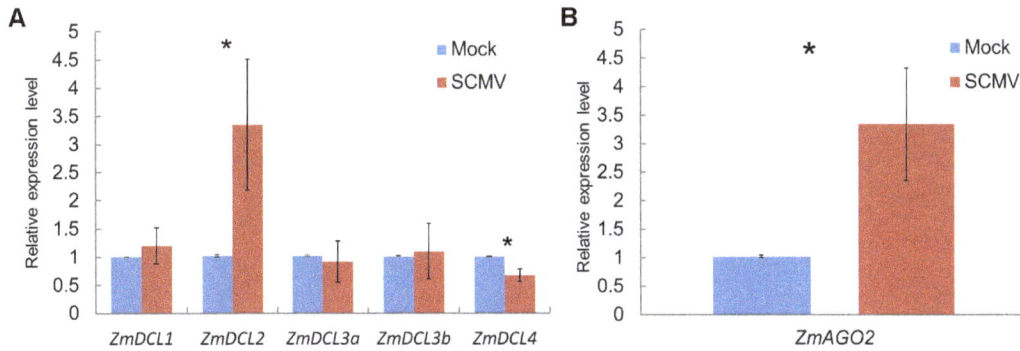

Figure 6. qRT-PCR analysis of the expression of maize *DCLs* and *AGO2* mRNAs in mock- (blue) and SCMV-inoculated (red) maize plants. For each gene, asterisk indicates significant differences (*$P<0.05$) of SCMV-inoculated versus mock-inoculated maize plants. The information and primer sequences used for amplification of *ZmDCLs* and *ZmAGO2* were listed in Table S6.

participated in antiviral defense. In addition, the presence of large amounts of vsiRNAs whose 5′-terminal nucleotides were G or C accumulated, revealing that other AGOs may also be recruited to form different RISCs and involved in antiviral defense.

Polarity distribution analysis of the sequenced vsiRNAs demonstrated the presence of approximately equal ratios of sense and antisense vsiRNAs (Figure 2B), indicating that most of vsiRNAs would be produced from dsRNA precursors comprised of sense and antisense strands. However, this could not explain the existence of hotspots and non-hotspots, because each position on the viral genome was a potential cleavage site in producing vsiRNA [40]. Moreover, the hotspots of sense and antisense strand were clustered in different regions of SCMV genome (Figure 3B). Although it had been suggested that dsRNA-like secondary structures within the single-stranded viral RNA were more likely to be the main source of vsiRNAs than dsRNA replication intermediates [4,12,13,41,60], it was not successful to find significant correlations between hotspots and regions predicted to adopt a potential hairpin structure in this study (data not shown). Recently, it was reported that (-) RNA was not accessible to antiviral RNA silencing, which could be another explanation for plants infected with different RNA viruses, e.g. the TBSV-related CymRSV, revealed a strong bias for the generation of vsiRNAs from the (+) RNA [25]. However, it is not clear whether this mechanism also functions in SCMV-infected maize plants. Our results suggested that most vsiRNAs of non-hotspots might be produced and subsequently degraded by unknown mechanisms, which need to be further investigated.

It is unclear if all the vsiRNAs produced in the host cell can be incorporated into AGO-containing RISCs, and it remains to be established whether vsiRNAs can be recruited into all the AGO family members [40]. Recently, the findings from a research seemed to give us an answer that the majority of the vsiRNAs derived from TBSV were inefficient in guiding the formed RISC and specific vsiRNAs could be recruited into AGOs 1, 2, 3, 5, 7 and 10 of *Arabidopsis*, which were demonstrated to exhibit *in vitro* slicer activity [25]. In the presence of vsiRNAs, only a distinct number rather than a broad variety of cleavage products were obtained, revealing that only some distinct vsiRNAs may be highly effective [25]. In another report, only a few cleavage sites were found in the viral genomes by degradome analysis, and vsiRNA hotspots were not directly associated with cleavage sites [9]. It has been reported that vsiRNAs generated from hotspots, in spite of their much greater abundance, do not exhibit a greater efficiency than those from non-hotspots regions [40,60]. Thus, we speculated that only some distinct vsiRNAs would be incorporated into

specific AGO-containing RISC and involved in the antiviral silencing. Some cleavage sites on the SCMV genome directed by vsiRNAs have been found (data not shown), but the functional vsiRNAs of antiviral response remains a subject of further investigation.

Previous studies suggested that vsiRNAs can target host mRNAs at post-transcriptional level [6,7,9,43]. In this study, most of the predicted targets were not down-regulated (Figure 4), inferring that many factors, for example, virus-encoded silencing suppressors and abundance of vsiRNAs, might affect the functionality of vsiRNAs and hence restrict their regulatory potential on host targets *in vivo* [40]. In addition, vsiRNAs might regulate host targets by translation inhibition, not only cleavage of mRNAs, similar to the characteristics of miRNAs [61-63]. Moreover, the possibility cannot be excluded that SCMV infection could induce over-expression of some transcripts in a non-RNA silencing-related pathway [9].

Materials and Methods

Ethics statement

No specific permission is required for these sampling locations in this study, and do not need to provide details on why this is the case. Also, we did not require ethical approval to conduct this study as we did not handle or collect any animal species considered in any animal welfare regulations and no endangered or protected species were involved in the samplings or the experiments.

Plant growth, virus source and small RNA sequencing

Maize (*Zea mays* L.) inbred line Zong 31 plants were grown in growth chambers (28 °C day and 22 °C night, 16 h light and 8 h dark cycles) for plant growth and virus inoculation. SCMV-BJ (accession number AY042184) were isolated from diseased maize in the northern suburbs of Beijing [46] and maintained at -80 °C. At 8 days post-inoculation (dpi), when the newly developed leaves started to show viral symptoms, the systemically infected leaves were harvested (16 days after maize germination). With each treatment, the systemic leaves of at least 15 maize seedlings were pooled for RNA extraction. Total RNA was extracted using Trizol reagent (Invitrogen, Carlsbad, CA, USA) for qRT-PCR, small RNA sequencing and Northern blotting. For deep sequencing, total RNA concentration was examined with a spectrophotometer (Nanodrop ND-2000, ThermoFisher Scientific, Wilmington, DE, USA), and RNA sample integrity was verified by a Bio-Analyzer 2100 (Agilent Technologies, Waldbronn, Germany). Then, in brief, total RNA was separated through 17% denaturing polyacrylamide gels and small RNAs of 15–36 nt were recovered. After that, RNA adaptors were ligated to these small RNAs followed by reverse transcription into cDNAs. These cDNAs were finally amplified by PCR and subjected to Solexa/Illumina sequencing by SBC (Shanghai Biotechnology Corporation, Shanghai, China).

Bioinformatic analyses of small RNA sequences

Small RNA sequences were computationally analysed by a set of Perl scripts from datasets generated from Illumina sequencing data. The adapter sequences were trimmed from raw reads and small RNAs between 18–28 nt in length were extracted. Only small RNA reads of sequences identical or complementary to SCMV genomic sequences within 2 mismatches were recognized as vsiRNAs (Figure 1A).

Target Gene Prediction and Analysis

In this study, we adopted MiRnada to predict maize mRNAs targeted by vsiRNAs derived from SCMV [57]. Briefly, the criteria used were as follows: 1) No more than four mismatches between vsiRNA and target (G-U bases count as 0.5 mismatches), 2) No more than two adjacent mismatches in the vsiRNA/target duplex, 3) No adjacent mismatches in positions 2–12 of the vsiRNA/target duplex (5′-terminus of vsiRNA), 4) No mismatches in positions 10-11 of vsiRNA/target duplex, 5) No more than 2.5 mismatches in positions 1–12 of the vsiRNA/target duplex (5′-terminus of vsiRNA), 5) The predicted complementary structure between vsiRNA and target has a high minimal folding free energy (MFE) that must be no fewer than 75% of the best complementary structure.

The predicted target genes were aligned using BLAST (http://blast.ncbi.nlm.nih.gov/) and were mapped and annotated by BLAST2GO (version 2.5.0) [58]. The genes were characterized using GO terms, i.e., molecular function, biological process and cellular component.

Northern blot analysis of vsiRNAs

Approximately 15 μg of total RNA (prepared as described above) was individually separated in a 15% urea polyacrylamide gel, electrophoretically transferred to Hybond-NX membrane (GE Healthcare, Buckinghamshire, UK) using a semi-dry transfer apparatus (Amersham Biosciences, Piscataway, NJ), and was chemically cross-linked via 1-ethyl-3-(3-dimethylaminopropyl) carbodiimide (EDC) [64]. For labeling reaction of probes, 1 μl of 10 μM probes, 2.5 μl of 10 x T4 PNK buffer (New England Biolabs), 3 μl of [γ-^{32}P] ATP (~10 μCi/μl), 17.5 μl of ddH$_2$O and 1 μl of T4 Poly Nucleotide Kinase (New England Biolabs) were added (a total volume of 25 μl reaction) and kept in a water bath for 1 hour at 37°C. Probe sequences used for Northern blot analysis were shown in Table S7. Blots were pre-hybridized and hybridized at 42°C overnight using hybridization buffer (Sigma, USA). Post-hybridization washes were performed using 2 x SSC and 0.2% sodium dodecyl sulfate (SDS) at 42°C for 20 min for twice. Hybridization signals were detected by exposing blots to autoradiographic film.

Quantitative Real-time RT-PCR

Total RNA was extracted from mock- and SCMV-inoculated maize leaves using TRIzol reagent (Invitrogen) and treated with 5 U of RNase-free DNAase I (TaKaRa Bio Inc., Dalian, China) at 37 °C for 30 min. The DNase I-treated total RNAs were recovered by ethanol precipitation. About 2 μg of total RNA was reverse-transcribed into cDNA and the qRT-PCR was performed as previously reported [49]. The sequence information of ZmDCLs and ZmAGO2 refers to the report by Qian et al. [65]. The sequences of the primers used in the qRT-PCR experiments were listed in Table S5 and S6. The qRT-PCR experiments were performed to explore the expression of predicted targets; qRT-PCR amplification was also performed to determine the expression levels of ZmDCL1, 2, 3a, 3b, 4 and ZmAGO2. The mean and standard errors were calculated over three biological and three technical replicates and the experimental data were subjected to t-test statistical analysis for these qRT-PCR experiments.

Supporting Information

Table S1 Single-base resolution maps of 21–24 nt vsiRNAs originated from sense (+) and antisense (-) stand of SCMV genome in SCMV-inoculated maize plants.

Table S2 The characteristics of vsiRNA hotspots.

Table S3 The information of vsiRNAs selected for target prediction.

Table S4 Predicted maize mRNA targets of the selected vsiRNAs.

Table S5 The primers used for qRT-PCR amplification for predicted target transcripts.

Table S6 The primers used for qRT-PCR of ZmDCLs and ZmAGO2 mRNAs.

Table S7 The probes used for Northern blotting of vsiRNAs.

Author Contributions

Conceived and designed the experiments: ZX ZF. Performed the experiments: ZX LC SL. Analyzed the data: ZX JP YL TZ ZF. Contributed reagents/materials/analysis tools: ZF. Wrote the paper: ZX ZF.

References

1. Hamilton AJ, Baulcombe DC (1999) A species of small antisense RNA in posttranscriptional gene silencing in plants. Science 286: 950–952.
2. Zamore PD, Tuschl T, Sharp PA, Bartel DP (2000) RNAi: double-stranded RNA directs the ATP-dependent cleavage of mRNA at 21 to 23 nucleotide intervals. Cell 101: 25–33.
3. Moissiard G, Voinnet O (2006) RNA silencing of host transcripts by cauliflower mosaic virus requires coordinated action of the four *Arabidopsis* Dicer-like proteins. Proc Natl Acad Sci USA 103: 19593–19598.
4. Molnar A, Csorba T, Lakatos L, Varallyay E, Lacomme C, et al. (2005) Plant virus-derived small interfering RNAs originate predominantly from highly structured single-stranded viral RNAs. J Virol 79: 7812–7818.
5. Baumberger N, Baulcombe DC (2005) *Arabidopsis* ARGONAUTE1 is an RNA slicer that selectively recruits microRNAs and short interfering RNAs. Proc Natl Acad Sci USA 102: 11928–11933.
6. Shimura H, Pantaleo V, Ishihara T, Myojo N, Inaba J, et al. (2011) A viral satellite RNA induces yellow symptoms on tobacco by targeting a gene involved in chlorophyll biosynthesis using the RNA silencing machinery. PLoS Pathog 7: e1002021.
7. Smith NA, Eamens AL, Wang MB (2011) Viral small interfering RNAs target host genes to mediate disease symptoms in plants. PLoS Pathog 7: e1002022.
8. Zhu H, Duan C, Hou W, Du Q, Lv D, et al. (2011) Satellite RNA-derived satsiR-12 targeting the 3′UTR of *Cucumber mosaic virus* triggers viral RNAs for degradation. J Virol 85: 13384–13397.

9. Miozzi L, Gambino G, Burgyan J, Pantaleo V (2013) Genome-wide identification of viral and host transcripts targeted by viral siRNAs in *Vitis vinifera*. Mol Plant Pathol 14: 30–43.

10. Donaire L, Barajas D, Martínez-García B, Martínez-Priego L, Pagán I, et al. (2008) Structural and genetic requirements for the biogenesis of *Tobacco rattle virus*-derived small interfering RNAs. J Virol 82: 5167–5177.

11. Ding SW (2010) RNA-based antiviral immunity. Nat Rev Immunol 10: 632–644.

12. Garcia-Ruiz H, Takeda A, Chapman EJ, Sullivan CM, Fahlgren N, et al. (2010) *Arabidopsis* RNA-dependent RNA polymerases and Dicer-like proteins in antiviral defense and small interfering RNA biogenesis during *Turnip mosaic virus* infection. Plant Cell 22: 481–496.

13. Wang XB, Wu Q, Ito T, Cillo F, Li WX, et al. (2010) RNAi-mediated viral immunity requires amplification of virus-derived siRNAs in *Arabidopsis thaliana*. Proc Natl Acad Sci USA 107: 484–489.

14. Bouché N, Lauressergues D, Gasciolli V, Vaucheret H (2006) An antagonistic function for *Arabidopsis* DCL2 in development and a new function for DCL4 in generating viral siRNAs. EMBO J 25: 3347–3356.

15. Deleris A, Gallego-Bartolome J, Bao J, Kasschau KD, Carrington JC, et al. (2006) Hierarchical action and inhibition of plant Dicer-like proteins in antiviral defense. Science 313: 68–71.

16. Fusaro AF, Matthew L, Smith NA, Curtin SJ, Dedic-Hagan J, et al. (2006) RNA interference-inducing hairpin RNAs in plants act through the viral defence pathway. EMBO Rep 7: 1168–1175.

17. Zhang X, Zhang X, Singh J, Li D, Qu F (2012). Temperature-dependent survival of *Turnip crinkle virus*-infected *Arabidopsis* plants relies on an RNA silencing-based defense that requires DCL2, AGO2, and HEN1. J Virol 86: 6847–6854.

18. Wang XB, Jovel J, Udomporn P, Wang Y, Wu Q, et al. (2011) The 21-nucleotide, but not 22-nucleotide, viral secondary small interfering RNAs direct potent antiviral defense by two cooperative Argonautes in *Arabidopsis thaliana*. Plant Cell 23: 1625–1638.

19. Diaz-Pendon JA, Li F, Li WX, Ding SW (2007) Suppression of antiviral silencing by cucumber mosaic virus 2b protein in *Arabidopsis* is associated with drastically reduced accumulation of three classes of viral small interfering RNAs. Plant Cell 19: 2053–2063.

20. Donaire L, Wang Y, Gonzalez-Ibeas D, Mayer KF, Aranda MA, et al. (2009) Deep-sequencing of plant viral small RNAs reveals effective and widespread targeting of viral genomes. Virology 392: 203–214.

21. Qu F (2010) Antiviral role of plant-encoded RNA-dependent RNA polymerases revisited with deep sequencing of small interfering RNAs of virus origin. Mol Plant Microbe Interact 23: 1248–1252.

22. Ding SW, Voinnet O (2007) Antiviral immunity directed by small RNAs. Cell 130: 413–426.

23. Hutvagner G, Simard MJ (2008) Argonaute proteins: key players in RNA silencing. Nat Rev Mol Cell Biol 9: 22–32.

24. Vaucheret H (2008) Plant ARGONAUTES. Trends Plant Sci 13: 350–358.

25. Schuck J, Gursinsky T, Pantaleo V, Burgyán J, Behrens SE (2013) AGO/RISC-mediated antiviral RNA silencing in a plant *in vitro* system. Nucleic Acids Res 41: 5090–5103.

26. Mi S, Cai T, Hu Y, Chen Y, Hodges E, et al. (2008) Sorting of small RNAs into *Arabidopsis* Argonaute complexes is directed by the 5′ terminal nucleotide. Cell 133: 116–127.

27. Takeda A, Iwasaki S, Watanabe T, Utsumi M, Watanabe Y (2008) The mechanism selecting the guide strand from small RNA duplexes is different among Argonaute proteins. Plant Cell Physiol 49: 493–500.

28. Azevedo J, Garcia D, Pontier D, Ohnesorge S, Yu A, et al. (2010) Argonaute quenching and global changes in Dicer homeostasis caused by a pathogen-encoded GW repeat protein. Genes Dev 24: 904–915.

29. Harvey JJ, Lewsey MG, Patel K, Westwood J, Heimstadt S, et al. (2011) An antiviral defense role of AGO2 in plants. PLoS ONE 6: e14639.

30. Jaubert MJ, Bhattacharjee S, Mello AF, Perry KL, Moffett P (2011) ARGONAUTE2 mediates RNA-silencing antiviral defenses against *Potato virus X* in *Arabidopsis*. Plant Physiol 156: 1556–1564.

31. Morel JB, Godon C, Mourrain P, Béclin C, Boutet S, et al. (2002) Fertile hypomorphic ARGONAUTE (*ago1*) mutants impaired in post-transcriptional gene silencing and virus resistance. Plant Cell 14: 629–639.

32. Qu F, Ye X, Morris TJ (2008) *Arabidopsis* DRB4, AGO1, AGO7, and RDR6 participate in a DCL4-initiated antiviral RNA silencing pathway negatively regulated by DCL1. Proc Natl Acad Sci USA 105: 14732–14737.

33. Zhang X, Yuan YR, Pei Y, Lin SS, Tuschl T, et al. (2006) *Cucumber mosaic virus*-encoded 2b suppressor inhibits *Arabidopsis* Argonaute1 cleavage activity to counter plant defense. Genes Dev 20: 3255–3268.

34. Scholthof HB, Alvarado VY, Vega-Arreguin JC, Ciomperlik J, Odokonyero D, et al. (2011) Identification of an ARGONAUTE for antiviral RNA silencing in *Nicotiana benthamiana*. Plant Physiol 156: 1548–1555.

35. Mourrain P, Béclin C, Elmayan T, Feuerbach F, Godon C, et al. (2000) *Arabidopsis* SGS2 and SGS3 genes are required for posttranscriptional gene silencing and natural virus resistance. Cell 101: 533–542.

36. Boutet S, Vazquez F, Liu J, Béclin C, Fagard M, et al. (2003) *Arabidopsis* HEN1: a genetic link between endogenous miRNA controlling development and siRNA controlling transgene silencing and virus resistance. Curr Biol 13: 843–848.

37. Llave C (2010) Virus-derived small interfering RNAs at the core of plant-virus interactions. Trends Plant Sci 15: 701–707.

38. Jakubiec A, Yang SW, Chua N-H (2012) *Arabidopsis* DRB4 protein in antiviral defense against *Turnip yellow mosaic virus* infection. Plant J 69: 14–25.

39. Zhu S, Jeong R, Lim G, Yu K, Wang C, et al. (2013) Double-stranded RNA-binding protein 4 is required for resistance signaling against viral and bacterial pathogens. Cell Rep 4: 1168–1184.

40. Zhu H, Guo H (2012) The role of virus-derived small interfering RNAs in RNA silencing in plants. Sci China Life Sci 55: 119–125.

41. Qi X, Bao FS, Xie Z (2009) Small RNA deep sequencing reveals role for *Arabidopsis thaliana* RNA-dependent RNA polymerases in viral siRNA biogenesis. PLoS ONE 4: e4971.

42. Wang MB, Bian XY, Wu LM, Liu LW, Smith NA, et al. (2004) On the role of RNA silencing in the pathogenicity and evolution of viroids and viral satellites. Proc Natl Acad Sci USA 101: 3275–3280.

43. Navarro B, Gisel A, Rodio ME, Delgado S, Flores R, et al. (2012) Small RNAs containing the pathogenic determinant of a chloroplast-replicating viroid guide the degradation of a host mRNA as predicted by RNA silencing. Plant J 70: 991–1003.

44. Shukla DD, Tosic M, Jilka JM, Ford R, Toler W, et al. (1989) Taxonomy of potyvirus infecting maize, sorghum, and sugarcane in Australia and the United States as determined by reactivities of polyclonal antibodies directed towards virus-specific N-termini of coat proteins. Phytopathology 79: 223–229.

45. Shi C, Thümmer F, Melchinger AE, Wenzel G, Lübberstedt T (2006) Comparison of transcript profiles between near-isogenic maize lines in association with SCMV resistance based on unigene-microarrays. Plant Sci 170: 159–169.

46. Fan Z, Chen H, Liang X, Li H (2003) Complete sequence of the genomic RNA of the prevalent strain of a potyvirus infecting maize in China. Arch Virol 148: 773–782.

47. He J, Dong Z, Jia Z, Wang J, Wang G (2010) Isolation, expression and functional analysis of a putative RNA-dependent RNA polymerase gene from maize (*Zea mays* L.). Mol Biol Rep 37: 865–874.

48. Zhang X, Du P, Lu L, Xiao Q, Wang W, et al. (2008) Contrasting effects of HC-Pro and 2b viral suppressors from *Sugarcane mosaic virus* and *Tomato aspermy cucumovirus* on the accumulation of siRNAs. Virology 374: 351–360.

49. Cao Y, Shi Y, Li Y, Cheng Y, Zhou T, et al. (2012) Possible involvement of maize Rop1 in the defence responses of plants to viral infection. Mol Plant Pathol 13: 732–743.

50. Cheng Y, Liu Z, Xu J, Zhou T, Wang M, et al. (2008) HC-Pro protein of sugar cane mosaic virus interacts specifically with maize ferredoxin-5 *in vitro* and *in planta*. J Gen Virol 89: 2046–2054.

51. Melchinger AE, Kuntze L, Gumber RK, Lübberstedt T, Fuchs E (1998) Genetic basis of resistance to sugarcane mosaic virus in European maize germplasm. Theor Appl Genet 96: 1151–1161.

52. Shi Y, Qin Y, Cao Y, Sun H, Zhou T, et al. (2011) Influence of an m-type thioredoxin in maize on potyviral infection. Eur J Plant Pathol 131: 317–326.

53. Tao Y, Jiang L, Liu Q, Zhang Y, Zhang R, et al. (2013) Combined linkage and association mapping reveals candidates for *Scmv1*, a major locus involved in resistance to sugarcane mosaic virus (SCMV) in maize. BMC Plant Biol 13: 162.

54. Užarowska A, Dionisio G, Sarholz B, Piepho HP, Xu M, et al. (2009) Validation of candidate genes putatively associated with resistance to SCMV and MDMV in maize (*Zea mays* L.) by expression profiling. BMC Plant Biol 9: 15.

55. Ho T, Rusholme Pilcher RL, Edwards ML, Cooper I, Dalmay T, et al. (2008) Evidence for GC preference by monocot Dicer-like proteins. Biochem Biophys Res Commun 368: 433–437.

56. Ho T, Wang H, Pallett D, Dalmay T (2007) Evidence for targeting common siRNA hotspots and GC preference by plant Dicer-like proteins. FEBS Lett 581: 3267–3272.

57. Enright AJ, John B, Gaul U, Tusch T, Sander C, et al. (2003) MicroRNA targets in *Drosophila*. Genome Biol 5: R1.

58. Conesa A, Götz S, García-Gómez JM, Terol J, Talón M, et al. (2005) Blast2GO: a universal tool for annotation, visualization and analysis in functional genomics research. Bioinformatics 21: 3674–3676.

59. Silva TF, Romanel EAC, Andrade RRS, Farinelli L, Osteras M, et al. (2011) Profile of small interfering RNAs from cotton plants infected with the polerovirus *Cotton leafroll dwarf virus*. BMC Mol Biol 12: 40.

60. Szittya G, Moxon S, Pantaleo V, Toth G, Rusholme Pilcher RL, et al. (2010) Structural and functional analysis of viral siRNAs. PLoS Pathog 6: e1000838.

61. Brodersen P, Sakvarelidze-Achard L, Bruun-Rasmussen M, Dunoyer P, Yamamoto YY, et al. (2008) Widespread translational inhibition by plant miRNAs and siRNAs. Science 320: 1185–1190.

62. Li S, Liu L, Zhuang X, Yu Y, Liu X, et al. (2013) MicroRNAs inhibit the translation of target mRNAs on the endoplasmic reticulum in *Arabidopsis*. Cell 153: 562–574.

63. Incarbone M, Dunoyer P (2013) RNA silencing and its suppression: novel insights from *in planta* analyses. Trends Plant Sci 18: 382–392.

64. Pall GS, Hamilton AJ (2008) Improved northern blot method for enhanced detection of small RNA. Nat Protoc 3: 1077–1084.

65. Qian Y, Cheng Y, Cheng X, Jiang H, Zhu S, et al. (2011) Identification and characterization of Dicer-like, Argonaute and RNA-dependent RNA polymerase gene families in maize. Plant Cell Rep 30: 1347–1363.

8

Plant Virology and Next Generation Sequencing: Experiences with a *Potyvirus*

Monica A. Kehoe[1,2]*, Brenda A. Coutts[1,2], Bevan J. Buirchell[1,2], Roger A. C. Jones[1,2]

1 School of Plant Biology and Institute of Agriculture, Faculty of Science, University of Western Australia, Crawley, WA, Australia, **2** Crop Protection and Lupin Breeding Branches, Department of Agriculture and Food Western Australia, Perth, WA, Australia

Abstract

Next generation sequencing is quickly emerging as the go-to tool for plant virologists when sequencing whole virus genomes, and undertaking plant metagenomic studies for new virus discoveries. This study aims to compare the genomic and biological properties of *Bean yellow mosaic virus* (BYMV) (genus *Potyvirus*), isolates from *Lupinus angustifolius* plants with black pod syndrome (BPS), systemic necrosis or non-necrotic symptoms, and from two other plant species. When one *Clover yellow vein virus* (ClYVV) (genus *Potyvirus*) and 22 BYMV isolates were sequenced on the Illumina HiSeq2000, one new ClYVV and 23 new BYMV sequences were obtained. When the 23 new BYMV genomes were compared with 17 other BYMV genomes available on Genbank, phylogenetic analysis provided strong support for existence of nine phylogenetic groupings. Biological studies involving seven isolates of BYMV and one of ClYVV gave no symptoms or reactions that could be used to distinguish BYMV isolates from *L. angustifolius* plants with black pod syndrome from other isolates. Here, we propose that the current system of nomenclature based on biological properties be replaced by numbered groups (I–IX). This is because use of whole genomes revealed that the previous phylogenetic grouping system based on partial sequences of virus genomes and original isolation hosts was unsustainable. This study also demonstrated that, where next generation sequencing is used to obtain complete plant virus genomes, consideration needs to be given to issues regarding sample preparation, adequate levels of coverage across a genome and methods of assembly. It also provided important lessons that will be helpful to other plant virologists using next generation sequencing in the future.

Editor: Boris Alexander Vinatzer, Virginia Tech, United States of America

Funding: This research was funded by an Australian Postgraduate Award (APA), and an Australian Grains Research and Development Corporation (GRDC) Studentship, Project number GRS10039. It was undertaken using the facilities at the Department of Agriculture and Food, Western Australia. This study forms part of a PhD project by the first author at the University of Western Australia. The funders had no role in study design, data collection and analysis, decision to publish, or preparation of the manuscript.

Competing Interests: The authors have declared that no competing interests exist.

* Email: monica.kehoe@agric.wa.gov.au

Introduction

Next generation sequencing (NGS) technologies are fast becoming a popular method to obtain whole plant virus genomes in a relatively short period of time [1]. Their uptake by plant virologists has been slower than by their counterparts in the medical sciences where the applications are extending much further, rapidly approaching the concept of personalized medicine. Such a situation was impossible before the advent of NGS and its' rapid evolution into an affordable and accessible technology now appearing on laboratory bench-tops throughout the world [2,3]. Because of the ability to use total RNA extractions for NGS, it is becoming increasingly common to use it to sequence complete genomes of plant viruses and still obtain excellent results [4–9]. The challenge now lies not in accessing and using NGS technology, but in analyzing and interpreting the very large datasets suddenly at our disposal [1].

Bean yellow mosaic virus (BYMV) (family *Potyviridae*, genus *Potyvirus*) is a single stranded positive sense RNA virus that occurs worldwide. It is a virus with an extensive natural host range that encompasses monocots and dicots, and both domesticated and wild plant species [10,11]. It is transmitted non-persistently by many different aphid species [12]. BYMV causes serious diseases and losses in many cultivated plant species worldwide. For example, early BYMV infection, which causes serious losses, normally results in systemic necrosis and plant death [13–15]. In contrast, late infection with BYMV causes black pod syndrome (BPS) in *Lupinus angustifolius* (narrow-leafed lupin) also resulting in damaging losses [16]. Plants with BPS develop characteristic flat, black pods that have little or no seed [17]. It seems likely that both the BPS and systemic necrosis responses are related to presence of hypersensitivy *Nbm-1* gene and another similar resistance gene [15,18–20].

Wylie *et al*. [21] provided evidence for existence of seven BYMV phylogenetic groupings based on coat protein (CP) sequences and the original hosts of the isolates sequenced: one generalist group with a broad host range including monocots and dicots called the general group, and six other specialist groups each named after the original hosts of the isolates within them (broad bean, canna, lupin, monocot, pea, W). Partial CP sequences from BYMV isolates originally from *L. angustifolius* plants with BPS, systemic necrosis or non-necrotic symptoms placed all of them into the general group [16,21].

This study aims to compare the genomic and biological properties of BYMV isolates from *L. angustifolius* plants with BPS, systemic necrosis or non-necrotic symptoms, and from two other plant species. NGS was used to sequence 22 BYMV isolates, obtained as part a study conducted in 2011 and from previous studies in south-west Australia [16,19]. Here, we present the results of genome comparisons with the resulting 23 new BYMV genomes and one *Clover yellow vein virus* (ClYVV) genome with 17 genomes retrieved from Genbank, and biological host range studies with seven BYMV and one ClYVV isolates. We also make recommendations based on the lessons learned from our NGS studies which will be useful to plant virologists employing this approach to obtain whole genomes of other plant viruses.

Materials and Methods

Isolates and host plants

Seventeen BYMV isolates were collected from *L. angustifolius* plants with BPS (i.e. systemic necrotic stem streaking with black pods) (11) and systemic necrosis (no black pods) (6), and two from *L. cosentinii* plants with mosaic and leaf deformation as part of a 2011 study in south-western Australia [16]. The remaining three BYMV isolates (FB, LMBNN and LP) were from previous studies [19]. They had been maintained as freeze-dried leaf material obtained from the West Australian Plant Pathogen Culture Collection (FB - WAC10051, LMBNN - WAC10094 and LP - WAC10059). The ClYVV isolate was from the same culture collection (WAC10102).

All plants were maintained at 18–22°C in an insect-proof, air conditioned glasshouse. Plants of *L. angustifolius* cvs Jenabillup (partially resistant to BPS), Mandelup (susceptible to BPS) and germplasm accession P26697 (*Nbm-1* gene absent) were grown in washed river sand. Plants of *Nicotiana benthamiana*, *Trifolium subterraneum* cv. Woogenellup (subterranean clover), *Chenopodium amaranticolor*, *C. quinoa*, *Pisum sativum* cv. Greenfeast (pea) and *Vicia faba* cv. Coles early dwarf (faba bean) were grown in steam-sterilised potting mix. Cultures of virus isolates were maintained by serial mechanical inoculation of infective sap to plants of *N. benthamiana* or *T. subterraneum*. For inoculations to maintain cultures, or as part of experiments, virus-infected leaves from systemically infected plants were ground in 0.1M phosphate buffer, pH 7.2, and the infective sap mixed with celite before being rubbed onto leaves.

For testing by ELISA, leaf samples were extracted (1 g per 20 ml) in phosphate-buffered saline (10 mM potassium phosphate, 150 mM sodium chloride, pH 7.4, Tween 20 at 5 ml/liter, and polyvinyl pyrrolidone at 20 g/liter) using a mixer mill (Retsch, Germany). Sample extracts were tested for BYMV or ClYVV by double-antibody sandwich ELISA based on a modified protocol described by Clark and Adams [22] and according to manufacturer's recommendations. For generic *Potyvirus* testing, samples were extracted in 0.05 M sodium carbonate buffer, pH 9.6, and tested using the antigen-coated indirect ELISA protocol of Torrance and Pead [23]. The polyclonal antiserum to BYMV was from DSMZ (AS-0717), Germany, to ClYVV from Neogen Phytodiagnostics – formerly Adgen, UK (1171-05) and to generic potyvirus from Agdia, USA (SRA27200). All samples were tested in duplicate wells in microtiter plates. Sap from BYMV or ClYVV infected and healthy *T. subterraneum* leaf samples was included in paired wells to provide positive and negative controls. The substrate was *p*-nitrophenyl phosphate at 1.0 mg/ml in diethanolamine, pH 9.8, at 100 ml/liter. Absorbance values at A_{405} were measured in a microplate reader (Bio-Rad laboratories, USA). Absorbance values of positive samples were always more than three times those of the healthy sap control.

Sequence data

Twenty two BYMV and one ClYVV sample were sent for NGS on an Illumina HiSeq 2000 (Table 1). For BYMV in total there were 11 samples from *L. angustifolius* plants with BPS, six from *L. angustifolius* plants with systemic necrosis and one from a *L. angustifolius* plant with non-necrotic symptoms. The remaining samples consisted of isolates from other *Lupinus* spp. or were isolates from other hosts representing other phylogenetic groups based on Wylie *et al.* [21], including two samples from *L. cosentinii*, one from *L. pilosus*, and one from *V. faba*. The single ClYVV sample was from *T. repens* (white clover). Total RNA was extracted from each sample using a Spectrum Plant Total RNA kit (Sigma-Aldrich, Australia). Following extraction, total RNA was sent to the Australian Genome Research Facility (AGRF) for library preparation and barcoding (24 samples per lane) before 100 bp paired-end sequencing on an Illumina HiSeq2000. For each sample, reads were first trimmed using CLC Genomics Workbench 6.5 (CLCGW) (CLC bio) with the quality scores limit set to 0.01, maximum number of ambiguities to two and removing any reads with <30 nucleotides (nt). Contigs were assembled using the *de novo* assembly function of CLCGW with automatic word size, automatic bubble size, minimum contig length 500, mismatch cost two, insertion cost three, deletion cost three, length fraction 0.5 and similarity fraction 0.9. Contigs were sorted by length and the longest subjected to a BLAST search [24]. In addition, reads were also imported into Geneious 6.1.6 (Biomatters) and provided with a reference sequence obtained from Genbank (JX173278 for BYMV and NC003536 for ClYVV). Mapping was performed with minimum overlap 10%, minimum overlap identity 80%, allow gaps 10% and fine tuning set to iterate up to 10 times. A consensus between the contig of interest from CLCGW and the consensus from mapping in Geneious was created in Geneious by alignment with Clustal W. Open reading frames (ORFs) were predicted and annotations made using Geneious. Finalized sequences were designated as "complete" based on comparison with the reference sequences used in the mapping process, "nearly complete" if some of the 5′ or 3′ UTR was missing but the coding region was intact, and "partial" if all of the 5′ or 3′ UTR and some of the P1 or CP genes were missing.

Phylogenetic analysis

The new sequences were aligned with the 17 retrieved from Genbank using Clustal W in MEGA 5.2.1, prior to phylogenetic analysis [25]. Phylogenetic analysis compared (i) coding regions of all BYMV genome sequences and (ii) coding regions of all BYMV genome sequences except seven with average coverage of 10 times or less. Neighbor-joining trees were made using the number of differences model with a bootstrap value of 1000, Maximum Likelihood trees using the Tamura-Nei model with a bootstrap value of 1000, and Minimum Evolution trees using the number of differences model with a bootstrap value of 1000. Tables of nucleotide (nt) percentage differences were calculated for the complete genomes using the pairwise comparison function with the number of differences model. Final sequences were submitted to the European Nucleotide Archive (ENA) with accession numbers HG970847–HG970870 (Table 1).

Biological data

For host range studies, seven isolates of BYMV and one of ClYVV were mechanically inoculated onto leaves of *L. angusti-*

Table 1. Next generation sequencing data from twenty two Bean yellow mosaic virus (BYMV) and one Clover yellow vein virus (ClYVV) samples.

Plant/Host ID	Symptoms[a]	No. of reads obtained	No. of reads after trimming	No. of Contigs produced (CLC)	Sample sequence ID	Accession number	Contig length (nt) (CLCGW[c])	Average coverage (CLCGW)	No. of reads mapped to contig of interest (CLCGW)	Length of consensus (nt) (Geneious)	Average coverage (Geneious)	No. of reads mapped to ref. sequence (Geneious)	Length of Geneious+ CLCGW consensus (nt) (Geneious)	Genome completenes
Lupinus cosentinii	M	12,684,310	12,402,361	387	MD1	HG970847	9,547	10,173	987,972	9,581	10,562	1,002,513	9,285	partial
L. angustifolius	SS, SC	31,131,660	29,497,124	1,851	MD5	HG970848	968–2,576 (5)[b]	11–14 (5)	111–359 (5)	9,541	9	894	9,287	partial[d]
L. angustifolius	SS, SC	14,342,828	13,995,123	887	MD6	HG970849	2,625; 1,430	6; 7	103; 202	9,544	7	713	9,285	partial[d]
L. cosentinii	M, LD	12,068,236	11,791,675	472	MD7	HG970850	9,563	1,780	173,242	9,636	1,821	175,858	9,530	nearly
L. angustifolius	SS, SC	10,841,138	10,582,250	802	SP1	HG970851	9,524	446	43,220	9,544	457	43,877	9,528	nearly
L. angustifolius	SS, SC	11,348,684	11,076,092	1,098	GB17A	HG970852	9,655	1,723	169,883	9,983	12,313	954,307	9,530	complete
L. angustifolius	SS, BP	30,708,142	29,877,487	2,498	GB42C	HG970853	701–1,557 (8)	6–14 (8)	49–191 (8)	9,544	10	1,045	9,390	nearly[d]
L. angustifolius	SS, BP	11,780,826	11,532,853	360	GB32A	HG970854	9,593	5,241	511,194	9,915	5,317	511,902	9,538	complete
L. angustifolius	VC	11,166,630	10,924,778	476	LMBNN	HG970855	9,533	2,869	278,099	9,544	2,925	279,855	9,531	nearly
L. angustifolius	BP, NVSS	14,710,190	14,415,378	820	ES69C	HG970856	554–2,360 (9)	3–11 (9)	19–211 (9)	9,544	8	850	9,369	partial[d]
L. angustifolius	MSS, BP	14,308,420	13,991,683	907	ES67C	HG970857	534–1,413 (9)	3–8 (9)	18–101 (9)	9,544	6	660	9,479	nearly[d]
L. angustifolius	MSS, BP	15,468,144	15,091,935	837	ES55C	HG970858	9,587	1,631	159,066	9,567	1,656	159,704	9,514	complete
L. angustifolius	SS, SC	13,676,576	13,370,007	798	ES11A	HG970859	709–1,417 (9)	767–2207 (9)	8,193–31,962 (9)	10,324	2,708	280,928	9,530	complete
L. angustifolius	SS, BP	14,890,148	14,565,374	855	PN83A	HG970860	9,556	7,239	704,254	10,262	7,382	715,978	9,530	complete
L. angustifolius	SS, BP	15,789,358	15,445,952	795	PN80A	HG970861	9,532	4,825	468,607	9,625	4,934	475,461	9,530	complete
L. angustifolius	SS, BP	16,826,134	16,427,970	919	PN77C	HG970862	646–926 (5)	4–5 (5)	27–50 (5)	9,544	4	471	9,274	partial[d]
L. angustifolius	SS, BP	14,713,078	14,380,894	1,643	AR87C	HG970863	9,561	2,600	252,909	9,912	2,587	256,835	9,530	complete
L. angustifolius	SS, BP	12,537,974	12,252,296	827	AR98C	HG970864	676–1,264 (5)	6–7 (5)	56–88 (5)	9,544	6	617	9,447	nearly[d]
L. angustifolius	SS, BP	16,337,836	15,965,779	1,023	AR93C	HG970865	9,547	5,856	568,049	9,990	6,004	576,414	9,530	complete
L. pilosus	VC, SM, LD, Y	17,060,028	16,402,537	889	LP	HG970866	9,521	1,159	113,369	9,034	737	6,866	9,520	nearly
Vicia faba	VC, SM, LD	16,293,018	15728187	519	FB	HG970867	9,464	1,035	100,666	-	-	-	9,417[e]	nearly
					LPexFB	HG970868	9,450	603	58,411	-	-	-	9,417[e]	nearly
L. angustifolius	SS, SC	15,974,728	15,402,955	720	NG1	HG970869	9,548	4,192	409,456	9,595	4,399	422,536	9,530	complete
Trifolium repens	Not recorded	11,385,986	11,121,357	149	ClYVV	HG970870	9,439	6,195	6,195	9,585	65	6,295	9,439	nearly

[a]Coded symptom description: BP, black pods; LD, distortion; M, mosaic; MSS, mild necrotic stem streaking; NVSS, no visible stem streaking; SS, necrotic stem streaking; SM, severe mosaic; SC, shepherds crook appearance (i.e. bending over and apical tip necrosis); VC, vein clearing; Y, yellowing.
[b]Numbers in parenthesis represent the total number of contigs for the sample with lengths indicated by the preceding range.
[c]CLC genomics workbench.
[d]Indicates that the genome is draft only, meaning less than or equal to ten times average coverage.
[e]Indicates that the final sequence is derived entirely from CLCGW de novo assembly.

folius, *N. benthamaniana*, *T. subterraneum*, *C. amaranticolor*, *C. quinoa*, *P. sativum* and *V. faba* plants (5 plants/isolate). For each experimental host, uninoculated and mock-inoculated controls were included at time of inoculation (five plants each). There were five isolates from *L. angustifolius*, one from a plant with BPS (AR93C), three from plants with systemic necrosis (MD5, GB17A and ES11A), and one from a plant with non-necrotic symptoms (LMBNN). The remaining isolates were from plants of *L. cosentinii* (MD7) and *L. pilosus* (LP) with non-necrotic symptoms. Symptoms were recorded and samples from inoculated and tip leaves tested by ELISA weekly beginning 7 days after inoculation for up to six weeks.

Results

Sequence data

From the single ClYVV and 22 BYMV samples, the numbers of raw reads obtained from NGS were 10,841,138–31,131,660, but these numbers were reduced to 10,582,250–29,877,478 after trimming (Table 1). Following *de novo* assembly of each individual sample using CLCGW, the numbers of contigs produced were 149–2498. Contig of interest lengths were 534–9,655 nt with average coverage 3–10,173 times and the numbers of reads mapped to each contig were 18–987,972. After mapping to a reference genome in Geneious, the lengths of the consensus sequences were 9,034–10,324 nt, with average coverage of 4–12,313 times and the numbers of reads mapped to the references sequence were 471–1,002,513. Final sequence lengths consisted of the consensus of the contig from CLCGW and the consensus from Geneious, and were 9,274–9,530 nt. All samples yielded one sequence of interest, with the exception of FB, which contained a second BYMV sequence which we called "LPexFB". In all cases, except for ClYVV, the contigs of interest were most closely related to BYMV after being subjected to Blastn analysis. ClYVV was most closely related to the only other ClYVV complete genome available on Genbank. In total, there were nine complete genomes, ten nearly complete genomes (including ClYVV) and five partial genomes.

Phylogenetic analysis

Phylogenetic analysis comparing the coding regions of 23 new complete or nearly complete BYMV genomes and one new nearly complete ClYVV genome with those of 17 BYMV and one ClYVV genome retrieved from Genbank provided 100% bootstrap support for eight of nine phylogenetic groups (I, II, IV–IX). The remaining group (III) had 98% bootstrap support. Seven of the new genomes had average coverages of less than or equal to ten times (MD5, MD6, GB42C, ES69C, ES67C, PN77C and AR98C) and five of these (MD5, MD6, ES67C, ES69C and PN77C) did not sit well within groups I and II. Although they appear to belong to them, genomes such as MD6 and PN77C sit out on their own, almost separate from the other sequences, leaving groups I and II poorly resolved (Figure 1a). In contrast, when sequences of the seven genomes with poor average coverage (≤10 times) were removed, phylogenetic analysis gave the same results but with much greater resolution between groups I and II and improved bootstrap support for groups I–IX (Figure 1b). Those removed were designated as "draft" genomes because all had low coverage and/or small gaps. When all the genomes, including those with poor coverage were analyzed using Maximum Likelihood or Minimum Evolution methods, the tree topologies shown were the same as the Neighbor-Joining method.

The range of original isolation hosts within each grouping varied (Table 2). Group I consisted of nine sequences from two

dicot, and two monocot species. Group II consisted of seven sequences from two dicot and one monocot species. Group III consisted entirely of three sequences from one monocot species. Group IV was made up of three sequences from an unknown original host or hosts, as well as two from a monocot and one from a dicot species. Groups V–IX consisted entirely of dicot species belonging to a single family, and were represented by up to three sequences. All dicot species were from families *Fabaceae* or *Gentianaceae*, and all monocot species were from families *Orchidaceae* or *Iridaceae*.

Sequence analysis

When the coding regions of the 16 new BYMV genomes (draft genomes excluded) and one ClYVV genome were analyzed against those retrieved from Genbank, the nt percentage identities within each phylogenetic group were ≥96.6% (I), ≥98.6% (II), ≥93.9% (III), ≥94% (IV), ≥90.7% (V), ≥99.8% (VI), ≥97.6% (VII) and ≥97.5% for ClYVV (Table S1). When the six sequences from *L. angustifolius* plants with BPS were compared to each other their percentage nt identites were ≥93.8%. When the sequences from all *L. angustifolius* plants were compared to each other their percentage nt identities were also ≥93.8%. Across all 33 BYMV sequences used in this analysis the nt identities were ≥75.6%. When the ClYVV sequences were compared to the BYMV sequences, overall the percentage nt identities were 66.4–67.9%.

Biological data

All seven BYMV isolates and one ClYVV isolate inoculated to plants caused systemic symptoms of varying severity in *N. benthamiana*, *T. subterraneum* and *V. faba* (Table 3). However, apart from ClYVV and BYMV isolate GB17A in *V. faba*, none of them induced systemic necrotic symptoms, which were severe only with ClYVV. In *C. amaranticolor*, ClYVV and five BYMV isolates caused obvious systemic symptoms, while infection was restricted to inoculated leaves with the isolate originally from *L. angustifolius* plants with BPS and another originally from an *L. angustifolius* plant with non-necrotic symptoms. In *C. quinoa*, although all isolates infected inoculated leaves, only ClYVV caused systemic invasion. In contrast, in *P. sativum*, only BYMV isolate LP caused any infection.

In *L. angustifolius* cvs Jenabillup and Mandelup, three BYMV isolates caused systemic non-necrotic symptoms. These were originally from plants of this species with non-necrotic symptoms (LMBNN) or systemic necrotic symptoms (ES11A), and *L. cosentinii* (MD7) from a plant with mosaic and leaf distortion. All other BYMV isolates and the ClYVV caused systemic necrotic symptoms in cvs Jenabillup and Mandelup. In accession P26697, with ClYVV and four BYMV isolates for which symptom data are available, the reactions resembled those in cv. Jenabillup, with the exception of MD5 which produced severe mosaic (i.e. non-necrotic) symptoms instead of systemic necrosis. Isolates LBMNN and ES11A caused non-necrotic symptoms, while ClYVV and LP caused systemic necrosis. Failure of isolates AR93C and MD7 to infect P26697 probably represents escapes, but there was no seed left of P26697 for further testing. Isolate LP did not infect *L. angustifolius* cv. Mandelup on two separate occasions by sap inoculation, but further inoculations using grafting or aphids would be needed to establish if this is a resistance reaction.

Discussion

Before this study was conducted, there were only 17 complete BYMV genomes on Genbank. The ten complete and eight nearly complete genomes from this study doubled available BYMV

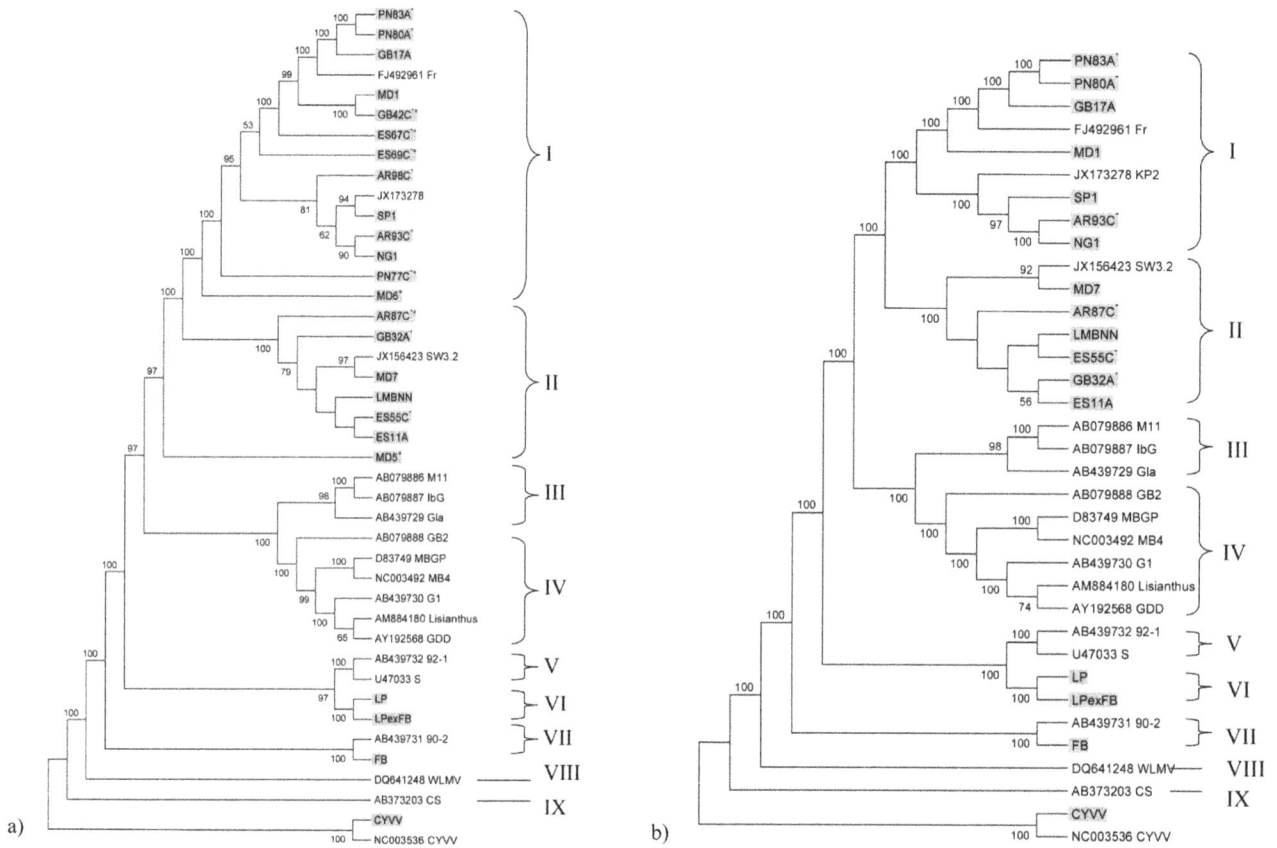

Figure 1. Neighbor-joining relationship phylograms obtained from alignment of the coding regions of *Bean yellow mosaic virus* **(BYMV) genomes.** The alignments were generated in MEGA 5.2.1 using ClustalW and tree branches were bootstrapped with 1000 replications. The trees were rooted with a sequence of *Clover yellow vein virus* (CIYVV), the closest relative to BYMV. New isolates from this study shown in grey, isolates obtained from *Lupinus angustifolius* plants with BPS are denoted by *, and isolates with genomes designated as "draft" are denoted by +. a) Complete coding regions of BYMV genomes, including draft sequences, with isolates retrieved from Genbank. b) The same sequences as in a) but with draft sequences removed from the analysis.

Table 2. Original hosts of isolates within each phylogenetic grouping.

Phylogenetic group (old name)	Accession numbers	Dicot	Monocot
I (general)	FJ492961, JX173278, HG970847, HG970851, HG970851-52, HG970856-57, HG970860-62, HG970864-65, HG970865	*Lupinus angustifolius*[a] (6)[b], *L. cosentinii* (1)	*Diuris magnifica* (1), *Freesia* sp. (1)
II (general)	JX156423, HG970848, HG970850, HG970854-55, HG970858-59, HG970863	*L. angustifolius* (5), *L. cosentinii* (1)	*Diuris* sp. (1)
III (monocot)	AB079886, AB079887, AB439729	-	*Gladiolus hybrid* (3)
IV (general)	AB079888[c], D83749[c], NC003492[c], AB439730, AM884180, AY192568	*Eustoma russellianum* (1),	*Gladiolus* sp. (1), *Gladiolus hybrid* (1)
V(faba bean)	AB439732, U47033	*Trifolium pratense* (1), *Vicia faba* (1)	-
VI (lupin)	HG970866, HG970868	*L. pilosus*(1), *Vicia faba* (1)	-
VII (faba bean)	AB439731, HG970867	*V. faba* (2)	-
VIII (W)	DQ641248	*L. albus* (1)	-
IX (pea)	AB373203	*Pisum sativum* (1)	-

[a]Species from *Lupinus*, *Vicia* and *Trifolium* are from family *Fabaceae*. *Eustoma* is from family *Gentianaceae*, *Gladiolus* and *Freesia* are from family *Iridaceae*, *Diuris* is from family *Orchidaceae*.
[b]Numbers in parentheses represent the numbers of genomes with from this original isolation host.
[c]Denotes an unknown original host for that accession number.

Table 3. Responses of seven plant species to inoculation with eight different isolates of *Bean yellow mosaic virus* (BYMV) and one of *Clover yellow vein virus* (CIYVV).

Isolate[a]	Original host	Symptoms in original host	Chenopodium amaranticolor	C. quinoa	Lupinus angustifolius cv. Jenabillup	L. angustifolius cv. Mandelup	L. angustifolius line P26697	Nicotiana benthamiana	Pisum sativum cv. Greenfeast	Trifolium subterraneum cv. Woogenellup	Vicia faba cv. Coles Early Dwarf	Phylogenetic grouping
CIYVV	T. repens	Not recorded	LNS[b], SCS, SVC	LNS, SCS, SVC	SS, N, St, Y, LD	SS, SN, DR	SSS, R, B, M	SM	NI	VC, SM, LD, St	SS, SM, LD, SN	n/a
LMBNN	L. angustifolius	VC	LCS	LMCS	SM, DR, B	B, M	B, M	MM	NI	MM	SM, LD	II
AR93C	L. angustifolius	SS, BP	LCS	LNS	MSS, SN, St, Y	SSS, DR, SN	NI[c]	M	NI	VC, M, LD, St	MM	I
MD5	L. angustifolius	SS, SC	LNS, SCS, SVC	LNS, LCS	SSS, SN	SSS, DR, SN	SM	MM	NI	VC, M, LD	M	II
GB17A	L. angustifolius	SS, SC	LCS, SCS	LCS, LNS	M, DR, Y, SSS, N	nt	nt	MM	NI	MVC, SM, LD	SS, M	I
ES11A	L. angustifolius	SS, SC	LCS, SCS, SVC	LVC, LCS, NSH	M, DR, Y	B, M	B, M	MM	NI	MVC, MM	SM	II
MD7	L. cosentinii	M, DR	LCS, SCS, SVC	LCS	SM, DR, B	PSN, M, B	NI[c]	MM	NI	MVC, M	SM	II
LP	L. pilosus	VC, SM, LD, Y	SCS, SVC	LNS	SSS, DR, SN	NI[d]	SSS, SN	M	MM	VC, M, LD, St	M	VI
Uninoculated	n/a	n/a	NI	NI	NI	NI	NI	NI	NI	NI	NI	n/a
Mock	n/a	n/a	NI	NI	NI	NI	NI	NI	NI	NI	NI	n/a

Leaves were inoculated with infective sap. Samples from inoculated and tip leaves were tested by ELISA for BYMV, CIYVV and potyviruses in general.

[a]Locations where isolates were collected from: CIYVV, New South Wales (NSW); LMBNN, Mt. Barker, Western Australia (WA); AR93C, Arthur River, WA; MD5 and MD7, Medina, WA; GB17A, Gibson, WA; ES11A, Esperance, WA; LP, South Perth WA.

[b]Coded symptom descriptions: B, bunchy new growth; BP, black pods; LCS, local chlorotic spots; LD, leaf drop; LMCS, local mild chlorotic spots; LNS, local necrotic spots; M, mosaic; MM, mild mosaic; MSS, mild necrotic stem streaking; MVC, mild local vein clearing; NI, not infected; NSH, local necrotic spots with halo; nt, not tested; PSN, partial necrotic stem streaking (no infection in uninoculated leaves); R, reddening; SC, shepherds crook appearance (i.e. bending over and apical tip necrosis); SS, severe necrotic stem streaking leading to plant death; SCS, systemic chlorotic spots; SM, severe mosaic; SN, severe necrosis of uninoculated leaves and new growth; St, stunting; SVC, systemic vein clearing; VC, localized vein clearing; Y, yellowing.

[c]Denotes that the initial round of inoculations failed to infect plants of this species, and due to a lack of seed the inoculations were not repeated.

[d]Denotes that two rounds of inoculation failed to infect L. angustifolius cv. Mandelup. Further testing involving sap, aphid and graft inoculations are required.

genomic data in the database. Moreover, the five additional partial genomes we obtained will be useful in future studies. Our genome results enabled the phylogenetic makeup of BYMV to be examined thoroughly, revealing presence of nine distinct groups, including the subdivision of the former generalist group into three new groups. We recommend replacing the phylogenetic groupings of Wylie et al. [21] with numbered group names (I–IX). We have not included one former specialist group based on CP genes, the canna group, in our analysis because it was not represented by any whole genome sequence. Use of whole genomes revealed that the previous phylogenetic grouping system based on partial genome sequences and original isolation hosts was unsustainable. This is because genome sequences from broad bean are present in two former specialist groups (now V and VII), from various *Lupinus* species in two former specialist groups (now VI and VIII), and two former generalist groups (now I and II). Moreover, although we have not re-analyzed sequences of CP genes, Wylie et al. [21] had previously placed a CP sequence from the dicot species *Eustoma russellianum* (family *Gentianaceae*), in the former monocot group (now III). Numbering of groups prevents such confusion arising from use of natural isolation host names. Our results highlight the importance of using complete genomes wherever possible to define phylogenetic groupings. The results also highlight the need for further sequencing and analysis of BYMV isolates likely to belong to former specialist phylogenetic groupings, which will provide greater insight into the genetic makeup of BYMV.

Close examination of the nt percentage sequence identities between BYMV and ClYVV genomes revealed that the divergence between them is greater than previously thought. Overall, BYMV percentage nt identities ranged from 75.6 to 99.5%. The species demarcation for potyviruses is currently 23–24% divergence at the nt level [26], and some of the BYMV isolates compared came close to this. The two ClYVV genomes shared 97.5% nt identity, but when compared to all the BYMV genomes, nt identities were 66.4–67.9%, well beyond the species demarcation point for potyviruses. ClYVV was originally considered an isolate of BYMV but was later shown to be a distinct virus [26,27,28]. Our percentage identities support that distinction. However, some BYMV phylogenetic groups were more closely related to ClYVV than others. For example when compared with all other BYMV sequences, the single sequences from groups VIII and IX had percentage identities of just 78.4–79.8% and 75.6–76.9% to BYMV respectively, whereas when compared to ClYVV their nt percentage identities were 67.0–67.7% (Table S1). Again, further genome sequences from these groups and ClYVV are required for a more conclusive analysis.

Based on our phylogenetic and sequence analyses, BYMV isolates associated with BPS in *L. angustifolius* were not different phylogenetically from other BYMV isolates we sequenced from *L. angustifolius*, *L. cosentinii*, or other hosts within the same phylogenetic groups (I and II). Also, from the host data from our inoculations, there was no host reaction that could be used to distinguish a particular isolate as causing BPS. However, there were some other interesting differences. Although isolate ES11A behaved in a similar manner to isolate LMBNN, which overcomes the *Nbm-1* hypersensitivity gene in *L. angustifolius* plants [19,20], it was isolated from a *L. angustifolius* plant originally displaying systemic necrosis. ClYVV behaved like isolate LP, but whether ClYVV interacts with both *Nbm-1* and the second putative BYMV hypersensitivity genes, or unknown ClYVV-specific genes in *L. angustifolius*, is not clear [19]. ClYVV and all group I and II isolates failed to infect *P. sativum* cv. Greenfeast although the group VI isolate LP did cause infection. This may be due to the fact that this cultivar, like many commercial pea cultivars, may

contain the BYMV resistance gene *mo* and ClYVV resistance genes *cyv* or *cyv-2* [29,30] and their responses are strain specific. Induction of severe necrotic symptoms in *V. faba* by ClYVV but not the BYMV isolates is expected, as this is the classical method for distinguishing BYMV from ClYVV [10,20].

In this study, we used NGS to obtain complete virus genomes and it proved both an advantage and a disadvantage over traditional sequencing methods. It allowed large amounts of data to be generated quickly, but analysis of the data proved a major challenge. Many free programs exist for the assembly of NGS data (e.g. Velvet, SOAP de novo, Abyss and bowtie) but they all require the researcher to be proficient in the use of command line driven applications. As so-called "benchtop biologists", the use of Geneious and CLCGW was easy to learn and their cost was acceptable in view of the time saved in learning the use of command line driven programs. That said, our success was probably attributable to the small genome sizes of plant viruses, particularly BYMV and ClYVV, which are both c. 9535 nt long. Larger genomes, from unpurified RNA samples would undoubtedly be much harder to piece together, but not impossible. We found in most cases (17 out of 23) there was sufficient average coverage to be confident of good genome representation for the isolate sequenced. These sequences had average coverages as low as 65 and 457 with remaining average coverages being greater than 737 and up to 12,313 times when mapped back to a reference sequence using Geneious. Currently, sequencing a human genome of approximately 300 MB on an Illumina platform requires 30 times coverage to be adequate [31]. Therefore, it seems reasonable to designate our virus genomes with less than 30 times coverage as draft sequences. Although not meeting minimum requirements for average coverage, they are still valuable data sets, particularly given the low numbers of complete or nearly complete BYMV genomes available (now 32 including those from this study).

The settings used in *de novo* assembly are sufficient to distinguish between more than one strain or group of a plant virus when present in the same sample, as previously demonstrated by Kehoe et al. [9]. In our case, the sample from a *V. faba* BYMV isolate (FB) retrieved from the culture collection also contained a nearly complete LP isolate genome. The contamination probably occurred more than ten years ago when they were maintained next to each other in the same glasshouse prior to freeze-drying and storage in the collection. In such instances, if we had only been using Geneious to map to a reference genome, we would have likely missed the second sequence. It is therefore important to perform *de novo* assembly, as well as mapping to a reference genome. In cases where either the mapping or the *de novo* sequence had a gap, it was usually resolved after alignment with the sequence from the second program. However, for genomes with coverage less than ten (i.e. the draft genomes) this method was ineffective.

The uptake of NGS amongst plant virologists is increasing as the cost associated with it decreases [1]. The relatively small genome size of plant viruses allows us the opportunity to extract complete or nearly complete genomes using commercial packages. Use of NGS does raise concerns regarding the consequences of an increase in the discovery of virus or virus-like sequences. As such, MacDiarmid et al. [32] made recommendations regarding the identification of plant viruses through NGS, and the potential biosecurity issues associated with this. One of the recommendations was that the term "uncultured virus" should be used with any plant virus sequence not associated with a recognized virus infection. We support this recommendation whole-heartedly.

We know of no recommendation regarding requirements for depth of coverage for plant virus genomes, particularly ones involving new virus discoveries. Until such time as an appropriate set of comparative studies are done, we would recommend following in the path of our human genome colleagues by requiring a minimum coverage of at least 30 times, but this would likely lead to many nearly complete or draft plant virus genomes. However as with BYMV for example, we required coverage well into the 1000's to ensure a complete genome (including 5' and 3' UTRs, a constant challenge for plant virologists). Our samples sent for sequencing were total RNA, so different methods of sample preparation might have increased the numbers of virus reads. For example, use of subtractive hybridization [4], or extracting for dsRNA first, followed by random cDNA synthesis [1,33]. Despite this, there is no doubt that NGS has been an exceedingly useful tool for our study.

Supporting Information

Table S1 Nucleotide percentage similarities of the coding regions of thirty three *Bean yellow mosaic virus* and two *Clover yellow vein virus* isolates, calculated in MEGA 5.2.1 using a pairwise comparison with the number of differences model.

Acknowledgments

We thank E. Gajda, S. Vincent and M. Banovic for glasshouse and laboratory support.

Author Contributions

Conceived and designed the experiments: MAK BAC BJB RACJ. Performed the experiments: MAK. Analyzed the data: MAK. Contributed reagents/materials/analysis tools: MAK. Contributed to the writing of the manuscript: MAK BAC BJB RACJ.

References

1. Boonham N, Kreuze J, Winter S, van der Vlugt R, Bergervoet J, et al. (2014) Methods in virus diagnostics: from ELISA to next generation sequencing. Virus Res 186: 20–31.
2. Mardis ER (2013) Next-generation sequencing platforms. Ann Rev Analyt Chem 6: 287–303.
3. Koboldt DC, Steinberg KM, Larson DE, Wilson RK, Mardis ER (2013) The Next-generation sequencing revolution and its impact on genomics. Cell 155: 27–38.
4. Adams IP, Glover RH, Monger WA, Mumford R, Jackeviciene E, et al. (2009) Next-generation sequencing and metagenomic analysis: a universal diagnostic tool in plant virology. Mol Plant Pathol 10: 537–545.
5. Roossinck MJ (2012) Plant virus metagenomics: biodiversity and ecology. Ann Rev Gen 46: 359–369.
6. Wylie SJ, Luo H, Li H, Jones MGK (2012) Multiple polyadenylated RNA viruses detected in pooled cultivated and wild plant samples. Arch Virol 157: 271–284.
7. Wylie SJ, Li H, Sivasithamparam K, Jones MGK (2014) Complete genome analysis of three isolates of narcissus late season yellows virus and two of narcissus yellow stripe virus: three species or one? Arch Virol 159: 1521–1525.
8. Jones RAC (2014) Trends in plant virus epidemiology: opportunities from new or improved technologies. Virus Res 186: 3–19.
9. Kehoe MA, Coutts BA, Buirchell BJ, Jones RAC (2014) *Hardenbergia mosaic virus*: crossing the barrier between native and introduced plant species. Virus Res 184: 87–92.
10. Bos L (1970) Bean yellow mosaic virus. Association of Applied Biologists. Descriptions of Plant Viruses No. 40.
11. Edwardson JR, Christie RG (1991) Bean yellow mosaic virus. In CRC Handbook of viruses infecting legumes pp137–148. CRC Press, Boca Raton, FL.
12. Berlandier FA, Thackray DJ, Jones RAC, Latham LJ, Cartwright L (1997) Determining the relative roles of different aphid species as vectors of cucumber mosaic and bean yellow mosaic viruses in lupins. Ann Appl Biol 131: 297–314.
13. Jones RAC, McLean GD (1989). Virus diseases of lupins. Ann Appl Biol 114: 609–637.
14. Jones RAC (2001) Developing integrated disease management strategies against non-persistently aphid-borne viruses: A model programme. Integ. Pest Man. Reviews 6: 15–46.
15. Jones RAC, Coutts BA, Cheng Y (2003) Yield limiting potential of necrotic and non-necrotic strains of *Bean yellow mosaic virus* in narrow-leafed lupin (*Lupinus angustifolius*). Aust J Agric Res 54: 849–859.
16. Kehoe MA, Buirchell BJ, Coutts BA, Jones RAC (2014) Black pod syndrome of *Lupinus angustifolius* is caused by late infection with *Bean yellow mosaic virus*. Plant Dis 98: 739–745.
17. Buichell B J (2008) Narrow-leafed lupin breeding in Australia – where to from here? Pages 226–230 in: Lupins for Health and Wealth, Proceedings of the 12th International Lupin Conference, 14–18 September 2008, Fremantle, Western Australia. JA . Palta and JB . Berger editors. International Lupin Association, Canterbury, New Zealand.
18. Cheng Y, Jones RAC (1999) Distribution and incidence of necrotic and non-necrotic strains of bean yellow mosaic virus in wild and crop lupins. Aust J Agric Res 50: 589–599.
19. Cheng Y, Jones RAC (2000) Biological properties of necrotic and non-necrotic strains of bean yellow mosaic virus in cool season grain legumes. Ann Appl Biol 136: 215–227.
20. Jones RAC, Smith LJ (2005) Inheritance of hypersensitive resistance to *Bean yellow mosaic virus* in narrow-leafed lupin (*Lupinus angustifolius*). Ann Appl Biol 146: 539–543.
21. Wylie SJ, Coutts B.A, Jones MGK, Jones RAC (2008) Phylogenetic analysis of *Bean yellow mosaic virus* isolates from four continents: relationship between the seven groups found and their hosts and origins. Plant Dis 92: 1596–1603.
22. Clark MF, Adams AN (1977) Characteristics of the microplate method of enzyme-linked immunosorbent assay for the detection of plant viruses. J Gen Virol 34: 475–483.
23. Torrance L, Pead MT (1986) The application of monoclonal antibodies to routine tests for two plant viruses. pp 103–118 in: Developments in applied biology. 1 Developments and applications on virus testing. Jones RAC and Torrance L, editors. Association of Applied Biologists, Wellesbourne, UK.
24. Altschul SF, Gish W, Miller W, Myers EW, Lipman DJ (1990) Basic local alignment search tool. J Mol Biol 215: 403–410.
25. Tamura K, Peterson D, Peterson N, Stecher G, Nei M, et al.(2011) MEGA5: Molecular evolutionary genetics analysis using maximum likelihood, evolutionary distance, and parsimony methods. Mol Biol Evol 28: 2731–2739.
26. Adams MJ, Antoniw JF, Fauquet CM (2005) Molecular criteria for genus and species discrimination within the family *Potyviridae*. Arch Virol 150: 459–479.
27. Uyeda I, Takahasi T, Shikata E (1991) Relatedness of the nucleotide sequence of the 3'-terminal region of clover yellow vein potyvirus RNA to bean yellow mosaic potyvirus RNA. Intervirol 1991: 234–245.
28. Hollings M, Stone OM (1974) Clover yellow vein virus. Association of Applied Biologists. Descriptions of Plant Viruses No. 131.
29. Schroeder WT, Provvidenti R (1971) A common gene for resistance to Bean yellow mosaic virus and Watermelon mosaic virus 2 in Pisum sativum. Phytopathology 61: 846–847.
30. Provvidenti R (1987) Inheritance of resistance to clover yellow vein virus in *Pisum sativum*. J Heredity 70: 126–128.
31. Wetterstrand KA (2014) DNA Sequencing Costs: Data from the NHGRI Genome Sequencing Program (GSP) Available: www.genome.gov/sequencingcosts. Accessed 2014 Feb 1.
32. MacDiarmid R, Rodoni B, Melcher U, Ochoa-Corona F, Roosinck M (2013) Biosecurity implications of new technology and discovery in plant virus research. Plos Pathogens 9: e1003337.
33. Al Rwahnih M, Daubert S, Úrbez-Torres JR, Cordero F, Rowhani A (2011) Deep sequencing evidence from single grapevine plants reveals a virome dominated by mycoviruses. Arch Virol 156: 397–403.

9

Complete Mitochondrial Genome of *Eruca sativa* Mill. (Garden Rocket)

Yankun Wang[1,9], Pu Chu[1,9], Qing Yang[1], Shengxin Chang[1], Jianmei Chen[1], Maolong Hu[2], Rongzhan Guan[1,3]*

1 State Key Laboratory of Crop Genetics and Germplasm Enhancement, Nanjing Agricultural University, Nanjing, Jiangsu, China, 2 Institute of Economic Crop, Jiangsu Academy of Agricultural Sciences, Nanjing, Jiangsu, China, 3 Nanjing Agricultural University, Jiangsu Collaborative Innovation Center for Modern Crop Production, Nanjing, Jiangsu, China

Abstract

Eruca sativa (Cruciferae family) is an ancient crop of great economic and agronomic importance. Here, the complete mitochondrial genome of *Eruca sativa* was sequenced and annotated. The circular molecule is 247 696 bp long, with a G+C content of 45.07%, containing 33 protein-coding genes, three rRNA genes, and 18 tRNA genes. The *Eruca sativa* mitochondrial genome may be divided into six master circles and four subgenomic molecules via three pairwise large repeats, resulting in a more dynamic structure of the *Eruca sativa* mtDNA compared with other cruciferous mitotypes. Comparison with the *Brassica napus* MtDNA revealed that most of the genes with known function are conserved between these two mitotypes except for the *ccmFN2* and *rrn18* genes, and 27 point mutations were scattered in the 14 protein-coding genes. Evolutionary relationships analysis suggested that *Eruca sativa* is more closely related to the *Brassica* species and to *Raphanus sativus* than to *Arabidopsis thaliana*.

Editor: Weijun Zhou, Zhejiang University, China

Funding: This work was supported by the National Natural Science Foundation of China (No. 31270386, 31301352 and 31101174), the National Key Technology R & D Program (No. 2010BAD01B02 and 2011BAD13B09) in China, the Open Research Fund of State Key Laboratory of State Key Laboratory of Crop Genetics and Germplasm Enhancement(ZW2011006), the Priority Academic Program Development of Jiangsu Higher Education Institutions (PAPD), the Special Fund for Independent innovation of Agricultural Science and Technology in Jiangsu province (Nos. CX (11) 1026), and the Science and Technology Support Program of Jiangsu Province (BE2012327). The funders had no role in study design, data collection and analysis, decision to publish, or preparation of the manuscript.

Competing Interests: The authors have declared that no competing interests exist.

* Email: guanrz@njau.edu.cn

9 These authors contributed equally to this work.

Introduction

Mitochondria supply energy in the form of ATP through oxidative phosphorylation in almost all eukaryotic cells [1]. In comparison to their counterparts in animals and fungi, plant mitochondrial (mt) genomes have unique features, such as large and dramatic variations in size [2], dynamic structure [3], extremely low rate of point mutations [4] and incorporation of foreign DNA [5]. The largest known mitochondrial genomes are those of seed plants, with sizes ranging from 208 kb for *Brassica hirta* [6] to over 11.3 Mb for *Silene conica* [4]. The dramatic variation may occur within closely related species [7]. Active recombination via repeated sequences appear to be responsible for the dynamic nature and multipartite organization of the mt genome in all angiosperms investigated [8], which may produce significantly different gene orders even among close relatives [9].

Mitochondria play an important role in plant growth and development. Genomic rearrangements involving substoichiometric shifting (SSS), a consequence of intermediate repeat DNA exchange [10], is often accompanied by changes in the plant's phenotype. SSS activity in plant mitochondria has been reported to be associated with cytoplasmic male sterility [11], nitrate sensing and GA-mediated pathways for growth and flowering

[12]. Plant mitochondria have also been associated with stress responses [13] and regulation of programmed cell death [14]. Therefore, determining mitochondrial genomes is important for determining specific metabolic activities of plants [15].

Eruca sativa Mill.or *Eruca vesicaria* subsp. *sativa* (Miller) (Garden rocket), a member of the Cruciferae family, has several desirable agronomic traits, such as resistance to salt, drought, white rust and aphids [16–18]. Introducing these beneficial genes of *E. sativa* into economically important cultivated species will promote crop improvement [19,20]. Crosses of *E. sativa* with other species of the family Cruciferae, including *B. rapa*, *B. juncea*, and *B. oleracea*, have been reported [20].

To date, several mt genomes from the *Cruciferae* family have been sequenced, including *Arabidopsis thaliana* (*tha*) [21], *Raphanus sativus* (*sat*) [22] and five species from the *Brassica* genus, i.e., *B. napus* (*pol*, *nap*), *B. rapa* (*cam*), *B. oleracea* (*ole*), *B. juncea* (*jun*), and *B. carinata* (*car*) [23–25]. In this study, we reported the complete mitochondrial genome sequences of *E. sativa* and provide a comparison with other sequenced *cruciferous* mt genomes. This research will help to characterize the *E. sativa* crop and further our understanding of the evolution of mitochondrial genomes within the *Cruciferae* family.

Materials and Methods

Mitochondrial DNA isolation and sequencing

A commercial cultivar of *E. sativa* was used in this study. Mitochondrial DNA was isolated from 7-day-old etiolated seedlings according to Chen's methods (Chen et al., 2011), and stored at −80°C until use. Genome sequencing was performed using the GS-FLX platform (Roche, Branford, CT, USA). The reads were assembled into contigs using Newbler v.2.6. Sanger sequencing of PCR products was used to join the contigs to form the complete genome.

Sequence data analysis

The NCBI database (http://www.ncbi.nlm.nih.gov/) was searched for mitochondrial sequences annotation, using previously annotated mitochondrial genes from angiosperms as query sequences. The tRNAs were identified using the tRNA scan-SE software (http://lowelab.ucsc.edu/tRNAscan-SE/). Putative open reading frames (ORFs) with a minimum size of 100 codons were predicted and annotated using ORF-Finder (http://www.ncbi.nlm.nih.gov/gorf/gorf.html). The circular map was drawn using OGDraw v1.2 (http://ogdraw.mpimp-golm.mpg.de/). Repeats analysis was performed as previously described [25].

Comparing mitochondrial genomes and evolutionary analysis

The *E. sativa* mitochondrial genome sequence presented here was compared with eight other reported Cruciferae mitotypes: *B. rapa* (GenBank: NC_016125), *B. oleracea* (GenBank: NC_016118), *B. juncea* (GenBank: NC_016123), *B. carinata* (GenBank: NC_016120), *B. napus* (GenBank: NC_008285), *B. napus* cultivar Polima (EMBL: FR715249), *R. sativus* (GenBank: JQ083668) and *A. thaliana* (GenBank: NC_001284), using NCBI-blastn. For comparison, the exons of 32 protein coding genes (atp1, atp4, atp6, atp8, atp9, ccmB, ccmC, ccmFc, ccmFN1, ccmFN2, cob, cox1, cox2-1, cox3, matR, nad1, nad2, nad3, nad4, nad4L, nad5, nad6, nad7, nad9, rpl2, rpl5, rpl16, rps3, rps4, rps7, rps12, tatC), which were shared by these nine species, were

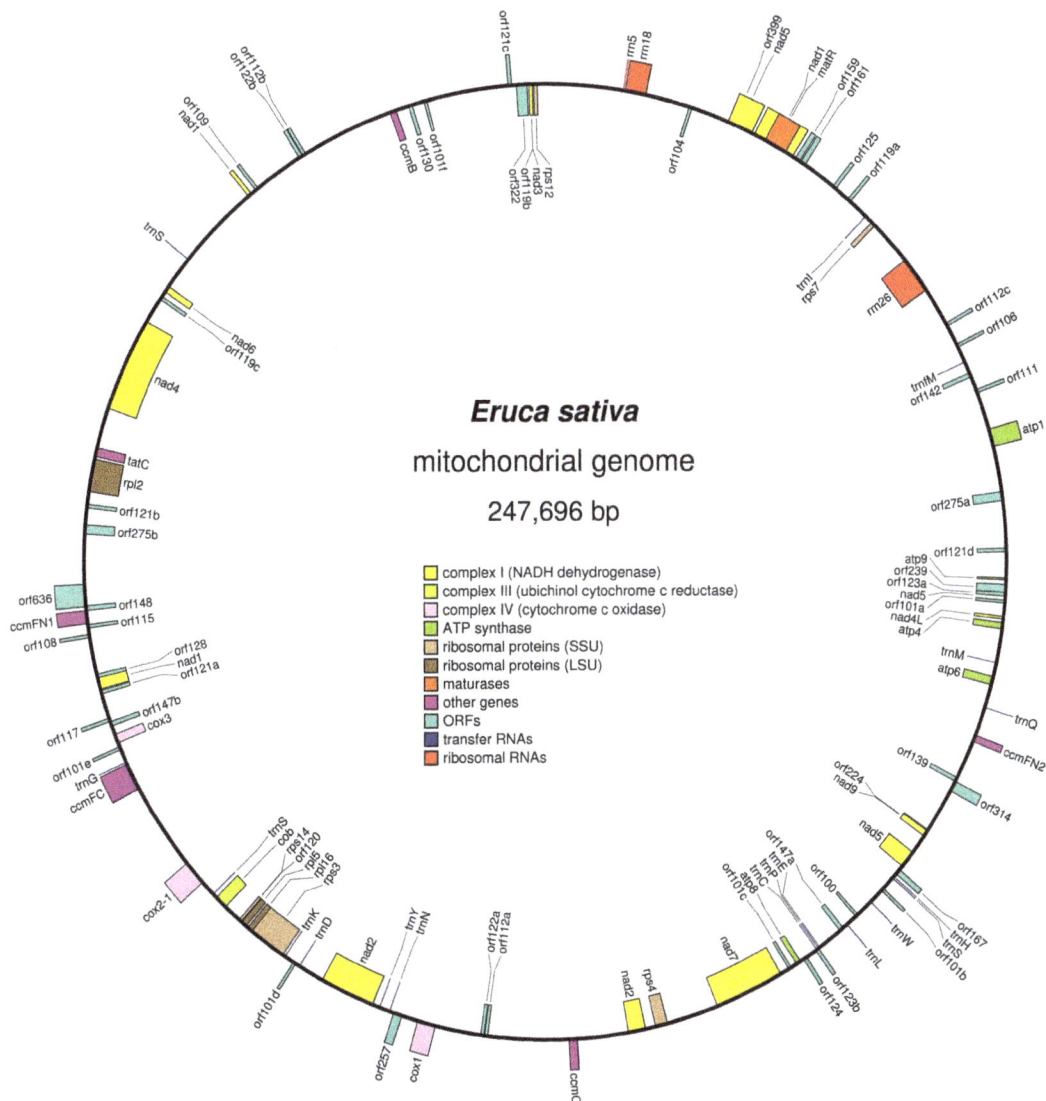

Figure 1. Mitochondrial genome map of *Eruca sativa*. Features on the clockwise- and counter-clockwise-transcribed strands are drawn on the inside and outside of the circle, respectively. The figure was drawn using OGDraw v1.2.

extracted and sequentially joined together. A neighbor-joining tree [26] was constructed with MEGA 5, using the Kumar method [27]. The number of bootstrap replications was set as 1000 [28].

Results

The mitochondrial genome of *E. sativa*

The mitochondrial genome of *E. sativa* was assembled as a single circular molecule of 247 696 bp (Figure 1, deposited in GenBank under the accession KF442616). The overall GC content of the mtDNA is 45.07%, which is comparable to those of other mtDNAs of *Cruciferae*. The largest part of the *E. sativa* mtDNA comprises the non-coding sequences (85.14%), which is slightly smaller than the average non-coding sequences content (89.4±3.1%) in other reported angiosperm mitochondrial genomes [29]. Genes account for 26.27% of the genome (65 070 bp in total length), 56.61% of which represent exons (36 837 bp) and 43.39% represent introns (28 233 bp).

Gene content and ORFs

Using BLAST and tRNA scan-SE, 54 genes were identified, including 33 protein coding genes, three rRNA genes (5S, 18S and 26S rRNAs) and 18 transfer RNA genes (Figure 1, Table 1). The 33 protein coding genes (PCGs) were in the range of 225 bp (*atp9*) to 7 979 bp (*nad4*), including 18 genes for components of the electron transport chain and ATP synthase: nine subunits of complex I (*nad1-7, 4L, 9*), one subunit of complex III (*cob*), three subunits of complex IV (*cox1-3*) and five subunits of complex V (*atp1, 4, 6, 8, 9*). In addition, there are five genes for cytochrome c biogenesis (*ccmB, ccmC, ccmFN1/ccmFN2* and *ccmFC*), eight genes for ribosomal proteins (*rpl2, 5, 16* and *rps3, 4, 7, 12, 14*), two genes for maturase (*matR*) and one gene for other functions (*tatC*). The total length of the 33 PCGs of *E. sativa* mtDNA is 58 569 bp, accounting for 23.64% of its total mtDNA genome length, which is lower than that of the *Brassica* and *R. sativus* mitotypes. Nine genes had an exon–intron structure. All exons of *ccmFC* (exons a, b), *nad2* (a-e), *cox2* (a, b), *nad4* (a-d), *nad7* (a-e), *rps3* (a, b) and *rpl2* (a, b) were cis-spliced, whereas some exons of *nad1* and *nad5* were trans-spliced as follows: *nad1a/nad1b-e; nad5a, b, d, e/nad5c* (the slash indicating trans-spliced exons). ATG is the most commonly used initiation codon for mitochondrial PCGs in *E. sativa*, except for *nad1* (start with ACG), *matR* (start with AGA) and *tatC* (start with ATT), as predicted by previous studies (Handa, 2003). Ten genes (*nad4, cob, ccmC, ccmFN1/ccmFN2, cox3, atp8, atp9, rpl2* and *rps12*) are predicted to terminate with TGA and six (*atp1, nad7, rps3, rps14, matR* and *tatC*) with TAG; other PCGs use TAA as their termination codon.

18 tRNA sequences (1 383 bp) were found in *E. sativa* mtDNA (Table 2), in the range of 71–88 bp in length. The A+T content of the tRNA genes is 48.81%, which is lower than the overall A+T composition of the mtDNA. Among these genes, tRNAs for 15 amino acids, including duplication of the methionine (Met) and triplication of the serine (Ser), are encoded. The genome lacks tRNAs for the amino acids alanine (Ala), valine (Val), phenylalanine (Phe), threonine (Thr) and arginine (Arg). To enable gene expression for protein synthesis in mitochondria, the missing tRNAs may be supplied by either the chloroplast or nuclear genomes [30].

Using ORF-Finder and BLAST searching, 50 ORFs longer than 100 codons were identified in the *E. sativa* mitochondrial genome. Among the 50 ORFs, only the *orf112, orf121, orf122*, and *orf275* have two copies. All others are single-copy ORFs. Most of the ORFs are between 300 and 500 bp in length, except for 10 ORFs that are longer than 500 bp, including the 1 200 bp *orf399* and the 1 911 bp *orf636*.

Subgenomic circles mediated by large repeats

Large repeats (>1 Kb) have been identified in most of the seed plants analyzed, except for white mustard (*Brassica hirta*) (Palmer and Herbo, 1987). The repeats in the *E. sativa* mitochondrial genome were analyzed. Three pairs of large repeats were identified, accounting for 13.48% of the genome. The large repeats were designated as R1, R2 and R3 (Table 3). R1 (10 320 bp) has a pair of large repeats in the opposite orientation, while R2 (4 864 bp) and R3 (1 513 bp) have a pair of large repeats in the same orientation. Large repeat R1 contains two ORFs, *orf112* and *orf122*, while R2 and R3 contain *orf275* and the *orf121*, respectively. No known protein coding gene was found in these large repeats.

Table 1. Gene content of the mitochondrial DNA of *Eruca sativa*.

Product group	Gene					
Complex I	*nad1*	*nad2*	*nad3*	*nad4*	*nad4L*	
	nad5	*nad6*	*nad7*	*nad9*		
Complex III	*cob*					
Complex IV	*cox1*	*cox2-1*	*cox3*			
Complex V	*atp1*	*atp4*	*atp6*	*atp8*	*atp9*	
Ribosome large subunit	*rpl2*	*rpl5*	*rpl16*			
Ribosome small subunit	*rps3*	*rps4*	*rps7*	*rps12*	*rps14*	
Cytochrome c biogenesis	*ccmB*	*ccmC*	*ccmFC*	*ccmFN1*	*ccmFN2*	
Intron maturase	*matR*					
Protein translocase	*tatC*					
rRNA genes	*rrn5*	*rrn18*	*rrn26*			
tRNA genes	*trnN*	*trnD*	*trnC*	*trnE*	*trnQ*	*trnG*
	trnH	*trnI*	*trnK*	*trnM*	*trnfM*	*trnP*
	trnW	*trnY*	*trnL*	*trnS*(3×)		

Table 2. Recognition of anticodons by tRNA genes found in the mtDNA of *Eruca sativa*.

Name	Type	Anticodon	Length(bp)	Orientation
chloroplast origin				
trnD	Asp	GTC	74	inverted
trnH	His	GTG	74	direct
trnL	Leu	CAA	81	direct
trnM	Met	CAT	73	direct
trnN	Asn	GTT	72	inverted
trnW	Trp	CCA	74	direct
trnS	Ser	GGA	87	inverted
mitochondrial origin				
trnfM	Met	CAT	74	inverted
trnG	Gly	GCC	72	direct
trnI	Ile	CAU	81	inverted
trnK	Lys	TTT	73	inverted
trnQ	Gln	TTG	72	direct
trnS	Ser	TGA	87	direct
trnS	Ser	GCT	88	direct
trnY	Tyr	GTA	83	inverted
trnC	Cys	GCA	71	inverted
trnE	Glu	TTC	72	inverted
trnP	Pro	TGG	75	inverted

Large repeats have been implicated in mediating high frequency, reciprocal DNA exchange that can result in subdivision of the genome into a multipartite configuration [31]. The formation of the multipartite structure of the *E. sativa* mitochondrial genome was predicted based on the assumptions of intra-molecular homologous recombination (Figure 2). Six isometric master circular (MC) genomic structures of the same length (including MC1 shown in Figure 1) could be produced by intra-molecular recombination between different repeat pairs. In addition, MC molecule 1 and 6 may generate four subgenomic circles, including two small circles of 129 447 bp (SC1) and 118 249 bp (SC2) via the pairwise large repeat R2, and another two small circles of 132 016 bp (SC3) and 115 680 bp (SC4) mediated by the pairwise large repeat R3. MC3 may produce SC1 and SC2, and MC4 may produce SC3 and SC4, mediated by the pairwise large repeat R1.

Sequence comparison between *E. sativa* and *B. napus* mtDNAs

We compared the sequences of the mtDNAs from *E. sativa* and *B. napus*. Most of the protein coding and RNA genes were conserved in length, except *ccmFN2* and *rrn18*. The 5′ portion of the coding region of *ccmFN2* in *E. sativa* mtDNA was quite different (Figure S1) and a 25-bp deletion in *rrn18* was found in *E. sativa* mtDNA (Figure S2) compared with that in *B. napus*. The *E. sativa* mitotype is devoid of *cox2-2*, compared with that of *B. napus*. 27 single nucleotide polymorphisms (SNPs) were detected in 14 genes when compared with *B. napus* (Table 4). Thirteen synonymous substitutions were found in *atp6*, *ccmB*, *cob*, *cox1*, *nad2*, *nad6*, *rpl2*, *rps3*, and *rps4*. Fourteen nonsynonymous mutants were found in 11 genes, including an S to N (199aa) switch in *atp1*, a V to I (18aa) and an H to F (51aa) switch in *atp6*, a P to L (107aa) switch in *ccmB*, an R to K (113aa) switch in *ccmFC*, an H to Y (285aa) switch in *cob*, a P to L (112aa) switch in *cox1*, an S to L (126aa) and an S to N (438aa) switch in *matR*, a C to R (72aa) switch in *nad2*, an S to L (29aa) switch in *rpl2*, an L to P (172aa) switch in *rpl5*, and an M to I (50aa) switch in *rps7*. Of these 27 SNPs, most were transitions and only three were transversion (G→T in *nad2*, T→A in *cox1*, and T→A in *atp6*). All tRNAs in the *B. napus* mitochondrial genome were detected in *E. sativa* mtDNA. However, the ORFs were quite different between these two mitotypes.

Table 3. Large repeats in the mtDNA of *Eruca sativa*.

No.	Type[a]	Size(bp)	Copy-1	Copy-2	Difference between copies	Identity
R1	IR	10320	77495-87814	176149-186468	identical	100%
R2	DR	4864	4083-8946	119763-124626	2 bp mismatch	99.95%
R3	DR	1513	1-1513	118250-119762	identical	100%

[a]DR and IR: direct and reverse repeats, respectively.

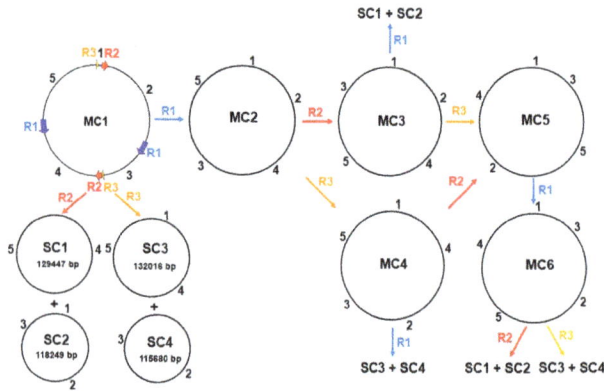

Figure 2. The multipartite mitochondrial genome structure of *Eruca sativa*. Schematic diagrams showing six master circles and four subgenomic circles. The three repeat pairs are shown in different colors. MC and SC mean master and subgenomic circles, respectively. Numbers outside circles indicate segments separated by repeat pairs.

Evolutionary relationships of the cruciferous mitotypes

To further illustrate the evolution of mitochondrial genomes within the Cruciferae family, the *E. sativa* mtDNA and other reported Cruciferous mtDNAs were compared using BLASTN [32]. *E. sativa* mtDNA was used as the reference sequence and similar regions in two or more mtDNA sequences were aligned. The alignable *E. sativa* sequence (93%) was 81% identical to that of *R. sativus* mtDNA. The sequence identity shared by the mtDNA of *E. sativa* and *Brassica* was more than 83%, with a coverage in the range of 83–85%. Only 63% of the *E. sativa* mtDNA matched those of *Arabidopsis thaliana*, with an identity of more than 68%, and the longest fragment was only 8.0 kb. This result suggested that the evolutionary relationship of mitochondrial genomes among *E. sativa*, the *Brassicas* and *R. sativus* is closer than that between *E. sativa* and *A. thaliana*.

In support of this hypothesis, a dot matrix analysis showed that the lengths of syntenic regions between *E. sativa* and *A. thaliana* are shorter than those between *E. sativa* and *Brassica* or *R. sativus*. Additionally, the distribution of syntenic regions between the mtDNAs of *E. sativa* and *A. thaliana* is more dispersed, and the identity is lower, than that between *E. sativa* and the *Brassica* mitotypes (Figure 3). Moreover, the phylogenetic relationships among the Cruciferae family (Figure 4) were inferred using the neighbor-joining method and 23 conserved genes among the

Table 4. SNP in protein-coding genes of mtDNA between *Eruca sativa* and *Brassica napus*.

Gene	Position from the start codon	nucleotide variation		Position from the first amino acid	amino acid change
		B.napus	*E.sativa*		
atp1	596	AGT	AAT	199	S→N
atp6	7	GAG	AAG	260	Synonymous
	559	GAA	AAA	76	Synonymous
	635	ATG	AAA	51	H→F
	735	GAC	GAT	18	V→I
ccmB	235	CCC	TCC	129	Synonymous
	303	AAG	AAA	107	P→L
	577	GGG	AGG	15	Synonymous
ccmFC exonB	337	GCG	ACG	113	R→K
cob	330	ATG	ATA	285	H→Y
	409	TCC	CCC	258	Synonymous
cox1	335	CCC	CTC	112	P→L
	702	TAC	TAT	234	Synonymous
	1466	CCT	CCA	489	Synonymous
matR	377	CGG	TGG	126	S→L
	1313	AGC	AAC	438	S→N
	1566	GGG	GGA	522	Synonymous
nad2 exonB	179	CAA	CGA	72	C→R
nad2 exonD	500	GAT	TAT	25	Synonymous
nad6	388	CGC	TGC	77	Synonymous
rpl2 exonB	47	CGA	CAA	29	S→L
rpl5	44	CAG	CGG	172	L→P
rps3 exonB	685	GGT	AGT	302	Synonymous
	823	CTT	TTT	256	Synonymous
rps4	391	CTT	TTT	233	Synonymous
rps7	298	CAT	TAT	50	M→I

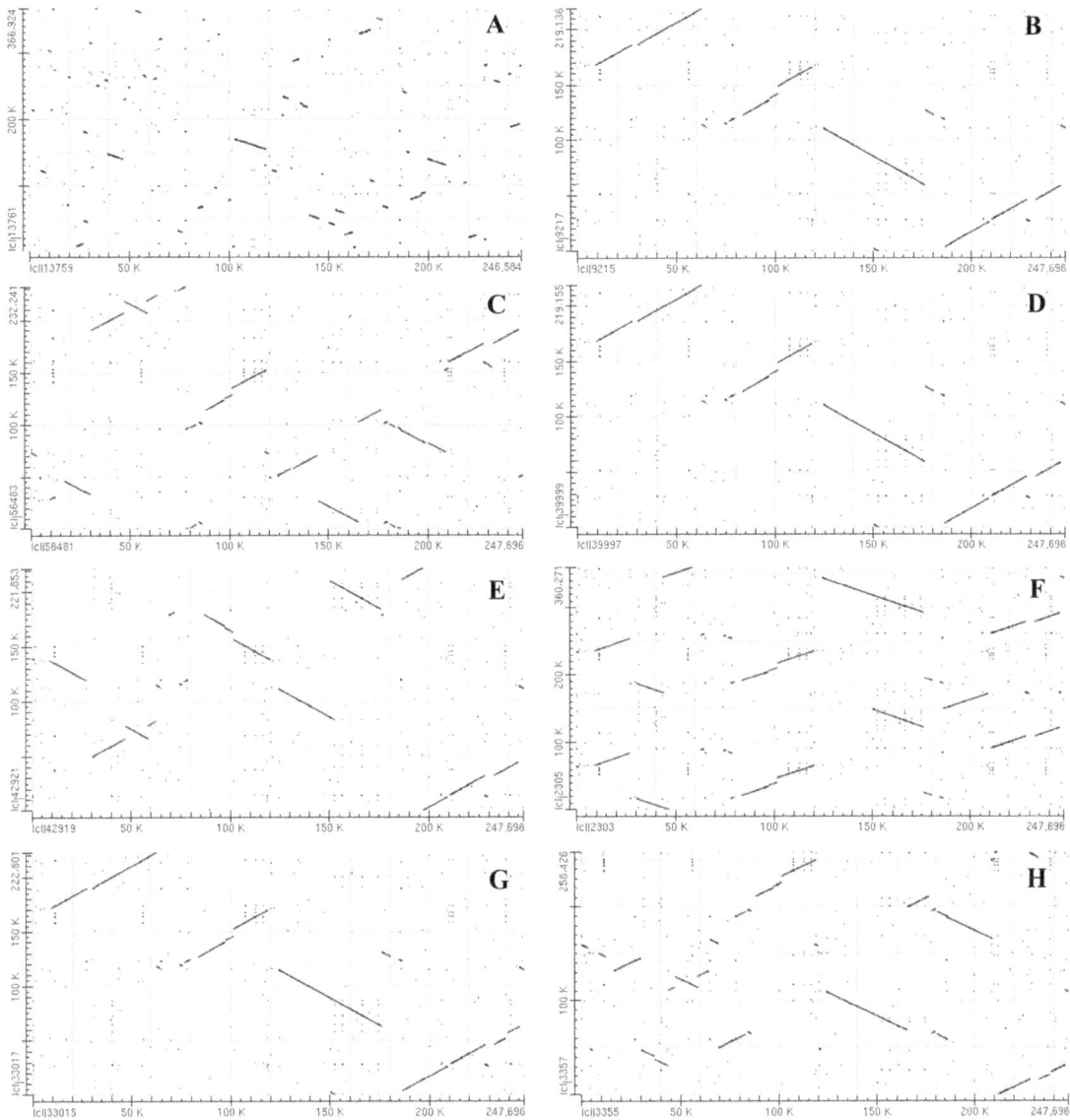

Figure 3. Dot matrix alignments of the *Eruca sativa* (x-axis) and other known cruciferous mtDNAs (y-axis). (A) *Arabidopsis thaliana* (tha), (B) *B. rapa* (cam), (C) *B. carinata* (car), (D) *B. juncea* (jun), (E) *B. napus* (nap), (F) *B. oleracea* (ole), (G) *B. napus* (pol), (G) *Raphanus sativus* (sat).

reported Cruciferae mitotypes. The results are mainly consistent with previous reports based on mitochondrial genome analysis [22] and strongly support the conclusion that *E. sativa* is more closely related to the *Brassica* species and *R. sativus* than to *A. thaliana*.

Discussion

The *Cruciferae* family is one of the largest dicot families of the flowering plant kingdom and includes several vegetable and oilseed crops, as well as several model species of great scientific, economic and agronomic importance [33]. Annotations for mitochondrial genomes from closely related species would improve the understanding of molecular evolution and phylogenetic relationships [34] in the *Cruciferae* family. *E. sativa*, a member of the Cruciferae family, is a conventional crop consumed as food and fodder. The economic potential of *E. sativa* lies in various other aspects, including the protein sources for edible

purposes, a potential source of industrial oil, an effective biological control of crop pests and traditional pharmacopoeia for various purposes [35]. To better understand this important crop, the mitochondrial genome of *E. sativa* was sequenced and annotated.

Cruciferae mitochondrial genomes are generally small (208–367 kb) compared with other seed plants. The *E. sativa* mt genome (248 kb) is larger than most *Brassica* mitotypes, but smaller than that of *B. oleracea* (360 kb) and *A. thaliana* (367 kb). Comparison of the *E. sativa* mtDNA with the *B. napus* mtDNA revealed that the *cox2-2* gene was absent from the *E. sativa* mt genome. This gene was also absent from the genomes of *B. oleracea*, *B. carinata*, and Ogura-cms-cybrid (oguC) rapeseed mitotypes [25,36]. A distinguishing feature of Cruciferae mitochondrial genomes is that the *ccmFN* genes are divided into two reading frames (*ccmFN1* and *ccmFN2*) [23]. The translation of *ccmFN2* has been confirmed in *A. thaliana* mitochondria, which demonstrated that *ccmFN2* was not a pseudo gene, although it lacks a classical ATG initiation codon [37]. Sequence alignments

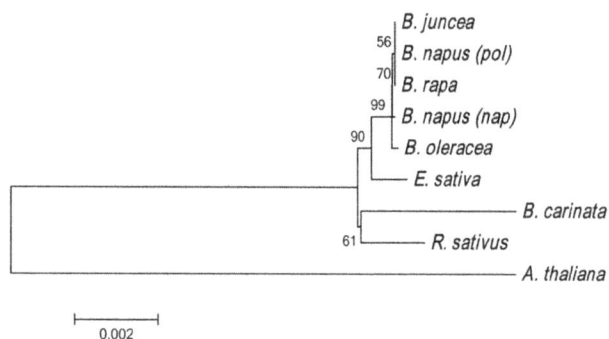

Figure 4. Phylogeny of nine cruciferous mitotypes. The phylogenetic tree was inferred using the neighbor-joining method based on the exons of 32 protein coding genes. The evolutionary distances were computed using the Kumar method, and the branch lengths are in units of synonymous substitutions per synonymous site. Evolutionary analyses were conducted in MEGA5.

of *ccmFN2* from reported Cruciferae mtDNAs showed that the first 45 bp of the putative *ccmFN2* gene in *E. sativa* mt genome is quite different from the *ccmFN2* gene in *Brassica* and *A. thaliana* mitotypes (Figure S1), suggesting that this non-conserved region may not be critical for gene function. However, the tryptophan-rich WWD domain in *ccmFN2*, which is responsible for heme binding [38], is conserved among these mitotypes.

The 5S and 18S rRNA genes in *E. sativa* mtDNA are closely linked, as they are in other plants, and the 26S rRNA gene is separated from the 18S and 5S by 26 459 bp. To elucidate the evolutionary origins of mitochondria, the ribosomal RNA genes have been extensively examined [39]. Sequence analysis of the *rrn18* gene from wheat, maize and soybean showed highly similarity between the plant mitochondrial *rrn18* genes and the eubacterial 16S rRNA, suggesting that there is a much slower rate of sequence change in plant mitochondria compared with their animal counterparts [40]. We compared the *rrn18* among the reported Cruciferae mitotypes and found a 25-bp deletion in *rrn18* in E. sativa mtDNA (Figure S2) compared with that in *Brassica* mitotypes. We also noticed a 46-bp deletion in *rrn18* within the same region of the *Brassica* mitotypes when compared with that in *A. thaliana* mtDNA. However, the overall nucleotide identities of the *rrn18* gene sequences were markedly high, from 89.50% between *E. sativa* and *A. thaliana* to 93.92% between *E. sativa* and the *Brassica* family. The nucleotide identity of the *rrn18* gene between *E. sativa* and *R. sativus* was 93.86% (Figure S2). This result is consistent with the results of the phylogenetic analysis based on 32 protein coding genes (Figure 4), which suggested that *E. sativa* is closer to *Brassica* and *R. sativus* than to *A. thaliana*.

18 tRNA genes were identified in *E. sativa* mtDNA, accounting for only 0.56% of the mitochondrial genome. Among them, six seem to be chloroplast derived, which exhibit high sequence identity (>99%) to their chloroplast counterparts. The chloroplast-derived *trnH-GTG, trnM-CAT, trnS-GGA, trnW-CCA, trnD-GUC*, and *trnN-GTT* genes, which are frequently found in mitochondrial genomes of angiosperms [15], were identified in the *E. sativa* mtDNA. An additional chloroplast-originating tRNA gene (*trnL-CAA*), which is found in the *R. sativus* and *Brassica* mitotypes [22], was also identified in *E. sativa* mitochondrial genome. This result indicated that mt tRNA genes are frequently transferred from chloroplast genomes during the evolution of angiosperms. However, another two gene (*trnP-GGG* and *trnQ-UUG*)

transfer events reported in dicots [41,42] were not found in *E. sativa*.

Genes with known functions are relatively conserved among the Cruciferae mitotypes, especially for the protein coding genes. However, the mitochondrial genomes structural differences are remarkable among the Cruciferae family. Multipartite structures of mtDNA mediated by large repeats have been commonly observed in plant species [43]. Direct electron-microscopic evidence of the coexistence of multipartite molecules in the plant mitochondrial genome has been found in tobacco [44]. The large repeat, RB, which is 2,427 bp in length and has been identified in most of the reported *Brassica* (except the oguC rapeseed) mitotypes, was not found in the *E. sativa* mtDNA. Instead, three pairwise large repeats were identified. Large repeat R1 in *E. sativa* mtDNA showed significantly high sequence similarity to the 6 580-bp large repeat R in *B. carinata* mitochondria (Figure S3). The 1 513-bp large repeat R3 showed 99% identity to the corresponding segments of the large repeat R2 in *B. oleracea* mitochondrial genome. Only 2% and 23% of R1 in *E. sativa* mtDNA showed high similarity (>83%) with the large repeats in *A. thaliana* and *R. sativus* mtDNA, respectively. The tripartite structure of the mitochondrial genome, including one master circle and two smaller subgenomic circles, has been reported in *Brassica* species (except the *ole* mitotype) and *R. sativus* [22,25,36]. The predicted multipartite structure of the mitochondrial genome in *E. sativa* is more complex than other Cruciferae species because of the three pairwise large repeats, including six master circles and four smaller subgenomic circles.

Conclusions

In this study, we reported the complete mitochondrial genome sequence of *E. sativa*, a member of the Cruciferae family. The *E. sativa* mtDNA is 247 696 bp and harbors 33 known protein coding genes, three rRNAs (5 S, 18 S, and 26 S rRNAs) and 18 tRNAs. In addition, the *cox2-2* gene is absent, the *ccmFN2* and *rrn18* genes have different lengths and 27 SNPs are involved in 14 protein coding genes in comparison with *B. napus* mtDNA. Reorganization of the genome may have occurred via three pairs of large repeats, resulting in a more dynamic structure of the *E. sativa* mtDNA compared with other cruciferous mitotypes. This may produce six master circles and four smaller subgenomic circles. The evolutionary relationships analysis among reported Cruciferous mitotypes revealed that the mitochondrial genome of *E. sativa* is divergent from *A. thaliana*, but closely related to those of *Brassica* and *R. sativus*. This study will improve our understanding of the *E. sativa* crop and the evolution of mitochondrial genomes within the Cruciferae family.

Supporting Information

Figure S1 Sequence alignments of *ccmFN2* from reported Cruciferae mtDNAs. The highly and partly conserved amino acids are shaded black or grey respectively. The black block diagram indicates the un-conserved region of *ccmFN2* in *E. sativa* mtDNAs compared to other reported Cruciferae mtDNAs.

Figure S2 Sequence alignments of *rrn18* from reported Cruciferae mtDNAs. The highly and partly conserved amino acids are shaded black or grey respectively. The black block diagram indicates the deletion region of *rrn18* in *E. sativa* mtDNAs compared to other reported Cruciferae mtDNAs.

Figure S3 Alignment of the large repeats in *Eruca sativa* mtDNA with the large 6.6 kb repeats in *car*. The alignment was made using Mauve. Blocks of the same color denote homologous regions; the *B. carinata* blocks above or below the middle line represent direct or inverted, respectively, compared with *E. sativa*. The extent to which a block is filled indicates the similarity of the syntenic region.

Author Contributions

Conceived and designed the experiments: RG. Performed the experiments: YW. Analyzed the data: YW PC SC QY. Contributed reagents/materials/analysis tools: JC MH RG. Contributed to the writing of the manuscript: PC.

References

1. Mower JP, Sloan DB, Alverson AJ (2012) Plant mitochondrial genome diversity: the genomics revolution. In: Jonathan FW, Johann G, Jaroslav D, Ilia JL, editors. Plant Genome Diversity Volume 1. Vienna: Springer. pp. 123–144.
2. Kubo T, Newton KJ (2008) Angiosperm mitochondrial genomes and mutations. Mitochondrion 8: 5–14.
3. Ogihara Y, Yamazaki Y, Murai K, Kanno A, Terachi T, et al. (2005) Structural dynamics of cereal mitochondrial genomes as revealed by complete nucleotide sequencing of the wheat mitochondrial genome. Nucleic Acids Res 33: 6235–6250.
4. Sloan DB, Alverson AJ, Chuckalovcak JP, Wu M, McCauley DE, et al. (2012) Rapid evolution of enormous, multichromosomal genomes in flowering plant mitochondria with exceptionally high mutation rates. PLoS Biol. 10: e1001241.
5. Tanaka Y, Tsuda M, Yasumoto K, Yamagishi H, Terachi T (2012) A complete mitochondrial genome sequence of Ogura-type male-sterile cytoplasm and its comparative analysis with that of normal cytoplasm in radish (*Raphanus sativus* L.). BMC genomics 13: 352.
6. Palmer JD, Herbo LA (1987) Unicircular structure of the *Brassica hirta* mitochondrial genome. Curr Genet 11: 565–570.
7. Alverson AJ, Wei X, Rice DW, Stern DB, Barry K, et al. (2010) Insights into the evolution of mitochondrial genome size from complete sequences of *Citrullus lanatus* and *Cucurbita pepo* (Cucurbitaceae). Mol Biol Evol 27: 1436–1448.
8. Woloszynska M (2010) Heteroplasmy and stoichiometric complexity of plant mitochondrial genomes—though this be madness, yet there's method in't. J Exp Bot 61: 657–671.
9. Palmer JD, Herbon LA (1988) Plant mitochondrial DNA evolved rapidly in structure, but slowly in sequence. J Mol Evol 28: 87–97.
10. Arrieta-Montiel MP, Shedge V, Davila J, Christensen AC, Mackenzie SA (2009) Diversity of the Arabidopsis mitochondrial genome occurs via nuclear-controlled recombination activity. Genetics 183: 1261–1268.
11. Sandhu APS, Abdelnoor RV, Mackenzie SA (2007) Transgenic induction of mitochondrial rearrangements for cytoplasmic male sterility in crop plants. Proc. Natl. Acad. Sci. U S A. 104: 1766–1770.
12. Pellny TK, Van Aken O, Dutilleul C, Wolff T, Groten K, et al. (2008) Mitochondrial respiratory pathways modulate nitrate sensing and nitrogen-dependent regulation of plant architecture in *Nicotiana sylvestris*. Plant J 54: 976–992.
13. Huang S, Millar AH, Taylor NL (2011) The plant mitochondrial proteome composition and stress response: conservation and divergence between monocots and dicots. In: Frank K, editors. Plant Mitochondria: Springer. pp. 207–239.
14. Diamond M, McCabe PF (2011) Mitochondrial regulation of plant programmed cell death. Plant Mitochondria. New York: Springer. pp. 439–465.
15. Chang S, Wang Y, Lu J, Gai J, Li J, et al. (2013) The mitochondrial genome of soybean reveals complex genome structures and gene evolution at intercellular and phylogenetic levels. PLoS One 8: e56502.
16. Tsunoda S, Hinata K, Gomez-Campo C (1980) Eco-physiology of wild and cultivated forms in Brassica and allied genera. Brassica crops and wild allies Biology and breeding. Tokyo: Japan Scientific Societies Press. 109–120 p.
17. Lakshmikumaran M, Negi MS (1994) Structural analysis of two length variants of the rDNA intergenic spacer from *Eruca sativa*. Plant Mol Biol 24: 915–927.
18. Ashraf M (1994) Organic substances responsible for salt tolerance in *Eruca sativa*. Biol Plantarum 36: 255–259.
19. Fahleson J, Lagercrantz U, Mouras A, Glimelius K (1997) Characterization of somatic hybrids between *Brassica napus* and *Eruca sativa* using species-specific repetitive sequences and genomic in situ hybridization. Plant Sci 123: 133–142.
20. Sastry ED (2003) Taramira (*Eruca sativa*) and its improvement-A Review. Agricultural reviews- Agricultural Reseach Communications Centre India 24: 235–249.
21. Unseld M, Marienfeld JR, Brandt P, Brennicke A (1997) The mitochondrial genome of *Arabidopsis thaliana* contains 57 genes in 366,924 nucleotides. Nat Genet 15: 57–61.
22. Chang S, Chen J, Wang Y, Gu B, He J, et al. (2013) A Mitochondrial Genome of Raphanus sativus and Gene Evolution of Cruciferous Mitochondrial Types. J Genet Genomics. 40:117–126.
23. Handa H (2003) The complete nucleotide sequence and RNA editing content of the mitochondrial genome of rapeseed (*Brassica napus* L.): comparative analysis of the mitochondrial genomes of rapeseed and *Arabidopsis thaliana*. Nucleic Acids Res 31: 5907–5916.
24. Chen J, Guan R, Chang S, Du T, Zhang H, et al. (2011) Substoichiometrically different mitotypes coexist in mitochondrial genomes of *Brassica napus* L. PLoS One 6: e17662.
25. Chang S, Yang T, Du T, Huang Y, Chen J, et al. (2011) Mitochondrial genome sequencing helps show the evolutionary mechanism of mitochondrial genome formation in Brassica. BMC genomics 12: 497.
26. Saitou N, Nei M (1987) The neighbor-joining method: a new method for reconstructing phylogenetic trees. Mol Biol Evol 4: 406–425.
27. Kumar S, Nei M, Dudley J, Tamura K (2008) MEGA: a biologist-centric software for evolutionary analysis of DNA and protein sequences. Brief Bioinform 9: 299–306.
28. Felsenstein J (1985) Confidence limits on phylogenies: an approach using the bootstrap. Evolution. 39: 783–791.
29. Chaw SM, Shih AC, Wang D, Wu YW, Liu SM, et al. (2008) The mitochondrial genome of the gymnosperm *Cycas taitungensis* contains a novel family of short interspersed elements, Bpu sequences, and abundant RNA editing sites. Mol Biol Evol 25: 603–615.
30. Fang Y, Wu H, Zhang T, Yang M, Yin Y, et al. (2012) A complete sequence and transcriptomic analyses of date palm (*Phoenix dactylifera* L.) mitochondrial genome. PLoS One 7: e37164.
31. Arrieta-Montiel MP, Mackenzie SA (2011) Plant mitochondrial genomes and recombination. In: Frank K, editors. Plant mitochondria. New York: Springer. pp. 65–82.
32. Zhang Z, Schwartz S, Wagner L, Miller W (2000) A greedy algorithm for aligning DNA sequences. J Comput Biol. 7: 203–214.
33. Anjum NA, Gill SS, Ahmad I, Pacheco M, Duarte AC, et al. (2012) The Plant Family Brassicaceae: An Introduction. In: Anjum N.A., Ahmad I, Pereira ME, Duarte AC,Umar S, Khan NA, editors. The Plant Family Brassicaceae. Berlin: Springer. pp. 1–33.
34. Yuan ML, Wei DD, Wang BJ, Dou W, Wang JJ (2010) The complete mitochondrial genome of the citrus red mite *Panonychus citri* (Acari: Tetranychidae): high genome rearrangement and extremely truncated tRNAs. BMC genomics 11: 597.
35. Slater SM (2013) Biotechnology of *Eruca Sativa* Mill. In: Shri MJ, Shourya DG., editors. Biotechnology of Neglected and Underutilized Crops. Netherlands: Springer. pp. 203–216.
36. Wang J, Jiang J, Li X, Li A, Zhang Y, et al. (2012) Complete sequence of heterogenous-composition mitochondrial genome (*Brassica napus*) and its exogenous source. BMC genomics 13: 675.
37. Rayapuram N, Hagenmuller J, Grienenberger JM, Bonnard G, Giegé P (2008) The three mitochondrial encoded CcmF proteins form a complex that interacts with CCMH and c-type apocytochromes in Arabidopsis. J Biol Chem 283: 25200–25208.
38. Goldman BS, Beck DL, Monika EM, Kranz RG (1998) Transmembrane heme delivery systems. Proc. Natl. Acad. Sci. U S A. 95: 5003–5008.
39. Gray MW (1982) Mitochondrial genome diversity and the evolution of mitochondrial DNA. Can J Biochem 60: 157–171.
40. Grabau EA (1985) Nucleotide sequence of the soybean mitochondrial 18S rRNA gene: evidence for a slow rate of divergence in the plant mitochondrial genome. Plant Mol Biol 5: 119–124.
41. Zhang T, Fang Y, Wang X, Deng X, Zhang X, et al. (2012) The complete chloroplast and mitochondrial genome sequences of *Boea hygrometrica*: insights into the evolution of plant organellar genomes. PLoS One 7: e30531.
42. Goremykin VV, Lockhart PJ, Viola R, Velasco R (2012) The mitochondrial genome of *Malus domestica* and the import-driven hypothesis of mitochondrial genome expansion in seed plants. Plant J 71: 615–626.
43. Park JY, Lee Y-P, Lee J, Choi B-S, Kim S, et al. (2013) Complete mitochondrial genome sequence and identification of a candidate gene responsible for cytoplasmic male sterility in radish (*Raphanus sativus* L.) containing DCGMS cytoplasm. Theor Appl Genet.126: 1763–1774.
44. Sugiyama Y, Watase Y, Nagase M, Makita N, Yagura S, et al. (2005) The complete nucleotide sequence and multipartite organization of the tobacco mitochondrial genome: comparative analysis of mitochondrial genomes in higher plants. Mol Genet Genomics 272: 603–615.

10

A *Picea abies* Linkage Map Based on SNP Markers Identifies QTLs for Four Aspects of Resistance to *Heterobasidion parviporum* Infection

Mårten Lind[1]*, Thomas Källman[2], Jun Chen[2], Xiao-Fei Ma[2], Jean Bousquet[3], Michele Morgante[4], Giusi Zaina[4], Bo Karlsson[5], Malin Elfstrand[1], Martin Lascoux[2], Jan Stenlid[1]

1 Department of Forest Mycology and Plant Pathology, Swedish University of Agricultural Sciences, Uppsala, Sweden, 2 Department of Ecology and Genetics, Evolutionary Biology Centre, Uppsala University, Uppsala, Sweden, 3 Institute for Systems and Integrative Biology, Université Laval, Québec City, Québec, Canada, 4 Dipartimento di Scienze Agrarie e Ambientali, Universita di Udine, Udine, Italy, 5 Skogforsk, Svalöv, Sweden

Abstract

A consensus linkage map of *Picea abies*, an economically important conifer, was constructed based on the segregation of 686 SNP markers in a F_1 progeny population consisting of 247 individuals. The total length of 1889.2 cM covered 96.5% of the estimated genome length and comprised 12 large linkage groups, corresponding to the number of haploid *P. abies* chromosomes. The sizes of the groups (from 5.9 to 9.9% of the total map length) correlated well with previous estimates of chromosome sizes (from 5.8 to 10.8% of total genome size). Any locus in the genome has a 97% probability to be within 10 cM from a mapped marker, which makes the map suited for QTL mapping. Infecting the progeny trees with the root rot pathogen *Heterobasidion parviporum* allowed for mapping of four different resistance traits: lesion length at the inoculation site, fungal spread within the sapwood, exclusion of the pathogen from the host after initial infection, and ability to prevent the infection from establishing at all. These four traits were associated with two, four, four and three QTL regions respectively of which none overlapped between the traits. Each QTL explained between 4.6 and 10.1% of the respective traits phenotypic variation. Although the QTL regions contain many more genes than the ones represented by the SNP markers, at least four markers within the confidence intervals originated from genes with known function in conifer defence; a leucoanthocyanidine reductase, which has previously been shown to upregulate during *H. parviporum* infection, and three intermediates of the lignification process; a hydroxycinnamoyl CoA shikimate/quinate hydroxycinnamoyltransferase, a 4-coumarate CoA ligase, and a R2R3-MYB transcription factor.

Editor: Mingliang Xu, China Agricultural University, China

Funding: Financial support was received from the European Community's Sixth Framework Programme, under the Network of Excellence Evoltree (www.evoltree.eu/), by the Seventh Framework Programme (FP7/2007-2013) under the grant agreement n° 211868 (Project Noveltree, www.noveltree.eu/), and by the Swedish Research Council FORMAS grant dnr 217-2007-433 (www.formas.se/en). Support also came from the Biodiversa projects Linktree (www.igv.fi.cnr.it/linktree/) and TipTree, grant ANR-12-EBID-0003 (www.agence-nationale-recherche.fr/en/anr-funded-project/?tx_lwmsuivibilan_pi2[CODE]=ANR-12-EBID-0003), and from the Swedish Foundation for Strategic Research (SSF), grant number R8b08-0011 (www.stratresearch.se). The funders had no role in study design, data collection and analysis, decision to publish, or preparation of the manuscript.

Competing Interests: The authors have declared that no competing interests exist.

* Email: marten.lind@slu.se

Introduction

Norway spruce [*Picea abies* (L.) Karst.] is ecologically one of the most important conifer species in Europe, naturally present in central Europe across the Alps, the Carpathians and the Balkans, and in northern Europe from the west coast of Norway to far into the Russian mainland [1]. It is also one of the most important species economically in European and Swedish forestry, constituting 41% of the standing Swedish tree volume [2]. This dominance means that vast areas of forest land are largely monocultural, covered by only *P. abies*. Albeit favouring a single species has been a sound strategy in terms of generating forest revenue, this approach has also been extremely beneficial to the forest pathogen *Heterobasidion parviporum*, which is a highly competitive early colonizer of fresh wounds and newly cut stumps. *H. parviporum* is a causal agent of annosum root rot, a serious and very common fungal disease in conifer forests of the Northern

Hemisphere [3] causing yearly losses to European forest owners exceeding € 790 million yearly in growth reduction and devaluation of timber [4]. In 1986, the average incidence of root rot in Swedish Norway spruce trees was estimated to 15% [5], and the frequency has later been reported to increase in managed forests with ~23% per decade [6].

Trees less susceptible to root rot would be highly coveted by the forest industry. Nevertheless, most breeding endeavours so far have been directed at high tree growth, with little or no effort targeted at breeding for trees resistant to *H. parviporum* infection. Still, resistance breeding is potentially fruitful, as it has been shown that *H. parviporum* resistance is a genetically variable trait and does not affect growth rate negatively [7]. It would be an important step towards efficient resistance breeding to have molecular resistance-associated markers as well as better knowledge of whether resistance is a multifactorial trait or instead, mainly controlled by a single locus. One way to shed light on this is through quantitative trait loci (QTL) mapping of *P. abies* resistance to *H. parviporum* infection.

Mapping of QTLs in genomes such as those of conifers of the *Pinaceae* family is fraught with various problems. The huge size of the *Pinaceae* genomes alone (~14–37 Gb) [8–9] has its drawbacks, but the extremely low gene density in combination with great regions of repeatable elements with low recombination rate [10] is an even greater obstacle in terms of converting identified QTLs to underlying sequence variation. In *P. abies*, the haploid genome is estimated to 19.6 Gb and the number of transcribed genes to 28,354 [11], making the average distance between two neighbouring genes 0.69 Mb. The proportion of various repetitive elements in the genome has been estimated to 70% [11], making whole genome assembly a difficult task where very few, if indeed any, scaffolds are likely to be large enough to contain more than a single gene. This makes it very hard to know the genome sequence between two given *P. abies* loci. In terms of making sense out of a QTL, this means that only markers originating from a known position are of potential use; anonymous markers are almost impossible to translate unless they originate from the same scaffold as a sequence-tagged marker. It also means that if several markers within a QTL are significantly associated with the trait, only the ones originating from the gene actually controlling the trait will carry any immediately useful information. This is a significant limitation compared to QTLs in maps based on less fragmented genome sequences since the search for potential candidate genes is more likely to be rewarding if the entire QTL is located to a single, large scaffold. On the other hand, the low gene density also means that there will be a limited number of possible candidate genes in the vicinity of each marker associated with the trait. Despite these potential obstacles, QTL analyses have been applied previously to map host resistance in similar pathosystems. For example, *Pinus radiata* resistance to Dothistroma needle blight has been found to be controlled by multiple QTLs [12], and two major QTLs explained 52% of the variance in *Eucalyptus globulus* resistance to *Mycosphaerella cryptica* infection [13]. Furthermore, Eucalyptus inter-specific hybrids have been used to show a multifactorial control of resistance to *Puccinia psidii* rust infection [14], and rust resistance in both *Populus* and *Salix* spp. is based on several loci [15–16].

Several linkage maps of *P. abies* have been constructed already, but only one [17–18] has been saturated enough to create a number of linkage groups corresponding to the haploid number of chromosomes. However, of the 768 markers in that map, only 32 were derived from ESTPs and thus originating from known genes, crucial for making use of QTLs in fragmented genome assemblies. Another *Picea* map, based on a *Picea mariana* x *Picea rubens* hybrid population, consisted of 835 positioned markers, of which 318 originated from known genes [19]. In terms of mapped genes, the most comprehensive conifer maps to date has been constructed for *Picea glauca* [20], containing 1743 SNP markers from transcribed genes, and twice for *Pinus taeda* [21–22], based on 2841 and 2393 segregating genes.

The goal of this study was to map resistance QTLs in *P. abies* involved in response to four aspects of *H. parviporum* infection. The map was constructed using 247 full-sib progenies from a cross between two *P. abies* parents, based on the segregation of 686 polymorphic SNP markers, all derived from known *P. abies* genes. All four traits were controlled by multiple loci, explaining between 4.6 and 10.1% of the phenotypic variance in the mapping population. If the effect of all loci is additive, the variation within the four traits would be explained by in total 12–25.7%. These are the first resistance QTLs identified against this fungus, and also the first identified against any pathogen in *P. abies*. Naturally, the

SNPs underlying these QTLs only represent a fraction of the actual genes and might not be the ones causal for the traits. Nevertheless, the SNPs originated from genes with homologs to at least four plausible candidate genes. One of these, a leucoanthocyanidine reductase, has previously been shown to be upregulated in *P. abies* in response to *H. parviporum* infection [23].

Materials and Methods

The mapping population and resistance assays

The biological material used to construct the linkage map and conduct the resistance assays was a family of originally 251 progenies (later adjusted to 247), with six ramet cuttings from each original ortet, from a cross between the *Picea abies* parents S21K7622162 and S21K7621678. The biological material and the virulence assays have been described previously [24]. In short, the resistance of the progeny was estimated by infecting four ramets of each progeny, as 2-year-old potted plants, with a *Heterobasidion parviporum* (strain Rb 175) infested wooden plug through a 5×5 mm cambial wound. The entire assay was conducted in the greenhouse. After 4 weeks, death of the inner bark (lesion) was noted and stems were cut aseptically into 5-mm lengths that were incubated in moist condition. Sapwood growth of *H. parviporum* was assessed from the presence of conidiophores on sections of the stem. These data were treated as best linear unbiased prediction (BLUP) values [24]. Resistance was estimated by measuring lesion lengths in the bark surrounding the wound, fungal growth within sapwood up- and down-stem from the wound and fungal exclusion, i.e. the ability of the host to exclude the fungus once it has entered the sapwood. The fungal exclusion concept was based on the observation that after 4 weeks infection, the fungus was not always continuously present in the sapwood all the way from the wound to its farthest frontline but rather seemed to lose substrate to the host, whose ability to reclaim sapwood thus can be seen as a resistance trait. Exclusion was calculated as the cleansed proportion of the infected sapwood: exclusion = x/y, where x is the distance of fungus-free sapwood between the wound and the fungus, and y is the total distance of sapwood between the wound and the fungal front. Both x and y were calculated as mean values across all four ramets.

Arnerup et al [24] observed no fungal material in 27% of the infected plants and argued that this might either be due to infections failing at an early stage and never reaching the sapwood, or due to a later, full exclusion of the infection (x = y). They further observed that excluding or including the potentially failed infections had little effect on the overall results. To investigate whether these 27% could be genetically explained as the ability of the host to stop the fungus from entering the wound altogether, this was also mapped as a separate trait. Infection prevention was quantified as the proportion of ramets of each progeny that did not contain any conidiophores after incubation and had no more than 2 mm inner bark lesion up- or down-stem from the wound after 4 weeks of infection.

DNA extraction and SNP marker design

Genomic DNA was extracted from spruce needles of the two ramets remaining from each progeny after the resistance assays, using the DNeasy Plant Mini Kit (QIAGEN, Germantown, MD). An Illumina 3072 SNP Golden Gate Assay was developed by merging SNPs from a number of different resequencing and genotyping projects. All markers included in the assay had a score higher than 0.6 in the Illumina OPA design. The majority of SNPs (1879) were originally identified in and designed for *Picea glauca* and later also tested on and found to be variable in a small number of *P. abies* individuals [25]. The rest of the SNPs came from three different sources; 250 from a previously developed 768 SNP Golden Gate Assay [26], 269 from sequencing pooled PCR products using Illumina next-generation sequencing technology (Ma XF, Zaina, G, Källman T, Chen J, Morgante M, Lascoux M, unpublished), and 674 SNPs identified in a single individual subjected to mRNA sequencing using Illumina technology [27].

Linkage analysis and map construction

The Illumina Golden Gate assay was used to genotype the mapping population according to standard procedures [28] at the SNP Technology Platform, Uppsala University (http://molmed. medsci.uu.se/SNP+SEQ+Technology+Platform/Genotyping/). The SNP markers formed three clearly distinguishable clusters (A/A, A/a, a/a), and those polymorphic between the parental trees were used as mapping data. Their segregation patterns within the 247 progenies were observed and analysed using JoinMap 3.0 (Kyazma) [29] and visualized using MapChart 2.1 (Plant Research International) [30]. To be accepted, a marker had to be scored as present or absent (i.e., not as missing data) in at least 80% of the individuals. Markers with a segregation pattern deviating strongly from the expected ratio ($P<0.005$, χ^2) were also omitted. The segregation data were coded as CP, i.e. a population resulting from a cross between two heterogeneously heterozygous and homozygous diploid parents. Linkage groups were determined using pairwise comparisons at minimum likelihood of odds (LOD) value of 4 and a recombination frequency threshold 0.4. As no previous information on marker order existed, the internal order within groups was accepted as presented by the Map 2-function (regression mapping, Kosambi's mapping function, adding markers one at a time, accepting the position that results in the best goodness-of-fit for the map, discarding markers that results in a *jump* in χ^2 for goodness-of-fit of 5 or more), after discarding any markers that contributed to the *mean* χ^2 for for goodness-of-fit of the group by more than 4. Finally, to screen for double recombination events, an average genotype probability (–Log10(P)) threshold value of 0.3 was set for each group, eliminating all markers with a higher average. Similarly, markers involved in highly improbable genotypes (-Log10(P) >2.0) in more than 10 of the 247 individuals were also removed.

From each gene carrying more than one marker, every marker but the best genotyped one (defined by least missing data and least distorted segregation) was deemed as redundant. Markers were deemed as originating from the same genes as another marker, if they shared identical accession numbers in GenBank, TAIR or the *Picea glauca* genome and mapped within 10 cM of each other. Even though recombination in *P. abies*, as in many plants, seem to occur mainly within the genes [31–33], the population of the present study is not big enough and the SNP markers not numerous enough to expect many recombination events between these redundant markers. As they are expected to have identical genotypes, the discordance can be used as a way of measuring genotyping error in the data. The effect of this genotyping error was estimated by summing the largest possible distance between two markers on every gene with redundant markers and divide by the number of such genes. The QTL ranges were subsequently adjusted at both ends by this mean in order to compensate for the genotyping error.

Genome length was estimated using method 4 of Chakravarti *et al.* [34], in which the observed length in cM of each linkage group is multiplied by $(m+1)/(m-1)$, with m being the marker count for each group. Genome coverage was calculated as (total observed genome length)/(total estimated genome length), and as the

proportion c of the genome within d cM of a marker, using the formula $c = 1 - e^{-2dn/L}$ [35] with L being the estimated genome length and n the number of mapped markers. As these methods assume normal distribution of data, we tested for this by dividing every linkage group into 10 cM fragments and counting the number of markers, x, in each, with x ranging from 0 to 15. A Poisson distribution function, $P(x) = \mu^x e^{-\mu}/x!$, was used to compare this to a normal distribution, with $P(x)$ being the probability of x markers per 10 cM fragment and μ the genome-wise average number of markers per fragment. The expected distribution of intervals with x markers was calculated by multiplying $P(x)$ for each x (0–15) with the total number of 10 cM intervals in the map. Finally, a Kolmogorov-Smirnov test for two populations was conducted to decide whether the observed distribution of markers was likely similar to the expected distribution [36].

QTL analyses

BLUP values for lesion lengths and fungal growth in sapwood, and mean values for excluded proportion of total infection length [24], were used in the QTL analyses, whereas raw proportion data was used for the infection prevention trait.

The likelihood of resistance QTLs was determined using the MapQTL (Kyazma) mixture model method [37] according to the following approach. First, interval mapping [38] (200 iterations) was employed for every 1 cM throughout the linkage map in order to identify markers significantly associated with the trait. The significance was calculated by a permutation test ($P<0.05$, χ^2) (5000 permutations). QTL peaks significant at either genome or linkage group level were picked for further study. Then, the marker closest to the respective regional QTL peak was tentatively picked as cofactor and tested using the restricted MQM-mapping algorithm (as there were multiple putative QTLs for each trait). If this altered the position of the QTL peak, a new cofactor was picked. This process was repeated until every designated cofactor was located as closely as possible to the peak of their respective QTL. The confidence interval for each resulting QTL was determined as a decrease in 1 LOD unit on both sides of the QTL LOD-peak, adjusted at both ends by the mean effect of the suspected genotyping error (see above). Finally, a Kruskal-Wallis test was conducted to determine the individual level of association between the markers within the confidence interval and the examined traits. A QTL was only deemed significant if it contained at least one significant marker according to the Kruskal-Wallis test ($P<0.05$, marker level). Such markers were noted as potentially causative for the trait along with the designated cofactor (Table 1). All calculations were performed using MapQTL 4.0 (Kyazma) [37].

After the significant QTLs were identified, the false discovery rate (FDR) [39] was estimated in order to determine the rate of type I errors (i.e. false discoveries). The individual *P*-values for every significant QTL, for the individual traits as well as all pooled together, was listed as $P(1)$, $P(2)$ … $P(m)$, where $P(1)$ was the smallest and $P(m)$ the largest. According to the Benjamini-Hochberg (BH) procedure [39], the FDR is controlled at the specified significance level q if $P(i) < q*i/m$, where $P(i)$ is the *P*-value of a given QTL in the list, i is the numerical listed position of that QTL (i.e., between 1 and 13 for the pooled traits, since 13 QTLs were found) and m is the number of QTLs considered.

Table 1. Numeric data and SNP content for each resistance QTL found in the *Picea abies* linkage map.

| Trait | Linkage group | QTL interval at -1LOD (cM) | | Cofactor | % explained[1] (at cofactor) | LOD thresholds P<0.05. χ2 | | SNPs in QTL interval | KW Significance[2] | Position | Accession/TAIR number | BlastX homology[3] |
		Interval	LOD peak (cofactor peak)			Linkage group level	Genome level					
Infection prevention												
	1	152.4-158.8	3.95	GQ03113-N13.1.1005	5.3 (5.1)	3.2	4.4	08pg07937c	0.05	158.4	BT101292	**hydroxycinnamoyl CoA shikimate/quinate hydroxycinnamoyltransferase**
								GQ03113-N13.1.1005	0.05	158.6	BT107879	uncharacterized protein
	2	64.0-82.5	3.37 (3.33)	FCL324Contig1-740	5.9 (5.8)	3.1	4.4	GQ03010-E07.1.207	0.05	69.9	BT106774	5-alpha-reductase
								FCL324Contig1-740	-	73.5	AT3G54260.1	TBL-gene family member
	11	96.2-119.4	3.83 (3.68)	PabiesFT1-1251	7.2 (6)	3.0	4.4	208PG15708n	0.005	96.8	BT100755	**pectin methylesterase**
								PabiesFT1-1251	0.01	114.2	BT115191	Mother of FT-like protein
								GQ0255-K05.2.1102	0.05	114.3	BT103501	**R2R3-MYB transcription factor MYB11**
Exclusion												
	1	157.5-161.8	3.59 (3.53)	02776-N18-3098	4.8 (4.6)	3.2	7.6	FCL3309Contig1-1061	0.05	161.3	AT3G44730.1	kinesin-like protein 1
								02776-N18-3098	0.01	161.3	AT1G09160.1	protein phosphatase 2C
	2	93.7-99.8	3.31 (3.31)	GQ0041-H19.1.1034	5.1 (5.1)	3.1	7.6	FCL1145Contig1-703	0.05	96.5	AT5G50920.1	ATP-dependent Clp protease ATP-binding subunit/ClpC
								GQ0041-H19.1.1034		97.7	BT100742	Cu/Zn Superoxide dismutase
	3	74.4-80.1	3.96 (3.96)	GQ04004-O18.1.247	5.7 (5.3)	3.2	7.6	GQ04004-O18.1.247	0.0005	75.7	BT18309	**4-coumarate:CoA ligase**
	6	133.4-158.9	5.24 (5.23)	GQ02803-C08.1.1246	10.1 (8)	3.2	7.6	GQ03404-I14.1.173	0.005	136.3	BT113308	uncharacterized protein
								GQ03302-B08.1.275	0.005	137.1	BT111676	uncharacterized protein
								GQ02803-C08.1.1246	0.005	140.3	BT103859	uncharacterized protein
								GQ03412-L23.1.1340	0.05	144.9	BT113543	Elongation factor Tu, chloroplastic
Lesion length												
	8	0.0-5.2	5.28 (5.15)	0-9749-01-9969-co	6.7 (6.2)	2.8	4.0	0-12329-01-11666-	0.01	0.0	AT1G49760.2	polyadenylate-binding protein
								0-9749-01-9969-co		0.9	AT5G49760.1	**Leucine-rich repeat protein kinase family protein**
								02739-B22-309	0.01	4.2	AT4G21450.3	PapD-like superfamily protein
	9	32.9-61.7	3.64 (3.64)	NODE-3044-length-2	5.3 (5)	2.9	4.0	NODE-3044-length-2	0.05	47.3	PUT-175a-Picea_glauca-12549270	uncharacterized protein
Growth in sapwood												
	2	135.8-142.7	3.42 (3.42)	GQ02827-E23.1.1092	4.6 (4.2)	3.1	4.3	GQ02827-E23.1.1092	0.05	141.2	BT105733	short chain dehydrogenase

Table 1. Cont.

Trait Linkage group	QTL interval at -1LOD (cM)			% explained[1] (at cofactor)	LOD thresholds P<0.05. χ^2		SNPs in QTL interval	KW Significance[2]	Position	Accession/TAIR number	BlastX homology[3]
	Interval	LOD peak (cofactor peak)	Cofactor		Linkage group level	Genome level					
6	42.9–53.0	5.79 (5.75)	WS-2.0-GQ0064.T	8.4 (8.1)	3.2	4.3	WS-2.0-GQ0064.T	0.005	46.3	BT105286	uncharacterized protein
	176.8–182.6	3.54 (3.54)	GQ03204-B13.1.1304	4.8 (4.8)	3.2	4.3	GQ03719-P11.1.1591	0.05	178.8	BT116508	magnesium-protoporphyrin IX monomethyl ester [oxidative] cyclase, chloroplastic [Vitis vinifera]
							GQ03204-B13.1.1304	0.01	182.6	BT109050	**leucoanthocyanidin reductase**
9	103.7–113.1	5.27 (5.27)	PabiesFT4pr-2046	7.3 (7)	3.1	4.3	PabiesFT4pr-2046	0.005	110.4	BT112861	flowering locus T-like/terminal flower1-like protein
							GQ03615-J11.1.673	0.05	112.0	BT115393	G2/mitotic-specific cyclin
							0276-A13-934	0.01	112.9	AT2G36970.1	UDP-glycosyltransferase superfamily

[1]Percentage of phenotypic variation explained by this QTL at its LOD peak. In parenthesis the percentage explained at the cofactor.

[2]Significance of association according to Kruskal-Wallis-test. $0.05 = P < 0.05$, χ^2

[3]**Bold** indicates homology (BlastX, MaxScore >378, E-value < 1e-125) to Pinaceae genes, *Italics* homology (BlastX, MaxScore >83.2, E-value < 1e-15) to angiosperm genes, with a suggested role in defence response

Table 2. Statistics for SNP markers segregating between the parental trees S21K76221 and S21K7621678.

	Heterozygous for both parents	for S21K7622162	for S21K7621678	Total
Polymorphic[1]	137	292	309	738
Distorted (0.005<P<0.05)	7	16	12	35
Positioned	118	278	290	686
Unlinked	19	14	19	52

[1]Remaining after removing heavily distorted loci (P<0.005) or missing data in >20% of the mapping populatio

Results

Polymorphic SNP markers segregating in the progeny

Screening the 247 progenies and the parental trees S21K7622162 and S21K7621678 against the Illumina 3072 SNP Golden Gate Assay discovered that 1014 assays failed to produce useful results. This resulted in 2058 (67%) successfully scored SNP markers. Of these, 874 (43%) SNPs proved to be polymorphic among the parental genomes and useful as genetic markers. The marker names, sequences and corresponding accession numbers in GenBank, TAIR and the *Picea glauca* genome are presented in Table S1. Analysing the SNP markers revealed that the 874 markers originated from 769 unique transcribed genes. Sixty-nine of these genes gave rise to more than one SNP marker each. In total, 174 of the 874 SNP markers came from such multi-SNP generating genes, i.e. the dataset had 105 redundant markers. The most widely spaced redundant markers of each corresponding gene mapped on average 0.97 cM of each other. Since not enough recombination events should occur within these genes to justify this figure, redundant markers were removed so that each of the 769 transcribed genes were represented by only one SNP marker.After further trimming the marker set by removing those that were heavily distorted (*P*< 0.005, χ^2) or poorly characterized (missing data in >20% of the population), 738 markers remained. Of these, 292 SNPs were heterozygous in S21K7622162 (segregating 1:1), 309 in S21K7621678 (1:1), and 137 were heterozygous in both parents (segregating 1:2:1) (Table 2). Thirty-five SNP markers (4.7%) segregated at a ratio distorted from the expected 1:2:1 (7 markers) or 1:1 ratio (28 markers) (0.005<*P*<0.05, χ^2).

The SNP analysis showed that eight individuals (four pairs) of the originally 251 members of the mapping population were between 97.5% and 99.3% identical in terms of SNP genotypes. As this probably was due to DNA contamination at some stage, one member of each pair was omitted, resulting in 247 trees.

Twelve well supported linkage groups

JoinMap and MapQTL input files are included as Tables S2 and S3.

The markers that proved to be involved in many improbable genotypes (average −Log10(P) for linkage with neighbouring markers above 0.3, or −Log10(P) above 2.0 in more than 10 individuals of the population) were deemed suspicious and omitted during the mapping process. After this final screening, which removed 20 markers, JoinMap 3.0 (Kyazma) organized 686 of the remaining 718 into a consensus map of twelve linkage groups (Figures 1-3) as per result of the Map 2-function. No marker in the resulting linkage groups contributed with more than 4 to the mean χ^2 of goodness-of-fit for the group. Although a minimum LOD score of 4 was arbitrarily determined as sufficient to accept a linkage group as significant, this threshold was not relevant as all

twelve groups remained stable up to a LOD score of 10. The groups varied in size from 110.7 to 186.2 cM and included from 32 to 73 markers (Figures 1-3, Table 3). The twelve linkage groups covered 1889.2 cM in total and had an average density of 2.8 cM/marker, varying across the groups from 2.2 to 3.5 cM/ marker. The genome wide physical size to genomic distance ratio was 10.4 MB/cM, if assuming the Norway spruce genome to be 19.6 GB large [11]. Across the groups, assuming each chromosome is 1.63 GB large, the ratio varied between 8.8 and 14.7 MB/ cM (Table 3).

Genome length was estimated as 1958 cM, using the (m +1)/(m − 1) method [34]. The map thus covers 96.5% of the estimated genome and has a 97.0% probability that any given locus in the genome lies at most 10 cM from a positioned marker (determined from c – see Material and Methods). Whether the observed distribution of markers across 10 cM blocks of the genome was different from the expected random distribution was tested with a Kolmogorov-Smirnov test. The difference was small (D = 0.08) and non significant (P = 0.53). The distributions are displayed in Figure 4.

Of the 35 markers displaying distorted segregation, 3 were unlinked while 32 (91.4%) were distributed over 9 of the 12 the linkage groups (Table 3). Six groups had just 1 or 2 distorted markers, while the other three groups (1, 2 and 6) had 7, 12 and 5 markers, respectively. In three cases two or more neighbouring distorted markers constitute a cluster. On LG 1, five distorted markers were positioned between 20.4 and 23.3 cM, whereas on LG 2, three others were clustered between 84.9 and 87.4 cM. Two distorted neighbouring markers were found on LG 6 (at 35.1 and 36.9 cM), while the other 14 on these groups were distributed without any other such marker as their closest neighbour.

Resistance variation detected in the population

Two-hundred and fifty-one full sibling plants stemming from a cross between the *P. abies* parents S21K7622162 and S21K7621678 were used for scoring resistance against *H. parviporum* infection. The resistance variation among the progeny, expressed as lesion lengths around the wound and fungal growth within the sapwood, has been described previously [24]. In addition, the exclusion of fungus from the host was calculated as the proportion of fungal growth in sapwood eliminated from the wound and outward, and the ability to prevent infection was calculated as the proportion of ramets completely lacking conidiophores after 4 weeks of infection. The distributions of the resistance over the progeny for all traits are displayed in Figure 5.

Resistance mapping reveal multiple QTLs

As visible from Figures 1-3 and Table 1, the traits exclusion (purple in Figures 1–3), lesion length (green), growth in sapwood (red) and infection prevention (orange), were associated with four, two, four, and three specific QTL regions, as per defined by a

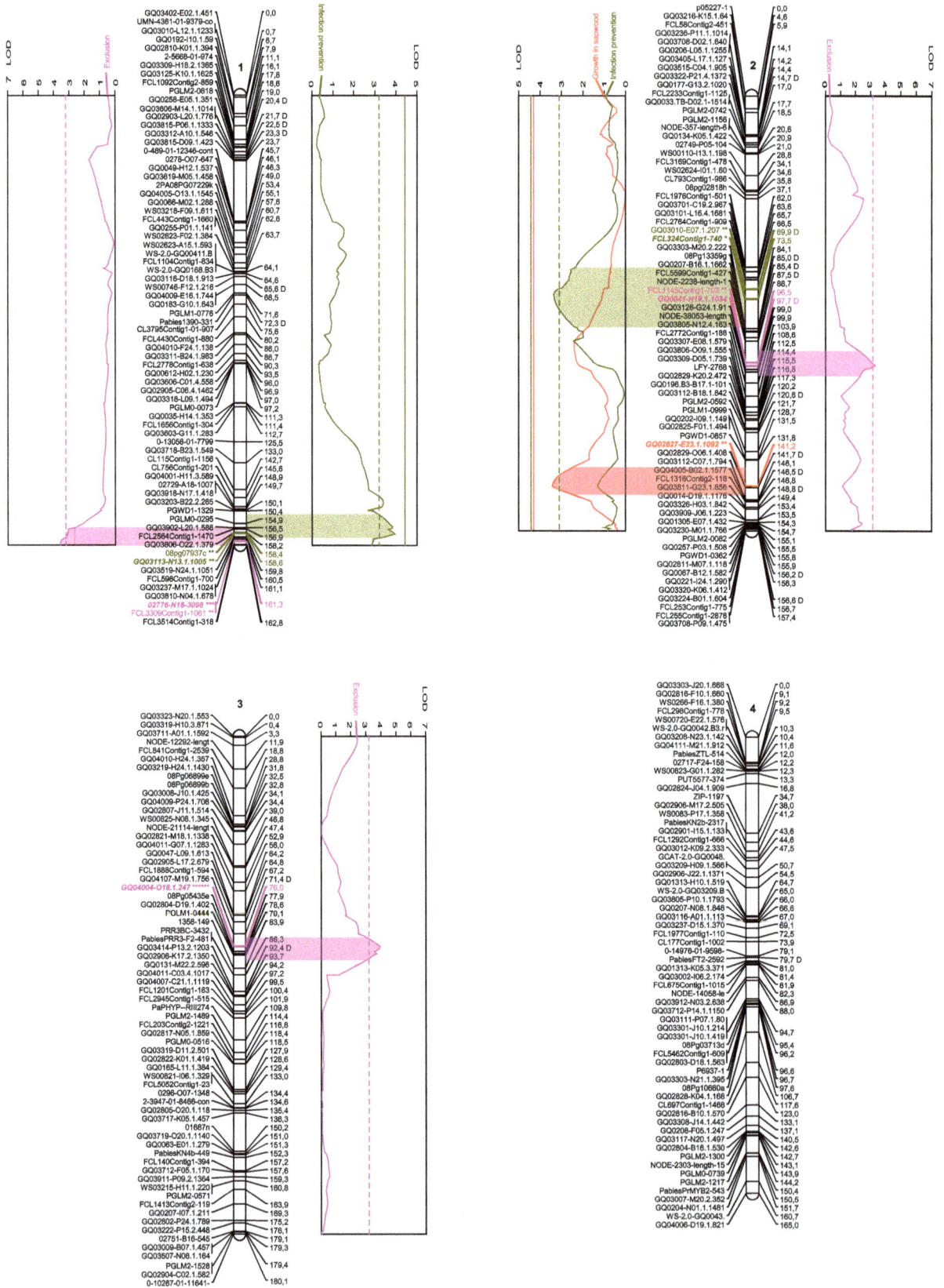

Figure 1. Linkage groups 1-4 of the *Picea abies* genome and the QTLs for various resistance traits. Names of the SNP markers are displayed on the left of the linkage groups. Genetic distance (cM) is indicated on the right. SNP markers are described in detail in Table S1. Graphs denote QTL effects for the current group. Red curves indicate fungal growth within sapwood, purple fungal exclusion and orange curves indicates infection prevention. Complete and dashed vertical lines describe 0.1% and 5% levels of significance for the individual trait and group. Wide colored

areas between curve and group show the QTL confidence interval based on a 1 LOD drop from the QTL peak. The colored marker denotes SNPs under the QTL confidence interval, the stars level of significance according to the Kruskal-Wallis test (ranging from p<0.1 (*) to p<0.0005 (******) and **bold** style the designated cofactors. "D" after a marker name indicates segregation pattern deviating from the expected Mendelian ratios of 1:1 or 1:2:1 (0.005<P<0.05, χ^2). Image created using MapChart 2.1 (Plant Research International). [28]

1-LOD interval around a significant QTL LOD-peak (permutation test, $P<0.05$, χ^2), containing at least one marker significant at $P<0.05$, χ^2 according to the Kruskal-Wallis test, and adjusted at both ends by 0.97 cM in order to compensate for genotyping error. The QTL regions contained nine, four, seven, and seven such SNP markers, respectively. The QTLs for each trait explained similar percentages of the phenotypic variation; 4.8–10.1% for the exclusion (25.7% in total, if additive), 5.3–6.7% for the lesion lengths (12% in total), 4.6–8.4% for the growth in sapwood (25.1% in total), and 5.3–7.2% for the infection prevention (18.4% in total), while the significant ($P<0.05$, χ^2) LOD peaks varied between 3.31 and 5.79 across the traits and QTLs. Three QTLs (1 for lesion lengths, 2 for growth in sapwood) were significant at the genome level; the others only at the chromosome level. The genes corresponding to each significant SNP and the corresponding cofactor can be seen in Table 1.

The BH-procedure [39] showed that for the QTLs from individual traits as well as for all QTLs in the study together, the false discovery rate was controlled at q = 0.05, since $P(i)<q*i/m$ for every i (QTL) at that level of significance (see Material and Methods).

Among these resulting 27 SNP markers, we identified four markers in genes with homology (BlastX, MaxScore >378, E-value < 1e-125) to *Pinaceae* genes with a suggested role in defense response: a hydroxycinnamoyl CoA shikimate/quinate hydroxycinnamoyltransferase [40], a 4-coumarate CoA-ligase [41], a R2R3-MYB transcription factor [42], a leucoanthocyanidine reductase [43], (Table 1). These are specifically involved in the phenylpropanoid and flavanoid pathways.

Discussion

One of the purposes of QTL mapping is to investigate the number of involved loci in complex traits and their individual importance. Another coveted benefit of QTL mapping is the identification of candidate genes behind these traits, which makes it essential to be able to decipher the mapped area into actual sequence. Because of the difficulties in assembling conifer DNA into scaffolds containing multiple genes, the number of known neighbouring genes to any given identified gene in the *Picea abies* genome is virtually zero. Genetic linkage can circumvent this obstacle, but only if the markers are fashioned from known sequence, i.e. not anonymous. An example of such markers is SNPs located in transcribed genes. Since the *P. abies* whole transcriptome has been sequenced and a large part of the genome assembled [11], and transcriptome profiles are available for *P. glauca* and *P. sitchensis* [44], the potential for SNP based linkage maps to identify causative genes for important traits is growing. Extreme high-density maps containing gene-based markers for every single gene seem possible in the foreseeable future.

In the present study, a full-sib *P. abies* family of 247 progenies, stemming from a cross between the parents S21K7622162 and S21K7621678 [24], was used to construct a consensus linkage map based on segregation patterns of 686 SNP markers. The markers were derived from an Illumina 3072 SNP Golden Gate Assay, resulting in 2058 successful assays and 874 informative markers. The success rate from SNP design to successful SNP core (67%) was consistent with what has been reported from conifers

earlier [19,45–47]. The final linkage map is one of the most saturated *P. abies* linkage map to date and the one most enriched in mapped genes, although well behind some other conifer maps such as the 1745 genes mapped in *Picea glauca* [20] or the 2841 in *Pinus taeda* [22]. Quantitative trait loci for four distinct traits of resistance against *Heterobasidion parviporum* infection were positioned on the map. This is the first report of resistance QTLs in any host to this economically important pathogen, which is a vital step towards the identification of the causal genes.

The twelve linkage groups of the present map correspond to the twelve chromosomes of *P. abies* and most other species of the *Pinaceae* family [48]. The linkage groups are all large, varying between 110.7 and 186.2 cM and between 32 and 73 unique loci. As visible from Table 3, this variation in size among the linkage groups corresponds to a relative size of 5.9 to 9.9% of the total map length. This mimics the variation in relative size among *P. abies* chromosomes, which span from 6.0 to 10.7% and 5.8 to 10.8% of the total genome size as per two previous estimations based on morphometrics of karyotype data [49–50]. Of course, this correlation is only circumstantial evidence that the largest linkage group reflects the largest chromosome, but still suggests not only that the linkage groups indeed represent the actual chromosomes, but also that the SNP markers have been derived from all parts of the genome. This assumption was strengthened by the fact that the map covers 96.5% of the estimated genome length, and that every loci is with 97% probability located within 10 cM of a mapped marker. Thus, the map should be saturated enough to allow QTL detection for every causative region for the respective trait, and even though adding more markers to the map certainly would make the existing groups denser, it would not substantially increase the map size in centimorgans. This assumption has been confirmed in a saturated *Picea glauca* map [20], which was increased from 1301 to 2211 markers, while decreasing from 2086.8 to 2065 centimorgans.

There is strong evidence that the marker organisation into groups is valid, since every group remained stable up to LOD 10 and no marker had an average genotype probability (–Log10(P)) higher than 0.3. In total, the twelve groups span 1889.2 cM, with a marker density of 2.8 cM/marker. This is fairly similar to the only other *P. abies* map of similar saturation available [17–18], which reported a size of 2035 cM and 2.6 cM/marker using 775 markers. As expected, the 174 SNPs sharing a mutual origin with another marker always did map very close to their intragenic neighbours - the average distance between such markers (measured on the most widely spaced two of each gene) was 0.97 cM. Recombination in *P. abies*, and plants in general, has indeed been suggested to occur mainly within genes [31–33], but for enough such events to occur within these genes to justify this average distance, the genes would have to be many times larger than the genome average. Instead, this might reflect the genotyping error rate for the experiment. Knowing that two of markers that were expected to locate very close to the same position actually were separated by on average 0.97 cM infers that every marker position in the map should be considered as potentially being misplaced by this much. To compensate for this, the QTL confidence interval defined by the 1-LOD drop was adjusted by 0.97 cM.

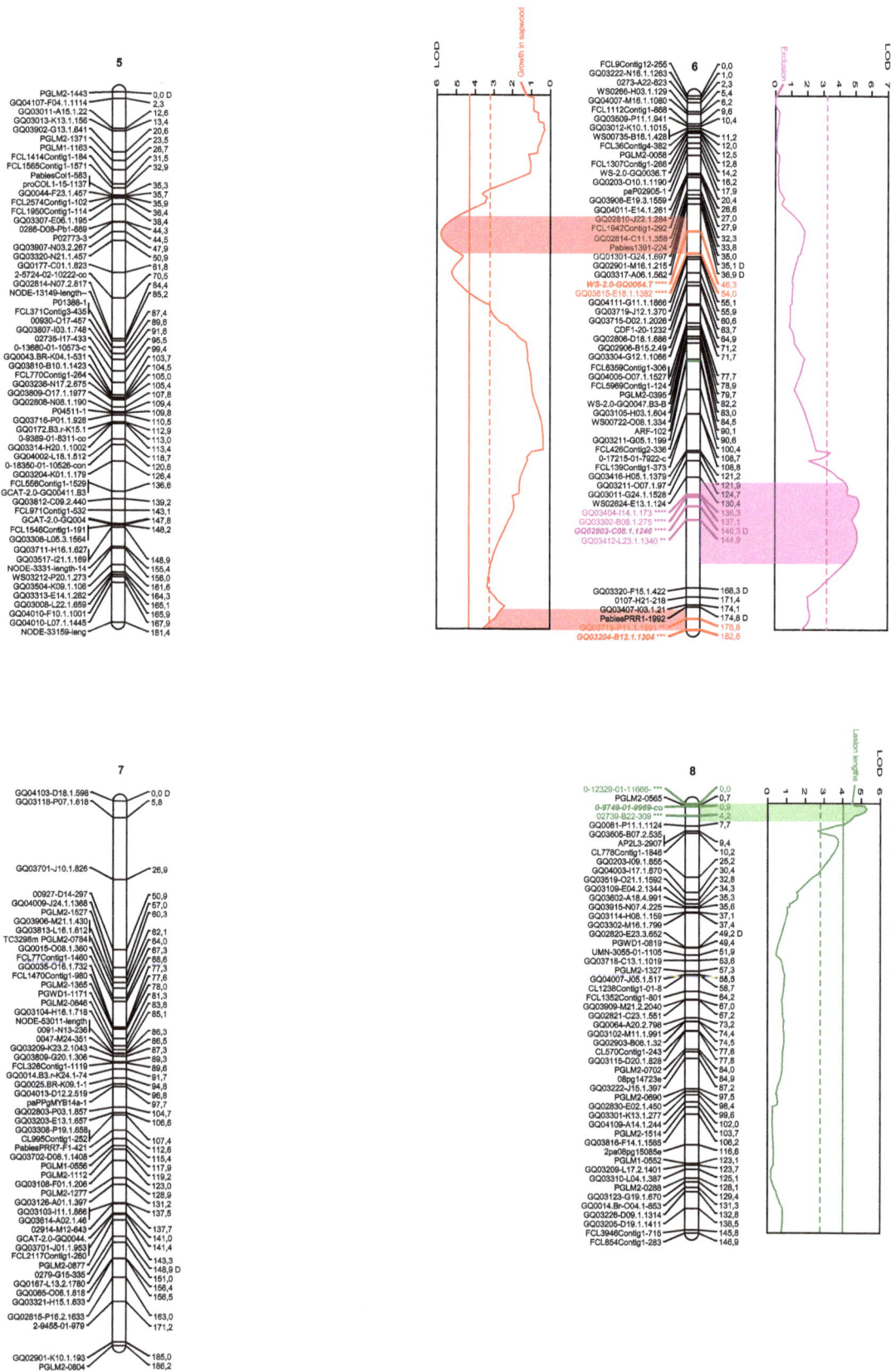

Figure 2. Linkage groups 5-8 of the *Picea abies* genome and the QTLs for various resistance traits. Names of the SNP markers are displayed on the left of the linkage groups. Genetic distance (cM) is indicated on the right. SNP markers are described in detail in Table S1. Graphs denote QTL effects for the current group. Red curves indicate fungal growth within sapwood, green lesion length, and purple fungal exclusion.

Complete and dashed vertical lines describe 0.1% and 5% levels of significance for the individual trait and group. Wide colored areas between curve and group show the QTL confidence interval based on a 1 LOD drop from the QTL peak. The colored marker denotes SNPs under the QTL confidence interval, the stars level of significance according to the Kruskal-Wallis test (ranging from p<0.05 (**) to p<0.005 (****) and **bold** style the designated cofactors. "D" after a marker name indicates segregation pattern deviating from the expected Mendelian ratios of 1:1 or 1:2:1 ($0.005<P<0.05$, χ^2). Image created using MapChart 2.1 (Plant Research International). [28]

Markers exhibiting segregation distortion can be a sign of erroneous genotyping data, especially if the markers are randomly dispersed across the map and do not aggregate into clusters [51]. On the other hand, clusters of distorted markers may weaken map structure [52]. In this study, 4.7% (32) of the 686 positioned SNP markers segregated at a ratio distorted from the expected Mendelian ratios of 1:1 or 1:2:1 ($0.005<P<0.05$, χ^2). This proportion is similar to the 6% reported for a *P. abies* map previously [17–18], but lower than the 12% reported for a *P. mariana* x *P. rubens* cross [52] and higher than the 1.9% reported for *P. glauca* [18]. As for other conifers, similar distortion levels have been reported for maps of *Pinus* species, such as 9% for *P. sylvestris* [53], 1–2.4% for various populations of *P. pinaster* [54], 7.4–7.5% for two pedigrees of *P. taeda* ($P<0.005$, χ^2) [55] and 12% for a *P. palustris* x *P. elliottii* cross [56], while 5.4% of 1364 markers has been reported as distorted ($P<0.01$, χ^2) in *Cryptomeria japonica* [57]. In this study, ten distorted markers formed three minor clusters (i.e. had at least one distorted immediate neighbour) while the other 22 were scattered across the map (had no such neighbour). However, since the distortion level was not extreme ($0.005<P<0.05$, χ^2) and their presence did not force any other markers out of the groups, they were included.

When distorted markers are clustered, the reason may be selection bias. As certain genotypes are lethal or otherwise detrimental to viability, there will be an inherent selection for these areas, so-called viability QTLs. Viability QTLs are of course interesting in their own right as they may provide insight into genes with a pivotal impact on fitness, but they are not necessarily involved in the biology of host resistance. Without selection, the segregation pattern in a family would be expected to follow Mendelian ratios all across the genome. At a viability QTL, these ratios would be distorted due to the genotype-dependent mortality [58]. Also, viability QTLs would be expected to be significant for all measurements, as the effect is based on selection prior to phenotyping. It is not likely that the QTLs of this study can be considered as viability QTLs, because only three of the 27 SNP markers found within a QTL confidence intervals in this study were distorted ($P<0.05$, χ^2), all in different QTLs, on two different linkage groups (2, 2 and 6), and none of the QTLs were significant for more than one measurement.

The finalized map was used to locate QTLs for four phenotypically distinct aspects of resistance against *H. parviporum* infection using interval mapping [38]. Interval mapping assumes that the measured traits follow a normal distribution across the mapping population. As visible from Figure 5, at least the exclusion and infection prevention traits do not follow a normal distribution but rather seem to have a spike at 0 effect (a null phenotype). Interval mapping can still provide reliable results, if the spike at 0 is small and the rest of the data set normally distributed and not much larger than 0 [59]. This is probably not the case of the exclusion and infection prevention traits. In order to verify the QTLs observed for these traits, a Kruskal-Wallis test was performed. A Kruskal-Wallis test is a non-parametric test that makes no assumption about the probability distributions of the traits, and it was performed on each locus separately without using any linkage information [37]. The standpoint was that loci deemed significant both by interval mapping and the Kruskal-Wallis test

would be considered as valid QTLs even if the data was not normally distributed. This approach was also employed for the normally distributed traits for Lesion length and Growth in sapwood (Figure 5).

To avoid scoring false positive QTLs while at the same time maintaining enough statistical power to detect as many true positive QTLs as possible, the false discovery rate (FDR) was estimated at the level of significance q. According to [60], all traits in a multitrait study should be tested simultaneously in a FDR-controlling approach. The BH-test [39] was used both for all traits together and for each trait individually. For both approaches, the FDR was controlled at $q<0.05$, χ^2. Thus, if accepting all 13 QTLs as true positives, it is simultaneously assumed that on average 5% of these would be false positives.

According to the QTLs, neither trait is controlled by a single gene or locus, but significantly affected by at least two to four regions of the genome. No overlap between QTLs for the separate traits was identified which suggests that the separate traits measure different aspects of host resistance. The percentages of phenotypic variation explained (PVE) levels detected for individual QTLs (4.6–10.1%), or combined for each trait (12–25.7%), suggest that the measured traits are complex in nature and probably controlled by a large number of loci. This is expected as defence against a necrotroph such as *H. parviporum* is not based on gene-for-gene interactions, but rather on the employment of a large battery of genes involved in systemic resistance. Similar PVE levels were found for the respective loci associated with bud set and height growth in a 283 individual population of *P. mariana*; 4–11.7% and 6.5–12.3% [61]. In a composite *P. glauca* map, two populations of 500 respective 200 individuals were used to detect QTLs for bud flush, bud set and height growth, explaining at the most 16.4, 22.2 and 10.5% of the respective variation of these traits [62], As for other conifers, the PVE of height growth, wood density and fibre length QTLs in *Pseudotsuga menziesii* peaked at 17.7, 14.9 and 15.7% [63] in a population of 320 trees (40 from each of 8 full-sib families), and resistance to *Dothistroma* needle blight in *Pinus radiata*, using 202 individuals from 6 full-sib families, was only explained to 4.8% by the strongest loci [12].

In every QTL mapping experiment, there is a danger of overestimating the importance of identified QTLs due to the so called Beavis effect [64]. According to Beavis, undetected QTLs with small effects inflate the estimated impact of closely located, detected QTLs. This bias has been reported to increase with LOD significance threshold and decrease with population size and map saturation [65]. Typically, large PVE-values in low density maps and small mapping populations run a greater risk of being affected by the Beavis effect. Indeed, Pelgas and coworkers [62] compared results from two mapping populations of 260 and 500 progenies respectively. They found that only 24% of the QTLs found in the larger population were also identified in the smaller, and that a randomized subset of 250 progenies out of the 500 only identified 29% of the QTLs. Furthermore, the PVE values of QTLs in the smaller population were twice as large as those obtained with the larger. This suggests that our population of 247 individuals was not large enough to disregard the Beavis effect and high PVE values should be regarded with caution as they might be the result of overestimation. However, this effect is a lesser factor for QTLs of

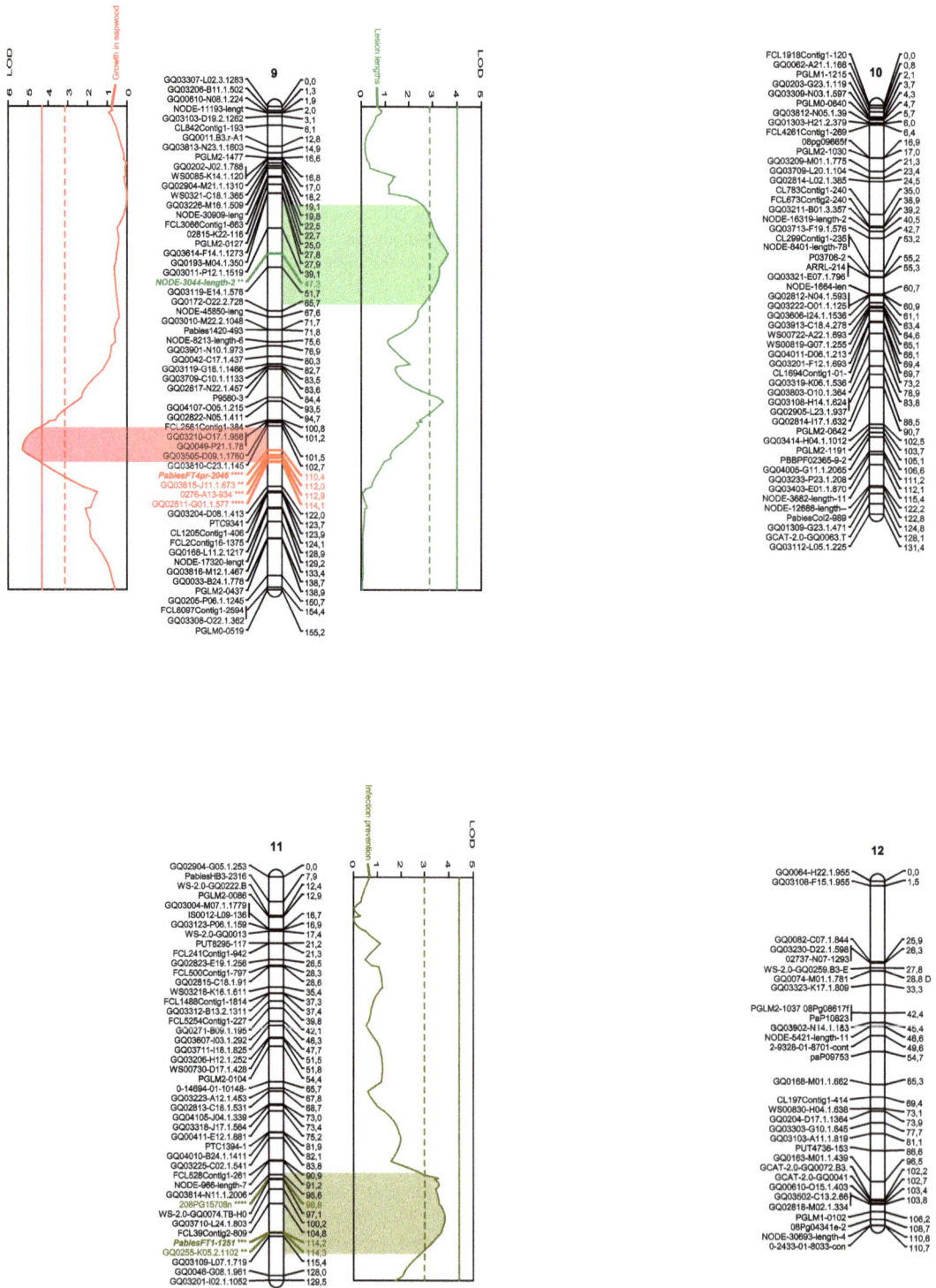

Figure 3. Linkage groups 9-12 of the *Picea abies* genome and the QTLs for various resistance traits. Names of the SNP markers are displayed on the left of the linkage groups. Genetic distance (cM) is indicated on the right. SNP markers are described in detail in Table S1. Graphs denote QTL effects for the current group. Red curves indicate fungal growth within sapwood, green lesion length, and orange curves indicates infection prevention. Complete and dashed vertical lines describe 0.1% and 5% levels of significance for the individual trait and group. Wide colored areas between curve and group show the QTL confidence interval based on a 1 LOD drop from the QTL peak. The colored marker denotes SNPs under the QTL confidence interval, the stars level of significance according to the Kruskal-Wallis test (ranging from p<0.05 (**) to p<0.005 (****) and **bold** style the designated cofactors. "D" after a marker name indicates segregation pattern deviating from the expected Mendelian ratios of 1:1 or 1:2:1 (0.005<P<0.05, χ²). Image created using MapChart 2.1 (Plant Research International). [28]

Table 3. Numeric data for each linkage group in the *Picea abies* linkage map.

Linkage group	Size (cM)	SNP markers. total	SNP markers. distorted segregation	cM/marker	Linkage group size (% of total length)	Chromosome size (% of total length)[1]	Chromosome size (% of total length)[2]	~Mb/cM[3]
1	162.8	72	7	2.3	8.6	8.7	8.6	10.0
2	157.4	73	12	2.2	8.3	8.6	8.1	10.4
3	180.1	67	2	2.7	9.5	9.0	9.1	9.1
4	165.0	64	1	2.6	8.7	8.8	8.9	9.9
5	181.4	60	1	3.0	9.6	9.3	9.7	9.0
6	182.6	59	5	3.1	9.7	9.3	9.8	8.9
7	186.2	54	2	3.4	9.9	10.7	10.8	8.8
8	146.9	51	1	2.9	7.8	7.4	7.5	11.1
9	155.2	58	0	2.7	8.2	8.2	7.8	10.5
10	131.4	52	0	2.5	7.0	7.0	7.3	12.4
11	129.5	44	0	2.9	6.9	6.9	6.7	12.6
12	110.7	32	1	3.5	5.9	6.0	5.8	14.7
Avg	157.4	57.2	2.7	-	8.3	8.3	8.3	-
Total	1889.2	686	32	2.8	100	100	100	10.4

[1]Relative chromosome size according to Lubaretz, 1996.
[2]Relative chromosome size according to Siljak-Yakovlev, 2002.
[3]Calculated on an average chromosome size of 1.63 Gb (Nystedt, 2013).

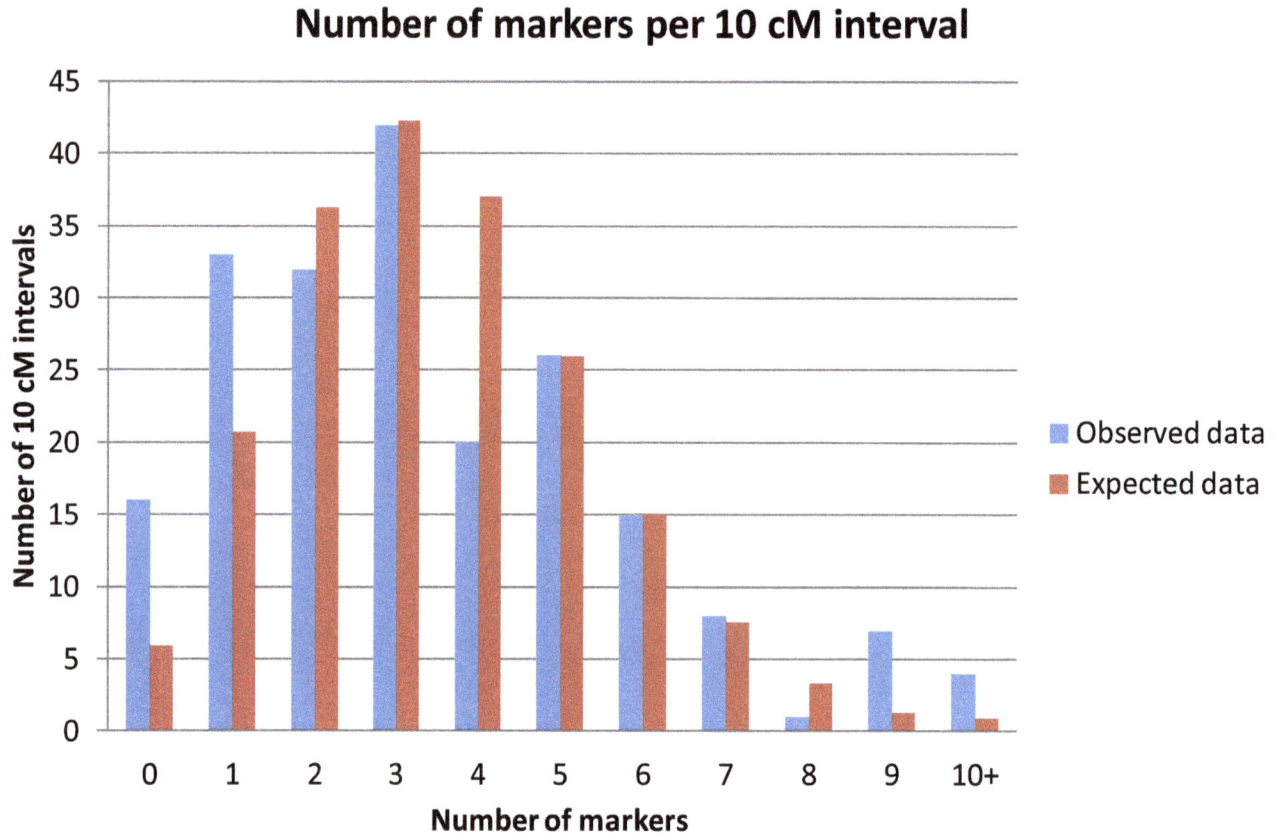

Figure 4. Distribution of markers. Observed and expected Poisson distribution of the markers frequencies for each 10 cM increment of the map.

lower PVE values. Jermstad et al [66–67] found that expanding a population from 98 to about 400 only changed the PVE interval for identified QTLs from 1.2–11.5% to 0.7–9.5%. Since all QTLs identified in this study fall within a similar range (PVE 4.6–10.1%), it is conceivable that the Beavis effect has played a minor role in determining their PVE.

When screening the infected trees for *H. parviporum* conidiophores, Arnerup *et al.* [24] concluded that 27% of the ramets contained no fungi. The reasons for this could be either technical, for example poorly infested wood blocks, or biological, i.e. some inherent trait that allows the host to prevent the establishment of the fungus in the wound. By mapping the proportion of immune ramets of each progeny as a trait, we discovered that 18.4% of this variation was explained by three QTLs on LG 1, 2 and 11 (Table 1). This result suggests that infection prevention at a very early stage is a heritable trait. As such a trait could potentially confer complete resistance to infection, rather than just limit the fungal spread within the tree, these QTLs might be of special interest for future *P. abies* breeding programmes.

The size of the *P. abies* genome, the low density of coding regions per MB and the difficulties with assembling genomes rich in transposable elements will complicate the investigation of the whole sequence corresponding to each QTL. In fact, as each assembled scaffold contains only one gene or less on average, we can at present only identify candidate genes from the QTLs if the map carries a SNP marker from them. The present linkage map contains 686 gene-based markers, a mere 2.4%, of the 28,354 genes of the *P. abies* genome [11]. However, some of the SNPs within the QTLs do originate from genes with known functions in plant response to wounding or microbial infection and have previously been shown to

be regulated during *H. parviporum* infection or belong to the same biosynthethic pathways. This warrants discussing them, despite the low proportion of represented genes.

Lignin plays an important part in both chemical and mechanical resistance against pathogens on *P. abies* [68]. Several of the SNPs found within the QTL confidence intervals are known intermediates in the lignification process. The QTL on LG 11 contained, at 114.3 cM, a SNP originating from the transcription factor R2R3-MYB11. The R2R3-MYB transcription factor family plays a role in lignin metabolism [42], and transcripts of the *MYB11* gene were shown to increase slightly but significantly in the lignin enriched compressed wood of *P. glauca* compared to opposite wood [69]. One of the other QTLs for infection prevention, at 158.4 cM on LG 1, contained a SNP from a gene for hydroxycinnamoyl CoA shikimate/quinate hydroxycinnamoyltransferase (HCT). HCT has been shown to be involved in lignin biosynthesis in several plants, including *Pinus radiata* [40] and *Populus euramericana* [70]. Furthermore, a SNP from a gene for 4-coumarate CoA-ligase (4CL) was located at 76.0 cM on LG 3, in a QTL for the fungal exclusion trait. This enzyme is involved in the conjugation of hydroxycinnamates with CoA, and thus also related to the biosynthesis of phenyl propanoid [41].

Another interesting observation was made in the QTL for fungal growth within sapwood on LG 6. A SNP from a gene for leucoanthocyanidin reductase (accession number BT10950 in Table 1, *LAR1* in [23]) was mapped to 182.6 cM. *LAR1* is an important intermediate in the flavonoid pathway, converting leucoanthocyanidins to catechins, which subsequently are converted into the defence-related phenolics flavonoid-derived proanthocyanidins [71]. Both catechin and *LAR1* mRNA-levels,

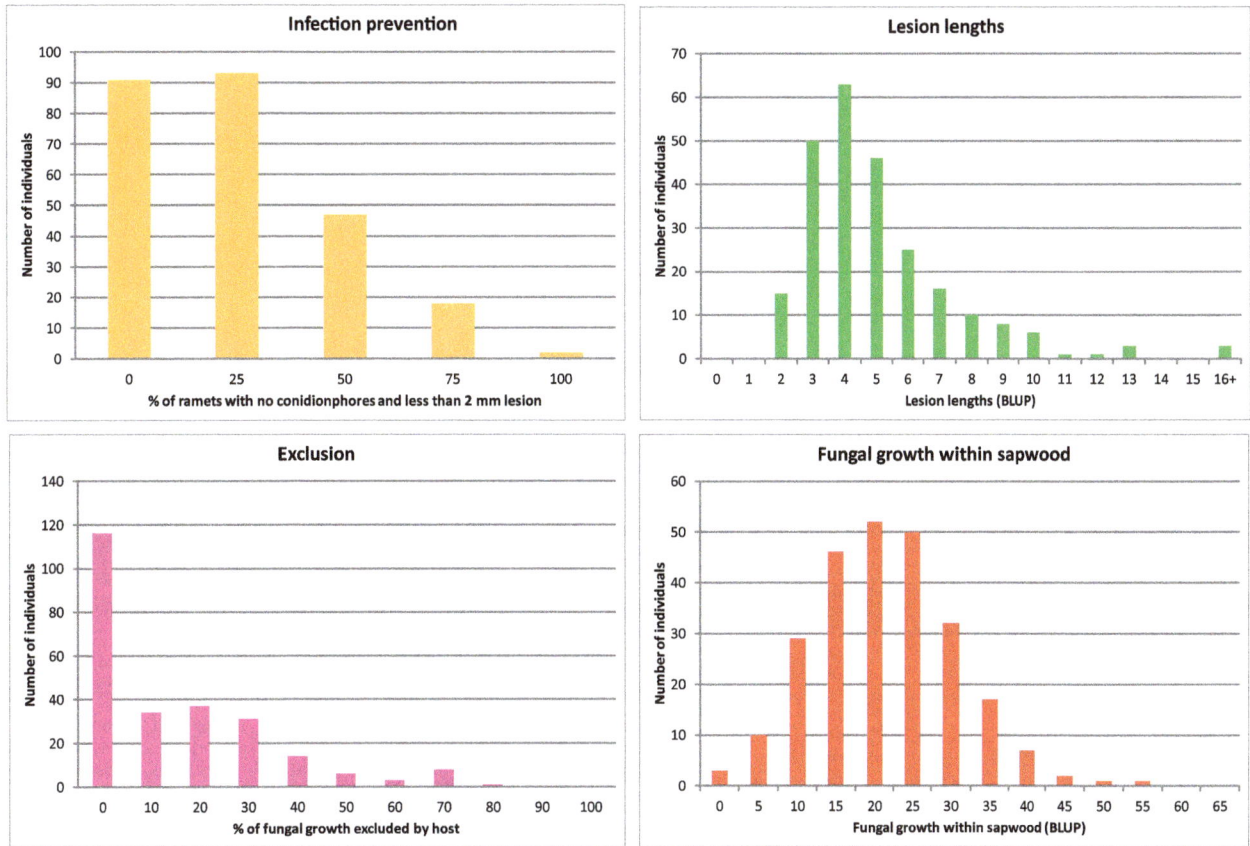

Figure 5. Distribution of resistance data. The distribution of observed resistance, measured as lesion length, fungal growth within sapwood, fungal exclusion, and ability to prevent infection throughout the mapping population, using all four ramets.

as well as other intermediates in the flavonoid pathway, have previously been shown to significantly increase in *P. abies* clones in response to *Heterobasidion*. The increase was greater in less susceptible clones than in more susceptible [24,72]. Still, this pathway is not unique for reactions against *Heterobasidion* attack but it is rather involved in general host response to many kinds of stress. For example, the very same *LAR* gene identified in this study is also induced in *P. abies* in response to *Ceratocystis polonica* inoculation and wounding (*LAR3* in [71]). Taken together, this strengthens the notion that the BT10950 locus is important for *P. abies* resistance to *H. parviporum* infections.

In this study, we used 247 full-sib progenies and 686 SNP markers derived from transcribed genes to construct one of the most saturated, and the most gene-enriched, *P. abies* maps to date. It constitutes a powerful tool for future mapping of important traits and understanding of recombination events in conifers. We used the map to identify QTL regions for four distinct resistance traits involved in host response to *H. parviporum* infection, explaining in total 12–25.7% of the variation for each trait. This showed that all traits are complex and controlled by several key genes on different chromosomes, as would be expected when measuring systemic resistance. When analysing the SNP markers within the QTLs, we identified four genes previously reported to play a part in plant defence against fungal attack, of which one has previously been shown to be regulated during infection and three play a role in the phenyl propanoid pathway. Albeit the significant SNPs likely represent only a few of the genes present in the sequence covered by the QTLs, they can be used as valuable starting points for further investigations.

Supporting Information

Table S1   SNP information. Short name, full name, accession information for GenBank and TAIR databases and from the *Picea glauca* genome, and full nucleotide sequence, for all 874 SNPs polymorphic between the parental trees.

Table S2   JoinMap files. Input files used to compute the linkage map using JoinMap 3.0.

Table S3   MapQTL files. Input files used to compute the QTL mapping using MapQTL 4.0.

Acknowledgments

The authors would like to thank Dr. Jenny Arnerup and Dr. Gunilla Swedjemark for their valued assistance in conducting the inoculation experiments.

Author Contributions

Conceived and designed the experiments: ME TK BK M. Lascoux JS. Performed the experiments: TK JC XFM M. Lascoux. Analyzed the data: M. Lind TK. Contributed reagents/materials/analysis tools: BK MM GZ JB. Wrote the paper: M. Lind.

References

1. Farjón A (1990) Pinaceae: drawings and descriptions of the genera Abies, Cedrus, Pseudolarix, Keteleeria, Nothotsuga, Tsuga, Cathaya, Pseudotsuga, Larix and Picea. Königstein: Koeltz Scientific Books.
2. Loman J (2007) Statistical Yearbook of Forestry Official Statistics of Sweden; Loman J, editor. Jönköping: Swedish Forestry Agency.
3. Asiegbu FO, Adomas A, Stenlid J (2005) Conifer root and butt rot caused by *Heterobasidion annosum* (Fr.) Bref. *s. l*. Molecular Plant Pathology 6: 395–409.
4. Woodward S, Stenlid J, Karjalainen R, Hûttermann A, editors (1998) *Heterobasidion annosum*. Biology, Ecology, Impact and Control. Cambridge, UK: CAB International.
5. Stenlid J, Wästerlund I (1986) Estimating the frequency of stem rot in *Picea abies* using an increment borer. Scandanavian Journal of Forest Research 1: 303–308.
6. Thor M, Ståhl G, Stenlid J (2005) Modeling root rot incidence in Sweden using tree, size and stand variables. Scandanavian Journal of Forest Research 20: 165–176.
7. Karlsson B, Swedjemark G (2006) Genotypic variation in natural infection frequency of *Heterobasidion* spp. in a *Picea abies* clone trial in southern Sweden. Scandanavian Journal of Forest Research 21: 108–114.
8. Ahuja MR, Neale DB (2005) Evolution of Genome Size in Conifers. Silvae Genetica 54: 126–137.
9. Williams CG (2009) Separate Female and Male Meioses. Conifer Reproductive Biology: Springer Netherlands. pp. 39–54.
10. Moritsuka E, Hisataka Y, Tamura M, Uchiyama K, Watanabe A, et al. (2012) Extended Linkage Disequilibrium in Noncoding Regions in a Conifer, *Cryptomeria japonica*. Genetics 190: 1145–1148.
11. Nystedt B, Street NR, Wetterbom A, Zuccolo A, Lin Y-C, et al. (2013) The Norway spruce genome sequence and conifer genome evolution. Nature.
12. Devey M, Groom K, Nolan M, Bell J, Dudzinsky M, et al. (2004) Detection and verification of quantitative trait loci for resistance to Dothistroma needle blight in *Pinus radiata*. Theoretical and Applied Genetics 108: 1056–1063.
13. Freeman J, Potts B, Vaillancourt R (2008) Few Mendelian genes underlie the quantitative response of a forest tree, *Eucalyptus globulus*, to a natural fungal epidemic. Genetics 178: 563–571.
14. Alves AA, Rosado CCG, Faria DA, Guimaraes LMdS, Lau D (2012) Genetic mapping provides evidence for the role of additive and non-additive QTLs in the response of inter-specific hybrids of Eucalyptus to *Puccinia psidii* rust infection. Euphytica 183: 27–38.
15. Jorge V, Dowkiw A, Faivre-Rampant P (2005) Genetic architecture of qualitative and quantitative *Melampsora larici-populina* leaf rust resistance in hybrid poplar: genetic mapping and QTL detection. New Phytologist 167: 113–127.
16. Samils B, Rönnberg-Wästljung A-C, Stenlid J (2011) QTL mapping of resistance to leaf rust in *Salix*. Tree Genetics & Genomes 7: 1219–1235.
17. Acheré V, Faivre-Rampant P, Jeandroz S, Besnard G, Markussen T, et al. (2004) A full saturated linkage map of *Picea abies* including AFLP, SSR, ESTP, 5S rDNA and morphological markers. Theoretical and Applied Genetics 108: 1602–1613.
18. Pelgas B, Beauseigle S, Acheré V, Jeandroz S, Bousquet J, et al. (2006) Comparative genome mapping among *Picea glauca*, *P. mariana* x *P. rubens* and *P. abies*, and correspondence with other Pinaceae. Theoretical and Applied Genetics 113: 1371–1393.
19. Pavy N, Pelgas B, Beauseigle S, Blais S, Gagnon F, et al. (2008) Enhancing genetic mapping of complex genomes through the design of highly-multiplexed SNP arrays: application to the large and unsequenced genomes of white spruce and black spruce BMC Genomics 9.
20. Pavy N, Pelgas B, Laroche J, Rigault P, Isabel N, et al. (2012) A spruce gene infers ancient plant genome reshuffling and subsequent slow evolution in the gymnosperm lineage leading to extant conifers. Bmc Biology 10: 84.
21. Martinez-Garcia PJ, Stevens KA, Wegrzyn JL, Liechty J, Crepeau M, et al. (2013) Combination of multipoint maximum likelihood (MML) and regression mapping algorithms to construct a high-density genetic linkage map for loblolly pine (*Pinus taeda* L.). Tree Genetics & Genomes 9: 1529–1535.
22. Neves LG, Davis JM, Barbazuk WB, Kirst M (2014) A High-Density Gene Map of Loblolly Pine (*Pinus taeda* L.) Based on Exome Sequence Capture Genotyping. G3 4: 29–37.
23. Danielsson M, Lundén K, Elfstrand M, Hu J, Zhao T, et al. (2011) Chemical and transcriptional responses of Norway spruce genotypes with different susceptibility to *Heterobasidion* spp. infection. BMC Plant Biology 11: 154.
24. Arnerup J, Swedjemark G, Elfstrand M, Karlsson B, Stenlid J (2010) Variation in growth of *Heterobasidion parviporum*. Scandanavian Journal of Forest Research 25: 106–110.
25. Pavy N, Gagnon F, Rigault P, Blais S, Deschênes A, et al. (2013) Development of high-density SNP genotyping arrays for white spruce (*Picea glauca*) and transferability to subropical and nordic congeners. Molecular Ecology Resources 13: 324–336.
26. Chen J, Källman T, Ma X, Gyllenstrand N, Zaina G, et al. (2012) Disentangling the roles of history and local selection in shaping clinal variation of allele frequencies and gene expression in Norway spruce (*Picea abies*). Genetics 191: 865–881.
27. Chen J, Uebbing S, Gyllenstrand N, Lagercrantz U, Lascoux M, et al. (2012) Sequencing of the needle transcriptome from Norway spruce (*Picea abies* Karst

28. Fan J-B, Oliphant A, Shen R, Kermani B, Garcia F, et al. (2003) Highly Parallel SNP Genotyping. Cold Spring Harbor Symposia on Quantitative Biology 68: 69–78.
29. Van Ooijen J, Voorrips R (2001) JoinMap 3.0, Software for the calculation of genetic linkage maps. Wageningen, the Netherlands: Plant Research International.
30. Voorrips R (2002) MapChart: Software for the graphical presentation of linkage maps and QTLs The Journal of Heredity 93: 77–78.
31. Gaut B, Wright S, Rizzon C, Dvorak J, Anderson LK (2007) Recombination: an underappreciated factor in the evolution of plant genomes. Nature Reviews Genetics 8: 77–84.
32. Larsson H, Källman T, Gyllenstrand N, Lascoux M (2013) Distribution of Long-Range Linkage Disequilibrium and Tajima's D Values on Scandnavian Populations of Norway Spruce *(Picea abies)*. G3 3: 795–806.
33. Schnable PS, Hsia A-P, Nikolau BJ (1998) Genetic recombination in plants. Current Opinion in Plant Biology 1: 123–129.
34. Chakravarti A, Lasher L, Reefer J (1991) A maximum likeliehood method for estimating genome length using genetic linkage data. Genetics 128: 157–162.
35. Lange K, Boehnke M (1982) How many polymoprhic marker genes will it take to span the human genome? American Journal of Human Genetics 34.
36. Massey F (1951) The Kolomogorov-Smirnov test for goodness of fit. Journal of the American Statistical Association 46: 68–78.
37. Van Ooijen J, Boer MP, Jansen RC, Maliepaard C (2002) MapQTL 4.0, Software for the calculation of QTL positions on genetic maps. Wageningen, the Netherlands: Plant Research International.
38. Lander ES, Botstein D (1988) Mapping Mendelian Factors Underlying Quantitative Traits Using RFLP Linkage Maps. Genetics 121: 185–199.
39. Benjamini Y, Hochberg Y (1995) Controlling the false discovery rate: a practical and powerful approach to multiple testing. Journal of the Royal Statistical Society, Series B 57: 289–300.
40. Wagner A, Ralph J, Akiyama T, Flint H, Phillips L, et al. (2007) Exploring lignification in conifers by silencing hydroxycinnamoyl-CoA:shikimate hydroxycinnamoyltransferase in *Pinus radiata*. Proceedings of National Academy of Science USA 104: 11856–11861.
41. Gross GG, Zenk MH (1974) Isolation and Properties of Hydroxycinnamate:CoA Ligase from Lignifying Tissue of *Forsythia*. European Journal of Biochemistry 42: 453–459.
42. Rogers L, Campbell M (2004) The genetic control of lignin deposition during plant growth and development. New Phytologist 164: 17–30.
43. Winkel-Shirley B (2001) Flavonoid Biosynthesis. A Colorful Model for Genetics, Biochemistry, Cell Biology, and Biotechnology. Plant Physiology 126: 485–493.
44. Raherison E, Rigault P, Caron S, Poulin P, Boyle B, et al. (2012) Transcriptome profiling in conifers and the PiceaGenExpress database show patterns of diversification within gene families and interspecific conservation in vascular expression. BMC Genomics 13: 434.
45. Chancerel E, Lepoittevin C, Le Provost G, Lin Y-C, Jaramillo-Correa JP, et al. (2011) Development and implementation of a highly-multiplexed SNP array for a genetic mapping in maritime pine and comparative mapping with loblolly pine. BMC Genomics 12.
46. Eckert AJ, Bower AD, Wegrzyn JL, Pande B, Jermstad KD, et al. (2009) Association Genetics of Coastal Douglas Fir *(Pseudotsuga menziesii* var. *menziesii*, Pinaceae). I. Cold-Hardiness Related Traits. Genetics 182: 1289–1302.
47. Eckert AJ, Pande B, Ersoz ES, Wright MH, Rashbrook VK, et al. (2009) High-throughput genotyping and mapping of single nucleotide polymorphisms in loblolly pine (*Pinus taeda* L.). Tree Genetics & Genomes 5: 225–234.
48. Sax H, Sax K (1933) Chromosome number and morphology in the conifers. Journal of the Arnold Arboretum 14: 356–375.
49. Lubaretz O, Fuchs J, Ahne R, Meister A, Schubert I (1996) Karyotyping of three Pinaceae species via fluorescent *in situ* hybridization and computer-aided chromosome analysis. Theoretical and Applied Genetics 92: 411–426.
50. Siljak-Yakovlev S, Cerbah M, Coulaud J, Stoian V, Brown S, et al. (2002) Nuclear DNA content, base composition, heterochromatin and rDNA in *Picea omorika* and *Picea abies*. Theoretical and Applied Genetics 104: 505–512.
51. Liu X, You J, Guo L, Liu X, He Y, et al. (2011) Genetic Analysis of Segregation Distortion of SSR Markers in F2 Population in Barley. Journal of Agricultural Science 3.
52. Pelgas B, Bousquet J, Beauseigle S, Isabel N (2005) A composite linkage map from two crosses for the species complex *Picea mariana* x *Picea rubens* and analysis of synteny with other Pinaceae. Theoretical and Applied Genetics 111: 1466–1488.
53. Yin T, Wang X, Andersson B, Lerceteau-Köhler E (2003) Nearly complete genetic maps of *Pinus sylvestris* L. (Scots pine) constructed by AFLP marker analysis in a full-sib family. Theoretical and Applied Genetics 106: 1075–1083.
54. Chancerel E, Lamy J-B, Lesur I, Noirot C, Klopp C, et al. (2013) High-density linkage mapping in a pine tree reveals a genomic region associated with inbreeding depression and provides clues to the extent and distribution of meiotic recombination. Bmc Biology 11.

55. Echt CS, Saha S, Krutovsky KV, Wimalanathan K, Erpelding JE, et al. (2011) An annotated genetic map of loblolly pine based on microsatellite and cDNA markers. BMC Genetics 12.

56. Kubisiak T, Nelson C, Nance W, Stine M (1995) RAPD linkage mapping in a longleaf pine x slash pine F1 family. Theoretical and Applied Genetics 90: 1119–1127.

57. Moriguchi Y, Ujino-Ihara T, Uchiyama K, Futamara N, Saito M, et al. (2012) The construction of a high-density linkage map for identifying SNP markers that are tightly linked to a nuclear-recessive major gene for male sterility in *Cryptomeria japonica* D. Don. BMC Genomics 13.

58. Plough LV, Hedgecock D (2011) Quantitative Trait Locus Analysis of Stage-Specific Inbreeding Depression in the Pacific Oyster *Crassostrea gigas*. Genetics 189: 1473–1486.

59. Broman KW (2003) Mapping Quantitative Trait Loci in the Case of a Spike in the Phenotype Distribution. Genetics 163: 1169–1175.

60. Benjamini Y, Yekutieli D (2005) Quantitative Trait Loci Analysis Using the False Discovery Rate. Genetics 171: 783–790.

61. Prunier J, Pelgas B, Gagnon F, Desponts M, Isabel N, et al. (2013) The genomic architecture and association genetics of adaptive characters using a candidate SNP approach in boreal black spruce. BMC Genomics 14.

62. Pelgas B, Bousquet J, Meirmans P, Ritland K, Isabel N (2011) QTL mapping in white spruce: gene maps and genomic regions underlying adaptive traits across pedigrees, years and environments. BMC Genomics 12: 145.

63. Ukrainetz NK, Ritland K, Mansfield SD (2008) Identification of quantitative trait loci for wood quality and growth across eight full-sib coastal Douglas-fir families. Tree Genetics & Genomes 4: 159–170.

64. Beavis W (1998) QTL analyses: power, precision, and accuracy. In: Paterson AH, editor. Molecular Dissection of Complex Traits. New York: CRC Press.

65. Xu S (2003) Theoretical Basis of the Beavis Effect. Genetics 165: 2259–2268.

66. Jermstad KD, Bassoni D, Jech K, Ritchie G, Wheeler N, et al. (2003) Mapping of quantitative trait loci controlling adaptive traits in coastal Douglas fir. III. Quantitative trait loci-by-environment interactions. Genetics 165: 1489–1506.

67. Jermstad KD, Bassoni D, Jech K, Wheeler N, Neale DB (2001) Mapping of quantitative trait loci controlling adaptive traits in coastal Douglas-fir. I. Timing of vegetative bud flush. Theoretical and Applied Genetics 102: 1142–1151.

68. Franceschi V, Krokene P, Christiansen E, Krekling T (2005) Anatomical and chemical defences of conifer bark against bark beetles and other pests. New Phytologist 167: 353–375.

69. Bedon F, Grima-Pettanati J, MacKay JJ (2007) Conifer R2R3-MYB transcription factors: sequence analyses and gene expression in wood-forming tissues of white spruce (*Picea glauca*). BMC Plant Biology 7.

70. Kim B-G, Kim I-A, Ahn J-H (2011) Characterization of Hydroxycinnamoyl-coenzyme A Shikimate Hydroxycinnamoyltransferase from *Populus euramericana*. Journal of Korean Society of Applied Biological Chemistry 54: 817–821.

71. Hammerbacher A (2011) Biosynthesis of polyphenols in Norway spruce as a defence strategy against attack by the bark beetle associated fungus *Ceratocystis polonica*. Jena, Germany: Max Planck Institute for Chemical Ecology. 127 p.

72. Arnerup J (2011) Induced defence reponses in *Picea abies* triggered by *Heterobasidion annosum s.l.* Uppsala: Swedish University of Agricultural Sciences. 56 p.

11

Genome-Wide Analysis of the bZIP Transcription Factors in Cucumber

Mehmet Cengiz Baloglu[1]*, Vahap Eldem[2], Mortaza Hajyzadeh[3], Turgay Unver[3]

1 Kastamonu University, Faculty of Engineering and Architecture, Department of Genetics and Bioengineering, Kastamonu, Turkey, 2 Istanbul University, Faculty of Science, Department of Biology, Istanbul, Turkey, 3 Cankırı Karatekin University, Faculty of Science, Department of Biology, Cankiri, Turkey

Abstract

bZIP proteins are one of the largest transcriptional regulators playing crucial roles in plant development, physiological processes, and biotic/abiotic stress responses. Despite the availability of recently published draft genome sequence of *Cucumis sativus*, no comprehensive investigation of these family members has been presented for cucumber. We have identified 64 bZIP transcription factor-encoding genes in the cucumber genome. Based on structural features of their encoded proteins, *CsbZIP* genes could be classified into 6 groups. Cucumber *bZIP* genes were expanded mainly by segmental duplication rather than tandem duplication. Although segmental duplication rate of the *CsbZIP* genes was lower than that of Arabidopsis, rice and sorghum, it was observed as a common expansion mechanism. Some orthologous relationships and chromosomal rearrangements were observed according to comparative mapping analysis with other species. Genome-wide expression analysis of *bZIP* genes indicated that 64 *CsbZIP* genes were differentially expressed in at least one of the ten sampled tissues. A total of 4 *CsbZIP* genes displayed higher expression values in leaf, flowers and root tissues. The *in silico* micro-RNA (miRNA) and target transcript analyses identified that a total of 21 *CsbZIP* genes were targeted by 38 plant miRNAs. *CsbZIP20* and *CsbZIP22* are the most targeted by miR165 and miR166 family members, respectively. We also analyzed the expression of ten *CsbZIP* genes in the root and leaf tissues of drought-stressed cucumber using quantitative RT-PCR. All of the selected *CsbZIP* genes were measured as increased in root tissue at 24th h upon PEG treatment. Contrarily, the down-regulation was observed in leaf tissues of all analyzed *CsbZIP* genes. *CsbZIP12* and *CsbZIP44* genes showed gradual induction of expression in root tissues during time points. This genome-wide identification and expression profiling provides new opportunities for cloning and functional analyses, which may be used in further studies for improving stress tolerance in plants.

Editor: Baohong Zhang, East Carolina University, United States of America

Funding: The study was supported by Ministry of Development of Turkey with grant no. DPT2010K120720. The funders had no role in study design, data collection and analysis, decision to publish, or preparation of the manuscript.

Competing Interests: Turgay Unver is currently serving as an academic editor for PLOS ONE and this does not alter the authors' adherence to all the PLOS ONE policies on sharing data and materials.

* E-mail: mcbaloglu@gmail.com

Introduction

Cucumber (*Cucumis sativus* L.), a major vegetable crop consumed worldwide, belongs to Cucurbitaceae family commonly known as cucurbits. Agricultural production of cucurbits utilizes nine million hectares of land, and yields 192 million tons of vegetables, fruits, and seeds annually (http://faostat.fao.org). Cucurbit family is a model system for the study of sex determination and plant vascular biology in which long-distance signaling events are examined using xylem and phloem sap [1–2]. In 2009, cucumber became the seventh plant to have its genome sequence published, following the well-studied model plant *Arabidopsis thaliana*, the poplar tree, grapevine, papaya, and the crops rice and sorghum. In 2013, a variation map of the cucumber genome at single-base resolution was generated by performing deep re-sequencing of all 115 lines with the wild cucumber genome, which was compared to the genome of cultivated cucumber [3]. These genomic resources provide new insights for understanding the genetic basis of domestication and diversity of this important crop. As a result, the released genome sequence of the cucumber encouraged the scientific research community for further study related with its

structural and functional genomics, which has resulted in crop improvement and ensuring food security [4]. Consequently, the substantial findings in the aspects of both structural [5–9], and functional genomics [10–15] were reported in the vegetable model crop, cucumber.

Transcription factors (TFs) consist of sequence-specific DNA-binding domain for binding to the promoter and/or enhancer regions of corresponding genes, thereby inducing or repressing transcription of downstream target genes. TFs can be grouped into 40–60 families based on their primary and/or three-dimensional structure similarities in the DNA-binding and multimerization domains [16–18]. Among them, the basic leucine zipper (bZIP) transcription factor family is one of the largest and most diverse families. bZIP transcription factors have conserved bZIP domain which is composed of two structural features; a basic region that binds DNA and a leucine zipper dimerization motif [14]. The basic region of 16 amino acid residues with an invariant N-x7-R/ K motif is highly conserved and responsible for nuclear localization and DNA binding. The leucine zipper is a less-conserved dimerization motif and composed of a heptad repeat of leucines or other bulky hydrophobic amino acids positioned exactly nine amino acids towards the C-terminus. Plant bZIP proteins preferentially bind to DNA sequences with an ACGT core. Currently, the bZIP family members have been identified or predicted in multiple eukaryotic genomes including plants, animals and yeasts [10], [13], [18–20].

It has been reported that bZIP TFs are involved in developmental and physiological processes as well as biotic/abiotic stress responses under normal and stressed growth conditions. So, they are important for various plants to withstand adverse environmental conditions [14], [21]. As developmental processes, bZIP TFs play crucial roles in organ and tissue differentiation [22–26], cell elongation [27–28], nitrogen/carbon and energy metabolism [29–31], unfolded protein response [32–33], seed storage protein gene regulation [34] somatic embryogenesis [35]. On the other hand, bZIP TFs have also been regarded as important regulators in response to various abiotic stresses such as drought, high salinity and cold stresses in Arabidopsis [36–41], rice [42–50], wheat [51], tomato [52–53], soybean [54–56], pepper [57], bean [58], barley [59] and maize [60]. However, little is known about the genome-wide survey and expression patterns of this gene family in cucumber. The genome-wide survey and identification studies from Arabidopsis [10], castor bean [11], maize [12], rice [13], sorghum [14], algae, mosses, ferns, gymnosperms and angiosperms [15] are a few examples for bZIP TFs.

On the other hand, a draft of the *Cucumis sativus* L. genome sequence was reported recently [1]. Cucumber has seven pairs of chromosomes and a haploid genome of 367 Mbp, which is smaller than other species in Cucurbitaceae family. Its small genome makes itself as a model for functional genomics of vegetable crop. However, to our knowledge, no *bZIP* gene has been identified and isolated in cucumber so far. In addition, only limited data are available on genome-wide identification and their characterizations in the cucumber genome. Therefore, the genome-wide identification and expression analysis of the cucumber *bZIP* gene family in cucumber is one of the important issues to study. Here, we identified the bZIP family members in cucumber based on the complete genome sequence analysis. Then, we have identified the genomic distribution and conserved motifs of *bZIP* gene family. Consequently, we analyzed the expression patterns of these family members by using the publicly available expression and experimental data. This extended analysis is the first comprehensive study of the *bZIP* gene family in cucumber and provides valuable information for further exploration into the functions of this significant gene family in cucumber. In addition, these results provide information about the relationship between evolution and functional divergence in the bZIP family.

Materials and Methods

Sequence Retrieval and Identification of bZIP-domain Proteins from Cucumber

Three different approaches were applied to identify putative bZIP proteins from *C. sativus* L. Initially, 741 amino acid sequences encoding bZIP transcription factors from seven plants (*A. thaliana, Carica papaya, Oryza sativa* subsp. *japonica, Populus trichocarpa, Sorghum bicolor, Vitis vinifera* and *Zea mays*) were retrieved from plant transcription factor database 3.0 (plntfdb.bio.uni-potsdam.de) [61]. These sequences were used to identify homologous peptides from cucumber by performing a BLASTP search at PHYTO-ZOME v9.1 database (www. phytozome.net) using default parameters, [62]. In addition, the database was searched using the keywords 'bZIP'. Moreover, The Hidden Markov Model (HMM) profiles of the bZIP domains in the Pfam database (http://pfam.sanger.ac.uk) were searched against the PHYTO-

ZOME database of *C. sativus*. Similarity searches were also performed through TBLASTN at NCBI database against the EST sequences of *C. sativus* genome to eliminate possible exclusions of any additional bZIP members. All hits with expected values less than 1.0 were retrieved and redundant sequences were removed using the decrease redundancy tool (web.expasy.org/decrease_redundancy). Each non-redundant sequence was checked for the presence of the conserved bZIP domain by SMART (http://smart.emblheidelberg.de) [63] and Pfam (http://pfam.sanger.ac.uk) searches.

Chromosomal Location, Gene-structure Prediction, and Estimation of the Genomic Distribution

Specific chromosomal positions for the genes encoding these CsbZIP proteins were determined by BLASTP search of the *C. sativus* sequences against the PHYTOZOME database by using default settings. The genes were plotted separately onto the seven cucumber chromosomes according to their ascending order of physical position (bp), from the short-arm telomere to the long-arm telomere and finally displayed using MapChart [64]. Segmental duplications were identified based on the method of Plant Genome Duplication Database [65]. Briefly, BLASTP search was executed against all predicted peptide sequences of *C. sativus* and top 5 matches with ≤1e-05 was identified as potential anchors. Collinear blocks were evaluated by MCScan, and alignments with ≤1e-10 were considered as significant matches [65–66]. Tandem duplications were characterized as adjacent genes of same sub-family located within 10 predicted genes apart or within 30 kbp of each other [66–67]. The exon-intron organizations of the genes were determined using Gene-Structure Display Server (gsds.cbi.pku.edu.cn) [68] through comparison of their full-length cDNA or predicted coding sequence (CDS) with their corresponding genomic sequence.

Sequence Alignment, Phylogenetic Analysis and Identification of the Conserved Motifs

The amino acid sequences were imported into MEGA5 [69] and multiple sequence alignments were performed using ClustalW with a gap open and gap extension penalties of 10 and 0.1, respectively [70]. The alignment file was then used to construct an unrooted phylogenetic tree based on the neighbor-joining method [71]. After bootstrap analysis for 1000 replicates, the tree was displayed using Interactive tree of life (iTOL; http://itol.embl.de/index.shtml) [72]. Protein sequence motifs were identified using the multiple EM for motif elicitation (MEME); (http://meme.nbcr.net/meme3/meme.html) [73]. The analysis was performed by keeping number of repetitions, any; maximum number of motifs, 20; and optimum width of the motif ≥50. Discovered MEME motifs with ≤1e-30 were searched in the InterPro database with InterProScan [74].

Gene Ontology (GO) Annotation

The functional annotation of bZIP sequences and the analysis of annotation data were performed using Blast2GO (http://www.blast2go.com) [75]. The amino acid sequences of bZIPs were imported into Blast2GO program to execute three steps viz, (i) BLASTp against the non-redundant protein database of NCBI, (ii) mapping and retrieval of GO terms associated with the BLAST results, and (iii) annotation of GO terms associated with each query to relate the sequences to known protein function. The program provides the output defining three categories of GO classification; namely biological processes, cellular components, and molecular functions.

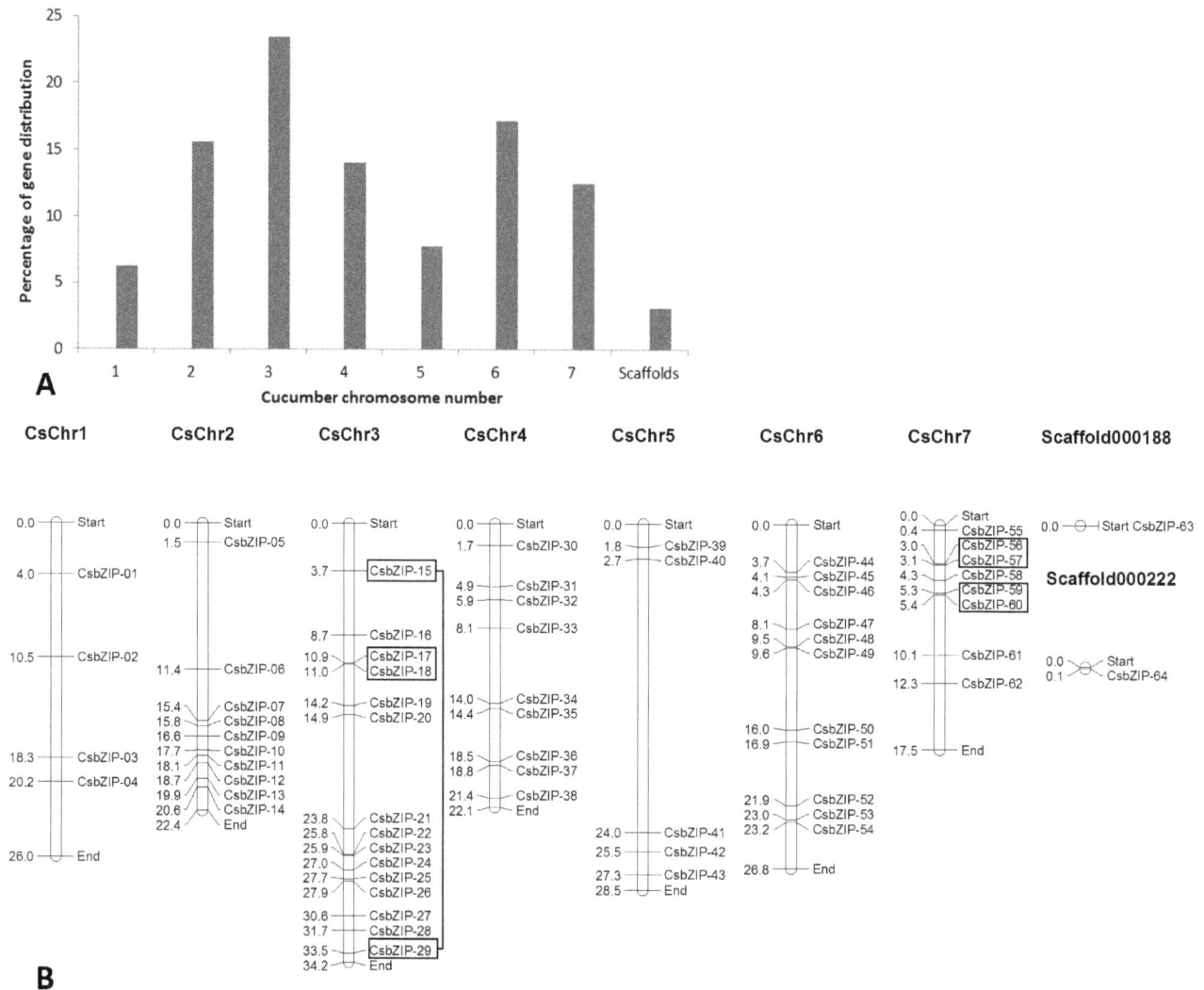

Figure 1. Distribution of 64 *CsbZIP* genes onto seven cucumber chromosomes. (A) Percentage of *bZIP* genes on each cucumber chromosome to show their distribution abundance. (B) Graphical (scaled) representation of physical locations for each *CsbZIP* gene on cucumber chromosomes (numbered 1–7). Tandem-duplicated genes on a particular chromosome are indicated in the box. Chromosomal distances are given in Mbp.

Comparative Physical Mapping of bZIP Proteins between Cucumber and other Species

For deriving orthologous relationship among the chromosomes of cucumber and three other species, amino acid sequences of cucumber bZIP were searched against peptide sequences of Arabidopsis, rice and poplar (www.phytozome.net) using BLASTP. Hits with ≤1e-5 and at least 80% identify were considered significant. The comparative orthologous relationships of *bZIP* genes among cucumber, Arabidopsis, rice and poplar chromosomes were finally visualized using MapChart.

Estimation of Synonymous and Non-synonymous Substitution Rates

The amino-acid sequences, duplicated protein-encoding bZIP genes, as well as orthologous gene-pairs between cucumber and Arabidopsis, rice and poplar were aligned using CLUSTALW based on multiple sequence alignment tool. The CODEML program in PAML interface tool of PAL2NAL (http://www.bork. embl.de/pal2nal) [76] was used to estimate the synonymous (Ks)

and non-synonymous (Ka) substitution rates by aligning the amino-acid sequences and their respective original cDNA sequences of *bZIP* genes. Time (million years ago, Mya) of duplication and divergence of each *bZIP* genes were estimated using a synonymous mutation rate of λ substitutions per synonymous site per year, as $T = Ks/2\lambda$ ($\lambda = 6.5 \times 10$ e-9) [77–78].

Homology Modeling of bZIP Proteins

All the CsbZIP proteins were searched against the Protein Data Bank (PDB) [79] by BLASTP (with the default parameters) to identify the best template having similar sequence and known three-dimensional structure. The data was fed in Phyre2 (Protein Homology/AnalogY Recognition Engine; http://www.sbg.bio.ic. ac.uk/phyre2) for predicting the protein structure by homology modeling under 'intensive' mode [80].

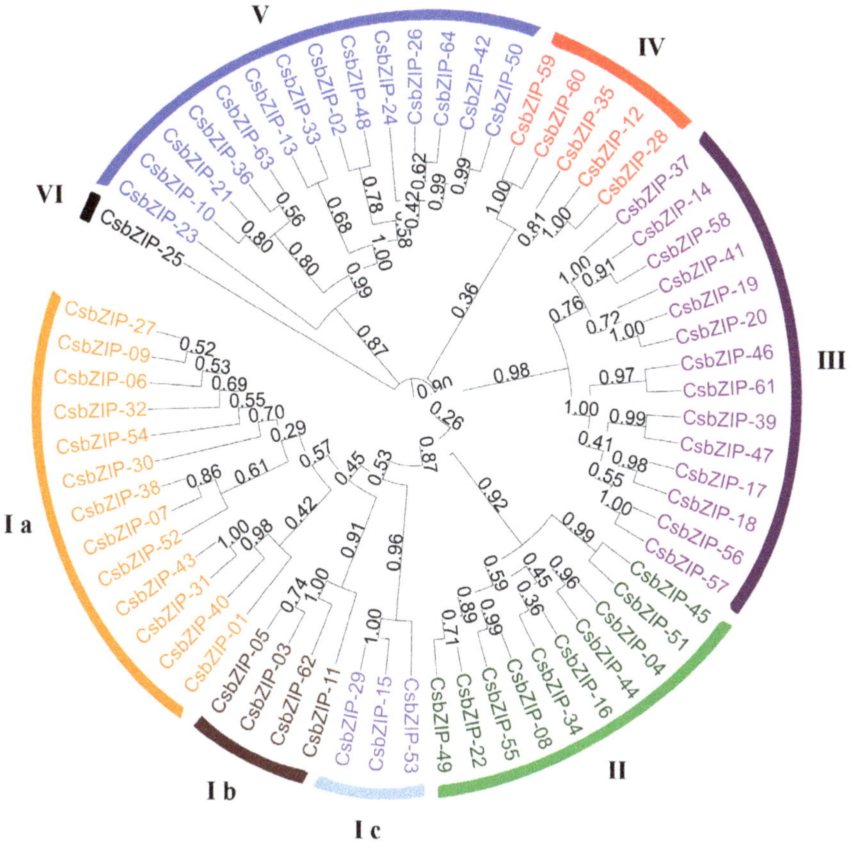

Figure 2. Phylogenetic relationships of cucumber bZIP proteins. The sequences were aligned by CLUSTALW at MEGA5 and the unrooted phylogenetic tree was deduced by neighbor-joining method. The proteins were classified into six distinct clusters. Each family was assigned a different color according to well-known members in other species.

Computational Identification of miRNAs Targeting the bZIP Genes

Identification of miRNA-regulated gene targets is crucial for understanding miRNA functions. Previously known plant pre-miRNA sequences obtained from miRBase v20.0 (http://www.mirbase.org) and plant miRNA database (http://bioinformatics.cau.edu.cn/PMRD) were used for identification of miRNAs targeting the *CsbZIP* genes. Therefore, the putative targets of all plant and *C. sativus* miRNAs were identified by aligning all known plant miRNAs with the assembled transcripts of CsbZIPs using the web-based psRNA Target Server (http://plantgrn.noble.org/psRNATarget) with default parameters. Alignment between all known plant miRNA and its potential target(s) was evaluated by the parameters described in Zhang et al. [81]. These computationally identified miRNA targets were further analyzed using BLASTX searches with ≤1e-10 against *C. sativus* EST sequences at NCBI database to identify putative gene homologous for confirmation.

Expression Profiling of *CsbZIP* Genes using Transcriptome Data

All Illimuna HiSeq reads for RNA-Seq analysis were retrieved from a public repository database (SRA, Sequence Read Archive) following accession numbers; SRR351476 (cucumber ovary tissue, unexpanded), SRR351489 (cucumber expanded ovary tissue, fertilized), SRR351495 (cucumber expanded ovary tissue, unfertilized), SRR351499 (cucumber root tissue), SRR351905 (cucum-

ber stem tissue), SRR351906 (cucumber leaf tissue), SRR351908 (cucumber male flower tissue), SRR351910 (cucumber tendril tissue), SRR351911 (cucumber tendril tissue basal) and SRR351912 (cucumber female flower tissue). All reads were downloaded in raw sequencing data ".sra" format and converted to "fastq" format by the NCBI SRA Toolkit's fastq-dump command. After discarding low-quality reads (Phred quality (Q) score <20) and trimming adapters by using FASTX toolkit, all clean reads were subjected to FastQC analysis for checking read qualities in terms of per-base sequence qualities, per-sequence quality scores, per-base nucleotide content and sequence duplication levels.

Following all preprocessing steps outlined above, the high-quality 75-b pair end reads were aligned to a Bowtie2-indexed *C. sativus* genome (v1.0) using the TopHat alignment software suite (http://tophat.cbcb.umd.edu). In RNA-Seq analysis, read counts have been found to be linearly related to the abundance of the transcripts. Therefore, before estimating read counts, the BAM (aligned read) file generated by TopHat must first be processed by appropriate software tools. SAMtools (v0.1.19) command tool was used for sorting, converting and indexing BAM files. For estimating raw counts, BEDTools (2.16.2) was used to estimate the number of raw reads by calculating mapped reads. The raw count data were normalized using DESeq's normalization step. Then, a variance-stabilizing transformation (VST) was performed on the normalized gene data set for downstream gene by applications, such as gene, gene expression measurement and hierarchical clustering. A two-way hierarchical clustering heat

Table 1. Amino acid composition of the cucumber bZIP motifs.

Motif No.	Sites	E-value	Amino acid sequence composition of motif	Width (aa)	Domain
Motif 1	50	3.4e-771	D[EPQ][KR][RK]Q[RK]R[MIL][IL][SAK] NR[EQ]SA[RA]RSR[LEM]RK[QK][AR] [YH][LI]x[EQ]L	29	bZIP_Basic
Motif 2	13	4,90E-201	[SG][KV]AAKAD[VI]FH[LIV][FL] [ST]G[MP]WKT[PS]AERC[FL] LW[IL]GGFR[PS]SEL	35	Transcription factor TGA like domain
Motif 3	20	8,60E-265	ER[KQ]VQTLQ[TA]E[AN][TS][TS]LSA [QRE][LV][TA][LFD]L[QS][QR][DQ] [TYR][LN]G[LA][TNS][VT][ED]N[SR] [EA]LK[LQ]R[LI][QEA][AT][LM][ER] [QA][QK][AVK][QH]L[RAK][DE]	50	NA
Motif 4	11	3,30E-148	LE[GS]F[IL]RQAD[NL] LRQQTLQ[QR][MV]HRILTTRQ[AS] ARALLAI[AG][ED]YFSRLRA	44	CAMP-response element binding protein
Motif 5	9	1,40E-160	F[DE][MV]EYARW[LV][ED]E[HQ][NHQ] R[LQ][IM][NC][ED]LRAA[VL][NQ] [SA]H[LA][SGPT]D[TI]ELRI[IL]V[DE] [GSN][CIV][LIM][AT]HYDE[LFIV]FR[LM]K	50	Transcription factor TGA like domain
Motif 6	9	5,40E-133	[LI]K[LIMV]L[VM][PSN]Q[LI][ED] [PT]LT[ED]QQ[LI][ML][GE]IC[NK]LQQSSQ[QE] [AT]EDALSQG[ML][ED][AKQ]L [QH]Q[SN]L[AIS][ED][TS][LIV][AS][SG]	50	bZIP_Basic
Motif 7	32	1,10E-90	A[EL]N[EA]ALK[EA]E[VI]QRL[KR]x[LA]Lx[QDE]Lx	21	NA
Motif 8	9	1,60E-59	SR[IL]KL[TA]QLEQ[ED]L[QEH]RAR[QS]QG[IL]F	21	NA
Motif 9	6	7,10E-43	GK[DNP][FL]GSMN[ML]DE[LF]LKNIWTAE[EA]NQ[TA][MV]	25	NA
Motif 10	10	6,30E-44	[PT][SRY][FW][RLS][MDV][DN][ISL][SEL] [NHKR][ME][PS][ED][NAT][PL][PHRV] [RP][RGN][SKV][GHA]HRR[SA] [HNS]S[DE][ISTV][SFLP][FAT]	29	NA
Motif 11	7	2,80E-39	[NSY]LQRQ[GA]S[LF][ST]LPR[APT]L [SC][GQ]KTVDEVW[KS][ED]I[HQ]	27	NA
Motif 12	7	2,20E-37	[RS]Q[PQ]TLGE[MV]TLE[DE]FL[IV][KR]AGVV[RA]E	22	NA

map was constructed based on these expression values, using euclidean distance metric and average linkage method. Heat maps and hierarchical clustering were computed with the gplots package in R.

Plant Materials, Growth Conditions, and Treatments

Cucumber seeds (Altay cultivar) were obtained from Monsanto Gıda ve Tarım (Antalya, Turkey). The seed coats were removed, and the seeds were washed with distilled water 3 times. Then, they were transferred to plastic containers and grown in hydroponic culture containing half-strength Hoagland's Solution [82] for 14 days in a plant growth chamber at $24 \pm 2°C$ with a 16 h light and 8 h dark photoperiod at a light intensity of 400 μmol m^{-2} s^{-1}. For drought stress, 10% polyethylene glycol 6000 (PEG-6000) was added to the half-strength Hoagland's solution. Stress treatment was initiated on the 14th day of normal growth. Both treated (stress) and non-treated (control) plants were kept in the growth chamber with the same growth conditions. Samples from the treated and control plants were harvested after 0, 3, 12, and 24 h of stress application. Time point zero (0 h) was used as a control. The roots and leaves of mature plants were collected separately and used for tissue-specific expression analysis. The tissues from 3 biological replicates were immediately frozen in liquid nitrogen.

RNA Extraction and Quantitative Real-time PCR Analysis

Total RNA extraction was performed with TRIzol reagent (Life Technologies Corporation, Grand Island, NY, USA). DNA contamination in samples was removed with DNase I (Fermentas, Thermo Fisher Scientific, Waltham, MA, USA) according to the manufacturer's instructions. The quality and integrity of the total RNA was checked with agarose gel electrophoresis and the NanoDrop 2000D (NanoDrop Technologies, Wilmington, DE, USA).

For RT-PCR, the specific primers were designed according to the *bZIP* gene sequences by Primer 5 software (http://www.primer-e.com/index.htm) (Table S8). Based on the literature search, highly expressed *bZIP* genes under the drought stress were selected for quantitative real-time-PCR. A cucumber *18S rRNA* gene (GenBank ID: X51542.1), amplified with primers 5'-GTGACGGGTGACGGAGAATT-3' and 5'- GACAC-TAATGCGCCCGGTAT-3', was used as a control. A suitable program was optimized according to primers Tm temperatures. Three biological replicates were carried out and triplicate quantitative assays for each replicate were performed using SYBR Green PCR Master mix kit (Roche Applied Science). The cucumber *18S rRNA* gene was used as an internal control. Relative gene expression was calculated. The ΔCT and $\Delta\Delta CT$ were calculated by the formulas $\Delta CT = CT$ target - CT reference and $\Delta\Delta CT = \Delta CT$ treated sample - ΔCT untreated sample (0 h treatment). For all chart preparations, selected RNA relative

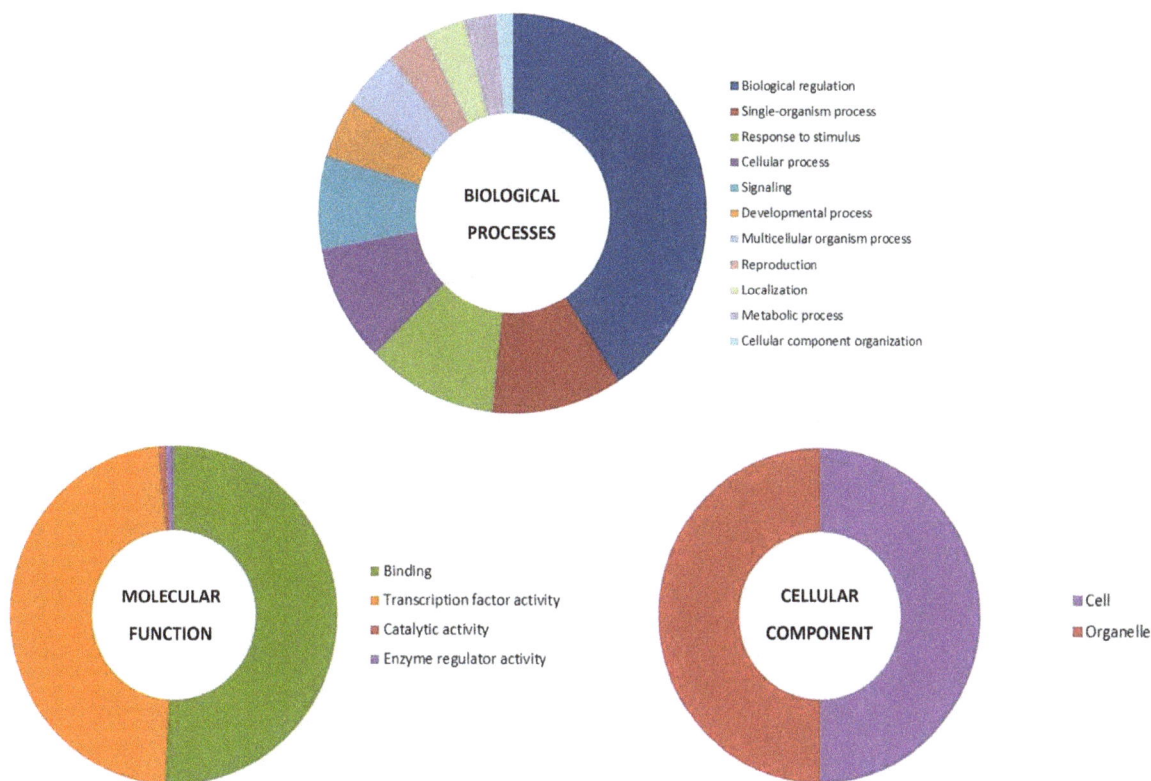

Figure 3. Gene Ontology (GO) distributions for the bZIP proteins. The Blast2Go program defines the gene ontology under three categories (A) biological processes, (B) molecular functions and (C) cellular component.

amount was evaluated for gene expression level using the 2-ΔΔCT. At the same time, the standard errors of mean among replicates were calculated. Student's t-test was used to obtain the statistical significance of the difference between treated samples and untreated samples (0 h treatment under abiotic stress). If P-values <0.01, we considered the *bZIP* genes as differentially expressed genes.

Results and Discussion

Genome-wide Identification of the bZIP Gene Family TFs in the Cucumber Genome

To identify *bZIP* TF genes in cucumber, both BLAST and Hidden Markov model (HMM) searches were performed. To better explore their expansion mechanisms, evolutionary history and expression divergence, seven plant genomes including Arabidopsis, grape, maize, papaya, poplar, rice, and sorghum were used for searches. After multiple cycles of searches, a total of 741 putative *bZIP* genes were detected in these plant genomes. These members were then subjected to the Pfam and SMART domain searches to validate the presence of the bZIP-related

Table 2. A summary of comparative mapping of cucumber *bZIP* genes on Arabidopsis, rice and poplar.

Cucumis sativus	Arabidopsis thaliana	Oryza sativa	Populus trichocarpa
Chr 1 (4)	Chr 5 (50%)	-	-
Chr 2 (10)	Chr 1 (90%)	Chr 2 (40%)	Chr 5 (30%)
Chr 3 (15)	Chr 1 (73%) Chr 7 (46%)	Chr 12 (27%) Chr 9 (20%)	Chr 6 (40%)
Chr 4 (9)	Chr 3 (55%)	-	Chr 2 (33%)
Chr 5 (5)	Chr 5 (60%)	Chr 3 (40%) Chr 7 (40%)	Chr 13 (40%)
Chr 6 (11)	Chr 3 (46%)	Chr 1 (45%)	Chr 5 (27%)
Chr 7 (8)	Chr 2 (88%) Chr 4 (50%)	Chr 2 (25%) Chr 6 (25%)	Chr 4 (75%) Chr 9 (50%)
Scaffolds (2)	-	-	-
Organism (64)	73%	56%	91%

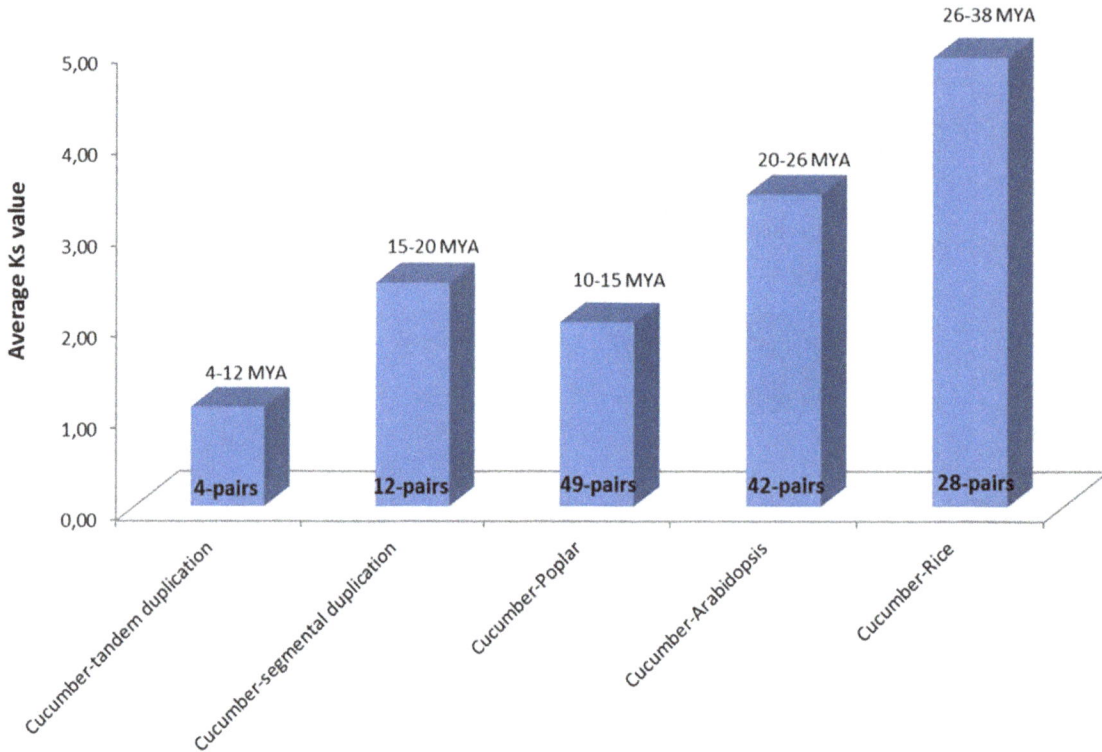

Figure 4. Time of duplication and divergence (MYA) based on synonymous substitution rate (Ks), which estimated using duplicated *bZIP* gene pairs of cucumber and orthologous *bZIP* gene pairs between cucumber and poplar (49) or Arabidopsis (42) or rice (28).

domains. By removal of different transcripts of the same gene, we identified 64 putative *CsbZIP* genes (Table S1). Both the search outputs showed the presence of bZIP-related domains in all the 64

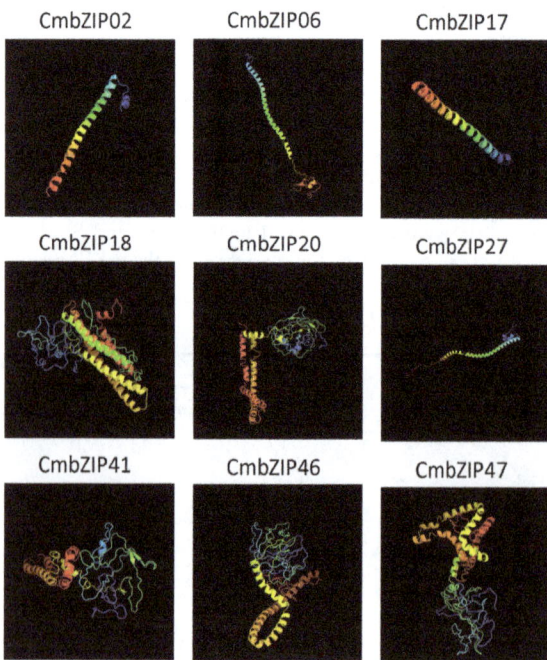

Figure 5. Predicated structures of bZIP proteins. The structure of 9 bZIP proteins with >90% confidence level are shown.

CsbZIP genes. For convenience, the 64 *CsbZIP* genes were named from *CsbZIP-01* to *CsbZIP-64* based on their order on the chromosomes, from chromosomes 1 to 7. Two *bZIP* genes (Cucsa.213060, Cucsa.365420) that could not be conclusively mapped to any chromosome were renamed *CsbZIP-63* and *CsbZIP-64*, respectively.

The *CsbZIP* genes vary substantially in the size and sequences of their encoded proteins, and their physicochemical properties. The location of the bZIP-related domains within the protein also differs. Protein length of CsbZIPs varied from 132 to 721 amino acids. EXPASY analysis suggested that the CsbZIP protein sequences had large variations in isoelectric point (pI) values (ranging from 4.6 to 9.9) and molecular weight (ranging from 15.308 kDa to 78.470 kDa). The details of other parameters of CsbZIP protein sequences were summarized in Table S1.

In Arabidopsis, a total of 75 and 77 *bZIP* gene family members have been identified by Jakoby et al. [10] and Correa et al. [15], respectively. However, Wang et al. [14] detected with the incomplete bZIP domain or lacked this domain based on the Pfam or SMART domain searches. So, they listed 72 members of *bZIP* gene in Arabidopsis [14]. Rice *bZIP* gene family was previously identified on a genome-wide level [13], [15]. Although these two studies reported the presence of 89 *bZIP* genes in rice, Wang et al. [14] made a detailed comparison with the sorghum bZIP family and found 88 bZIP TFs. Due to lack of bZIP domain, they eliminated some of the *bZIP* genes. Also, it has been reported that castor bean [11], maize [12], sorghum [14] and poplar [15] genomes encode 49, 125, 92 and 89 members of the *bZIP* gene family, respectively.

Chromosomal Distribution and Structure of CsbZIP

All cucumber *CsbZIP* gene members were physically mapped in all the 7 chromosomes of cucumber. Among all, chromosome 3 contains the highest number of *CsbZIP*s (23.4%), while minimum genes were distributed on chromosome 1 (6.3%) (Figure 1A). The exact position (in bp) of each *CsbZIP* on cucumber chromosome is given in Table S1. Distribution pattern of the *CsbZIP* genes on individual chromosomes also indicated certain physical regions with a relatively higher accumulation of gene clusters. For example, *CsbZIP* genes located on chromosomes 2, 3 and chromosomes 7 appear to be congregate at the lower end and upper end of the arms, respectively (Figure 1B).

Tandem duplication of *CsbZIP* gene members was also determined. Tandem duplicated genes on a particular chromosome were indicated in the box as shown in Figure 1B. Totally, we detected only four pairs of tandem duplicates (Table S2), indicating the limited contribution of tandem duplication to the gene family expansion. A similar result has been observed in the sorghum, rice and Arabidopsis genomes [14]. The distance between these genes ranged from 1 kb to 33.4 kb. We then carried out a genome-wide identification of segmentally duplicated *bZIP* genes in cucumber. Totally, 12 segmental duplicated cucumber *bZIP* genes have been detected, accounting for around 19% (12/64) of total *CsbZIP* genes (Table S3). Segmental duplication has been regarded as a major driver to contribute to the expansion of gene families. Segmental duplication rate of the *bZIP* genes was also examined in other plant species such as Arabidopsis, rice and sorghum, which ranging between 53% and 59% [14]. They showed higher segmental duplication rate, when compared to rate of cucumber *bZIP* genes.

Exon-intron organization of the 64 cucumber *bZIP* genes was also investigated to obtain some insight into their gene structures (Figure S1). We have detected a total of 12 *bZIP* genes with no intron, accounting for 18.75% of total *CsbZIP* genes. Similar cases have also been observed in Arabidopsis, castor bean, rice and sorghum [10–11], [13–14], suggesting the evolutionary conservation. Most of these intronless genes were clustered into the Cluster I a (Figure S1). Among the intron-containing *bZIP* genes in cucumber, the number of introns in their open reading frames varied from 1 to 12. They were distributed into different classes of the bZIP family. In castor bean, there were 11 introns [11], being 12 in Arabidopsis and rice, [10], [13] and 14 in sorghum [14].

Phylogenetic Classification of CsbZIPs and Identification of Domain Conservation

The comprehensive phylogenic analysis was performed to understand the evolutionary significance of domain architecture in CsbZIP proteins. The phylogenetic tree was constructed with 64 CsbZIP proteins by Neighbour-Joining (NJ) method. The phylogenetic analysis categorized all the CsbZIPs into six discrete groups (Cluster I to VI) comprising of 20, 10, 14, 5, 14, and 1 proteins, respectively (Figure 2). Since a good number of the internal branches were observed to have high bootstrap values, it was clearly through bootstrap analysis of 1000 replicates. A good number of the internal branches had high bootstrap values, reflecting derivation of statistically reliable pairs of possible homologous. Phylogeny-based function prediction has been applied for prediction of bZIP proteins in other species like rice, Arabidopsis and soybean. A total of 7 and 10 groups of bZIP TFs have been classified in sorghum and Arabidopsis, respectively. They were named as group A to I and S [10], [14]. Compared with the nomenclature of Arabidopsis and sorghum, both group I and E have been combined into the class 1; class 2 consists of group B, D, F and H; and group S has been divided into classes 6

and 7. In addition, rice OsbZIP transcription factors were subdivided into 10 clades, designated A to J, with well-supported bootstrap values [13]. In a different study, totally 333 sequences were analyzed to indicate phylogenetic relationship of bZIP transcription factors among maize (170 sequences), rice (89 sequences) and Arabidopsis (74 sequences) [12]. The three plant species had some divergences against the rice OsbZIP proteins in the phylogenetic tree. Certain members of groups were separated from their clusters which have also been observed by Nijhawan et al [13]. So, it can be concluded that the interspecies clustering indicates a parallel evolution of bZIP transcription factors in three plants.

Additionally, reliability of the phylogeny was further evidenced by parameters like motif compositions. MEME analysis identified 12 motifs according to their domain compositions of the cucumber bZIPs (Table 1, Figure S2). Except for CsbZIP10, CsbZIP21, CsbZIP36, and CsbZIP63, all CsbZIP proteins contain Basic-leucine zipper domain (Motif 1). Besides bZIP basic domain, two other known functional domains were classified into the CsbZIPs. Fourteen members containing the transcription factor TGA-like domain (Motif 2) were identified as Cluster V in which CsbZIP10, CsbZIP21, CsbZIP36, and CsbZIP63 were found. Eleven members contain CAMP-Response Element Binding Protein (Motif 4). Further, seven unidentified conserved motifs were found. It was observed that a majority of the members, predicted to have similar DNA-binding properties, clustered together. However, certain members of Clusters Ib, III, and V were exceptions because they clustered apart into different clades. All CsbZIP proteins in Cluster V have carried transcription factor TGA like domain. In addition, CsbZIP proteins belonging to Cluster Ib and Cluster III contain an unidentified conserved motif, which was named as Motif 3 (Table 1). Most of the members belonging to one cluster also shared one or more conserved motifs outside the bZIP domain. In addition, 4 tandemly duplicated and 12 segmentally duplicated CsbZIP proteins were located on the same cluster. For example, tandemly duplicated CsbZIP17-CsbZIP18 and CsbZIP56-CsbZIP57 were found in Cluster III; segmentally duplicated CsbZIP proteins (CsbZIP2-CsbZIP48, CsbZIP13-CsbZIP33, CsbZIP26-CsbZIP42, CsbZIP42-CsbZIP50 and CsbZIP50-CsbZIP64) were located on the Cluster V. Such motif-sequence conservation or variation between the proteins specifies the functional equivalence or diversification, respectively, with respect to the various aspects of biological functions [83]. Apart from the bZIP domain region, bZIP proteins usually contain additional conserved motifs, which might indicate potential function sites or participate in activating the function of bZIP proteins. As reported in some earlier studies, diverse conserved motifs outside of the bZIP domain region have been identified in Arabidopsis, castor bean, maize and rice [10–13]. Compared to those conserved motifs identified within bZIP transcription factors in other plants, eight motifs (2, 4, 5, 6, 8, 9, 14 and 19) were commonly shared by cucumber, castor bean, Arabidopsis, rice and maize, indicating that these additional motifs outside of bZIP domain might be conserved among plant species. However, some of the other motifs were variable among species and might be species-specific in plants.

Gene Ontology Annotation

The GO slim analysis conducted through Blast2Go showed the putative participation of 64 CsbZIP proteins in diverse biological processes (Figure 3, Table S4). Of the 11 categories of biological processes defined by Blast2Go, predominant of CsbZIPs were predicted to function in response to biological regulation (~41%) [60], followed by single-organism process and response to stimulus

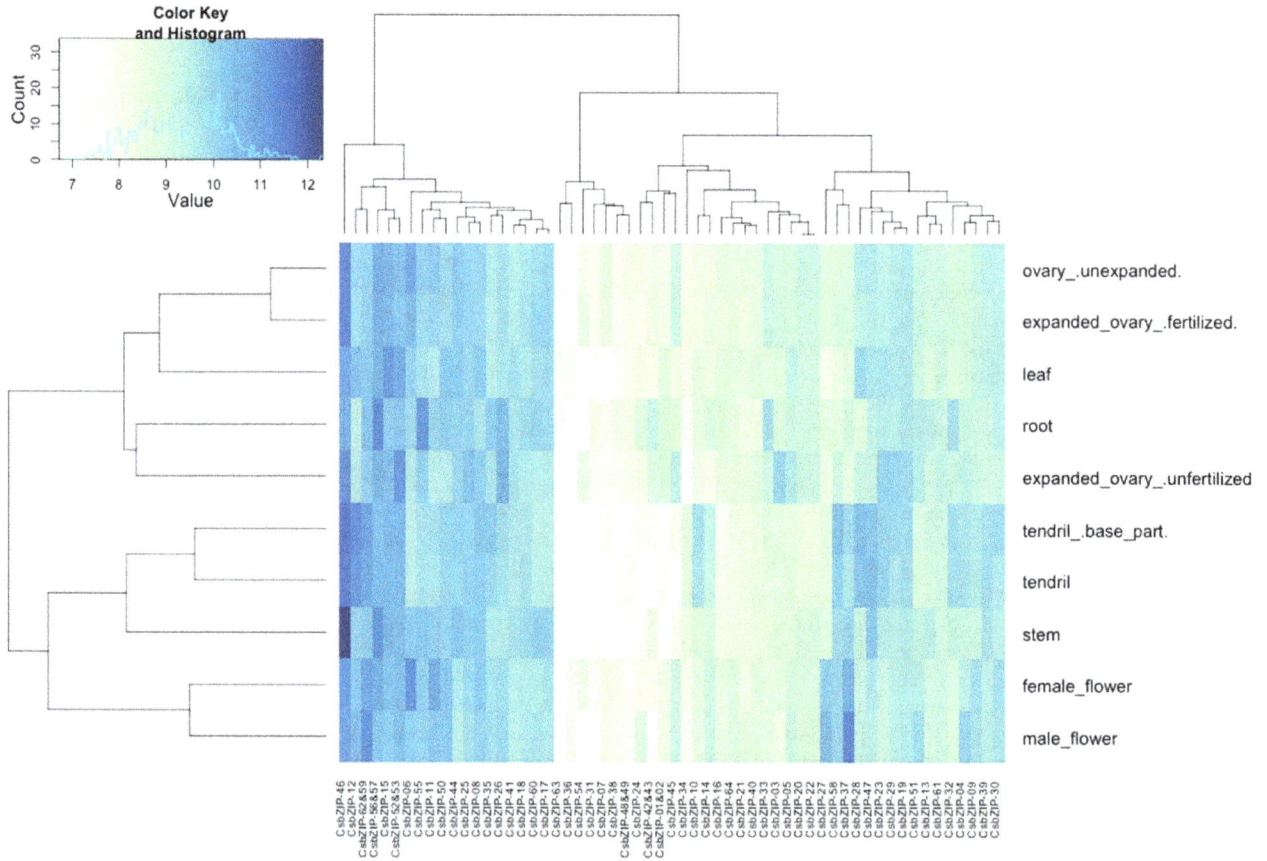

Figure 6. Heatmap of the differentially expressed *CsbZIP* genes in different tissues. The image summarizes the tissue-specific expression pattern of 64 bZIP transcription factors. Note that expression values mapped to a color gradient from low (plain green) to high expression (dark blue).

(~11%) [16]. Molecular-function prediction showed that about 64 (~51%) CsbZIP were evidenced to participate in small molecule binding, which concords with the molecular role of bZIP proteins in assisting protein-protein interactions. Regarding molecular function, about 60 (~48%) CsbZIP showed transcription factor activity, which correlates with abiotic stress tolerance behavior of

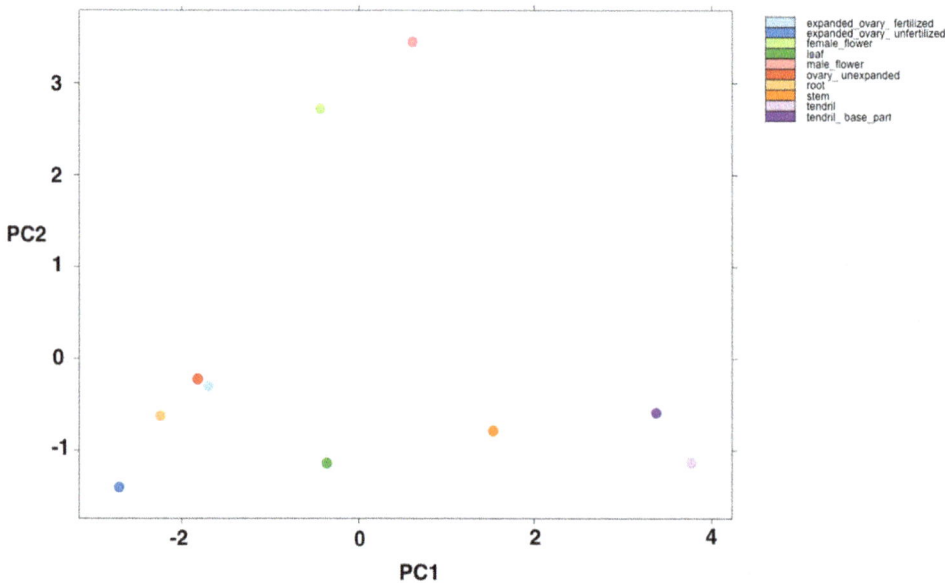

Figure 7. PCA score plots of different tissues. The graph shows a clear separation between flower and other tissues.

cucumber. Cellular localization prediction showed that 17 (~50%) CsbZIP proteins are localized in the cell part, of which 17 (~50%) are organelle-localized (Figure 3; Table S4).

Orthologous Relationships of bZIP Genes between Cucumber and other Species

For comparative mapping to derive orthologous relationships of CsbZIPs, the physically mapped CsbZIP genes were compared with those in chromosomes of other related genomes, namely Arabidopsis, rice and poplar. (Table 2, Figure S3). Of the identified 64 CsbZIP protein-encoding genes in cucumber, the specific orthologous relationships could be derived on an average for ~73.3% proteins. Maximum orthology of CsbZIP genes annotated on the cucumber chromosomes was obtained with poplar (91%), followed by Arabidopsis (73%) and rice (56%). The extensive gene-level synteny shared among cucumber, poplar and Arabidopsis supports their close evolutionary relationships. Interestingly, most of CsbZIP genes revealed syntenic biases towards particular chromosomes of Arabidopsis and poplar. For instance, maximum orthology was obtained between CsbZIP genes on cucumber chromosome 2 and Arabidopsis chromosome 1 (90%). In addition, the CsbZIP genes on cucumber chromosome 7 showed 88% orthology and colinearity with Arabidopsis chromosome 2 and poplar chromosome 4 (75%) (Table 2). A similar result was observed between Arabidopsis, rice and sorghum [14]. Total 72 sorghum bZIP genes could find their corresponding 66 rice orthologs. Also, six Arabidopsis bZIP genes were detected with seven orthologs in the rice genome [14]. The results indicated that chromosomal rearrangements like duplications and inversions were predominant in shaping the distribution and organization of CsbZIP genes in cucumber, Arabidopsis, rice and poplar genomes. The information from comparative mapping provides a useful preface for understanding the evolutionary process of bZIP genes among cucumber and other plant species. This can be also useful for isolation of orthologous bZIP genes from cucumber, using the map-based genomic information of other related plant-species for genetic enhancement.

Duplication and Divergence Rate of the CsbZIP Genes

Multiple copies of genes in a gene family could possibly evolve due to the flexibility provided by events of whole-genome tandem and segmental duplications. Gene duplication, either segmental or tandem, has been documented in several plant TF gene-families such as NAC, MYB, F-box, as well as in bZIP [12–14], [83–85]. We thus explored association of Darwinian positive selection in divergence and duplication of bZIP genes to understand the expansion of this gene family. To interpret this, the ratios of non-synonymous (Ka) versus synonymous (Ks) substitution rates (Ka/Ks) were estimated for four tandem and 12 segmentally duplicated gene-pairs, as well as between orthologous gene-pairs of CsbZIP with those of poplar (49-pairs), Arabidopsis (42) and rice (28). The ratios of Ka/Ks for tandem duplication varied from 0.09 to 0.20 with an average of 0.13 (Table S2), whereas Ka/Ks for segmentally duplicated gene-pairs ranged from 0.04 to 0.23 with an average of 0.11 (Table S3). It suggested that the duplicated CsbZIP genes are under strong purifying selection pressure, since their Ka/Ks ratios were estimated as <1. Additionally, the duplication event of these tandemly and segmentally duplicated genes may be estimated to have occurred around 4–12 and 15–20 Mya, respectively (Figure 4). Among the orthologous gene-pairs of CsbZIP with those of other plant species, the average Ka/Ks values were maximum between poplar and cucumber (0.14) and minimum for Arabidospsis and rice-cucumber gene-pairs (0.10; Table S5). Although synonymous substitution rates between rice-

cucumber and Arabidopsis-cucumber bZIP genes were the same, the earlier divergence was observed from rice-cucumber around 26–38 Mya, as compared to Arabidopsis-cucumber bZIP genes (20–26 Mya). Conversely, the bZIP gene-pairs between poplar and cucumber seem to have largely encountered intense purifying selection, as compared to rice-cucumber and Arabidopsis-cucumber bZIP genes. It agreed well with their recent time of divergence around 10–15 Mya (Figure 4). The estimation of tandem and segmental-duplication time (average of 8.3 Mya and 18.8 Mya, respectively) of cucumber bZIP genes in between the divergence time of cucumber-rice (37.8 Mya) and -Arabidopsis (26.2 Mya) and -poplar (15.5 Mya) orthologous bZIP gene-pairs are comparable to evolutionary studies, involving the protein-coding genes annotated from the recently released draft genome sequences of cucumber [1]. Interestingly, the CsbZIP gene-pairs showing tandem and segmental duplication events are under similar evolutionary pressure (Ka/Ks = 0.13, Ka/Ks = 0.11, respectively), of which the tandemly-duplicated genes revealed much recent duplication events (average 8.3 Mya), in contrast to that estimated for segmentally duplicated gene-pairs (average 18.8 Mya). It overall suggests that the segmental and tandem duplication events have played a predominant role in evolution, for shaping such gene family in foxtail millet. A total of three pairs of tandem duplicates were detected in the sorghum and a similar result have been also observed in the rice and Arabidopsis genomes [14]. Genome-wide analysis of segmentally-duplicated bZIP genes showed that in rice, sorghum and Arabidopsis, a total of 52, 49 and 39 bZIP genes have been also detected, respectively [10], [13–14]. It can be concluded that these segmentally-duplicated genes have contributed to the expansion of multiple classes of bZIP genes in different plant species.

Identification of miRNAs Targeting CsbZIP Transcripts

Two important parameters have been taken into consideration for identification of targets. To score the complementarity between miRNA and their target transcript, we have applied the scoring schema according to miRU [81]. The maximum expectation is the first one, which is the threshold of the score. A miRNA/target site pair has been discarded if its score is greater than the threshold. The default cut-off threshold was adjusted to 3.0. The second one is an UPE defining as maximum energy to unpair the target site. The accessibility of mRNA target site to miRNA has been identified as one of important factors that are involved in target recognition. The psRNATarget server employs RNA to calculate target accessibility, which is represented by the energy required to open (unpair) secondary structure around target. The less energy means the more possibility that small RNA is able to contact (and cleave) target mRNA.

A total of 21 CsbZIP genes (CsbZIP4-8-13-15-20-22-23-24-37-39-40-45-46-47-48-49-52-53-55-56-57) targeted by 38 plant miR-NAs were identified in cucumber by psRNATarget: A Plant Small RNA Target Analysis Server. However, some plant miRNAs could not indicate any gene target. Among the target genes, CsbZIP20 and CsbZIP22 are the most abundant transcripts, which were targeted by all 38 plant miRNAs (Table S6). According to the results of BLASTX analysis of the identified miRNA targets, many of the targets were homologous to conserved target genes existing in other plants species; these targets included bZIP transcription factors (also called as abscisic acid-insensitive 5-like protein, transcription factor TGA7-like, transcription factor RF2b-like) transcription factor HBP-1a-like, G-box-binding factor 4-like, transcription factor VIP1-like, TGACG-sequence-specific DNA-binding protein and some uncharacterized proteins. Most of these targets were found to be responsible for plant growth and response

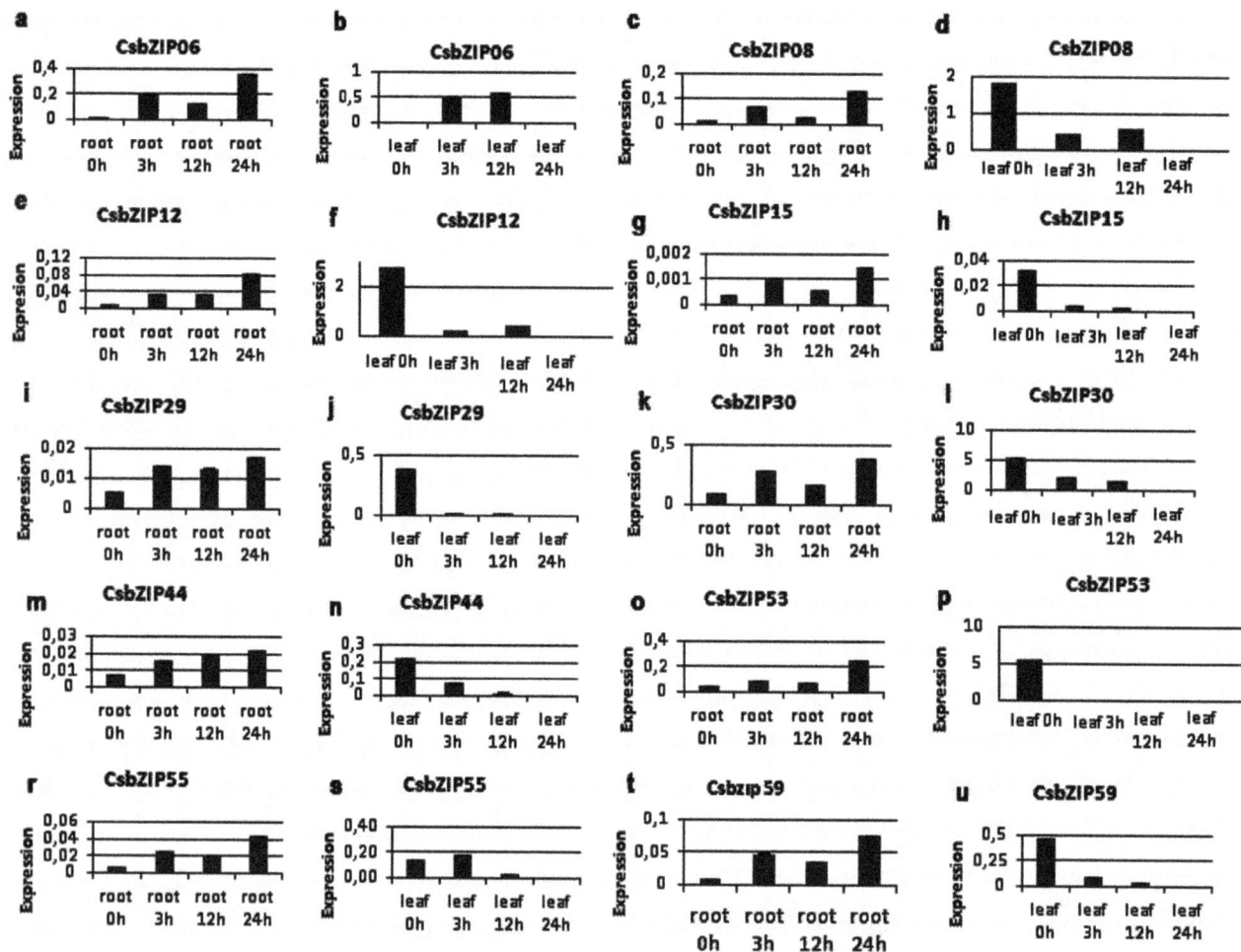

Figure 8. Expression profiles of selected cucumber *bZIP* genes under drought stress. Relative expression levels of the genes upon 0, 3, 12, and 24 h of PEG treatment in root (a, c, e, g, i, k, m, o, r and t) and leaf (b, d, f, h, j, l, n, p, s and u) tissues are shown.

to environmental changes. For example, the target transcript of miR166 was ABA-insensitive gene (*ABI5*), which functions in plant development [86] and in response to stress stimulus, such as NaCl, drought, ABA and cold stress in Arabidopsis [87].

Homology Modeling of CsbZIP Proteins

BLASTP search was performed against the PDB in order to build the homology model. Nine CsbZIP proteins (CsbZIP2-6-17-18-20-27-41-46-47) having higher homology were selected. Detection rate was utilized for prediction of homology modeling in Phyre2, which uses the alignment of hidden Markov models via HMM-HMM search [88] to significantly improve the accuracy of alignment. The intensive mode of Phyre 2 uses the multi-template modeling for higher accuracy. Furthermore, it integrates a new *ab initio* folding simulation termed as Poing [89] to model regions of proteins with no noticeable homology to known structures. The protein structure of all the nine CsbZIP are modelled at >90% confidence and the percentage residue varied from 80 to 100 (Figure 5). The secondary structure predominantly comprised of α helices, with rare occurrence of β sheets. Hence, all the predicted protein structures are considered highly reliable, and this offers a preliminary basis for understanding the molecular function of CsbZIP proteins.

Genome-wide Tissue-specific Expression Profile of bZIP Transcription Factors

One of the main purposes of a gene expression profiling on a genomic scale is to determine those genes that are differentially expressed within the organism being studied. Thus, to gain insight into the tissue-specific gene expression patterns of bZIP transcription factors in cucumber, a RNA-Seq approach was applied to data sets obtained from SRA. After normalization using DEseq and variance-stabilizing transformation for gene expression, *bZIP* genes were ranked from highest to lowest according to their differential expression across main tissues. Hierarchically clustered heat map in Figure 6 showed that *CsbZIP-46* gene was highly expressed in ovary (unexpanded, unfertilized and fertilized), stem, tendril, tendril base tissues, whereas *CsbZIP-06*, *CsbZIP-15*, *CsbZIP-37* and *CsbZIP-56/57* genes were found to be highly expressed in female flower, leaf, male flower and root tissues, respectively. Moreover, while *CsbZIP-36* and *CsbZIP-63* genes had their lowest expression level in reproductive tissues (all ovary tissues and female flower), the lowest expression level of *CsbZIP-54* was observed particularly in vegetative tissues (root, stem, leaf and tendril) and male flower. But when considering all of *CsbZIP* genes with expression level, we calculated and compared the coefficient of variation (CV) of *bZIP* genes for determining the most variable expression phenotypes among the tissues (Table S7). Although

CsbZIP-54 exhibited very low expression in many tissues (Figure 6), its expression was most variable among tissues according to the coefficient of variation value, contrary to *CsbZIP-61*, *CsbZIP-18* and *CsbZIP-41* whose expressions showed minimal variation among tissues. According to the dendrogram (hierarchical clustering of 64 *CsbZIP* genes) above the heat maps, the expression level of *CsbZIP-27*, *CsbZIP-58* and *CsbZIP-37* genes in female flower tissue showed remarkable similarity to those of *CsbZIP* genes in male flower. These transcription factors might exhibit a unique expression pattern in these reproductive tissues. Moreover, PCA score plots (Figure 7) and dendrogram (left side of the heat map) also showed a clear separation between flower and other tissues. Despite the partial similarity among genes, *CsbZIP* TFs have also a different expression pattern in ovary tissue of cucumber. Namely, the *CsbZIP* TFs from unexpanded ovary shared a higher degree of similarity with ovary-fertilized tissues than with ovary-unfertilized tissue in terms of gene expression level. Therefore, unexpanded ovary and fertilized ovary were clustered together in both PCA-space and dendrogram.

Drought-responsive *CsbZIP* Genes

To identify the drought-responsive cucumber bZIP transcription family members, the qRT-PCR analyses were performed. Expression levels of 10 distinct bZIP family members (*CsbZIP06*, *CsbZIP08*, *CsbZIP12*, *CsbZIP15*, *CsbZIP29*, *CsbZIP30*, *CsbZIP44*, *vbZIP53*, *CsbZIP55*, *CsbZIP59*) were measured in the drought-stressed cucumber leaves and roots. It was observed that all ten selected genes were up-regulated in cucumber roots upon 24 h PEG treatment. Among them, expression of the *CsbZIP12* and *CsbZIP44* showed gradual induction in root tissues with progressing drought stress condition. On the contrary, all selected *CsbZIP* genes were measured as down-regulated in drought-stressed leaf tissue. The expression levels of *CsbZIP29*, *CsbZIP30* and *CsbZIP44* genes were observed as gradually decreased with increasing drought exposure. Other measured *CsbZIP* family genes displayed various expression levels between 3 and 12 h of drought stress (Figure 8). According to the qRT-PCR results, the cucumber bZIP transcription factors were notably affected and showed changing expression levels in response to water deficiency in tissue-specific manner. Especially, relatively rapid increase in the accumulation of the bZIP transcripts in roots and suppression in leaf tissues were observed upon water deficiency. Such stresses as drought cause to huge crop looses all around the world. Therefore, understanding the plant responses at molecular level is crucial to improve the stress tolerance and productivity. The bZIP TFs are found in all eukaryotes, being considered as one of the largest families of TFs in plants, having diverse responsibilities such as abiotic stress response, seed maturation, flower development and pathogen defense [90]. However, it is not fully understood how they cope with abiotic stresses, particularly tolerating drought. Therefore, several functional studies have been done on the role of bZIP TFs against water-deficit stresses [91–93]. In this study, we measured the expression of ten *CsbZIP* TF genes to analyze their possible drought-responsive roles. The accumulation of the measured CsbZIP transcripts in root tissue was observed, whereas the suppression was detected for all of them in leaf upon water deficiency. It supports the general idea that the signals for water deficiency are generally realized first in roots [94]. Differential

expression profiles of the *bZIP* genes under drought stress suggest that some other genes functioning in water deficiency might also be regulated by this family. In other biotic and abiotic stress studies, similar results were reported [93], [95].

Supporting Information

Figure S1 Exon-intron organization of 6 classes of cucumber *bZIP* genes. The bZIP family was classified according to Figure 2. The values in parentheses indicate the number of correponding classes of *bZIP* genes. Exons and introns are represented by green boxes and black lines, respectively.

Figure S2 Variation in motif clades for the bZIP proteins. The MEME motifs are shown as differently-colored boxes at the N-terminal and C-terminal region for the transcription regulatory region.

Figure S3 Comparative physical mapping revealed high degree of orthologous relationships of *bZIP* genes located on seven chromosomes of cucumber with (A) Arabidopsis, (B) rice and (C) poplar.

Table S1 A catalog of 64 *Cucumis sativus* bZIP proteins.

Table S2 The Ka/Ks ratios and estimated divergence-time for tandemly-duplicated bZIP proteins.

Table S3 The Ka/Ks ratios and estimated divergence time for segmentally-duplicated bZIP proteins.

Table S4 Blast2Go annotation details of bZIP protein sequences.

Table S5 The Ka/Ks ratios and estimated divergence time for orthologous bZIP proteins between cucumber, Arabidopsis, rice and poplar.

Table S6 miRNA targets identified by psRNATarget.

Table S7 Expression patterns of the *CsbZIP* genes analyzed from eight RNA-Seq libraries.

Table S8 List of primers used in quantitative real-time-PCR expression analysis of bZIP genes.

Author Contributions

Conceived and designed the experiments: MCB TU. Performed the experiments: MCB MH. Analyzed the data: MCB VE. Wrote the paper: MCB TU.

References

1. Huang S, Li R, Zhang Z, Li L, Gu X, et al. (2009) The genome of the cucumber, *Cucumis sativus* L. Nat Genet 41: 1275–1281.

2. Lough TJ, Lucas WJ (2006) Integrative plant biology: role of phloem long-distance macromolecular trafficking. Annu Rev Plant Biol 57: 203–232.

3. Qi J, Liu X, Shen D, Miao H, Xie B, et al. (2013) A genomic variation map provides insights into the genetic basis of cucumber domestication and diversity. Nat Genet 45: 1510–1515.

4. Muthamilarasan M, Theriappan P, Prasad M (2013) Recent advances in crop genomics for ensuring food security. Curr Sci 105: 155–158.
5. Yang X, Li Y, Zhang W, He H, Pan J, et al. (2014) Fine mapping of the uniform immature fruit color gene *u* in cucumber (*Cucumis sativus* L.). Euphytica 196: 341–348.
6. Innark P, Khanobdee C, Samipak S, Jantasuriyarat C (2013) Evaluation of genetic diversity in cucumber (*Cucumis sativus* L.) germplasm using agro-economic traits and microsatellite markers. Sci Hort 162: 278–284.
7. Lv J, Qi J, Shi Q, Shen D, Zhang S, et al. (2012) Genetic Diversity and Population Structure of Cucumber (*Cucumis sativus* L.). PLoS ONE 7: e46919.
8. Zhang WW, Pan JS, He HL, Zhang C, Li Z, et al. (2012) Construction of a high density integrated genetic map for cucumber (*Cucumis sativus* L.). Theor Appl Genet 124 (2): 249–259.
9. Miao H, Zhang S, Wang X, Zhang Z, Li M, et al. (2011) A linkage map of cultivated cucumber (*Cucumis sativus* L.) with 248 microsatellite marker loci and seven genes for horticulturally important traits. Euphytica 182 (2): 167–176.
10. Jakoby M, Weisshaar B, Droge-Laser W, Vicente-Carbajosa J, Tiedemann J, et al. (2002) bZIP transcription factors in Arabidopsis. Trends Plant Sci 7(3): 106–111.
11. Jin Z, Xu W, Liu A (2014) Genomic surveys and expression analysis of bZIP gene family in castor bean (*Ricinus communis* L.). Planta 239: 299–312.
12. Wei K, Chen J, Wang Y, Chen Y, Chen S, et al. (2012) Genome-wide analysis of bZIP-encoding genes in maize. DNA Res 19: 463–476.
13. Nijhawan A, Jain M, Tyagi AK, Khurana JP (2008) Genomic survey and gene expression analysis of the basic leucine zipper transcription factor family in rice. Plant Physiol 146: 333–350.
14. Wang J, Zhou J, Zhang B, Vanitha J, Ramachandran S, et al. (2011) Genome-wide expansion and expression divergence of the basic leucine zipper transcription factors in higher plants with an emphasis on sorghum. J Integr Plant Biol 53(3): 212–231.
15. Correa LG, Riano-Pachon DM, Schrago CG, dos Santos RV, Mueller-Roeber B, et al. (2008) The role of bZIP transcription factors in green plant evolution: adaptive features emerging from four founder genes. PLoS One 3: e2944.
16. Wingender E, Chen X, Fricke E, Geffers R, Hehl R, et al. (2001) The TRANSFAC system on gene expression regulation. Nucleic Acids Res 29: 281–283.
17. Warren AJ (2002) Eukaryotic transcription factors. Curr Opin Struct Biol 12: 107–114.
18. Yilmaz A, Nishiyama MY Jr, Fuentes BG, Souza GM, Janies D, et al. (2009) GRASSIUS: a platform for comparative regulatory genomics across the grasses. Plant Physiol 149: 171–180.
19. Riechmann JL, Heard J, Martin G, Reuber L, Jiang C, et al. (2000) Arabidopsis transcription factors: genome-wide comparative analysis among Eukaryotes. Science 290: 2105–2110.
20. Iida K, Seki M, Sakurai T, Satou M, Akiyama K, et al. (2005) RARTF: database and tools for complete sets of Arabidopsis transcription factors. DNA Res 12: 247–256.
21. Zou M, Guan Y, Ren H, Zhang F, Chen F (2008) A bZIP transcription factor, OsABI5, is involved in rice fertility and stress tolerance. Plant Mol Biol 66: 675–683.
22. Walsh J, Waters CA, Freeling M (1998) The maize gene liguleless2 encodes a basic leucine zipper protein involved in the establishment of the leaf bladesheath boundary. Genes Dev 12: 208–218.
23. Chuang CF, Running MP, Williams RW, Meyerowitz EM (1999) The PERIANTHIA gene encodes a bZIP protein involved in the determination of floral organ number in *Arabidopsis thaliana*. Genes Dev 13: 334–344.
24. Abe M, Kobayashi Y, Yamamoto S, Daimon Y, Yamaguchi A, et al. (2005) FD, a bZIP protein mediating signals from the floral pathway integrator FT at the shoot apex. Science 309: 1052–1056.
25. Silveira AB, Gauer L, Tomaz JP, Cardoso PR, Carmello-Guerreiro S, et al. (2007) The Arabidopsis AtbZIP9 protein fused to the VP16 transcriptional activation domain alters leaf and vascular development. Plant Sci 172: 1148–1156.
26. Shen H, Cao K, Wang X (2007) A conserved proline residue in the leucine zipper region of AtbZIP34 and AtbZIP61 in *Arabidopsis thaliana* interferes with the formation of homodimer. Biochem Biophys Res Commun 362: 425–430.
27. Yin Y, Zhu Q, Dai S, Lamb C, Beachy RN (1997) RF2a, a bZIP transcriptional activator of the phloem-specific rice tungro bacilliform virus promoter, functions in vascular development. Embo J 16: 5247–5259.
28. Fukazawa J, Sakai T, Ishida S, Yamaguchi I, Kamiya Y, et al. (2000) Repression of shoot growth, a bZIP transcriptional activator, regulates cell elongation by controlling the level of gibberellins. Plant Cell 12: 901–915.
29. Ciceri P, Locatelli F, Genga A, Viotti A, Schmidt RJ (1999) The activity of the maize Opaque2 transcriptional activator is regulated diurnally. Plant Physiol 121: 1321–1328.
30. Weltmeier F, Ehlert A, Mayer CS, Dietrich K, Wang X, et al. (2006) Combinatorial control of Arabidopsis proline dehydrogenase transcription by specific heterodimerisation of bZIP transcription factors. Embo J 25: 3133–3143.
31. Baena-Gonzalez E, Rolland F, Thevelein JM, Sheen J (2007) A central integrator of transcription networks in plant stress and energy signalling. Nature 448: 938–943.
32. Iwata Y, Koizumi N (2005) An Arabidopsis transcription factor, AtbZIP60, regulates the endoplasmic reticulum stress response in a manner unique to plants. Proc Natl Acad Sci USA 102: 5280–5285.

33. Liu JX, Srivastava R, Che P, Howell SH (2007) Salt stress responses in Arabidopsis utilize a signal transduction pathway related to endoplasmic reticulum stress signaling. Plant J 51: 897–909.
34. Lara P, Onate-Sanchez L, Abraham Z, Ferrandiz C, Diaz I, et al. (2003) Synergistic activation of seed storage protein gene expression in Arabidopsis by ABI3 and two bZIPs related to OPAQUE2. J Biol Chem 278: 21003–21011.
35. Guan Y, Ren H, Xie H, Ma Z, Chen F (2009) Identification and characterization of bZIP-type transcription factors involved in carrot (*Daucus carota* L.) somatic embryogenesis. Plant J 60: 207–217.
36. Uno Y, Furihata T, Abe H, Yoshida R, Shinozaki K, et al. (2000) Arabidopsis basic leucine zipper transcription factors involved in an abscisic acid-dependent signal transduction pathway under drought and high-salinity conditions. Proc Natl Acad Sci USA 97: 11632–11637.
37. Kim S, Kang JY, Cho DI, Park JH, Kim SY (2004) ABF2, an ABRE binding bZIP factor, is an essential component of glucose signaling and its overexpression affects multiple stress tolerance. Plant J 40: 75–87.
38. Liu JX, Srivastava R, Howell SH (2008) Stress-induced expression of an activated form of AtbZIP17 provides protection from salt stress in Arabidopsis. Plant Cell Environ 31: 1735–1743.
39. Weltmeier F, Rahmani F, Ehlert A, Dietrich K, Schutze K, et al. (2009) Expression patterns within the Arabidopsis C/S1 bZIP transcription factor network: availability of heterodimerization partners controls gene expression during stress response and development. Plant Mol Biol 69: 107–119.
40. Yang O, Popova OV, Suthoff U, Luking I, Dietz KJ, et al. (2009) The Arabidopsis basic leucine zipper transcription factor AtbZIP24 regulates complex transcriptional networks involved in abiotic stress resistance. Gene 436: 45–55.
41. Yoshida T, Fujita Y, Sayama H, Kidokoro S, Maruyama K, et al. (2010) AREB1, AREB2, and ABF3 are master transcription factors that cooperatively regulate ABRE-dependent ABA signaling involved in drought stress tolerance and require ABA for full activation. Plant J 61: 672–685.
42. Aguan K, Sugawara K, Suzuki N, Kusano T (1993) Low-temperature dependent expression of a rice gene encoding a protein with a leucine-zipper motif. Mol Gen Genet 240: 1–8.
43. Gupta S, Chattopadhyay MK, Chatterjee P, Ghosh B, SenGupta DN (1998) Expression of abscisic acid-responsive element-binding protein in salt-tolerant indica rice (*Oryza sativa* L. cv. Pokkali). Plant Mol Biol 37: 629–637.
44. Shimizu H, Sato K, Berberich T, Miyazaki A, Ozaki R, et al. (2005) LIP19, a basic region leucine zipper protein, is a Fos-like molecular switch in the cold signaling of rice plants. Plant Cell Physiol 46: 1623–1634.
45. Mukherjee K, Choudhury AR, Gupta B, Gupta S, Sengupta DN (2006) An ABRE-binding factor, OSBZ8, is highly expressed in salt tolerant cultivars than in salt sensitive cultivars of indica rice. BMC Plant Biol 6: 18.
46. Xiang Y, Tang N, Du H, Ye H, Xiong L (2008) Characterization of OsbZIP23 as a key player of the basic leucine zipper transcription factor family for conferring abscisic acid sensitivity and salinity and drought tolerance in rice. Plant Physiol 148: 1938–1952.
47. Lu G, Gao C, Zheng X, Han B (2009) Identification of OsbZIP72 as a positive regulator of ABA response and drought tolerance in rice. Planta 229: 605–615.
48. Hossain MA, Cho JI, Han M, Ahn CH, Jeon JS, et al. (2010a) The ABRE-binding bZIP transcription factor OsABF2 is a positive regulator of abiotic stress and ABA signaling in rice. J Plant Physiol 167: 1512–1520.
49. Hossain MA, Lee Y, Cho JI, Ahn CH, Lee SK, et al. (2010b) The bZIP transcription factor OsABF1 is an ABA responsive element binding factor that enhances abiotic stress signaling in rice. Plant Mol Biol 72: 557–566.
50. Yun KY, Park MR, Mohanty B, Herath V, Xu F, et al. (2010) Transcriptional regulatory network triggered by oxidative signals configures the early response mechanisms of japonica rice to chilling stress. BMC Plant Biol 10: 16.
51. Kobayashi F, Maeta E, Terashima A, Takumi S (2008) Positive role of a wheat HvABI5 ortholog in abiotic stress response of seedlings. Physiol Plant 134: 74–86.
52. Yanez M, Caceres S, Orellana S, Bastias A, Verdugo I, et al. (2009) An abiotic stress-responsive bZIP transcription factor from wild and cultivated tomatoes regulates stress-related genes. Plant Cell Rep 28: 1497–1507.
53. Hsieh TH, Li CW, Su RC, Cheng CP, Sanjaya, et al. (2010) A tomato bZIP transcription factor, SlAREB, is involved in water deficit and salt stress response. Planta 231: 1459–1473.
54. Kim JC, Lee SH, Cheong YH, Yoo CM, Lee SI, et al. (2001) A novel cold-inducible zinc finger protein from soybean, SCOF-1, enhances cold tolerance in transgenic plants. Plant J 25: 247–259.
55. Liao Y, Zhang JS, Chen SY, Zhang WK (2008a) Role of soybean GmbZIP132 under abscisic acid and salt stresses. J Integr Plant Biol 50: 221–230.
56. Liao Y, Zou HF, Wei W, Hao YJ, Tian AG, et al. (2008b) Soybean *GmbZIP44*, *GmbZIP62* and *GmbZIP78* genes function as negative regulator of ABA signaling and confer salt and freezing tolerance in transgenic Arabidopsis. Planta 228: 225–240.
57. Lee SC, Choi HW, Hwang IS, Choi du S, Hwang BK (2006) Functional roles of the pepper pathogen-induced bZIP transcription factor, CAbZIP1, in enhanced resistance to pathogen infection and environmental stresses. Planta 224: 1209–1225.
58. Rodriguez-Uribe L, O'Connell MA (2006) A root-specific bZIP transcription factor is responsive to water deficit stress in tepary bean (*Phaseolus acutifolius*) and common bean (*P. vulgaris*). J Exp Bot 57: 1391–1398.

59. Xue GP, Loveridge CW (2004) HvDRF1 is involved in abscisic acidmediated gene regulation in barley and produces two forms of AP2 transcriptional activators, interacting preferably with a CT-rich element. Plant J 37: 326–339.

60. Kusano T, Berberich T, Harada M, Suzuki N, Sugawara K (1995) A maize DNA-binding factor with a bZIP motif is induced by low temperature. Mol Gen Genet 248: 507–517.

61. Zhang H, Jin JP, Tang L, Zhao Y, Gu XC, et al. (2011) PlantTFDB 2.0: update and improvement of the comprehensive plant transcription factor database. Nucleic Acids Res 39: D1114–D1117.

62. Goodstein DM, Shu S, Howson R, Neupane R, Hayes RD, et al. (2012) Phytozome: a comparative platform for green plant genomics. Nucleic Acids Res 40: D1178–D1186.

63. Letunic I, Doerks T, Bork P (2012) SMART 7: recent updates to the protein domain annotation resource. Nucleic Acids Res doi:10.1093/nar/gkr931.

64. Voorrips RE (2002) MapChart: software for the graphical presentation of linkage maps and QTLs. J Hered 93: 77–78.

65. Tang H, Bowers JE, Wang X, Ming R, Alam M, et al. (2008) Synteny and collinearity in plant genomes. Science 320: 486–488.

66. Du D, Zhang Q, Cheng T, Pan H, Yang W, et al. (2013) Genome-wide identification and analysis of late embryogenesis abundant (LEA) genes in Prunus mume. Mol Biol Rep 40(2): 1937–46.

67. Shiu SH, Bleecker AB (2003) Expansion of the receptor-like kinase/pelle gene family and receptor-like proteins in Arabidopsis. Plant Physiol 132: 530–543.

68. Guo AY, Zhu QH, Chen X, Luo JC (2007) GSDS: a gene structure display server. Yi Chuan 29: 1023–1026.

69. Tamura K, Peterson D, Peterson N, Stecher G, Nei M, et al. (2011) MEGA5: Molecular evolutionary genetics analysis using maximum likelihood, evolutionary distance, and maximum parsimony methods. Mol Biol Evol 28: 2731–2739.

70. Thompson JD, Gibson TJ, Plewniak F, Jeanmougin F, Higgins DG (1997) The CLUSTAL_X windows interface: Flexible strategies for multiple sequence alignment aided by quality analysis tools. Nucleic Acids Res 25: 4876–4882.

71. Saitou N, Nei M (1987) The neighbor-joining method: a new method for reconstructing phylogenetic trees. Mol Biol Evol 4: 406–425.

72. Letunic I, Bork P (2011) Interactive Tree Of Life v2: online annotation and display of phylogenetic trees made easy. Nucleic Acids Res 39: W475–8.

73. Bailey TL, Elkan C (1994) Fitting a mixture model by expectation maximization to discover motifs in biopolymers. Proceedings of the Second International Conference on Intelligent Systems for Molecular Biology, AAAI Press, Menlo Park, California, 28–36.

74. Quevillon E, Silventoinen V, Pillai S, Harte N, Mulder N, et al. (2005) InterProScan: protein domains identifier. Nucleic Acids Res 33: W116–W120.

75. Conesa A, Götz S (2008) Blast2GO: a comprehensive suite for functional analysis in plant genomics. Int J Plant Genomics 2008: 619832.

76. Suyama M, Torrents D, Bork P (2006) PAL2NAL: robust conversion of protein sequence alignments into the corresponding codon alignments. Nucleic Acids Res 34: W609–W612.

77. Lynch M, Conery JS (2000) The evolutionary fate and consequences of duplicate genes. Science 290: 1151–1155.

78. Yang Z, Gu S, Wang X, Li W, Tang Z, et al. (2008) Molecular evolution of the cpp-like gene family in plants: insights from comparative genomics of Arabidopsis and rice. J Mol Evol 67: 266–277.

79. Berman HM, Westbrook J, Feng Z, Gilliland G, Bhat TN, et al. (2000) The protein data bank. Nucleic Acids Res 28: 235–242.

80. Kelley LA, Sternberg MJE (2009) Protein structure prediction on the Web: a case study using the Phyre server. Nat Protoc 4: 363–371.

81. Zhang Y (2005) miRU: an automated plant miRNA target prediction server. Nucleic Acids Res 33 (Web Server issue): W701–4.

82. Hoagland DR, Arnon DI (1950) The water-culture method for growing plants without soil. Calif Agric Exp Stn Circ 347: 1–32.

83. Puranik S, Sahu PP, Srivastava PS, Prasad M (2012) NAC proteins: regulation and role in stress tolerance. Trends Plant Sci 17: 1360–1385.

84. Cannon SB, Mitra A, Baumgarten A, Young ND, May G (2004) The roles of segmental and tandem gene duplication in the evolution of large gene families in Arabidopsis thaliana. BMC Plant Biol 4: 10.

85. Jain M, Nijhawan A, Arora R, Agarwal P, Ray S, et al. (2007) F-Box Proteins in Rice. genome-wide analysis, classification, temporal and spatial gene expression during panicle and seed development, and regulation by light and abiotic stress. Plant Physiol 143: 14671483.

86. Tang X, Bian S, Tang M, Lu Q, Li S, et al. (2012). MicroRNA–mediated repression of the seed maturation program during vegetative development in Arabidopsis. PLoS Genet. 8: e1003091.

87. Eldem V, Okay S, Unver T (2013) Plant microRNAs: new players in functional genomics. Turk J Agric For 37: 1–21.

88. Söding J (2005) Protein homology detection by HMM-HMM comparison. Bioinformatics 21: 951–960.

89. Jefferys BR, Kelley LA, Sternberg MJE (2010) Protein folding requires crowd control in a simulated cell. J Mol Biol 397: 1329–1338.

90. Jakoby M, Weisshaar MJB, Dröge-Laser W, Vicente-Carbajosa J, Tiedemann J, et al. (2002) bZIP transcription factors in Arabidopsis. Trends Plant Sci. 7: 106–111.

91. Huang XS, Liu JH, Chen XJ (2010) Overexpression of PtrABF gene, a bZIP transcription factor isolated from Poncirus trifoliata, enhances dehydration and drought tolerance in tobacco via scavenging ROS and modulating expression of stress-responsive genes. BMC Plant Biol 10: 230.

92. Kim JS, Mizoi J, Yoshida T, Fujita Y, Nakajima J, et al. (2011) An ABRE promoter sequence is involved in osmotic stress-responsive expression of the DREB2A gene, which encodes a transcription factor regulating drought inducible genes in Arabidopsis. Plant Cell Physiol 52: 2136–2146.

93. Rodriguez-Uribe L, O'Connell MA (2006) A root-specific bZIP transcription factor is responsive to water deficit stress in tepary bean (Phaseolus acutifolius) and common bean (P. vulgaris). Journal of Experimental Botany 57 (6): 1391–1398.

94. Neill SJ, Burnett EC (1999) Regulation of gene expression during water deficit stress. Plant Growth Regulation 29: 23–33.

95. Lee SC, Choi HW, Hwang IS, Choi DS, Hwang BK (2006) Functional roles of the pepper pathogen-induced bZIP transcription factor, CAbZIP1, in enhanced resistance to pathogen infection and environmental stresses. Planta 1209–1225.

12

Appearances Can Be Deceptive: Revealing a Hidden Viral Infection with Deep Sequencing in a Plant Quarantine Context

Thierry Candresse[1,2], Denis Filloux[3], Brejnev Muhire[4], Charlotte Julian[3], Serge Galzi[3], Guillaume Fort[3], Pauline Bernardo[3], Jean-Heindrich Daugrois[3], Emmanuel Fernandez[3], Darren P. Martin[4], Arvind Varsani[5,6,7], Philippe Roumagnac[3]*

1 INRA, UMR 1332 Biologie du Fruit et Pathologie, CS 20032, 33882 Villenave d'Ornon Cedex, France, 2 Université de Bordeaux, UMR 1332 Biologie du Fruit et Pathologie, CS 20032, 33882 Villenave d'Ornon Cedex, France, 3 CIRAD, UMR BGPI, Campus International de Montferrier-Baillarguet, 34398 Montpellier Cedex-5, France, 4 Computational Biology Group, Institute of Infectious Diseases and Molecular Medicine, University of Cape Town, Cape Town, South Africa, 5 School of Biological Sciences and Biomolecular Interaction Centre, University of Canterbury, Christchurch, New Zealand, 6 Department of Plant Pathology and Emerging Pathogens Institute, University of Florida, Gainesville, Florida, United States of America, 7 Electron Microscope Unit, Division of Medical Biochemistry, Department of Clinical Laboratory Sciences, University of Cape Town, Observatory, South Africa

Abstract

Comprehensive inventories of plant viral diversity are essential for effective quarantine and sanitation efforts. The safety of regulated plant material exchanges presently relies heavily on techniques such as PCR or nucleic acid hybridisation, which are only suited to the detection and characterisation of specific, well characterised pathogens. Here, we demonstrate the utility of sequence-independent next generation sequencing (NGS) of both virus-derived small interfering RNAs (siRNAs) and virion-associated nucleic acids (VANA) for the detailed identification and characterisation of viruses infecting two quarantined sugarcane plants. Both plants originated from Egypt and were known to be infected with Sugarcane streak Egypt Virus (SSEV; Genus *Mastrevirus*, Family *Geminiviridae*), but were revealed by the NGS approaches to also be infected by a second highly divergent mastrevirus, here named Sugarcane white streak Virus (SWSV). This novel virus had escaped detection by all routine quarantine detection assays and was found to also be present in sugarcane plants originating from Sudan. Complete SWSV genomes were cloned and sequenced from six plants and all were found to share >91% genome-wide identity. With the exception of two SWSV variants, which potentially express unusually large RepA proteins, the SWSV isolates display genome characteristics very typical to those of all other previously described mastreviruses. An analysis of virus-derived siRNAs for SWSV and SSEV showed them to be strongly influenced by secondary structures within both genomic single stranded DNA and mRNA transcripts. In addition, the distribution of siRNA size frequencies indicates that these mastreviruses are likely subject to both transcriptional and post-transcriptional gene silencing. Our study stresses the potential advantages of NGS-based virus metagenomic screening in a plant quarantine setting and indicates that such techniques could dramatically reduce the numbers of non-intercepted virus pathogens passing through plant quarantine stations.

Editor: Hanu Pappu, Washington State University, United States of America

Funding: This work was supported by the ANR French National Research Agency (Netbiome Program, SafePGR Project). AV and DPM are supported by the National Research Foundation of South Africa. BM is funded by the University of Cape Town. The funders had no role in study design, data collection and analysis, decision to publish, or preparation of the manuscript.

Competing Interests: The authors have declared that no competing interests exist.

* Email: philippe.roumagnac@cirad.fr

Introduction

When attempting to prevent the spread of plant diseases, comprehensive inventories of viral diversity are fundamental both for effective quarantine and sanitation efforts, and to ensure that plant materials within biological resource centres (BRCs) can be safely distributed [1,2]. Detection of pathogens is one of the most critical quarantine and BRC operations. Ideally, the tools used for this purpose must be both sensitive enough to accurately detect the presence of even extremely low amounts of pathogen nucleic acids or proteins, and provide sufficient specific information to identify the genetic variants/strains of whatever pathogens are present.

The major challenge of using classical nucleic acid sequence-informed detection tools such as polymerase chain reaction (PCR) or Southern hybridisation assays, is that despite being highly sensitive, these techniques are generally either species or, at best, genus-specific. In addition, such tools lack the capacity to detect, let alone identify, pathogens that are unknown, poorly characterized or highly variable. Although it might be argued that the most economically important pathogens tend to be well characterized and that it is therefore not a serious issue that many of the more obscure pathogens go undetected, it is becoming better appreciated that the "importance" of any particular pathogenic microbe is very difficult to define. Specifically, the environmental and economic impacts of a particular pathogen can vary widely with varying climatic and ecological conditions and there are large numbers of microbes that are presently not classified as pathogens (or at least which are not noticeably pathogenic to humans or to domesticated plants and animals), which will eventually emerge as important future pathogens [3]. Also, since non-domesticated plant and animal species and countless numbers of microbes which contribute to natural terrestrial ecosystems [4–6] can also potentially be threatened by exotic pathogens, the unconstrained global dissemination of apparently harmless fungi, viruses and bacteria could have serious environmental and economic impacts.

Whereas "sequence-dependent" microbial detection methods, which are generally based on PCR or nucleic acid hybridisation, can only be used to target known pathogens, sequence-independent next generation sequencing (NGS) based approaches can potentially provide an ideal platform for identifying almost all known and unknown microbes present in any particular host organism [5,7–9]. Coupled with innovative sample processing procedures, "metagenomics" applications of NGS [10] have already enabled the identification of novel pathogens through the rapid and comprehensive characterization of microbial strains and isolates within environmental and host tissue samples [9,11].

In addition to numerous applications in the study of animal infecting viruses, NGS-based metagenomics approaches have also been used to detect plant infecting viruses [12]. Three main classes of nucleic-acids have been targeted by such analyses: (1) virion-associated nucleic acids (VANA) purified from viral particles [13,14]; (2) double-stranded RNAs (dsRNA) [15]; and (3) virus-derived small interfering RNAs (siRNAs) [16]. Large numbers of both known and new plant and fungus infecting DNA and RNA viruses and viroids have been detected using these approaches [12,17–21].

A major shortcoming of these metagenomic approaches is, however, that they remain technically cumbersome and too expensive for routine diagnostic applications on collections of eukaryotic hosts - even if barcoded primers are used to bulk-sequence pooled samples from multiple sources [15]. Although prohibitive for high throughput diagnostics, the costs of NGS in the context of viral diversity research are often offset by the vast volumes of useful data that can be generated on viral population dynamics, co-infections, mutation frequencies and genetic recombination [22–24].

Here we describe the application of siRNA- and VANA-targeted NGS approaches to the analyses of two Egyptian sugarcane plants maintained for a number of years at the CIRAD Sugarcane Quarantine Station in Montpellier, France. These plants were both known to be infected with *Sugarcane streak Egypt Virus* (SSEV; Family *Geminiviridae*, Genus *Mastrevirus*) and were maintained for use as positive controls during the application of diagnostic tools for SSEV detection in sugarcane plants passing through the quarantine station. Using the siRNA- and VANA-targeted NGS approaches, we discovered and characterized a novel highly divergent mastrevirus from these two plants. This novel virus was also identified in other sugarcane plants originating from Sudan that exhibited white spots on the base of their leaf blades that become fused laterally, so as to appear as chlorotic stripes. Accordingly, we have proposed naming this virus Sugarcane white streak Virus (SWSV). In addition, we present a detailed analysis of siRNAs derived from the SWSV and SSEV variants infecting the two analysed sugarcane plants.

Materials and Methods

Plant material and sugarcane quarantine DNAs collection

Leaves presenting typical symptoms of sugarcane streak disease were sampled from two sugarcane plants that had previously been found to be infected with *Sugarcane streak Egypt virus* (SSEV) and had been kept in a quarantine greenhouse at the CIRAD Sugarcane Quarantine Station, in Montpellier, France. The two sugarcane plants, VARX and USDA (which was initially maintained at USDA-APHIS Plant Germplasm Quarantine before being transferred to CIRAD in 2007), were both initially collected in Egypt during two independent sampling surveys in the late 1990s [25,26]. These sampling surveys were both carried out on experimental stations and commercial lands in close collaboration with Egyptian authorities (Sugar Crop Research Institute (SCRI), Dr Abdel Wahab I. Allam (Director of SCRI) regarding VARX; and Agricultural Genetic Engineering Research Institute, Dr N.A. Abdallah, and Dr M.A. Madkour regarding USDA). In addition, leaves from six sugarcane plants originating from Sudan (B0065, B0067, B0069, D0002, D0003 and D0005, Table S1) and maintained at the CIRAD Sugarcane Quarantine Station were also used (Material Transfer Agreements between CIRAD and Kenana Sugar Co. Ltd). DNAs from these six plants were extracted using the DNeasy Plant Mini Kit (Qiagen). In addition, DNA was extracted from an additional 18 frozen leaf samples (−20°C), including 17 samples originating from Sudanese sugarcane plants, which had passed through the Montpellier Quarantine station between 2000 and 2009 and one which had been obtained from a sugarcane seedling grown from sugarcane true seeds [fuzz] developed in Guadeloupe from a biparental cross involving plants H70-6957 and B86-049 using the DNeasy Plant Mini Kit (Table S1).

VANA extraction from viral particles, cDNA amplification and sequencing

One gram of leaf material from the VARX and USDA plants were ground in Hanks' buffered salt solution (HBSS) (1:10) with four ceramic beads (MP Biomedicals, USA) using a tissue homogeniser (MP biomedicals, USA). The homogenised plant extracts were centrifuged at 3,200×g for 5 min and 6 ml of the supernatants were further centrifuged at 8,228×g for 3 min. The resulting supernatants were then filtered through a 0.45 µm sterile syringe filter. The filtrate was then centrifuged at 148,000×g for 2.5 hrs at 4°C to concentrate viral particles. The resulting pellet was resuspended overnight at 4°C in 200 µl of HBSS. Non-encapsidated nucleic acids were eliminated by adding 15 U of bovine pancreas DNase I (Euromedex) and 1.9 U of bovine pancreas RNase A (Euromedex, France) followed by incubation at 37°C for 90 min. Total nucleic acids were finally extracted from virions using a NucleoSpin 96 Virus Core Kit (Macherey-Nagel, Germany) following the manufacturer's protocol. The amplification of extracted nucleic acids was performed as described by Victoria et al. [14] and aimed at detecting both RNA and DNA viruses. Reverse transcriptase priming and amplification of nucleic acids were used for detecting RNA viruses. A Klenow Fragment step was included in the protocol in order to detect DNA viruses as demonstrated by Froussard [27]. Briefly, viral cDNA synthesis was performed by incubation of 10 µl of extracted viral nucleic acids with 100 pmol of primer DoDec (5′-CCT TCG GAT CCT CCN NNN NNN NNN NN-3′) at 85°C for 2 min. The mixture was immediately placed on ice. Subsequently, 10 mM dithiothreitol, 1 mM of each deoxynucloside triphosphate (dNTP), 4 µl of 5× Superscript buffer, and 5 U of SuperScript III (Invitrogen, USA) were added to the mixture (final volume of 20 µl), which was then incubated at 25°C for 10 min, followed by 42°C incubation for 60 min and 70°C incubation for 5 min before being placed on ice for 2 min. cDNAs were purified using the QiaQuick PCR cleanup kit (Qiagen). Priming and extension was then performed using Large (Klenow) Fragment DNA polymerase (Promega). First, 20 µl of cDNA in the presence of 4.8 µM of primer DoDec were heated to 95°C for 2 min and then cooled to 4°C. 2.5 U of Klenow Fragment, 10X Klenow reaction buffer and 0.4 mM of each dNTP (final volume of 25 µl) were added. The mixture was incubated at 37°C for 60 min followed by 75°C enzyme heat inactivation for 10 min. PCR amplification was carried out using 5 µl of the reaction described above in a 20 µl reaction containing 2 µM primer (LinkerMid50 primer for VARX: 5'-ATC GTA GCA GCC TTC GGA TCC TCC-3' and LinkerMid52 primer for USDA: 5'-ATG TGT CTA GCC TTC GGA TCC TCC-3'), and 10 µl of HotStarTaq Plus Master Mix Kit (Qiagen). The following cycling conditions were used: one cycle of 95°C for 5 min, five cycles of 95°C for 1 min, 50°C for 1 min, 72°C for 1.5 min, 35 cycles of 95°C for 30 sec, 50°C for 30 sec, 72°C for 1.5 min +2 sec at each cycle. An additional final extension for 10 min at 72°C was then performed. DNA products were pooled (VARX and USDA products and 94 additional products obtained from other quarantine samples), cleaned up using the Wizard SV Gel and PCR Clean-Up System (Promega) and sequenced on 1/8th of a 454 pyrosequencing plate using GS FLX Titanium reagents (Beckman Coulter Cogenics, USA).

siRNA extraction and sequencing

The nucleic acid extraction and sequencing approach of Kreuze et al. [16] was used with slight modifications. Total RNAs were extracted from 100mg of VARX fresh leaf material using Trizol (Invitrogen) following the manufacturer's instructions. Small RNA libraries were directly generated from total RNAs. Small RNAs ligated with 3' and 5' adapters were reverse transcribed and PCR amplified (30 sec at 98°C; [10 sec at 98°C, 30 sec at 60°C, 15 sec at 72°C] ×13 cycles; 10 min at 72°C) to create cDNA libraries selectively enriched in fragments having adapter molecules at both ends. The last step was an acrylamide gel purification of the 140–150 nt amplified cDNA constructs (corresponding to cDNA inserts from siRNAs +120 nt from the adapters). Small RNA libraries were checked for quality and quantified using a 2100 Bioanalyzer (Agilent). The library was then sequenced on one lane of a HiSeq Illumina as single-end 50 base reads.

Sequence assembly

Analyses of reads produced by either Illumina (siRNA sequencing) or 454 GS FLX Titanium (Amplified-VANA sequencing) were performed using CLC Genomics Workbench 5.15. De novo assemblies of contigs were performed with a minimal contig size set at 100 bp and 200 bp for Illumina and 454 GS FLX Titanium reads, respectively. A posteriori mapping of reads against the complete genomes of SWSV (once the full genome had been cloned and sequenced) or SSEV or against parts of these genomes were also performed using CLC Genomics Workbench 5.15. Primary sequence outputs have been deposited in the sequence read archive of GenBank (accession numbers: VANA_USDA dataset: SRR1207274; VANA_VARX dataset: SRR1207275; siRNA_VARX dataset: SRR1207277).

SWSV genome amplification, cloning and sequencing

Two partially overlapping SWSV specific PCR primer pairs were designed so as to avoid any potential cross-hybridization to 63 representative species of the family Geminiviridae, including SSEV. These two primer pairs (pair1: SWSV_F1 forward primer

5'-GCT GAA ACC TAT GGC AAA GA-3' and SWSV-R1 reverse primer 5'-AGC CTC TCT ACA TCC TTT GC-3'; and pair2 ECORI-1F forward primer 5'-GAA TTC CCA GAG CGT GGT A-3' and ECORI-2R reverse primer 5'-GAG TTG AAT TCC GGT ACC AAG GAC-3') were complementary to sequences within the rep gene of SWSV. Total DNAs from the two sugarcane plants described above (VARX and USDA) were extracted using the DNeasy Plant Mini Kit (Qiagen) and screened for SWSV using the two pairs of primers and GoTaq Hot Start Master Mix (Promega) following the manufacturer's protocol. Amplification conditions consisted of an initial denaturation at 95°C for 2 min, 35 cycles at 94°C for 10 sec, 55°C for 30 sec, 68°C for 3 min, and a final extension step at 68°C for 10 min. Amplification products of ~2.8 Kbp were gel purified, ligated to pGEM-T (Promega) and sequenced by standard Sanger sequencing using a primer walking approach.

Reverse transcriptase priming and amplification of nucleic acids were carried out in order to detect the intron of the rep gene. Total RNAs from VARX were extracted using the RNeasy Plant Mini Kit (Qiagen). DNase treatment of extracted RNAs was carried out using RQ1 RNase-Free DNase (Promega) following the manufacturer's protocol. Viral cDNA synthesis was performed by incubation of 1 µl of DNase treated RNAs with 15 µl of RNase free water, 0.6 µM of each primers (SWSV_F2: 5'-ACC ATG TGC TGC CAG TAA TT-3' and ECORI-2R: 5'-GAG TTG AAT TCC GGT ACC AAG GAC-3'), and 0.4 mM of mixed deoxynucloside triphosphate (dNTPs), 5 µl of 5X Qiagen OneStep RT-PCR Buffer and 1 µl of Qiagen OneStep RT-PCR Enzyme Mix. Tubes were first placed at 50°C for 30 min for cDNA synthesis. PCR amplification was then carried out using the following cycling conditions: One cycle of 95°C for 15 min, 35 cycles of 94°C for 1 min, 55°C for 1 min, 72°C for 1 min. An additional final extension for 10 min at 72°C was then performed. Amplification products were gel purified, ligated to pGEM-T (Promega) and sequenced by standard Sanger sequencing.

PCR detection tests

DNAs extracted from 17 sugarcane plants originating from Sudan kept at −20°C or six freshly extracted from plants maintained at the CIRAD Sugarcane Quarantine Station were screened for SWSV. DNA extracted from one sugarcane seedling grown from true seeds (fuzz) was also screened for SWSV. PCR amplification was carried out using the two pairs of primers described above (SWSV_F1 and SWSV_R1; ECORI-1F and ECORI-2R) using GoTaq Hot Start Master Mix (Promega) following the manufacturer's protocol. Amplification products of ~2.8 Kbp were gel purified, ligated to pGEM-T (Promega) and sequenced as described above. Plants infected with SWSV were also screened for all known sugarcane-infecting mastreviruses:Sugarcane streak Egypt Virus, Sugarcane streak virus, Maize streak virus, Sugarcane streak Reunion virus, Eragrostis streak virus and Saccharum streak virus. PCR amplification was carried out using 1 µl of DNA template in a 25 µl reaction containing 0.2 µM of each broad spectrum primer (SSV_1732F: 5'-CAR TCV ACR TTR TTY TGC CAG TA-3' and SSV_2176R: 5'-GAR TAC CTY TCH ATG MTH CAG A-3') and GoTaq Hot Start Master Mix (Promega) following the manufacturer's protocol. The following cycling conditions were used: One cycle of 95°C for 2 min, 35 cycles of 94°C for 1 min, 53°C for 1 min, 72°C for 1 min. An additional final extension for 10 min at 72°C was then performed.

Sequence analyses

Six complete genomes of the novel mastrevirus were recovered from plants VARX, USDA, A0037, B0069, D0005 and E0144 (Table S1) and were aligned with the genomes of representative mastreviruses using MUSCLE (with default settings) [28]. Similarly, the predicted replication associated protein (Rep) and capsid protein (CP) amino acid sequences encoded by the viruses within the full-genome dataset were also aligned using MUSCLE. Maximum likelihood phylogenetic trees were inferred for the full genomes (TN93+G+I nucleotide substitution model chosen as the best-fit using jModelTest [29]), Rep (WAG+G+F amino acid substitution model chosen as the best-fit using ProtTest [30]) and CP (rtREV+G+F amino acid substitution model chosen as the best-fit using ProtTest) datasets with PHYML [31]. Approximate likelihood ratio tests (aLRT) were used to infer relative supports for branches (with branches having <80% support being collapsed). All pairwise identity analysis of the full genome nucleotide sequences, capsid protein (CP) amino acid sequences, replication associated protein (Rep) amino acid sequences and movement protein (MP) amino acid sequences were carried out using the MUSCLE-based pairwise alignment and identity calculation approach implemented in SDT v1.0 [32]. The full genome sequence alignment of representative mastrevirus genome sequences together with SWSV was used to detect evidence of recombination in SWSV using RDP 4.24 with default settings [33]. Sequences are deposited in GenBank under accession numbers (SWSV-A [SD-VARX-2013] - KJ187746; SWSV-A [SD -USDA-2013] - KJ187745; SWSV-B [SD -B0069-2013] – KJ210622; SWSV-B [SD -D0005-2013] - KJ187747; SWSV-B [SD -E0144-2013] - KJ187748 and SWSV-C [SD-A0037-2013] - KJ187749).

Test for associations between siRNAs and SWSV/SSEV genomic and transcript secondary structures

The SWSV/SSEV full genome sequences and predicted unspliced complementary and virion strand transcripts were separately folded using Nucleic Acid Secondary Structure Predictor [34], with the sequence conformation set as circular DNA, at a temperature of 25°C. NASP generates a list of all secondary structures detectable within given DNA or RNA sequences and through simulations it demarcates a set of structures referred to as a "high confidence structure set" (HCSS), that confers a higher degree of thermodynamic stability (lower free energy) to the sequences than what would be expected to be achievable by randomly generated sequences with the same base composition (with a $p <= 0.05$).

Given the genomic coordinates of pairing nucleotides within the HCSS, we investigated whether there was any significant trend for more reads (looking both at all reads collectively and at the 21 nt, 22 nt, 23 nt and 24 nt long reads separately) occurring within secondary structures predicted to occur within (i) the full genomes, (ii) the virion-strand transcripts and (iii) the complementary-strand transcripts. The reads were mapped to the secondary structures and we counted how many nucleotides were located at paired and unpaired sites. While Kolmogorov-Smirnov tests (implemented in R; www.r-project.org) were used to determine whether the distribution of reads between paired and unpaired sites were different, Wilcoxon rank-sum tests (also implemented in R, www.r-project.org) indicated whether there were significantly more reads at paired sites compared to unpaired sites and *vice versa*. Whereas the Kolmogorov-Smirnov tests were used to indicate whether any associations existed between siRNA locations and base pairing within nucleic acid secondary structures, the Wilcoxon rank-sum tests were used to determine whether detected associations were

positive (siRNAs tended to occur at structured sites) or negative (siRNAs tended to occur outside of structured sites).

Results

454-based sequencing of VANA from the VARX and USDA sugarcane samples

This approach was used in an attempt to detect both RNA and DNA viruses that may be present in the two sugarcane plants [27]. A total of 2612 and 1635 reads were respectively obtained from the VARX and USDA plants following length and quality filtering. One hundred and eight and 18 contigs were produced by *de novo* assembly from the VARX- and USDA-derived reads, respectively. Two contigs from the VARX plant (2706 nt and 412 nt) and two from the USDA plant (2706 nt and 649 nt), encoded proteins with between 91 and 100% sequence identity with previously described SSEV proteins (Table1). BLASTx analysis revealed that an additional two contigs from the VARX plant (2122 and 127 nt) and three contigs from the USDA plant (1836, 196 and 312 nt) were homologous with known mastreviruses but were nevertheless only distantly related to mastrevirus sequences currently deposited in GenBank (Table1).

A posteriori mapping of VANA 454 reads obtained from the VARX and USDA plants against the complete SWSV genome (see below), revealed that 23.9% (625/2612) and 16.1% (264/1635) of the total reads were derived from this genome and that these yielded complete genome coverage at an average depth of 81X and 29X, respectively. Interestingly, a ∼120 nt long region of very low coverage (<4X) was identified, which mapped to the large intergenic region (LIR) of the SWSV genome (Figure 1).

A mapping analysis performed with the genome of SSEV indicated that the corresponding values were 53.5% of reads (1398/2612, 159X average coverage depth) and 75% of reads (1227/1635, 138X coverage) for the VARX and USDA plants, respectively (Figure S1).

siRNA Illumina sequencing from the VARX sugarcane plant

A total of 15,275,640 raw reads were generated from the VARX sugarcane sample, which were then filtered down to 3,945,108 high quality reads in the 21 to 24 nt size range of siRNAs. From these reads, 226 contigs were obtained by *de novo* assembly, six of which showed significant degrees of similarity to mastreviruses based on BLASTx [35] searches (Table2). Of these six contigs, two (contigs #121 and #176) had a high degree of identity to SSEV while the remaining four were more distantly related to known mastreviruses. Three of these four contigs (contigs #44, #86 and #101) apparently corresponded with a mastrevirus capsid protein (CP) gene and the other one (contigs #79) with a movement protein (MP) gene, while the cumulative contig length of 761 bp corresponded to slightly more than a quarter of a typical mastrevirus genome (Table2).

Following the cloning and sequencing of the full genome of the new mastrevirus (SWSV; see below) it was determined that 0.59% of the Illumina reads obtained from the VARX plant could be mapped to this genome (Figure 1) to generate contigs that covered 96.3% of the genome at an average depth of 185X with only seven gaps of between three and 40 nucleotides. These gaps were located within the large intergenic region (three gaps) and within the probable replication associated protein (Rep) gene (four gaps; Figure 1) encoded by the C1 ORF. It is noteworthy that the ∼120 nt long region of very low coverage (<4X) identified using the VANA approach mapped to the same part of the LIR region of the SWSV genome that remained uncovered during the

Table 1. Lengths, numbers of reads and BlastX analysis results for VANA 454 de novo contigs from sugarcane plants VARX and USDA with detectable homology to mastreviral sequences.

Sample	Contig	Contig length (bp)	Number of reads	BlastX Virus	BlastX Locus	BlastX e-value	Percent identity
VARX	#1	2706	1387	SSEV (NP_045945)	RepA	0.00	100%
	#2	2122	470	DDSMV (YP_003915158)	CP	3.84E-56	70%
	#3	412	11	SSEV (AAC98076)	MP	2.07E-8	95.2%
	#7	127	1	BCSMV (YP_004089628)	RepA	1.72e-10	71%
USDA	#1	2706	1128	SSEV (NP_04945)	RepA	9.20E-177	99.2%
	#2	1836	82	DDSMV (YP_003915158)	CP	1.04E-56	48.8%
	#3	649	12	SSEV (NP_04945)	RepA	2.80E-66	91.1%
	#4	196	37	MSV (CAA10092)	RepA	1.34E-8	56.2%
	#13	312	83	SSEV (AAF76868)	RepA	1.65E-30	84.3%

Acronyms used are as follows: SSEV (Sugarcane streak Egypt virus), DDSMV (Digitaria didactyla striate mosaic virus), BCSMV (Bromus catharticus striate mosaic virus), MSV (Maize streak virus).

Illumina-based siRNA sequencing (Figure 1). As has been previously observed for other viruses, genome coverage was highly heterogeneous (Figure 1). However, a clear general trend could be observed, with the region corresponding to the virion sense V1 and V2 ORFs (encoding CP and MP proteins, respectively), showing an average coverage depth of ~436X and the complementary sense C1 ORF showing an average coverage of only ~38X. Coverage of the non-coding large and small intergenic regions and the presumed C1 ORF intron were even lower at 17.5X and 6.8X, respectively.

It is also noteworthy that besides differences in coverage depth, these various genomic regions of SWSV also showed differences in the siRNA size classes that they yielded. While there was an enrichment of the 21 and 22 nt siRNA size classes amongst the total siRNA reads mapping to the V1 and V2 ORFs, there was a depletion of the 21 nt siRNA size classes and an enrichment of the 24 nt size class amongst total siRNA reads mapping to the C1 ORF (Figure 2). The LIR and, to a lesser extent, the SIR showed a pattern similar to the C1 ORF region (data not shown). The C1 intron, however, had an extreme over-representation of the 24 nt size class with the other size classes being either nearly (22 nt) or totally (21 and 23 nt) absent (Figure 2).

Since the VARX plant was also infected with SSEV, a similar analysis of SSEV-derived siRNAs was performed. Mapping against the genome of SSEV (NC_001868) demonstrated that 0.17% of total reads (6572) were derived from it and that these reads covered 98.6% of the SSEV genome at an average depth of 55X, leaving only 4 gaps of between 5 and 15 nucleotides (Figure S1). Although showing some high degrees of local heterogeneity, genome coverage of SSEV was less biased when comparing the different genomic regions. Nevertheless a similar trend to that associated with SWSV was observed with a higher depth of coverage for the virion sense V1–V2 ORFs (76.5X) than for both the complementary sense C1 ORF (37.6X) and the non-coding regions (46X). Also, as for SWSV, the 21–22 nt siRNA size classes were enriched amongst those mapping to the virion sense ORFs and the 24 nt, siRNA size class was enriched amongst those mapping to the complementary sense C1 ORF (Figure S1). However, unlike for SWSV, no strong siRNA size-class biases were observed for the non-coding regions (data not shown).

By collectively using the Illumina siRNA reads and the 454 VANA reads it was possible to assemble a single genome of the novel mastrevirus from both the VARX and USDA plants. SWSV

Associations between siRNAs and SWSV/SSEV genomic and transcript secondary structures

It has been previously determined that nucleic acid structures can have an appreciable impact on both the distribution of siRNA targets [36,37], and the operational efficiency of small RNA mediated anti-viral and anti-viroid defences [37,38]. We detected strong evidence for the presence of ssDNA secondary structures in both the SWSV (30 high confidence structure set (HCSS) identified) and SSEV (29 HCSS structures identified) genomes (Table 3). The distributions of the HCSS structural elements were, however, different in the predicted virion and complementary strand transcripts of the two viruses, with only two HCCS structures detected in the SWSV complementary strand transcript and none being detected in the SSEV virion strand transcript (so that this particular transcript was not analysed further).

We detected a strong association between the absence of predicted secondary structures within the ssDNA SWSV genome and increased frequencies of corresponding 22, 23 and 24 nt long siRNAs (p-values <0.008; Table 3). Curiously, we found a

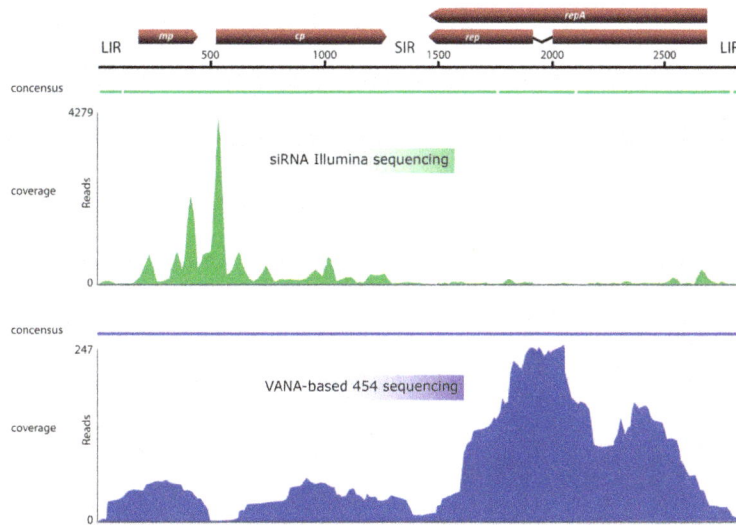

Figure 1. SWSV genome coverage following NGS. The genomic organization of SWSV is schematically shown above the graph. While relative degrees of coverage achieved after *a posteriori* mapping of reads produced by Illumina-based siRNA sequencing against the SWSV genome is indicated in green, the coverage achieved after mapping reads produced by 454 GS FLX Titanium-based VANA sequencing is indicated in blue.

different association when considering the predicted SWSV RNA transcripts with 21 nt siRNA reads displaying a strong tendency to correspond with nucleotide sites that were predicted to be base paired in both the virion and complementary strand transcripts (p-values $<2.49\times10\text{-}6$) and the 22, 23 and 24 siRNA size classes displaying a similar tendency with respect to the virion strand transcript (p-values $<6.07\times10\text{-}13$).

Similar to SWSV, for the SSEV full genome there was an association between the absence of ssDNA structural elements and increased frequencies of 22 nt siRNAs. Also similar to SWSV the 22, 23 and 24 nt long siRNAs display a significant tendency to correspond with transcript nucleotides that are base-paired within secondary structural elements.

A novel sugarcane-infecting mastrevirus originating from the Nile region

The complete genome of SWSV, as recovered from the VARX and USDA plants, is most similar to that of *Wheat dwarf India virus* (WDIV, Accession number NC_017828), with which it shares 61% genome-wide identity. Whereas the Rep and MP amino acid sequences of SWSV are also most similar to those of WDIV (54.4% and 44.8% identity, respectively), the CP is most similar to that of *Panicum streak virus* (PanSV, NC_001647, 51.4–53.9%). Based on the 78% species demarcation threshold set by the Geminivirus study group of the ICTV [32], it is clear that the novel mastrevirus should be considered a new species within the genus *Mastrevirus* of the Family *Geminiviridae* (Figure S2). This is further confirmed by phylogenetic analyses performed on both the full genome (Figure 3) and on the amino acid sequences of its encoded proteins (Figure 4). The new virus clearly clusters with mastreviruses on a branch that is not closely associated with any other species classified within this genus. Whereas the CP of SWSV clusters within the virus clade including the various African streak viruses, Australasian striate mosaic viruses, *Digitaria streak virus* (DSV) and WDIV, the Reps cluster with the African streak viruses and WDIV (Figure 4).

SWSV was not detected in sugarcane seedlings derived from sugarcane true seeds under sterile insect-proof conditions, in agreement with the fact that seed transmission of geminiviruses has

not so far been reported. The novel mastrevirus was, however, detected in five sugarcane plants originating from Sudan (A0037, B0065, B0069, D0005 and E0144) out of the 23 screened (Table S1).

Complete SWSV genomes from four sugarcane plants (A0037, B0069, D0005 and E0144) were cloned and sequenced. The genomes of these isolates have >91% genome-wide identity with those recovered from the VARX and USDA sugarcane plants. Phylogenetic analyses of the full genomes (Figure 3) and of the amino acid sequences that they likely encode (Figure 4) confirmed that the isolates obtained from the Sudanese sugarcane plants also belong to the SWSV species. The six isolates can be further classified into 3 strains, SWSV-A (VARX, USDA), -B (B0069, D0005, E0144) and -C (A0037) (Figure S3) based on the proposed classification of mastrevirus strains outlined by Muhire et al. [32].

Using primers that allow the amplification of all sugarcane-infecting mastreviruses, including Sugarcane streak Egypt Virus, Sugarcane streak virus, Maize streak virus, Sugarcane streak Reunion virus, Eragrostis streak virus and Saccharum streak virus, the five sugarcane plants originating from Sudan were shown to be free of co-infection with other known mastreviruses. Three of them are still maintained at the CIRAD sugarcane quarantine station (B0065, B0069 and D0005) and exhibit white spots on the base of their leaf blades, around the blade joint where the two wedge shaped areas called "dewlaps" are located (Figure S4). These spots can become fused laterally, so as to appear as chlorotic stripes (Figure S4). It is noteworthy that the SWSV infected D0005 plant displayed very little evidence of these spots (only one leaf out of eight exhibited tiny white spots that resembled thrip damage) and it is therefore very likely that SWSV infections could escape visual inspection (Figure S4). Given that three of the infected sugarcane varieties exhibited mild foliar symptoms, i.e. white spots on the base of their leaf blades that become fused laterally, so as to appear as chlorotic stripes, we propose naming the new species Sugarcane white streak Virus.

Genome analysis of SWSV

The SWSV genomes recovered from the various sugarcane plants were between 2828 and 2836 nt and are, in almost all

Table 2. Lengths, numbers of reads and BlastX analysis results for siRNA *de novo* contigs from sugarcane plant VARX with detectable homology to mastreviral sequences.

Virus	Contig	Contig length (bp)	Number of reads	BlastX Virus	BlastXLocus	BlastX e-value	Percent identity
SSEV	#121	101	270	SSEV (AAF76871)	CP	2.60E-9	100%
	#176	133	520	SSEV (AAC98080)	MP	7.22E-15	100%
SWSV	#44	117	914	SacSV (YP_003288767)	CP	7,16E-06	68%
	#79	275	7640	WDIV (YP_006273068)	MP	3,29E-11	70%
	#86	258	1649	WDIV (YP_006273069)	CP	6,78E-20	50%
	#101	111	1284	SSRV (ABZ03975)	CP	1,94E-05	64%

Acronyms used are as follows: SSEV (Sugarcane streak Egypt virus), SWSV (Sugarcane white streak virus), SacSV (Saccharum streak virus), WDIV (Wheat dwarf India virus), SSRV (Sugarcane streak Reunion virus).

respects, very similar to those of all other previously described mastreviruses. The one exceptional feature of the SWSV genomes is that in case of the VARX and USDA isolates alternative splicing of complementary sense transcripts likely results in the expression of both a standard Rep (which is predicted to be 396 amino acids long), and a rather unusual RepA of 418 amino acids long. This is the only known occurrence in any geminivirus of a RepA that is larger than Rep.

Given the uniqueness of this apparent genome organisation in the USDA and VARX isolates the correct identification of the intron within the complementary sense transcript was verified. RT-PCR reactions targeting the complementary sense transcript clearly indicated the presence of a mixture of spliced and non-spliced complementary sense mRNA transcripts, and confirmed that the correct locations of the acceptor and donor sites of the 66 nt long SWSV intron had been identified (Figure S5).

Analysis of recombination

All the SWSV genome sequences determined here share evidence of the same ancestral recombination event in the short intergenic region - corresponding to genomic coordinates 1419–1468 in the USDA isolate ($p = 3.821 \times 10$-7 for the GENECONV, MAXCHI and RDP methods implemented in RDP4.24). Corresponding coordinates are known to be very common sites of recombination in mastreviruses [39] and the fragment that they delimit in SWSV has apparently been derived from something resembling an African streak virus.

Discussion

We have performed NGS-based analyses of both siRNA and VANA isolated from sugarcane plants originating from Egypt. Both sequence-independent NGS approaches revealed the presence of a novel mastrevirus, SWSV, which had so far escaped routine quarantine detection assays, possibly because it was present in mixed infection with SSEV. The procedures used for the discovery of SWSV pave the way towards the application of NGS-based quarantine detection procedures. Such procedures would likely be hierarchical with a first stage sequence-independent NGS step followed by sequence-dependent secondary assays. Whereas the first step would be to identify novel viruses within a single plant (perhaps one displaying apparent disease symptoms), the second step would be to use sequence dependent approaches to both confirm the presence of any novel virus(es) identified in the original host, and identify the presence of this(ese) virus(es) in larger plant collections. A major strength of such an approach is that it would also yield complete genome sequences.

The present study also confirms that both VANA [13] and siRNA [16] can be successfully targeted by metagenomics approaches for the discovery and characterization of plant-infecting DNA viruses. The VANA-based 454 pyrosequencing approach has several advantages as it initially combines reverse-transcriptase priming and a Klenow Fragment step, which potentially enables the detection of both RNA and DNA viruses. Additionally, up to 96-tagged amplified DNAs (cDNA and DNAs amplified using the Klenow Fragment step) can be pooled and sequenced in multiplex format [15] making this approach very useful for routine diagnostic screening of plants within BRCs and quarantine stations. However, validation using plants infected or co-infected with RNA and DNA viruses needs to be carried out in order to determine the sensitivity and specificity levels of this 454-based VANA sequencing approach.

Virus-derived siRNAs naturally accumulate in virus-infected plants as a consequence of the action of Dicer enzymes as part of

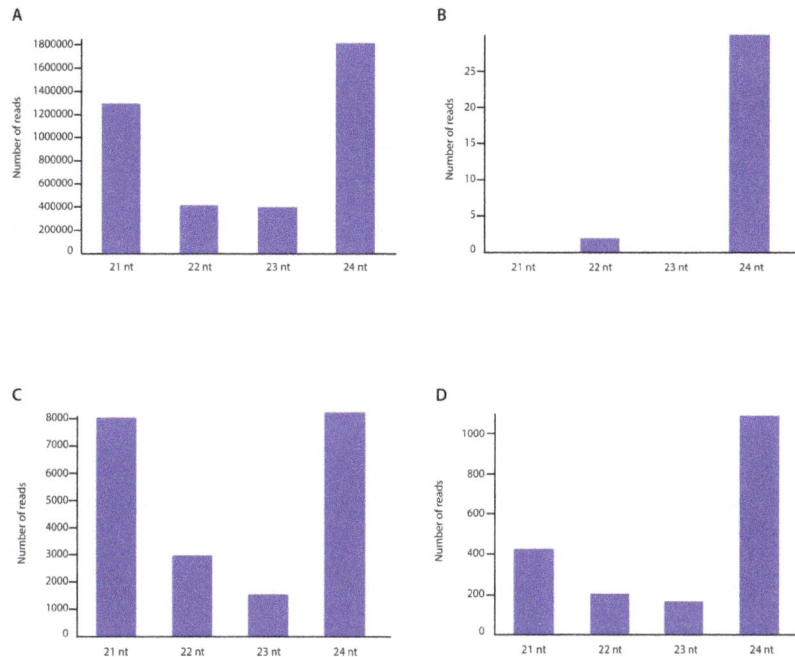

Figure 2. Size distribution of sequenced siRNAs obtained from the VARX plant. The histograms represent the numbers of siRNA reads in each size class. (A) The size distributions of total reads, (B) The size distributions of reads mapping to the *rep* gene C-sense intronic region of SWSV, (C) The size distributions of reads mapping to the V1–V2 ORFs region of SWSV and (D) The size distributions of reads mapping to the C1 ORF region of SWSV.

the RNA silencing-based plant antiviral defences [40]. Adopting a metagenomic approach and randomly sequencing these siRNAs is therefore an extremely powerful way to discover and characterise previously unknown plant viruses and viroids [16,41]. In addition to providing evidence for the presence of the two mastreviruses co-infecting the VARX sugarcane plant, this approach provided information on the interaction of the plant antiviral silencing machinery and these two viruses. Although these aspects have been studied previously in geminiviruses in the *Begomovirus* genus (Blevins et al., 2006; Akbergenov et al., 2006; Rodríguez-Negrete et al., 2009; Yang et al., 2011; Aregger et al., 2012), very little comparable information has previously been available for mastreviruses.

The siRNA distributions observed here for SWSV and SSEV, perhaps unsurprisingly, seem to largely parallel those previously reported for begomoviruses. In particular, the differences in size classes observed between different genome regions suggest that mastreviruses are subject both to transcriptional gene silencing, based on 24 nt long siRNAs produced through the action of DCL3, and to post-transcriptional gene silencing (PTGS) mediated by the 21–22 nt long siRNAs produced through the action of the antiviral Dicers DCL4 and DCL2 (Rodríguez-Negrete et al., 2009; Aregger et al., 2012). The action of the former mechanism is particularly evident in the siRNAs mapping to the SWSV intron but is also, to a lesser extent, evident in the siRNAs mapping to both the non-coding regions and the complementary sense ORFs of SWSV and SSEV. On the other hand, the 21–22 nt siRNA size classes associated with PTGS are particularly evident in the siRNAs mapping to the two virion sense ORFs which are known to be more actively transcribed in mastreviruses than their complementary sense counterparts [42].

For both SWSV and SSEV we detected a significant association between the frequencies of siRNAs and the presence/absence of

predicted secondary structures within both the single stranded DNA (ssDNA) genomes of these viruses and their predicted single stranded RNA (ssRNA) complementary and virion strand transcripts. However, whereas significantly more siRNAs corresponded with unstructured regions of the ssDNA genome, for the transcripts significantly more siRNAs corresponded with structured regions of ssRNA. It is plausible that base-paired nucleotides within transcript RNA molecules are protected from siRNA binding and that the secondary structures evident both in transcripts produced by SWSV, SSEV and in mastrevirus genomes in general [43] may represent an evolutionary adaptation for viral persistence. In mammalian RNA viruses there is an association between degrees of genomic secondary structure and infection duration with viruses having highly structured genomes tending to cause chronic infections and viruses with unstructured genomes tending to cause acute infections [44,45].

In both analysed Egyptian sugarcane accessions, VARX and USDA, SWSV was found to be present in co-infections with SSEV. Both sugarcane plants were independently collected in Egypt which suggests that SWSV infection of Egyptian sugarcane plants may not be a rare phenomenon. SWSV was also detected in SSEV-free plants that originated from Sudan. It is noteworthy that one of the Sudanese plants from which SWSV was isolated, E0144, was initially grown in Sudan in 1992 before being transferred to Barbados in 1998 and subsequently sent back to the CIRAD Sugarcane Quarantine Station in 2009 (unpublished data, CIRAD Sugarcane Quarantine Station). Assuming that SWSV did not infect this plant in Barbados between 1998 and 2009, it is plausible that SWSV was present along the Nile basin at least from the late 1980s. Interestingly, as a consequence of indel polymorphisms in the 66 nt long SWSV intron, the Egyptian SWSV isolates VARX and USDA have a highly unusual genome organization and likely express a RepA protein that, while having

Table 3. Associations between siRNAs and SWSV/SSEV genomic and transcript secondary structures in the HCSS.

Sequence name	Component	Length	Number of structures	siRNA type	Probability if association between siRNAs and secondary structure (KS Test)	Probability of no association between siRNAs and base-paired nucleotides (WRS test)	Probability of no association between siRNAs and unpaired nucleotides (WRS test)
SWSV	Full genome	2830	30	All	5.69×10^{-5}	0.999	6.20×10^{-4}
				21	0.205	0.590	0.410
				22	3.35×10^{-6}	0.992	0.008
				23	0.0005	0.999	6.81×10^{-4}
				24	3.48×10^{-9}	0.999	2.67×10^{-6}
	V-strand transcript	1222	16	All	0	1.80×10^{-15}	1
				21	0.022	4.93×10^{-17}	1
				22	0	4.1×10^{-17}	1
				23	8.23×10^{-14}	1.10×10^{-16}	1
				24	0	6.07×10^{-13}	1
	C-strand transcript	1446	2	All	2.00×10^{-4}	0.100	0.899
				21	1.78×10^{-7}	2.49×10^{-6}	0.999
				22	0.409	0.122	0.877
				23	0.008	0.925	0.075
				24	0.003	0.970	0.029
SSEV	Full genome	2706	29	All	0.007	0.889	0.111
				21	0.016	0.737	0.263
				22	0.0001	0.999	0.001
				23	0.140	0.861	0.139
				24	0.209	0.499	0.501
	V-strand transcript	1131	0	NA	NA	NA	NA
	C-strand transcript	1406	13	All	0.002	0.019	0.981
				21	0.670	0.465	0.535
				22	0.006	0.025	0.975
				23	0.002	0.001	0.999
				24	0.041	0.023	0.977

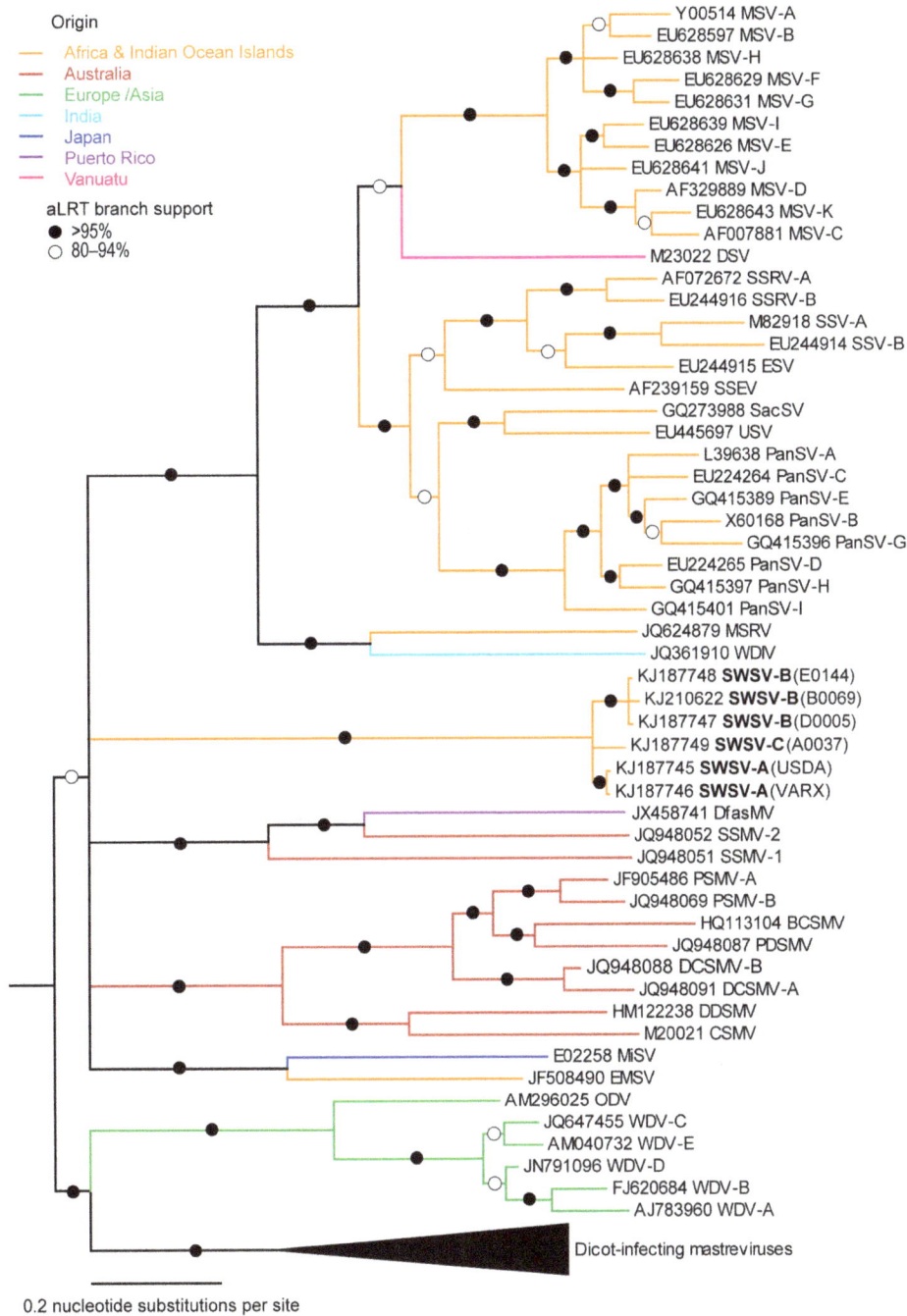

Figure 3. Maximum-likelihood phylogenetic tree of 63 virus isolates representing each known mastrevirus species (including major strains) and the 6 SWSV isolates determined in this study. Tree branches are coloured according to the geographical origins of the viruses. Branches marked with filled and open circles respectively have >95% and 80–94% approximate likelihood ratio test support; branches having <80% support were collapsed. The phylogenetic tree is rooted using the full genome sequence of Dicot-infecting mastreviruses.

the same N- and C-terminus sequences as Rep, is 22 amino acids longer than Rep.

The recent discoveries of SWSV and other highly divergent mastreviruses [46,47] suggest that this geminivirus genus likely encompasses a far greater diversity and has a greater global distribution than has been previously appreciated. The SWSV isolate from the Sudanese sugarcane plant that had been propagated in Barbados represents only the third instance of discovery of mastreviruses in the New World [16,48], and suggests

that there may have been other undetected recent introductions of mastreviruses to the Americas. Although insect transmission of mastreviruses in the New World remains to be reported, it is noteworthy that one of the three mastrevirus species that has so far been detected in the Americas was isolated from a dragonfly which had possibly eaten a plant feeding insect that was carrying the virus [48]. The presence of SWSV in Barbados offers an opportunity to investigate possible natural transmission of the virus by screening sugarcane planted near the SWSV infected

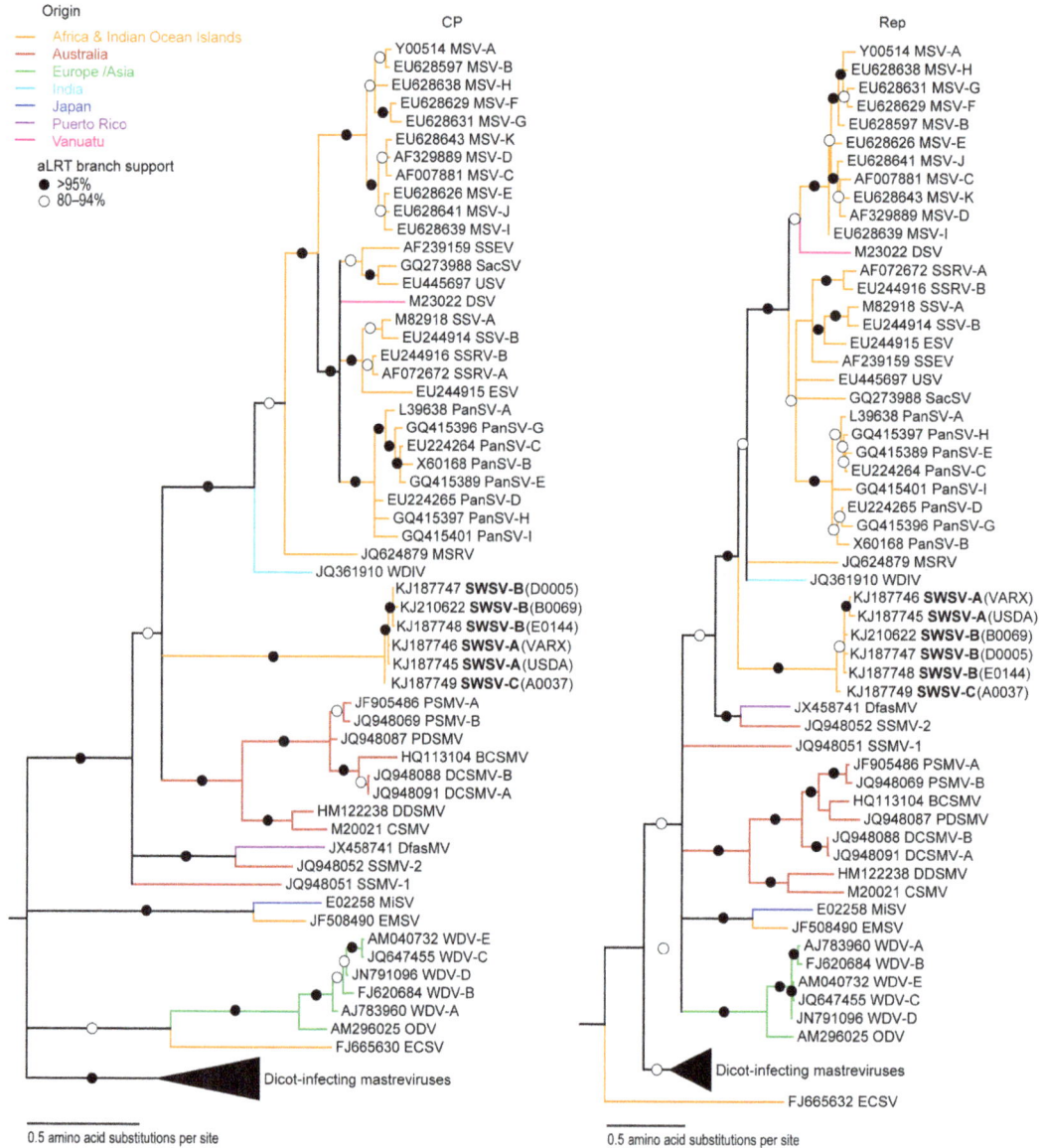

Figure 4. Maximum-likelihood phylogenetic tree of Rep (A) and CP (B) proteins. Tree branches are coloured according to the geographical origins of the viruses. Branches marked with filled and open circles are respectively have >95% and 80–94% approximate likelihood ratio test support; branches having <80% support were collapsed.

E0144 accessions. Phylogenetic analyses of any SWSV genomes sampled from such plants should reveal their likely recent transmission histories.

Given the relatively high degrees of sequence divergence observed between the different SWSV isolates described here (~9%), it is plausible that the natural geographical range of SWSV is broader than just the Nile basin. Also, the global dissemination of sugarcane cuttings, the absence of SWSV diagnostic tools, and the fact that SWSV induces, at least in one case, extremely mild symptoms in sugarcane imply that SWSV may have already been unknowingly distributed throughout the sugarcane growing regions of the world. The failure of established sugarcane quarantine diagnostics in this regard provides a dramatic example of how potentially pathogenic viruses can evade the screening procedures of quarantine facilities and may spread worldwide through international plant material exchanges. In this regard the

situation with SWSV might closely match that of *Sugarcane yellow leaf virus* (SCYLV), which remained unnoticed for at least 30 years during its spread throughout the world [49]. In order to accurately determine the potential economic impacts of the dissemination of SWSV, additional studies assessing the pathogenicity of this virus are certainly warranted.

Our study stresses both the potential advantages of NGS-based virus metagenomic screening in a plant quarantine setting, and the need to better assess viral diversity within plants that are destined for exotic habitats. It indicates that a combination of sequence-independent NGS-based partial viral genome sequencing coupled with sequence-dependent Sanger-based full genome cloning and sequencing is likely to reduce the number of non-intercepted virus pathogens passing through plant quarantine stations, while at the same time alerting authorities to the presence and potential spread of viruses with unknown pathogenic potentials.

Supporting Information

Figure S1 (A) Genome coverages obtained after *a posteriori* mapping against the complete genome of SSEV of reads produced by Illumina (siRNA sequencing). The genomic organization of SSEV is schematically shown at the top of the figure. (B) Size distribution of sequenced siRNAs obtained from the VARX plant mapping on the V1–V2 ORFs region of SSEV. Histograms represent the number of siRNA reads in each size class. and (C) Size distribution reads mapping on C1–C2 ORFs region of SSEV.

Figure S2 Two-dimensional genome-wide percentage pairwise nucleotide identity plot of monocot-infecting mastreviruses including the six novel SWSV isolates from this study.

Figure S3 (A) Maximum-likelihood phylogenetic tree of six SWSV isolates. The six isolates can be classified into 3 strains, SWSV-A (VARX, USDA), -B (B0069, D0005, E0144) and -C (A0037). (B) Genome-wide pairwise nucleotide similarity score matrix, the 94% strain demarcation threshold set by the Geminivirus study group of the ICTV (Muhire et al. 2013) is indicated (green coloured below 94% and pink-red coloured above 94%).

Figure S4 Symptoms caused by SWSV on B0065, B0069 and D0005 plants.

Figure S5 Reverse transcriptase priming and amplification of nucleic acids were carried out in order to detect the *rep* gene C-sense intronic region. (A) Agarose gel detection of presence of a mixture of spliced and non-spliced complementary sense mRNA transcripts. 1: 1 Kb ladder; 2: Reverse transcriptase priming and amplification of nucleic acids without DNase treatment of extracted RNAs; 3: Reverse transcriptase priming and amplification of nucleic acids with DNase treatment of extracted RNAs. (B) 66 nt long SWSV intron nucleotidic sequence and splice donor and acceptor sites. The sequence of the intron (in lower case) and its flanking exons (upper case) are shown. The 5′ (donor) and 3′ (acceptor) splice sites are underlined (lower case).

Acknowledgments

The authors are grateful to Sébastien Theil for depositing NGS datasets in GenBank.

Author Contributions

Conceived and designed the experiments: TC DF PR. Performed the experiments: BM CJ SG GF JHD EF PB. Analyzed the data: TC DF BM DPM AV PR. Wrote the paper: TC DPM AV PR.

References

1. Anderson PK, Cunningham AA, Patel NG, Morales FJ, Epstein PR, et al. (2004) Emerging infectious diseases of plants: pathogen pollution, climate change and agrotechnology drivers. Trends Ecol Evol 19: 535–544.
2. Jones KE, Patel NG, Levy MA, Storeygard A, Balk D, et al. (2008) Global trends in emerging infectious diseases. Nature 451: 990–993.
3. Jones RAC (2009) Plant virus emergence and evolution: Origins, new encounter scenarios, factors driving emergence, effects of changing world conditions, and prospects for control. Virus Research 141: 113–130.
4. Roossinck MJ (2011) The good viruses: viral mutualistic symbioses. Nat Rev Microbiol 9: 99–108.
5. Rosario K, Breitbart M (2011) Exploring the viral world through metagenomics. Curr Opin Virol 1: 1–9.
6. van der Heijden MG, Bardgett RD, van Straalen NM (2008) The unseen majority: soil microbes as drivers of plant diversity and productivity in terrestrial ecosystems. Ecol Lett 11: 296–310.
7. Li L, Delwart E (2011) From orphan virus to pathogen: the path to the clinical lab. Curr Opin Virol 1: 282–288.
8. Mokili JL, Rohwer F, Dutilh BE (2012) Metagenomics and future perspectives in virus discovery. Curr Opin Virol 2: 63–67.
9. Willner D, Hugenholtz P (2013) From deep sequencing to viral tagging: recent advances in viral metagenomics. Bioessays 35: 436–442.
10. Thurber RV, Haynes M, Breitbart M, Wegley L, Rohwer F (2009) Laboratory procedures to generate viral metagenomes. NatProtoc 4: 470–483.
11. Lipkin WI (2010) Microbe hunting. Microbiol Mol Biol Rev 74: 363–377.
12. Roy A, Shao J, Hartung JS, Schneider W, Brlansky RH (2013) A case study on discovery of novel *Citrus Leprosis* virus cytoplasmic type 2 utilizing small RNA libraries by Next Generation Sequencing andbioinformatic analyses. Journal of Data Mining in Genomics & Proteomics 4: 1000129.
13. Melcher U, Muthukumar V, Wiley GB, Min BE, Palmer MW, et al. (2008) Evidence for novel viruses by analysis of nucleic acids in virus-like particle fractions from Ambrosia psilostachya. Journal of Virological Methods 152: 49–55.
14. Victoria JG, Kapoor A, Dupuis K, Schnurr DP, Delwart EL (2008) Rapid identification of known and new RNA viruses from animal tissues. PLoS Pathog 4: e1000163.
15. Roossinck MJ, Saha P, Wiley G, Quan J, White J, et al. (2010) Ecogenomics: using massively parallel pyrosequencing to understand virus ecology. Mol Ecol 19 81–88.
16. Kreuze JF, Perez A, Untiveros M, Quispe D, Fuentes S, et al. (2009) Complete viral genome sequence and discovery of novel viruses by deep sequencing of small RNAs: a generic method for diagnosis, discovery and sequencing of viruses. Virology 388: 1–7.
17. Al Rwahnih M, Daubert S, Golino D, Rowhani A (2009) Deep sequencing analysis of RNAs from a grapevine showing Syrah decline symptoms reveals a multiple virus infection that includes a novel virus. Virology 387: 395–401.
18. Kraberger S, Stainton D, Dayaram A, Zawar-Reza P, Gomez C, et al. (2013) Discovery of Sclerotinia sclerotiorum Hypovirulence-Associated Virus-1 in Urban River Sediments of Heathcote and Styx Rivers in Christchurch City, New Zealand. Genome Announc 1.
19. Li R, Gao S, Hernandez AG, Wechter WP, Fei Z, et al. (2012) Deep sequencing of small RNAs in tomato for virus and viroid identification and strain differentiation. PLoS One 7: e37127.
20. Martinez G, Donaire L, Llave C, Pallas V, Gomez G (2010) High-throughput sequencing of Hop stunt viroid-derived small RNAs from cucumber leaves and phloem. Mol Plant Pathol 11: 347–359.
21. Sikorski A, Massaro M, Kraberger S, Young LM, Smalley D, et al. (2013) Novel myco-like DNA viruses discovered in the faecal matter of various animals. Virus Res 177: 209–216.
22. Borucki MK, Chen-Harris H, Lao V, Vanier G, Wadford DA, et al. (2013) Ultra-Deep Sequencing of Intra-host Rabies Virus Populations during Cross-species Transmission. PLoS Negl Trop Dis 7: e2555.
23. Simmons HE, Dunham JP, Stack JC, Dickins BJ, Pagan I, et al. (2012) Deep sequencing reveals persistence of intra- and inter-host genetic diversity in natural and greenhouse populations of zucchini yellow mosaic virus. J Gen Virol 93: 1831–1840.
24. Wang H, Xie J, Shreeve TG, Ma J, Pallett DW, et al. (2013) Sequence recombination and conservation of Varroa destructor virus-1 and deformed wing virus in field collected honey bees (Apis mellifera). PLoS One 8: e74508.
25. Bigarre L, Salah M, Granier M, Frutos R, Thouvenel J, et al. (1999) Nucleotide sequence evidence for three distinct sugarcane streak mastreviruses. Arch Virol 144: 2331–2344.
26. Shamloul AM, Abdallah NA, Madkour MA, Hadidi A (2001) Sensitive detection of the Egyptian species of sugarcane streak virus by PCR-probe capture hybridization (PCR-ELISA) and its complete nucleotide sequence. J Virol Methods 92: 45–54.
27. Froussard P (1993) rPCR: a powerful tool for random amplification of whole RNA sequences. PCR Methods Appl 2: 185–190.
28. Edgar RC (2004) MUSCLE: a multiple sequence alignment method with reduced time and space complexity. BMC Bioinformatics 5: 113.
29. Posada D (2008) jModelTest: Phylogenetic model averaging. Molecular Biology and Evolution 25: 1253–1256.
30. Abascal F, Zardoya R, Posada D (2005) ProtTest: selection of best-fit models of protein evolution. Bioinformatics 21: 2104–2105.

31. Guindon S, Delsuc F, Dufayard JF, Gascuel O (2009) Estimating maximum likelihood phylogenies with PhyML. Methods Mol Biol 537: 113–137.

32. Muhire B, Martin DP, Brown JK, Navas-Castillo J, Moriones E, et al. (2013) A genome-wide pairwise-identity-based proposal for the classification of viruses in the genus Mastrevirus (family Geminiviridae). Archives of Virology 158: 1411–1424.

33. Martin DP, Lemey P, Lott M, Moulton V, Posada D, et al. (2010) RDP3: a flexible and fast computer program for analyzing recombination. Bioinformatics 26: 2462–2463.

34. Semegni JY, Wamalwa M, Gaujoux R, Harkins GW, Gray A, et al. (2011) NASP: a parallel program for identifying evolutionarily conserved nucleic acid secondary structures from nucleotide sequence alignments. Bioinformatics 27: 2443–2445.

35. Altschul SF, Gish W, Miller W, Myers EW, Lipman DJ (1990) Basic Local Alignment Search Tool. Journal of Molecular Biology 215: 403–410.

36. Donaire L, Wang Y, Gonzalez-Ibeas D, Mayer KF, Aranda MA, et al. (2009) Deep-sequencing of plant viral small RNAs reveals effective and widespread targeting of viral genomes. Virology 392: 203–214.

37. Itaya A, Zhong X, Bundschuh R, Qi Y, Wang Y, et al. (2007) A structured viroid RNA serves as a substrate for dicer-like cleavage to produce biologically active small RNAs but is resistant to RNA-induced silencing complex-mediated degradation. J Virol 81: 2980–2994.

38. Westerhout EM, Ooms M, Vink M, Das AT, Berkhout B (2005) HIV-1 can escape from RNA interference by evolving an alternative structure in its RNA genome. Nucleic Acids Res 33: 796–804.

39. Varsani A, Monjane AL, Donaldson L, Oluwafemi S, Zinga I, et al. (2009) Comparative analysis of Panicum streak virus and Maize streak virus diversity, recombination patterns and phylogeography. Virol J 6: 194.

40. Voinnet O (2005) Induction and suppression of RNA silencing: insights from viral infections. Nat Rev Genet 6: 206–220.

41. Kashif M, Pietila S, Artola K, Jones RAC, Tugume AK, et al. (2012) Detection of Viruses in Sweetpotato from Honduras and Guatemala Augmented by Deep-Sequencing of Small-RNAs. Plant Disease 96: 1430–1437.

42. Morris-Krsinich BA, Mullineaux PM, Donson J, Boulton MI, Markham PG, et al. (1985) Bidirectional transcription of maize streak virus DNA and identification of the coat protein gene. Nucleic Acids Res 13: 7237–7256.

43. Muhire BM, Golden M, Murrell B, Lefeuvre P, Lett JM, et al. (2013) Evidence of pervasive biologically functional secondary-structures within the genomes of eukaryotic single-stranded DNA viruses. J Virol.

44. Davis M, Sagan SM, Pezacki JP, Evans DJ, Simmonds P (2008) Bioinformatic and physical characterizations of genome-scale ordered RNA structure in mammalian RNA viruses. J Virol 82: 11824–11836.

45. Simmonds P, Tuplin A, Evans DJ (2004) Detection of genome-scale ordered RNA structure (GORS) in genomes of positive-stranded RNA viruses: Implications for virus evolution and host persistence. RNA 10: 1337–1351.

46. Kraberger S, Harkins GW, Kumari SG, Thomas JE, Schwinghamer MW, et al. (2013) Evidence that dicot-infecting mastreviruses are particularly prone to inter-species recombination and have likely been circulating in Australia for longer than in Africa and the Middle East. Virology 444: 282–291.

47. Kraberger S, Thomas JE, Geering AD, Dayaram A, Stainton D, et al. (2012) Australian monocot-infecting mastrevirus diversity rivals that in Africa. Virus Res 169: 127–136.

48. Rosario K, Padilla-Rodriguez M, Kraberger S, Stainton D, Martin DP, et al. (2013) Discovery of a novel mastrevirus and alphasatellite-like circular DNA in dragonflies (Epiprocta) from Puerto Rico. Virus Res 171: 231–237.

49. Komor E, ElSayed A, Lehrer AT (2010) Sugarcane yellow leaf virus introduction and spread in Hawaiian sugarcane industry: Retrospective epidemiological study of an unnoticed, mostly asymptomatic plant disease. European Journal of Plant Pathology 127: 207–217.

13

Using Genotyping-By-Sequencing (GBS) for Genomic Discovery in Cultivated Oat

Yung-Fen Huang[1], Jesse A. Poland[2], Charlene P. Wight[1], Eric W. Jackson[3], Nicholas A. Tinker[1]*

1 Eastern Cereal and Oilseed Research Centre, Agriculture and Agri-Food Canada, Ottawa, Ontario, Canada, 2 Department of Plant Pathology, Kansas State University, Manhattan, Kansas, United States of America, 3 General Mills Crop Biosciences, Manhattan, Kansas, United States of America

Abstract

Advances in next-generation sequencing offer high-throughput and cost-effective genotyping alternatives, including genotyping-by-sequencing (GBS). Results have shown that this methodology is efficient for genotyping a variety of species, including those with complex genomes. To assess the utility of GBS in cultivated hexaploid oat (*Avena sativa* L.), seven bi-parental mapping populations and diverse inbred lines from breeding programs around the world were studied. We examined technical factors that influence GBS SNP calls, established a workflow that combines two bioinformatics pipelines for GBS SNP calling, and provided a nomenclature for oat GBS loci. The high-throughput GBS system enabled us to place 45,117 loci on an oat consensus map, thus establishing a positional reference for further genomic studies. Using the diversity lines, we estimated that a minimum density of one marker per 2 to 2.8 cM would be required for genome-wide association studies (GWAS), and GBS markers met this density requirement in most chromosome regions. We also demonstrated the utility of GBS in additional diagnostic applications related to oat breeding. We conclude that GBS is a powerful and useful approach, which will have many additional applications in oat breeding and genomic studies.

Editor: Xinping Cui, University of California, Riverside, United States of America

Funding: This work was supported by the Canadian Crop Genomics Initiative as part of Agriculture and Agri-Food Canada research grant 1885. The funders had no role in study design, data collection and analysis, decision to publish, or preparation of the manuscript.

Competing Interests: The authors have declared that no competing interests exist.

* Email: Nick.Tinker@AGR.GC.CA

Introduction

Cultivated oat (*Avena sativa* L.) is an allohexaploid ($2n = 6x = 42$) crop species that is grown as a source of food and feed. Oat and other crop species require continuous genetic improvement to meet the agronomic and nutritional needs of modern agriculture and food production. Major crop species such as corn, rice, wheat, canola, and soybean are benefiting considerably from advances in genome science and molecular breeding. These advances include the discovery and marker-assisted selection of single genes and quantitative trait loci (QTL), as well as the use of genomic selection (GS; [1,2]) to identify genotypes with superior performance and breeding value. Oat has also benefited from a long history of genomic and nutritional research [3] and from recent advances provided by a SNP platform and consensus map [4]. However, further advancements and applications of genomic technologies that can be integrated into traditional breeding strategies to accelerate and improve the development of superior oat cultivars are needed.

GS can be more efficient than phenotypic selection or marker-assisted selection for improving complex traits [5], and this has been demonstrated in oat [6]. Beyond GS, genomic characterization of breeding material offers many additional opportunities, including: the ability to monitor, maintain and expand germplasm diversity; the ability to diagnose identity or parentage of unknown material; and the ability to discover and deploy specific beneficial alleles [7,8]. Opportunities also exist for gene discovery, since species such as oat have unique biochemical pathways and adaptations not found in model plant species [9]. There are many technologies that can be applied routinely to whole genome characterization. These include parallel assays that target semi-random polymorphisms, such as Amplified Fragment Length Polymorphism (AFLP) and Diversity Array Technology (DArT), as well as parallel assays for specific single nucleotide polymorphisms (SNPs), such as the Golden Gate assays (Illumina, San Diego, CA). All technologies have strengths and weaknesses. Those that identify semi-random polymorphisms may not provide an adequate density of markers throughout the genome, and the technology may not transfer well between laboratories and different germplasm sets. The application of DArT technology in oat has been successful [10], but only a few hundred of the currently developed markers will segregate in a given population (unpublished results). The recently-developed SNP assay for oat [4] has a similar marker density to the DArT assay, but provides more precise and well-characterized gene-based predictions that may be more uniformly distributed throughout the genome and amenable to comparative mapping.

In oat, as in other polyploid species, genotyping is complicated because of the presence of homoeologous sub-genomes. Markers must be filtered to eliminate those that are confounded by multiple loci. This problem is less prevalent in pre-filtered SNP assays than it is in untargeted assays, although there are still SNP markers known to target different loci in different populations [4]. In all technologies, cost remains a critical factor. Currently, costs for DArT analysis and Illumina-based SNP assays range from $US 50–60 per sample, which can be prohibitive for routine genomics-assisted breeding where a large number of the lines will be discarded following genotyping and selection.

Recently, a robust genotyping method based on the sequencing of partial genome representations has been developed for parallel high-throughput genotyping. This is referred to as genotyping-by-sequencing (GBS) [11]. GBS utilizes one or more restriction enzymes [12] to digest the genome into fragments that are then sequenced by parallel high-throughput methods. Based on the sequencing data, SNP calling can be done using various bioinformatics pipelines [12–22]. Low per sample cost in GBS is achieved by multiplexing samples from many (*e.g.*, 48, 96, or 384) different genetic entities (hereafter 'lines') simultaneously through the use of short specific 'barcodes' ligated to each sample prior to sequencing. Thus, it is possible to reduce the cost per sample by multiplexing more lines for sequencing. For example, if 96 samples are sequenced in a reaction costing $960, the cost per sample would be $10 over and above the costs of sample preparation and bioinformatic analysis. The cost will also be affected by the choice of single-end or paired-end sequencing. GBS has been useful in a variety of applications in crop plants including: saturating an existing genetic map [19], genome characterization in wheat and barley [12], genomic selection in wheat [23], the genetic ordering of a draft genome sequence in barley [24,25], and the characterization of germplasm diversity in maize and switchgrass [8,16]. These results suggest that GBS could be utilized for basic and applied genomic studies in oat.

Here, we report the development and application of GBS in oat mapping populations and a diverse set of oat germplasm. Our objectives were: (1) to evaluate the effects of different factors that influence the quality and quantity of GBS SNP calls and establish a baseline of operating parameters and expectations for GBS in oat; (2) to compare alternate methods of bioinformatics analysis and establish a pragmatic workflow and nomenclature for GBS data analysis in oat; (3) to saturate a consensus linkage map with GBS loci and establish a positional reference for future work; and (4) to investigate the utility of GBS to address a variety of questions that are typical of potential uses, including: *de novo* linkage mapping, characterizing population structure and linkage disequilibrium, and solving diagnostic issues in breeding germplasm. We discuss these results in the context of where GBS is likely to be most useful in crop development.

Materials and Methods

Genetic materials

Sets of germplasm used in this study are listed in Table 1. Additional diverse oat lines not reported in this study were prepared and sequenced in parallel with this work, which led to a total number of 2,664 oat lines being genotyped with GBS. These samples are mentioned because their presence may have had a minor influence on the parallel sequencing results or global allele-calling pipelines. These effects would be marginal, since more stringent filters were applied within sub-populations.

DNA sample preparation

The isolation of DNA was performed using a variety of methods, as some samples were available from previous studies. The preparation of DNA stocks from the CxH, HxZ, OxT, OxP, and PxG populations was described by Oliver *et al.* [4], while stocks from the KxO population were prepared as described by Wight *et al.* [26]. The latter stocks still contained RNA, which was removed using a standard RNase procedure followed by phenol/chlorofrom extraction and ethanol precipitation.

For the diversity population (IOI panel), eight seeds of each line were germinated in cyg growth pouches (Mega International, Minneapolis, MN, USA). Leaf tissue was harvested in bulk as the second leaves emerged and was put into paper envelopes containing a 5:1 mix of non-indicating and indicating silica gel desiccant. The paper envelopes were then placed in sealed containers for drying. For the VxL population, leaf tissue was harvested from plants growing in the field, then dried in the same manner. DNA was extracted from the VxL and IOI leaf samples using DNeasy Plant Maxi kits (Qiagen Inc., Mississauga, ON, Canada).

GBS library preparation and sequencing

The GBS libraries were constructed in 95-plex using the P384A adapter set (Table S2 in [12]). For each plate, a single random blank well was included for quality control to ensure that libraries were not switched during construction, sequencing, and analysis. Genomic DNA was co-digested with the restriction enzymes *Pst*I (CTGCAG) and *Msp*I (CCGG) and barcoded adapters were ligated to individual samples. Samples were pooled by plate into libraries and polymerase chain reaction-amplified. Detailed protocols can be found in [23]. Each 95-plex library was sequenced to 100 bp on a single lane of Illumina HiSeq 2000 or HiSeq 2500 by the DNA Technologies core facility at the National Research Council, Saskatoon, SK, Canada.

UNEAK GBS pipeline

Sequence results were analysed using the UNEAK GBS pipeline [16], which is part of the TASSEL 3.0 bioinformatics analysis package [27]. This method does not require a reference sequence, since SNP discovery is performed directly within pairs of matched sequence tags and filtered through network analysis. In this method, tags (a tag is an unique sequence representing a group of reads) belonging to complex multi-locus families (as determined by network analysis) are ignored. Parameters in the UNEAK pipeline were set for maximum number of expected reads per sequence file (300,000,000), restriction enzymes used for library construction (*Pst*I-*Msp*I), minimum number of tags required for output (10), maximum tag number in the merged tag counts (200,000,000), option to merge multiple samples per line (yes), error tolerance rate (0.02), minimum/maximum minor allele frequencies (MAF, 0.02 and 0.5), and minimum/maximum call rates (0 and 1). Call rate is defined as the proportion of samples that are covered by at least one tag. The MAF and call rate were set at a low value for global analysis because these parameters were filtered within sub-populations in later steps.

GBS pipeline using population-level filter

A second SNP calling pipeline was employed as described by Poland *et al.* [23]. This pipeline is implemented in TASSEL 3 and was functionally identical to UNEAK to the point of developing a binary presence/absence matrix for each tag across multiple lines. To identify putative SNPs, tags were internally aligned allowing up to 3 bp mismatch in a 64 bp tag. From aligned tags, SNP alleles were identified and the number of lines in the population with each respective tag was tallied in a 2×2 table, counting the number of lines with one or the other tag, both, or neither [23]. A Fisher Exact Test was then used to determine if the two alleles were independent, as would be expected for a single locus, bi-allelic SNP in a population of inbred lines. If the null hypothesis of independence for the putative SNP was rejected ($p<0.001$), we assumed that the tags were allelic in the population (and, therefore, that the putative SNP was a true single locus, bi-allelic SNP). A significance threshold of $p<0.001$ was selected for the size of

Table 1. Populations and germplasm samples used in this study.

Genetic material	Abbreviation	Number of lines	Reference*	No. of SNP**
Bi-parental mapping populations				
Otana x PI269616 (F_6)	OxP	98	[4]	17,137
Provena x GS7 (F_8)	PxG	98	[4]	11,755
Ogle x TAMO-301 ($F_{6:7}$)	OxT	53	[48]	30,726
CDC SolFi x HiFi (F_7)	CxH	52	[4]	8,324
Hurdal x Z-597 (F_6)	HxZ	53	[4]	4,219
Kanota x Ogle (F_7)	KxO	52	[49]	2,582
VAO-44 x Leggett ($F_{4:5}$)	VxL	145	This study	280 (373)
Diversity panels				
Oat diversity panel	IOI	340	[31]	2,155

*First publication that refers to the population
**No. of SNP filtered for subsequent analyses. Please refer to the text for filtering criteria. For VxL, two sets of filtering criteria were used. The only difference between the filtering criteria sets is the heterozygosity level: 8% or 13% (SNP number is between brackets). For IOI, only markers passing filtering criteria with a map position were reported in the table.

population, based on previous work testing false discovery rates in duplicate samples.

Filtering and merging GBS SNP calls

Both of the above pipelines were applied globally to all available sequencing data, except where we deliberately tested SNP identification in partial datasets. This global strategy reduced the need to access large sequencing files repeatedly. However, there was then a need to generate genotype data for specific sub-populations, and to apply population-specific filters for allele frequency, heterozygosity, and data completeness (data completeness is defined as 100% - % missing data; *e.g.*, for a marker genotyped on 100 individuals with 10 individuals showing missing data points, the completeness of the marker is 90%). Furthermore, for genotypes of mapping progeny, it was necessary to recognize the parental phase of alleles, and to represent alleles using conventions required by the mapping software. These filters and secondary analyses were applied using in-house software ('CbyT') written in the Pascal programing language (Text S1). This software provided the additional feature of maintaining a cumulative index of unique SNPs with a consistent naming convention, such that data from different pipelines or subsequent assays could be merged to remove redundancy and to index matching SNPs with the same unique name. Each subsequent analysis required specific filtering criteria, which can be found at the beginning of the method section for each type of analysis.

Linkage mapping

For bi-parental mapping populations, parental lines were genotyped together with the progeny. GBS loci called using both pipelines across six bi-parental RIL populations (OxP, PxG, OxT, CxH, HxZ, and KxO) were filtered at ≥50% completeness, MAF ≥35%, and heterozygosity ≤8% inside each population, which gave a total of 45,117 GBS markers. The SNP data from the six mapping populations reported by Oliver *et al.* [4] were filtered to the same standards as the GBS SNP data, and the two data types concatenated. Marker phases were determined using parental genotypes when the latter were available and not monomorphic. Monomorphic parental genotypes can result from genotyping errors or genetic variation within the lines used to make the cross. Markers for which there were no good parental data were

converted into both parental phases for further analysis. For each mapping population, the phase of parental alleles was re-checked across the concatenated data by enumerating, for each SNP, the number of linked loci in the same phase (recombination fraction, $r<20\%$) vs. the number in opposite phase ($r>80\%$). Loci having a greater number of out-of-phase matches than in-phase matches were rescored in the opposite phase, or were eliminated through a recursive process if this did not improve the in-phase/out-of-phase ratio.

An updated version of the oat consensus map developed by Oliver *et al.* [4] was generated by placing each new candidate locus (GBS or other non-framework SNP) relative to framework SNPs from the existing map. Pair-wise recombination fraction (*rf*) was first calculated for all marker pairs, including both framework and non-framework markers. Marker placements were then made relative to the two framework loci showing the smallest *rf* among any of the six populations. The approximate map position of each placed marker was subsequently estimated by interpolating the cM position proportional to the recombination fraction with the closest two framework loci. When the closest framework locus was at the end of a linkage group, and the recombination with the next-closest framework locus was greater than that between the two framework loci, the candidate was placed distal to the end of the linkage group. In addition to this crude approximation of marker position, a detailed report of each placed marker was produced to show the actual recombination frequencies within each population and across populations between a given marker and all other loci that were within 20% recombination in any of the component populations. Marker data used for marker placement on the oat consensus map are in Table S1.

De novo linkage map construction was performed using MSTMap [28] for the VxL population. GBS loci for *de novo* map construction were called using the UNEAK GBS pipeline and filtered at high stringency (MAF ≥35%, completeness ≥90%) at two different levels of heterozygosity (8% and 13%). The resulting data contained 858 (heterozygosity ≥8%) and 1053 (heterozygosity ≥13%) GBS markers. The choices of 8% and 13% corresponded to the expected heterozygosity at F_5 and F_4, respectively, factoring in sequencing error and out-crossing rate. For MSTMap, a *p*-value equal to 10^{-11} was used for the marker clustering threshold. Markers were excluded as unlinked if they were 15 cM away from any other locus or if they belonged to a group containing only two loci. A simple recombination count was used for the objective function. Since map distances estimated by MSTMap are inflated (based on simulated data, result not shown), we re-estimated the recombination fractions between pairs of loci based on the marker order from MSTMap and converted them to map distances using the Kosambi mapping function.

Population structure and LD analysis

For population structure and LD analysis, GBS markers called by the UNEAK pipeline were filtered at $\geq$90% completeness, MAF $\geq$5%, and heterozygosity $\leq$5%. Population structure was investigated using principal component analysis (PCA). PCA was performed with the 'smartpca' function implemented in EIGEN-SOFT [29]. This function takes into account marker dependency (*i.e.*, markers in LD blocks) through the use of multiple-regression on adjacent markers prior to PCA. The maximum interval distance between markers (ldlimit) was set to 0.001. The number of adjacent markers included in LD adjustment (ldregress) was set at 0, 10, or 50 (designated k0, k10, and k50), such that k0 provided no LD correction and k10 and k50 corresponded to the median and maximum LD block sizes in the IOI dataset. Eigenvalues and Eigenvectors were transferred to the R statistical package [30] for scree plot drawing and other analyses.

A model-based approach was used to investigate the clustering pattern among lines in the diversity panel further, because it determines simultaneously the number of clusters and cluster membership and does not have underlying genetic assumptions that are rarely met [31]. Model-based clustering was based on the first ten PC and conducted using the clustCombi function of the R package mclust [32]. The purpose of clustCombi is to represent a non-Gaussian cluster by a mixture of two or more Gaussian distributions [33]. It first uses the Bayesian information criterion (BIC) to identify the number of Gaussian mixture components and then hierarchically combines components according to an entropy criterion. The final decision concerning the number of clusters to use was made based on an entropy plot; *e.g.*, if six components were identified by BIC and successive component combinations showed no large entropy decrease after four clusters, then the data were represented by four clusters.

Linkage disequilibrium (LD) between two loci was estimated as squared allele-frequency correlations (r^2) by an optimized version (Stéphane Nicolas, personal communication) of LD.Measure in the R package LDcorSV [34]. Four r^2 estimates were calculated: conventional r^2 based on raw genotype data, r^2 with population structure included in the calculation (r_s^2), r^2 with relatedness included in the calculation (r_v^2), and r^2 with both population structure and relatedness included (r_{sv}^2). Population structure was represented by the first four PC after scaling the coordinate identifiers across a range of zero to one. A matrix of relatedness was calculated by A.mat, implemented in the rrBLUP package [35].

The relationship between LD and genetic distance was modeled by fitting two alternate non-linear regression models: a drift-recombination equilibrium model [36] or a modified recombination-drift model including low level of mutation and an adjustment for sample size [37]. Both models were summarized in [38].

Other statistical analyses

We wished to examine how GBS technology could be used to solve diagnostic problems that arise occasionally in any plant breeding program. Germplasm diagnostics were performed using DARwin software [39] to generate clusters based on genetic distances among lines, estimated using simple allele-matching for bi-allelic diploid loci:

$$d_{ij} = 1 - \frac{1}{L}\sum_{l=1}^{L}\frac{m_l}{2}$$

where d_{ij} is the dissimilarity between lines i and j, L is the number of informative loci shared by those lines, and m_l is the number of matching alleles for locus l. Cluster analysis was performed using the un-weighted paired group mean analysis (UPGMA) method.

Results

Library construction, sequencing and coverage

For this study, a total of 38 libraries were generated, each multiplexing 95 lines. A single lane of Illumina HiSeq 2000 or HiSeq 2500 was used to sequence each library. The GBS libraries were constructed as previously described for wheat, with the exception that the forward barcode adapter concentration was reduced to 0.06 pmol for 200 ng of genomic DNA (*vs.* 0.1 pmol used for wheat in [12]). This adapter concentration was found to improve the oat libraries, reducing adapter dimers.

A complete set of short read archives for all GBS oat samples analysed to date has been made available for download from the NCBI short read archive (http://www.ncbi.nlm.nih.gov/sra/) under project accession number SRP037730. Details of these archives, including number of reads and number of good barcoded reads at the level of each flow-cell, single lane, and individual taxon are available in Table S4. Table S4 also provides the key file needed to support re-analysis of the raw short read archives by either of the GBS pipelines reported here.

From a total of 6.3×10^9 reads, 84.4% (5.3×10^9) included the barcode sequence and enzyme cut-site, and had no unreadable base ('N') in the sequence. The UNEAK pipeline found an average of 732,396 tags per oat line in the samples reported here, a total merged tag count across all samples of 358,177,647 tags, and a filtered tag count (tags appearing >10 times) of 17,700,128 that were covered by 564,946,411 matching reads.

Each sequencing lane generated approximately 2×10^8 100 bp reads. After discarding reads that did not have an exact match to one of the barcodes, there were approximately 2×10^6 100 bp reads per sequenced DNA sample. We designated the 2×10^6 100 bp-base reads/DNA sample as one unit of 'depth index'. To test the influence of plexity and sequencing depth on GBS data completeness, we used data from 53 OxT mapping progeny sequenced in three separate lanes. We split the raw sequencing data from two lanes in half and added these incrementally to the un-split lane, which contained a lower number of reads. This provided five different sequencing depths with mixed levels of plexity, from which we computed average depth indices of 0.58, 0.95, 1.33, 1.85, and 2.37. For example, the depth index of 0.58 means there were, on average, $0.58\times2\times10^6 = 1.16\times10^6$ reads/sample for that experiment. The exact read depth for individual samples varied because of sample quality and/or variations in barcode efficiencies. The UNEAK GBS pipeline was run on these five data subsets and SNPs were filtered at four levels of completeness (25%, 50%, 75%, and 90%). The results (Figure 1) showed that an increasing number of SNPs were called as the depth index was increased at all four completeness levels. The response to sequencing depth appears to be linear within the range tested. One of the sequencing runs, added at the second and third levels, had a higher variation in read depth among samples, which explains the lower slope at these levels and the fact that almost no SNPs had a completeness of 90%.

Number of GBS loci called

Level of completeness

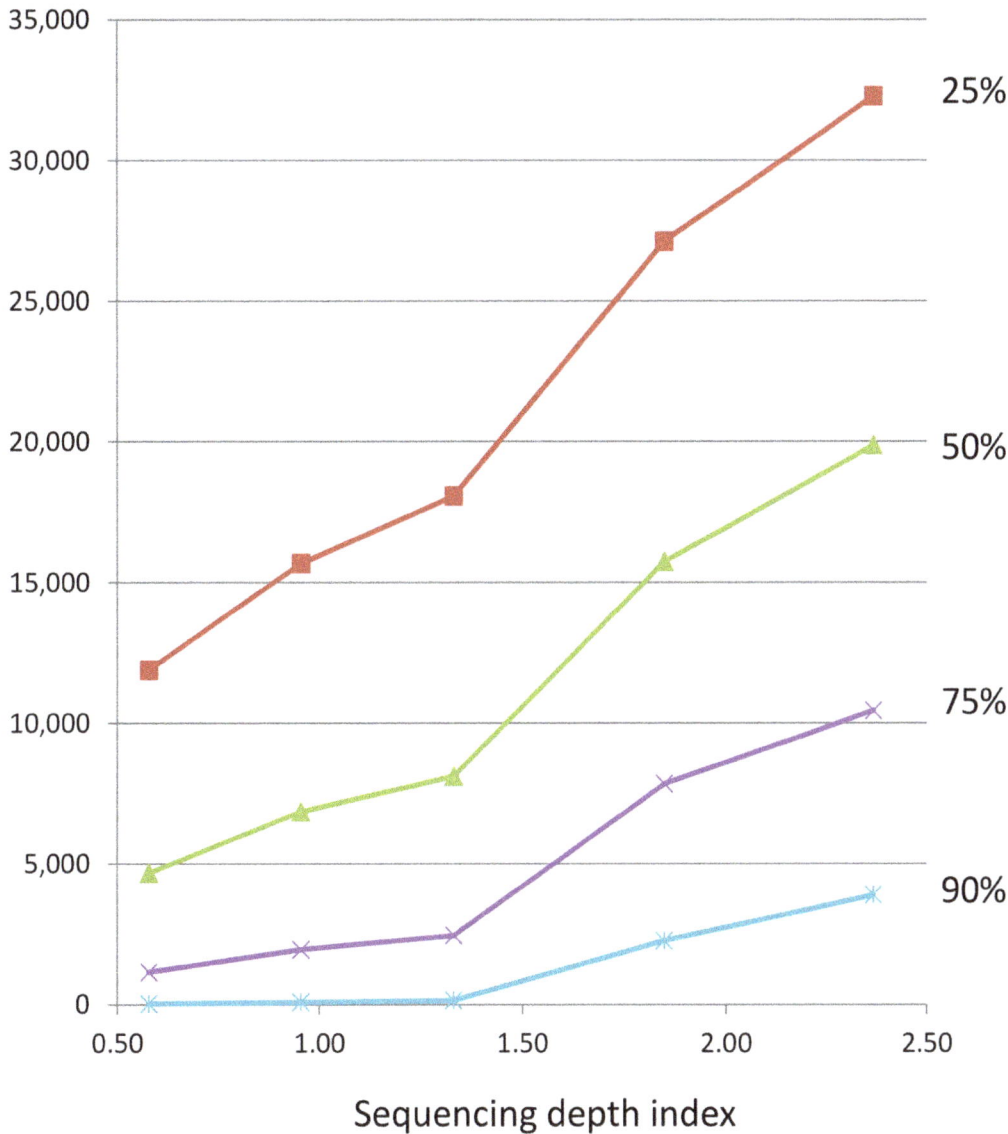

Figure 1. Number of GBS loci *vs.* sequencing depth. Number of GBS SNP loci called in 53 OxT mapping progeny at increasing sequencing depth, filtered at four levels of completeness (25%, 50%, 75%, and 90%). Other filtering parameters were constant, with heterozygosity ≤10% and minor allele frequency ≥30%. A sequencing depth index of 1 represents the average read depth that would be achieved with 95 samples multiplexed in a standard Illumina sequencing run giving approximately 2×10^8 short reads. Thus, an index of 2 would be equivalent to twice this number of reads or half of this plexity.

Population size *vs.* number of GBS SNPs

We examined random subsets of 366 diverse oat varieties, including the IOI set and 26 additional winter oat varieties, to determine how sample size would affect the number of GBS SNPs called at differing levels of completeness. Sample size was varied between 10 and 360 at increments of 10, with two randomly chosen subsets as replicates for each sample size. These data were filtered at a maximum heterozygosity of 10%, MAF of 5%, and minimum completeness of 25%, 50%, 75%, or 90%. At each sample size, the number of SNPs passing these filters was counted. At a low threshold for completeness (25%), the number of SNPs increased with sample size, plateauing at approximately 50,000 SNPs once 250 of the 360 oat lines had been included (Figure 2). At higher thresholds for completeness (50%, 75%, and 90%), the number of SNPs plateaued at approximately 20,000, 10,000, and 5,000 SNPs, respectively. These plateaus occurred at increasingly smaller sample sizes, and the number of SNPs appeared to decrease slightly as sample size increased beyond the initial plateau.

Number of GBS loci called

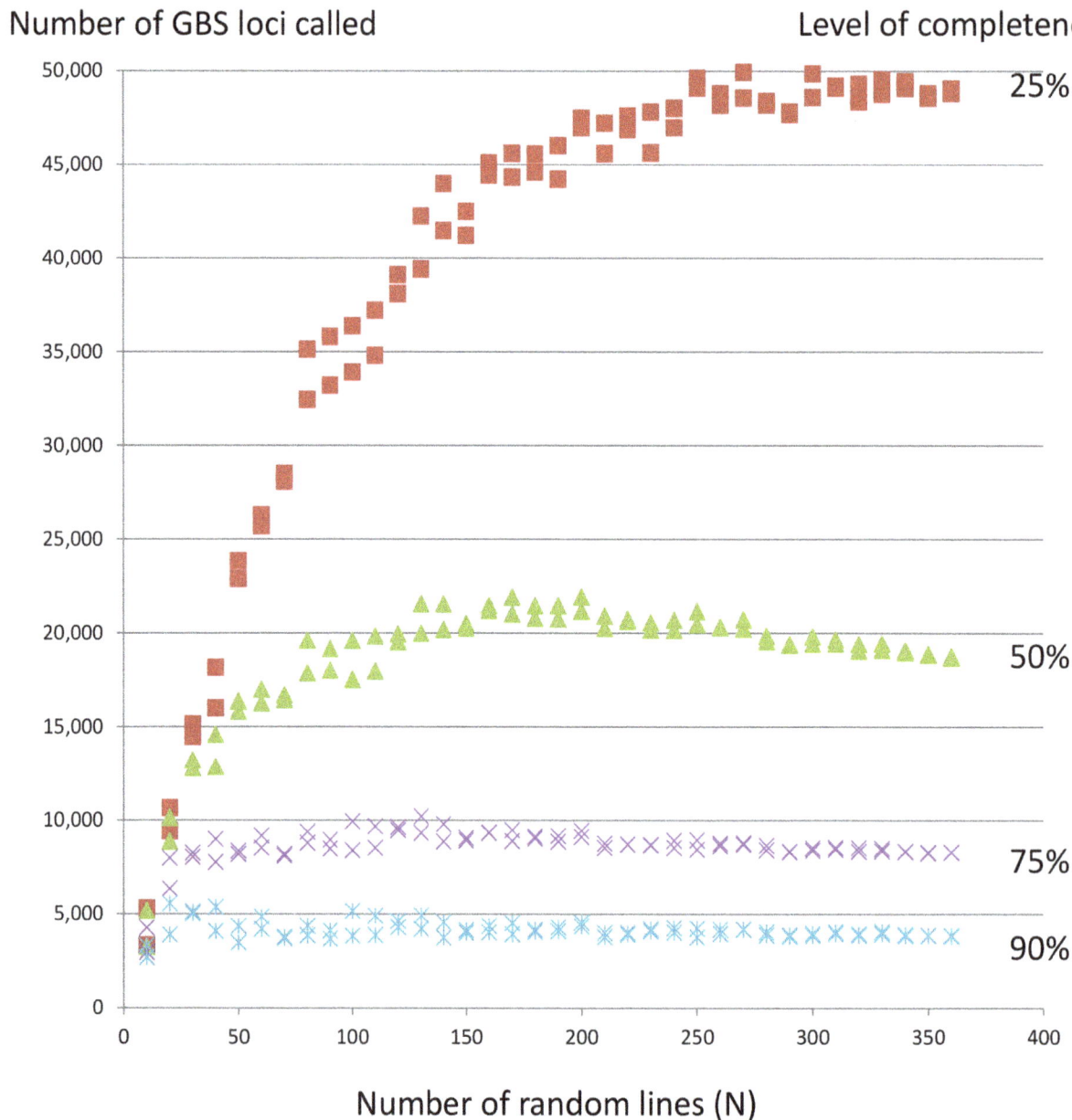

Figure 2. Number of GBS loci *vs.* sample size. Number of GBS loci called in samples of size *N* from a set of 360 diverse oat lines filtered at four levels of completeness (25%, 50%, 75%, and 90%) is shown. Other filtering parameters were constant, with heterozygosity ≥10% and MAF ≥5%.

Multi-allelic SNPs

Although multi-allelic loci were not specifically identified by either of the two GBS pipelines employed in this work, it was apparent that the population-based method would frequently call separate pairs of bi-allelic SNPs at the same site, providing evidence for third and, occasionally, fourth alleles. We used this result to make an approximate estimate of the frequency at which multi-allelic SNPs can be identified from existing pipelines. Of the 355,731 unique bi-allelic SNPs from all oat projects called using both pipelines, there were 343 sets of tri-nucleotide SNPs (*i.e.*, two bi-allelic SNPs having identical context sequence except for a third allele at the SNP position), and 21 sets of tetra-nucleotide SNPs (*i.e.* two bi-allelic SNPs having identical context sequence except for the SNP position). This analysis was not expected to give

comprehensive access to all multi-allelic SNPs, since most of the multi-allelic SNPs would have been filtered out during SNP calling. However, this result suggests that tri- and tetra-nucleotide SNPs are extremely rare, which is expected when SNPs arise primarily as random neutral mutations.

Integration of GBS SNPs with an existing genetic consensus map

45,117 GBS loci, filtered across six bi-parental RIL populations, were placed on the oat consensus map of Oliver *et al.* [4] based on simple counts of recombination fractions. Starting from the initial consensus map, each additional population provided from 2,535 (KxO) to 30,369 (OxT) more loci (Figure S16). As more populations were used for marker placement, fewer new markers were added to the map, but the number of new markers was always proportional to the number of markers available from the source population; *e.g.*, there were always more new loci from OxT (21,894–30,369) than KxO (1,065–2,535) (Figure S16). The complete report for this map is available as a supplementary HTML file in Text S2 (or online at: http://ahoy.aowc.ca/html_link_gbs_text_S2.html). The report format is described in Figure S1. Each individual marker on this HTML map is linked to a separate, detailed report that shows a complete matrix of recombination fractions between the reported marker and the neighbouring loci from the consensus map, as well as other placed markers across all populations.

To investigate the approximate global distribution of GBS loci relative to the SNP consensus map, we divided the map into bins of 5 cM intervals and produced a density histogram which showed the number of placed GBS loci in each bin along the genome (Figure 3). Some bins contained no markers, while some contained a much larger number of markers. In general, GBS loci tended to cluster in the same locations as array-based SNPs. Some clusters probably reflect centromeres, where suppressed recombination causes genetic clustering. However, some chromosomes contained multiple regions of clustering, especially 3C, 4C, 5C, 16A, 19A, 12D, and 21D. This may be caused in part by cytogenetic differences among the parents of the mapping populations, whereby individual maps contain underlying differences in the structure and order of genetic markers. The consensus map would have compressed these differences into a single 'average' map, but the underlying differences among populations remain, and placed markers may appear to cluster at points where the consensus has averaged these differences.

The inclusion of GBS markers appeared to fill gaps within the consensus map. For example, 112 marker intervals larger than 5 cM were present on the original consensus map [4], while only 25 are present on the same map once GBS markers are placed, and the maximum gap size decreased from 26.98 cM to 15.74 cM (Figure S2). These results do need to be interpreted with caution, because the interpolated positions of markers with miss-scored alleles may appear to fill some gaps. An accurate re-interpretation of the consensus map can only be achieved by a complete reanalysis and reinterpretation of the component maps. This work is in progress and will be reported elsewhere together with a complete report on additional SNP loci. Preliminary results of this work (unpublished data) indicate that GBS markers fill some gaps, but that their greatest benefit is to increase the number of loci that are mapped in multiple populations.

Examination of orthology with other crops

The high density of approximately-placed GBS markers provides a new opportunity to examine orthologous relationships between oat and model genomes. The orthology analysis was performed to determine whether matches of short GBS sequence to model genomes would be sufficient to identify major regions of genome co-linearity. Using dot-plots, we explored the locations of sequence similarity between the oat consensus map and pseudo-molecule sequence assemblies from *Brachypodium distachyon* L. (Bd), rice, and barley (Figures S3 to S6). As in a previous analysis using only array-based SNPs [4], we observed that *Brachypodium* had a greater number of matches and better colinearity with oat than did rice. Several stretches of colinearity were observed, such as those on oat 19A (similar to parts of chromosomes Bd1, Bd2, and rice1) and oat 20D (similar to Bd5 and rice4). In the current analysis, it is clear that, in regions of collinear sequence-based matching, both the GBS and other SNPs are contributing similar information. In some cases, the GBS loci appeared to extend the

regions of colinearity (for an example, see Bd1 and Bd4 in Figure S3). It was also clear from the higher density of GBS matches to *Brachypodium* sequences that a greater number of non-coding sequences were similar between *Brachypodium* and oat than between rice and oat, despite the larger genome of rice. This is probably because of the closer ancestral relationship between oat and *Brachypodium*. Barley showed very poor colinearity with oat, based on sequence matching to the newly-available ordered shotgun assembly [25]. We suspect this to be a result of the incomplete nature of the barley assembly. It was also notable that many GBS SNPs showed highly repetitive matches to the barley genome (Figure S5), which were mostly eliminated by removing GBS loci determined to have multiple matches to other GBS loci within oat (Figure S6). This suggests that there are many repetitive elements that are shared between the oat and barley genomes.

GBS SNP annotation

In order to give an approximation of the number and distribution of genic and intergenic GBS SNPs in oat, we compared a complete set of 355,731 context tags of GBS SNPs from all oat projects available at the time of analysis to the chromosome-based genome assembly and accompanying gene predictions from *Brachypodium* (release 2.1; http://www.brachypodium.org) by BLAST. Of these, 19,656 tags (5.5%) showed protein matches (BLASTx) with expectation <0.1, a level that corresponds approximately to a minimum 60% identity over the full tag length or 100% identity over half the tag length. Although there will be oat genes that do not have *Brachypodium* orthologues, it is still likely that fewer than 5% of GBS tags are within transcribed oat genes, because many of the protein signatures matching *Brachypodium* will likely represent vestigial genes in oat. Of the tags with protein matches, 16,712 showed DNA similarity within the transcribed region at the threshold expectation of <0.1. However, we noted that the average BLASTn expectation corresponding with a BLASTx expectation of 0.1 was approximately 0.001; therefore, we conducted further nucleotide matches at this level. This allowed us to identify a total of 46,370 tags (13% of total) with nucleotide matches in *Brachypodium*, among which 30,713 (66% out of matched tags or 8.6% out of total tags) were in intergenic regions, and 15,657 (34% of matched tags or 4.4% of total tags) inside gene regions, of which only 300 did not match a protein. Since gene regions in *Brachypodium* correspond to approximately 43% of the genome, it appears that there is some bias toward GBS nucleotide matches outside of gene regions.

Out of the 46,370 tags matched to *Brachypodium* sequences, 9,684 were positioned somewhere on the consensus map (*cf.* Results/Integration of GBS SNPs with an existing genetic consensus map). On average, 20.5% of mapped loci were positioned inside a gene, ranging from 14.2% to 29.08%, which is a slightly smaller than the proportion in overall tags having BLAST matches to the *Brachypodium* genome. The distribution of *Brachypodium* orthologous SNPs along the oat genome is similar to that of the overall oat GBS SNPs (Figure S17 and Figure 3). Genic and intergenic SNP counts per chromosome and their genome distribution can be found in Figure S17. While this result gives an approximation of the proportion and distribution of genic and intergenic SNPs along the oat genome, care should be taken in interpreting this result to form a general conclusion about oat, not just because of the partial coverage of the present consensus map, but also because of the approximate nature of the annotations made through the use of orthologues.

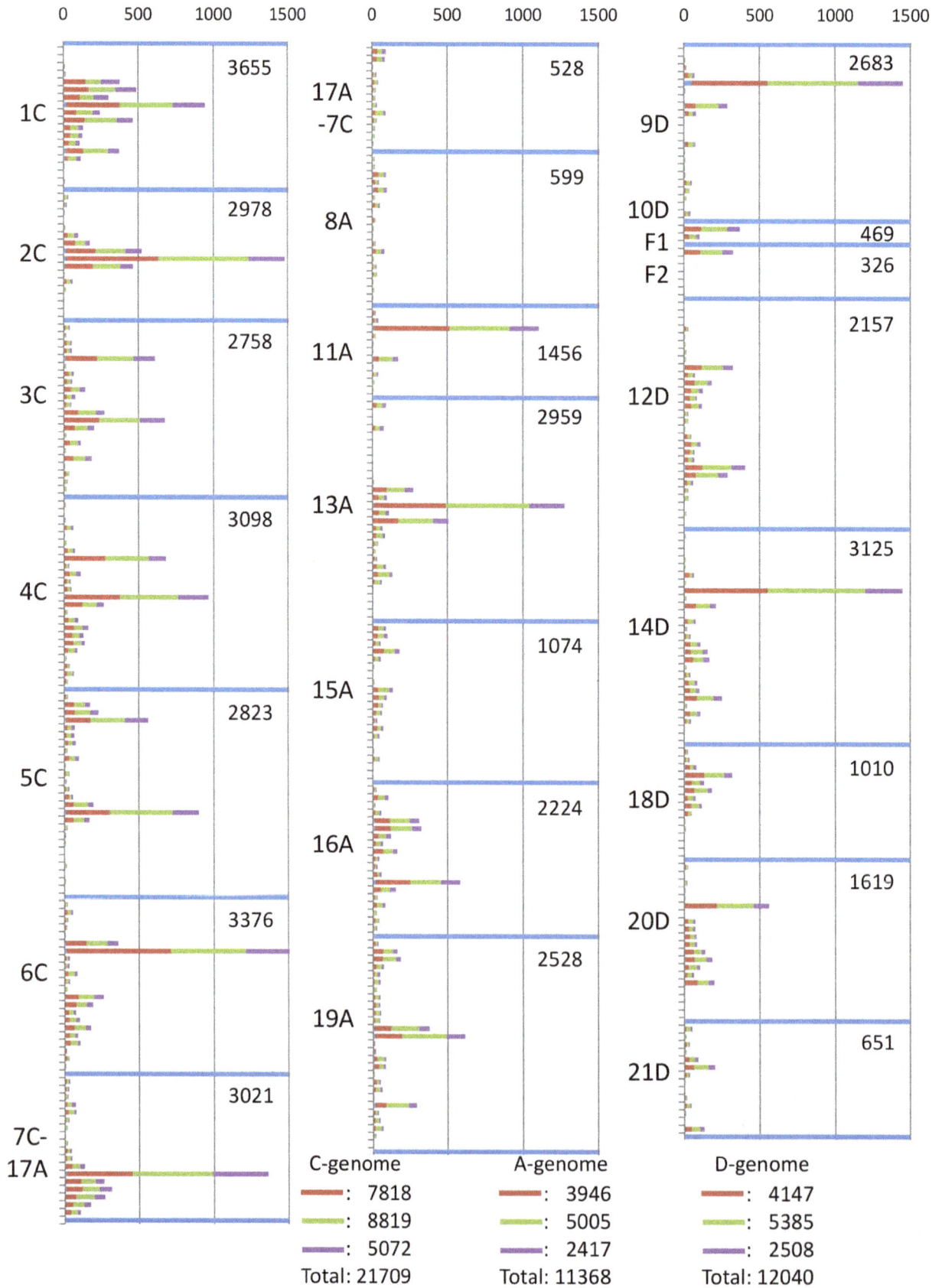

Figure 3. Distribution of GBS loci across the oat genome. Maps of each chromosome (delineated by blue lines and labeled on left) are divided into 5 cM bins with 0 cM starting at the top. Red bars show numbers of loci detected by two pipelines, green shows those detected only by the population-filtering pipeline, and violet shows those detected only by the UNEAK pipeline. Numerals inside boxes show total GBS loci by chromosome. A summary of placed GBS loci by pipeline and by sub-genome is shown.

A *de novo* genetic linkage map using GBS

In the above work, mapping was performed in six populations of reduced size, primarily to approximate the positions of a large inventory of GBS loci relative to an oat consensus map. We also wished to examine the utility of GBS markers for generating a *de novo* linkage map in the absence of other marker types, and to evaluate how well this map could be matched to the current consensus. For this work, we used the VxL population, composed of 145 $F_{4:5}$ RIL families. The GBS loci for map construction were called using the UNEAK GBS pipeline and filtered at high stringency (MAF $\geq$35%, completeness $\geq$90%) at two different levels of heterozygosity (8% and 13%). The resulting data contained 858 (heterozygosity $\geq$8%) and 1053 (heterozygosity $\geq$13%) GBS loci. From this, a map with 35 linkage groups having a total length of 1713 cM was constructed. A comparison between this map and the consensus (Figure S7) showed that 280 (heterozygosity $\geq$8%) or 373 (heterozygosity $\geq$13%) markers were present in VxL but not in the six mapping populations used for consensus map saturation. Most VxL linkage groups corresponded to single consensus chromosomes, and the relative positioning of loci within groups was approximately linear. Several sets of VxL linkage groups (*e.g.*, LG07 and LG20) likely represent single oat chromosomes (in this case, 12D). This suggests that there is good opportunity to perform comparative mapping of traits that are identified in new populations using only GBS technology. The filtering of loci at different levels of heterozygosity provided an opportunity to observe that certain regions of the VxL genetic map are more highly heterozygous than others (red dots in Figure S7, and graphical genotypes in Table S2), as would be expected in an F_4 population. In addition, most of the heterozygous loci were clustered at what are likely centromeres (Figure 3 and Figure S7), explained by the fact that low recombination in centromeric regions has been found to contribute to the retention of residual heterozygosity [40]. This provides good evidence that heterozygous genotype calls in the GBS pipeline are generally accurate and genetically consistent.

Use of GBS to evaluate population structure

PCA and model-based clustering were used to examine the effectiveness of GBS markers to identify population structure in 340 oat lines of global origin. Of these, 41% originated from North American breeding programs (81 lines from Canada and 59 from USA) and the remainder originated elsewhere (Table S3). GBS markers called by the UNEAK pipeline were filtered at $\geq$90% completeness, MAF $\geq$5%, and heterozygosity $\leq$5%. Of the filtered SNPs, only those that were placed on the consensus map (2155 loci) were considered. Because of genetic clustering, a large number of these SNP loci (1159, or 54%) co-segregated at identical positions, and 1755 (81%) were within 1 cM intervals (Figure S8). Since this dependency is also reflected in LD (see next section), it was likely to distort the Eigenvector/Eigenvalues and to bias the interpretation of population structure [29]. Therefore, we applied two levels of LD correction (k10 and k50), in addition to using an uncorrected analysis (k0) for PCA. At k0, no obvious reflection point was observed in the scree plot, while we could distinguish slight two-stage plateaus in k10 and k50, where the drop of Eigenvalues slowed at approximately PC5 and PC10 (Figure S9). The first ten PC explained 37.6, 32.1, and 31.6

percent of the total variation for k0, k10, and k50, respectively, whereas 25.0, 21.2, and 20.9 percent of the total variation was explained by the first four PC (the approximate point of the first plateau).

Since no obvious groups were separated by the first two PCs, model-based clustering was performed to explore the grouping of oat lines based on the first ten PC. The best solutions for k0, k10 and k50 were four, two, and two clusters, although the entropy plot of k50 could be understood as three clusters (Figure S10). The clusterings from k0, k10, and k50 were generally in agreement and reflected the geographic origins of the oat lines, with European lines tightly clustered together and lines from elsewhere spread out across the plot (Figures 4 and S11). The possible third cluster in k50 was positioned between European and North American lines and was comprised of oat lines from Eastern or Northern Europe, as well as some North American lines. One set of five lines was separated by PC4 and this separation is particularly obvious in the k0 data set (Figure S12). The separation of this cluster seemed to be related to growth habit (three of the five lines are winter oats, Figure S12), but because the number of lines was so small, a definitive conclusion could not be made. Our results showed that this diversity panel does not show substantial structural stratification, and this is in agreement with previous work based on DArT markers [31].

Linkage disequilibrium analysis

From the original 2155 markers, we retained r^2 estimates from 51,850 marker pairs with an average or minimum map distance less than 30 cM. Plotting these r^2 estimates against map distance (Figure 5, Figure S13, and Table 2) showed that LD decays such that, at 0.1, conventional r^2 is equal to an average distance of 21.5 cM (Hill-Weir model) or 13.6 cM (Sved model), while r_v^2 is equal to 2.8 cM (Hill-Weir model) or 2.5 cM (Sved model). The fact that r_v (corrected for relatedness) is much smaller than r^2 (uncorrected) illustrates the necessity of removing the effect of coancestry to reduce the inflation of r^2 estimates, and probably reflects that the IOI panel contained groups of related lines originating from the same breeding programs. Estimates of r_s^2 (accounting for population structure) did not reduce the bias in r^2 as substantially as did the models accounting for coancestry, which is consistent with earlier observations that this population is not highly structured. These results suggest that good genome coverage for GWAS will require a marker spacing of approximately 2.0 cM (r_{sv}^2 k0, min rf, Sved model) to 2.8 cM (r_v^2, average rf, Hill-Weir model, Table 2). Non-linear model fitting enabled us to estimate the effective population size required for GWAS, which varied from 68 to 110 lines, depending on the choice of r^2 estimates and evolution models (Table 2).

Germplasm diagnostics using GBS

We wished to examine how GBS technology could be used to solve two common diagnostic problems that arise occasionally in any plant breeding program. The first example involved a suspected error in the planting of one replication in an oat variety registration test. The questionable replication could have been discarded, but it had been grown and harvested at a cost that was more than double that required for genotyping the unknown samples, and discarding the replication would jeopardize the

A

B

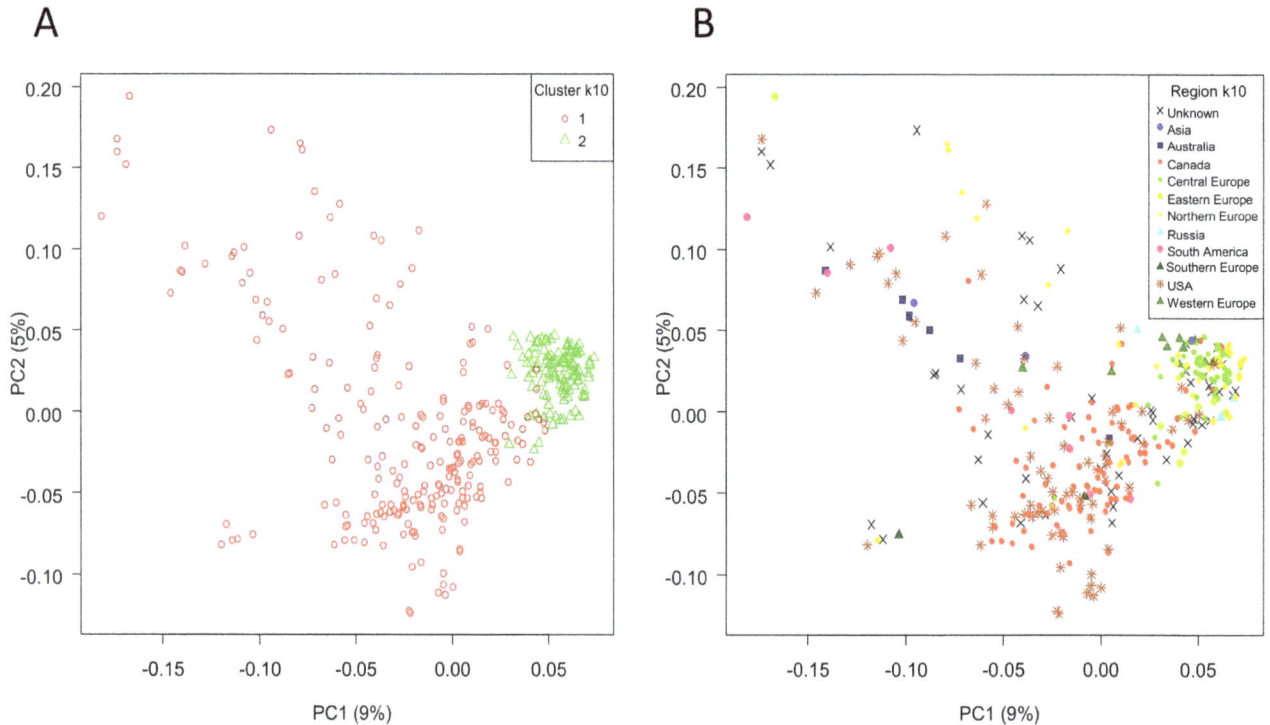

Figure 4. Scatter plots of PC1 *vs.* PC2. The k10 correction is shown: (A) coloured based on clustering from genotypic data, (B) coloured based on geographic origins.

statistical power of the experiment. Problems of this nature can have great economic impact if they delay or improperly influence the release of improved plant varieties. We filtered a set of 1518 diagnostic GBS loci that were polymorphic among known and unknown lines from this test. We then applied UPGMA cluster analysis to the simple allele dissimilarity (*d*) among samples. The results (Figure S14) illustrate that each unknown sample could be paired with one or more known samples. It was then clear that the source of error was a simple reversal of seed envelopes in one of the planting trays. The phenotypic data could then be reassigned to complete the analysis. Although the unknown experimental units could be unambiguously corrected based on the genetic data, a few of the distances between samples known to originate from the same variety (*e.g.*, samples 128 and 232 in Figure S14) were larger than expected. We expect that this is because the DNA samples were prepared from a few seeds sampled randomly from bulks that were harvested by combine from the registration test. If so, this draws attention to the fact that seed harvested from yield trials is often impure and should be used with caution in genetic studies.

The second diagnostic problem was to determine whether an F_2 population originated from a true cross or from selfed seed of a parent. In this case, the progeny appeared very homogeneous and an error was suspected. However, resources had been invested in the cross, and it seemed worthwhile to address this issue before discarding F_3 seed. We filtered GBS loci for ten F_2 progeny and the intended male parent of the cross, together with 340 additional progeny from the IOI diversity panel. The intended female parent was unavailable, but a maternal grandparent (Leggett) was available from the IOI set. Filtering at 10% heterozygosity, 10% MAF, and 90% completeness gave 2205 locus calls. All 2205 loci were completely homogeneous among the suspected progeny and the male parent, with the exception of minor variants that

appeared random and fell within a 1% tolerance for scoring error. A cluster analysis (Figure S15) supported this result. Thus, it was concluded that these seeds were not from a true cross, and that they probably represented a harvesting error in the crossing block.

Discussion

Factors that influence the quality and quantity of GBS SNP calls

The completeness of GBS SNP calls and the number of SNPs filtered at a given completeness are influenced by several factors, including: (1) the actual number of genomic fragments produced by the complexity reduction, (2) the sequencing depth, as determined by the number of reads and the number of samples that are multiplexed, and (3) the underlying density of SNPs, which is related to the diversity and structure of the population [41]. Although other restriction enzymes can be used for complexity reduction (*e.g.*, [11,17]), we limited our present investigation in oat to the *Pst*I-*Msp*I combination which had previously been optimized for the similar-sized genome of wheat [12,23]. This method provided suitable results in oat with minor modifications, and identified tens of thousands of sequence polymorphisms. Although the results in Figure 1 show a somewhat linear response in the number of SNPs identified as sequencing depth increases, the number of SNP calls for all levels of completeness would eventually plateau at a limit (possibly more than 100,000) determined by the complexity reduction and the population. The depth of sequencing required to obtain complete data for all SNP-containing fragments is currently not practical nor cost effective, nor is it required to obtain meaningful genetic data and results. However, we strongly recommend that additional replicate samples be used for the parents of mapping populations or other material that is critical to an experiment to achieve

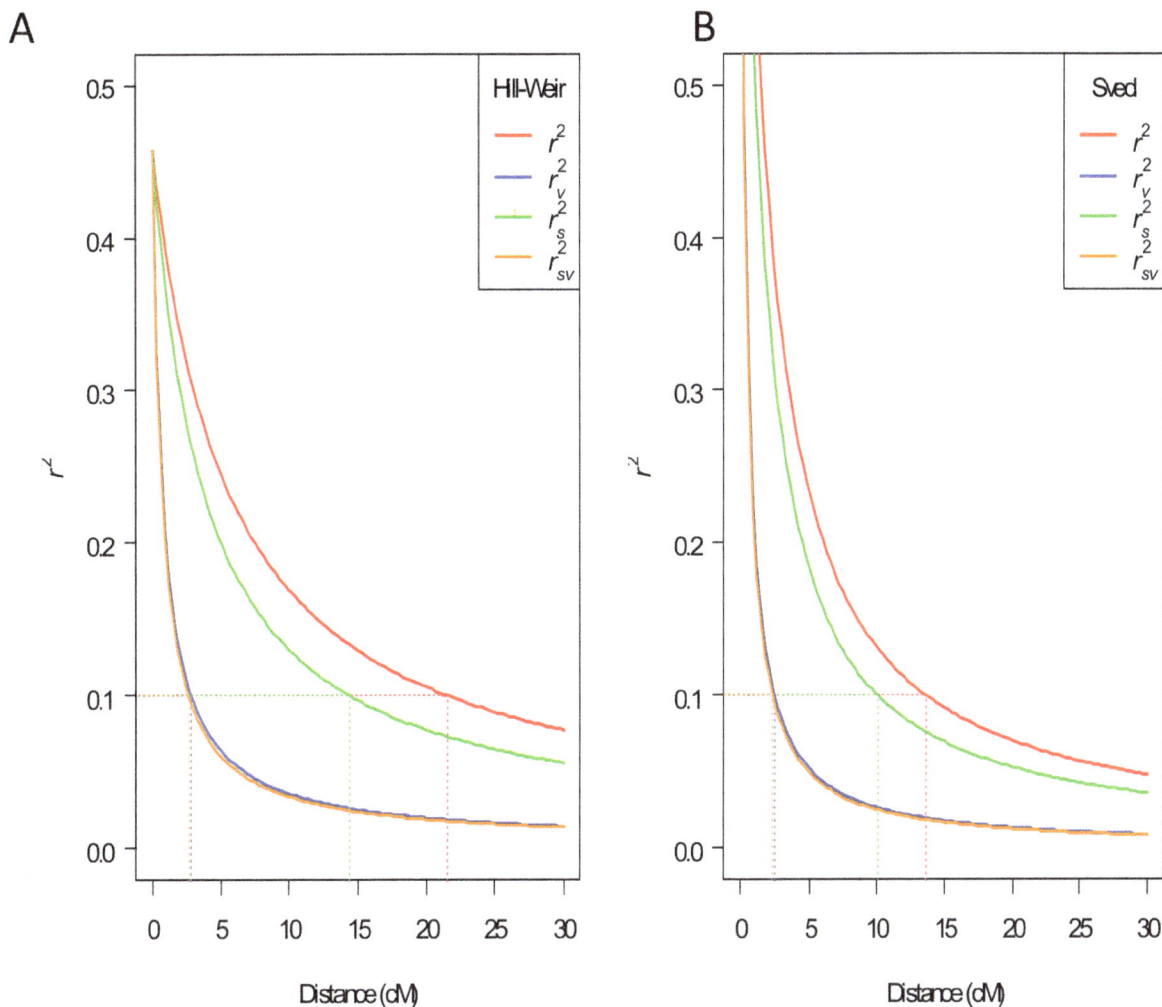

Figure 5. LD decay plot. r^2 estimates were plotted against the average map distance (recombination frequency expressed in cM): (A) relationship fit using the mutation model (Hill-Weir), (B) relationship fit using the recombination-drift model (Sved). Population structure was estimated using the k10 correction.

greater sequencing depth for these lines. In addition, there may be opportunities to produce and analyze reduced complexities generated using selective bases added to primers of existing enzyme systems [18]. Use of selective bases may have value for certain investigations where a more complete and consistent genotyping is required, and would have the advantage of providing data at a subset of loci already characterized in the present work. An example of an application where this might be useful is the routine diagnosis of variety identity or other diagnostic applications that do not rely on a high density of genetic markers. One can also "complete" the missing data points through data imputation. Various imputation methods have been developed which can be highly effective for subsequent analyses (reviewed in [42]). Although most methods require known marker order, often provided by a reference genome sequence, imputation can also be applied to unordered markers and could increase the accuracy of further analyses.

As the number of diverse lines was increased, the number of GBS loci also increased, but this number plateaued at a relatively small number of lines. This result is consistent with the notion that each additional line added an increasing likelihood of identifying rare alleles, such that an increasing number of loci will meet the

allele frequency threshold. Meanwhile, larger samples may make it increasingly unlikely that loci will meet filtering criteria and/or that spurious loci will be included. This is possibly why the plateau occurs earlier as data are filtered for higher completeness, and provides a possible explanation for the decrease in SNP number beyond the plateau (Figure 2). In addition, completeness is stochastically variable among samples and among loci. Thus, loci that are included in two diversity subsets are not necessarily those that are the most complete across the entire panel. This phenomenon may make it difficult to obtain consistent loci among different experiments if thresholds for completeness are set too high. For this reason, we recommend that filtering be performed at multiple levels, depending on the purpose of the experiment. For example, the VxL population was filtered at high stringency to develop an initial *de-novo* map, then at low stringency to identify additional GBS loci that mapped across populations.

One of the factors of concern in oat and other polyploid species is that the analysis of genetic loci may be confounded by duplicated homoeologous regions. This factor has made it difficult to discover and apply gene-based SNP loci, and has resulted in some assays where SNPs must be scored with dominant alleles [4,43]. Based on the relative scarcity of BLAST matches in related

Table 2. LD decay estimated in IOI.

| | Hill-Weir | | | | Sved | | | |
| | Average rf | | Min rf | | Average rf | | Min rf | |
r^2	Map distance (cM)	$N_e \pm$ SE	Map distance (cM)	$N_e \pm$ SE	Map distance (cM)	$N_e \pm$ SE	Map distance (cM)	$N_e \pm$ SE
r^2	21.5	9±0.1	19.5	10±0.1	13.6	17±0.1	12.3	18±0.1
r_s^2 k0	14.2	14±0.1	12.7	15±0.2	9.9	23±0.2	8.9	25±0.2
r_s^2 k10	14.4	13±0.1	12.9	15±0.2	10	22±0.2	9	25±0.2
r_s^2 k50	14.5	13±0.1	13	15±0.2	10.1	22±0.2	9.1	25±0.2
r_{sv}^2 k0	2.6	75±0.8	2.2	87±1	2.3	99±1.5	2	111±2
r_{sv}^2 k10	2.6	73±0.8	2.3	85±1	2.3	97±1.5	2.1	109±1.9
r_{sv}^2 k50	2.6	73±0.8	2.3	84±1	2.3	96±1.5	2.1	108±1.9
r_v^2	2.8	68±0.7	2.5	79±0.9	2.5	91±1.3	2.2	102±1.7

The relationship between LD and map distance was modeled by fitting two alternate non-linear regression models: a drift-recombination equilibrium model [36] or a modified recombination-drift model including low level of mutation and an adjustment for sample size [37]. Map distance at $r^2 = 0.1$ was shown. Both average distance across six bi-parental mapping population and minimum distance from available mapping populations were used. N_e effective population size; SE, standard error.

species, the majority of the GBS loci from the *Pst*I-*Msp*I complexity reduction appear to be located in non-genic regions. Furthermore, because the GBS marker calling is based on counts of specific allele variants, all GBS loci are scored with co-dominant alleles. For these reasons, and also because of intense filtering to remove loci with non-diploid inheritance, the GBS method provides good representation of single genetic loci. This would also tend to remove SNPs that fall in conserved genic regions, as these GBS tags would align to all three genomes and the resulting SNPs would not segregate as single loci. Although heterozygote calls are more subject to genotyping errors, our results show the interest of including heterozygous genotypes in certain applications. In particular, we observed that the overall quality of heterozygote determination in the VxL F_4 population was good, and that it enabled meaningful characterization of heterozygous regions in the graphical genotypes (Table S2). However, we have not thoroughly investigated the use of GBS markers in F_2 populations, where it is possible to confuse single loci with duplicated loci having similar genetic ratios. For this reason, GBS should be used with caution in F_2 mapping populations, unless a reference genome or an ordered scaffold is available such that heterozygous regions can be imputed.

Annotation of oat GBS SNPs

We estimated that 5% of oat GBS SNPs were within protein coding sequences. This estimate should be considered preliminary because of the current lack of public oat gene sequences. Using an automated SNP annotation pipeline, Kono *et al.* [44] estimated that only 1.3% of a preliminary subsample of 5000 oat GBS SNPs were genic SNPs. Most of their matches were also based on *Brachypodium*. Their estimate may be lower because of sampling bias and because they used a higher stringency in protein matching ($e < 5 \times 10^{-5}$). Using the same annotation pipeline, Kono *et al.* [44] identified 10.6% of a sample of barley GBS tags as being genic SNPs. The rate in barley may be substantially higher because of the greater availability of barley gene sequences.

Methods of bioinformatics analysis and workflow

We used two bioinformatics pipelines to perform SNP calling. The motivation for using two pipelines was that they were the only two non-reference-genome GBS pipelines of which we were aware. While the UNEAK pipeline gave clear, predictable results, the number of loci passing secondary filtering was low compared to those called by the population-level pipeline. In addition, as observed elsewhere [22], the GBS SNPs identified can vary considerably with different methods. Across all SNPs placed on the oat consensus map, 43% (19,209 out of 45,117) were called only by the population-based method, 22% (9997 out of 45,117) were called only by the UNEAK method, and 36% were called by both methods (Figure 3). Although the population filtering method called more SNPs, these SNPs contained a higher redundancy, and multiple SNPs (linked or unlinked) were sometimes assigned to the same context sequence in the report (data not shown). This is a result of calling SNPs in tags that belong to complex gene families and/or in context sequences of haplotypes that contain multiple SNPs. Such loci are usually ignored by the UNEAK pipeline, which reports only tags with single SNPs. However, in some cases, the direct use of SNPs from the population-filtering report would prevent the correct development of secondary allele assays, if no detailed validation of the assembly was performed to generate unambiguous diagnostic probes for correct SNP interrogation. For these reasons, we performed some of this work using only the UNEAK pipeline. Nevertheless, it was apparent that a much greater number of SNPs were called by the population-filtering method, and these additional SNPs may provide important information which would otherwise be discarded. Until bioinformatics methods can be further refined, we recommend using a combined data set composed of SNPs called by both pipelines. We also recommend the preferential use of SNPs called by the UNEAK pipeline in the development of secondary assays, and the use of appropriate statistical methods to reduce the influence of marker redundancy in subsequent applications such as association mapping.

Saturation of a consensus linkage map with GBS loci

High density maps are required for the precise mapping of important agronomic traits to be targeted in marker-assisted breeding. The high-throughput GBS technology enabled the placement of 45,117 GBS loci identified across six bi-parental mapping populations on the oat consensus map [4]. This high-density map showed marker clustering along chromosomes, similar to a barley map saturated with GBS markers [12]. Since many of these clusters represent centromeres, GBS loci are likely to be more evenly distributed along the physical map. However, gaps of up to 15 cM were still observed in the high-density map. Some gaps may result from lack of polymorphism in the mapping populations, which can be further improved by integrating other mapping populations. Gaps could also be filled in by using GBS libraries produced using different restriction enzymes, as shown in wheat [17]. Gaps and multiple clusters may also be related to the construction of the initial consensus map and it is possible that the consensus map will be improved once GBS markers are fully integrated with additional gene-based SNP loci.

Utility of GBS markers for genetic analysis in oat

We have investigated different GBS applications relevant to oat breeding. In addition to saturating an existing consensus map, GBS markers were suitable for building a *de-novo* linkage map with good genome coverage that revealed colinearity with the consensus. These results were successful because GBS provided a large number of markers that were polymorphic in multiple populations. This will facilitate comparative mapping to validate and refine the location of target alleles in diverse genetic backgrounds, and will increase the options available for molecular breeding.

Analyses of diverse germplasm showed weak population structure in our sample. This weak structure was observed previously when DArT markers were used across a larger oat diversity panel that included the IOI set used in this study [31]. While rice, barley, and maize are known for having strong population structure [45–47], oat, despite having four recognizable types (naked, covered, spring, and winter), has not shown obvious population structure within the samples analysed to date. Although the majority of lines in the IOI set are the spring type (318 out of 340) and covered-seeded (312 out of 340), the remaining naked or winter lines did not form distinct sub-clusters. Instead, the scatterplot tended to reflect the geographic locations of breeding programs and (by inference) the degree of coancestry among lines. A possible explanation for these results could be that, while oat breeders tend to make most crosses among parents that are locally adapted, they have also exchanged elite germplasm with some regularity. The relatively small effective population size compared to the number of IOI lines also supports this interpretation.

The use of GWAS is widely considered to be an attractive alternative to the use of structured (*e.g.*, bi-parental) populations for identifying adaptive alleles for use in molecular breeding. However, effective GWAS requires prior knowledge of LD decay and an awareness of the population under investigation. Our results highlight that there is a strong gradient of coancestry that needs to be accounted for through the use of an appropriate model, but that other sources of population structure are not important in the population investigated. Although GBS appears to provide a much higher density of markers than required for GWAS, it is possible that target loci are within gaps that do not contain a suitable marker density. For this reason, it may be useful to test additional enzyme combinations for GBS for use in a large association panel when a large investment has been made in phenotyping.

Conclusion

The choice of marker technologies is critical to the success and future application of genetic and genomic research. GBS is attractive because it provides thorough genome coverage and can be applied at low cost with or without a reference genome sequence. However, GBS requires intense bioinformatic analysis, an awareness of the need to filter data, and a tolerance for incomplete data. In this work, we have shown that GBS is an effective method to discover and apply SNPs in the large and complex oat genome, and that GBS integrates and compares favourably with an established SNP technology. The resulting data have provided whole-genome coverage at a density that enables detailed analysis of genetic diversity and high power to detect specific genetic variants.

Our overall conclusion and recommendation is that GBS be used as a cost-effective primary tool in any application similar to those that we have explored in oat. Other applications of GBS in oat, including QTL analysis, genomic selection, and the ordering of genome sequence scaffolds, remain to be fully tested, but are expected to be successful based on indications from this work and from similar use in other species. In future work, we intend to apply GBS routinely to genotype and select among advanced oat breeding lines. As a side benefit to improved selection, we hope to provide new information about the sources and locations of alleles for better adaptation in oat, and to integrate this information with the existing genomic knowledge for oat.

Supporting Information

Figure S1 HTML map format. Instructions for using the HTML-formatted map. The map can be found locally in Text S2 as "HTML_Local_text_S2.html" or online at: http://ahoy.aowc.ca/html_link_gbs_text_s2.html.

Figure S2 Distribution of map gaps in the original and updated oat consensus maps. Empty bars show the distribution of map gaps in the first oat consensus map (Oliver *et al.*, 2013); solid bars show the distribution of map gaps in the consensus map with the GBS markers placed on it (this study). Only gaps larger than 5 cM are shown.

Figure S3 Orthology between oat and *Brachypodium distachyon*. Each dot represents the position of a sequence match (BLASTn, $E < 10^{-12}$) between the oat consensus map (blue dots for GBS loci, red dots for array-based SNPs) and the assembled *Brachypodium distachyon* (Bd) pseudomolecule (release 2.1; http://www.brachypodium.org).

Figure S4 Orthology between oat and rice. Each dot represents the position of a sequence match (BLASTn, $E < 10^{-12}$)

between the oat consensus map (blue dots for GBS loci, red dots for array-based SNPs) and the genome sequence of rice (*Oryza sativa* L., release 6.1 from http://rice.plantbiology.msu.edu).

Figure S5 Orthology between oat and barley. Each dot represents the position of a sequence match (BLASTn, $E<10^{-12}$) between the oat consensus map (blue dots for GBS loci, red dots for array-based SNPs) and barley (*Hordeum vulgare* L., cv. Morex, release 2.0 from ftp://ftp.ensemblgenomes.org/pub/plants/release-20/fasta/hordeum_vulgare/dna/ non repeat-masked versions). Barley pseudomolecules are assembled according to chromosome arm (long (2HL to 7HL) or short (2HS to 7HS)), except for chromosome 1H.

Figure S6 Orthology between oat and barley (multiple matches removed). Each dot represents the position of a sequence match (BLASTn, $E<10^{-12}$) between the oat consensus map (blue dots for GBS loci, red dots for array-based SNPs) and barley (*Hordeum vulgare* L., cv. Morex, release 2.0 from ftp://ftp.ensemblgenomes.org/pub/plants/release-20/fasta/hordeum_vulgare/dna/ non repeat-masked versions). A subset of matches from Figure S5 is shown: oat sequences that matched other *Hv* sequences more than 6 times at the same BLASTn expectation have been removed.

Figure S7 Comparison between the VxL map and the consensus map. Each dot represents a marker shared by the two maps. Red dots highlight markers of higher heterozygosity (between 8 and 13%).

Figure S8 Distribution of distances between adjacent markers used for LD analysis. Markers were first sorted according to map position, then the distances were calculated as Position $_m$ minus Position $_{m-1}$. For the first marker of each chromosome, the interval was calculated as the difference between its position and the position of the second marker.

Figure S9 Scree plots of principal components of IOI genotypic data. The first 20 components at three levels of LD correction were used to draw the plots: k0 (without correction), k10 (using 10 adjacent markers for LD adjustment), and k50 (using 50 adjacent markers for LD adjustment). No obvious "elbows" were observed but there was a two-stage decay: at PC5 and at PC10.

Figure S10 IOI population structure scatter plot (PC1 vs. PC2) based on genetic clustering. Three levels of LD correction are shown: k0 (A), k10 (B), and k50 (C and D).

Figure S11 IOI population structure scatter plot (PC1 vs. PC2) coloured based on the geographical origins of the lines. Three levels of LD correction are shown: k0 (A), k10 (B), and k50 (C).

Figure S12 IOI population structure scatter plot (PC3 vs. PC4) coloured based on genetic clustering (left) or plant habitat (right). Three levels of LD correction are shown: k0 (up), k10 (middle), and k50 (bottom).

Figure S13 LD decay plot. r^2 estimates were plotted against both minimum and average map distance (recombination frequency expressed in cM): (A) relationship fit using the mutation model (Hill-Weir), (B) relationship fit using the recombination-drift model (Sved).

Figure S14 Using GBS markers to resolve an issue in a field experiment. UPGMA cluster analysis of simple allele-matching metric (d) based on 1518 GBS loci with heterozygosity <8%, MAF >20%, and completeness >95%. This evidence was used to correct a planting error in a field experiment. The samples in one replication (red samples) were out of order compared to those in a second, correct replication (green samples), and a set of known controls (blue samples). Analyzing the sub-clusters in the above dendrogram and assigning corrected identities (entry numbers 1–32, above) to the samples in replication 1 made it obvious that the planting order of the first replication had been reversed.

Figure S15 Using GBS markers to resolve an issue with breeding materials. UPGMA cluster analysis of simple allele-matching metric (d) based on 2205 GBS loci with >90% completeness. GBS calls were made across samples from 343 diverse oat varieties plus ten putative F_2 segregants (green) from a putative cross between SA060123 (red) and a progeny of Leggett (blue). Eight closely related oat cultivars are also shown in this partial cluster dendrogram. Of the 2205 loci, only 131 (6%) showed any variation among the ten progeny plus SA060123, and this variation was within the expectations of heterozygous miscalls. This evidence was used to conclude that the ten progeny were actually from selfed seed of SA060123 rather than true segregants from a hybrid.

Figure S16 Effect of adding populations on the number of markers placed on the oat consensus map. Using the consensus map [4] as a framework and starting with a different population each time, markers from the six populations were placed sequentially in all possible combinations (C_k^6, k = 1 to 6). The number of additional markers contributed by the final map at each step is represented by different colours and shapes. The box represents the range between the first and third quartiles and the thick horizontal bar represents the median.

Figure S17 Distribution of annotated GBS markers across the oat consensus map. Maps of each chromosome were divided into 5 cM bins and the number of intergenic/genic markers counted for each bin. Some markers are in the negative range because they are placed off the beginning of the linkage group.

Acknowledgments

This work was made possible through excellent technical and professional assistance from the following: Shuangye Wu, for constructing GBS libraries; Rebeca Oliver, Biniam Hizbai, Annick Gauthier, Muriel Jatar, Sophie Ménard, and Paul Gillespie for preparing and handling DNA samples; Andrew Sharpe and Darrin Klassen for performing DNA sequencing; Stéphane Nicolas for sharing R scripts used for statistical analysis; Weikai Yan for sharing breeding material that was used to evaluate genotyping procedures; and Brad de Haan, Steve Thomas, Matthew Hayes, and Kathie Upton for professional assistance with field and greenhouse procedures.

Author Contributions

Conceived and designed the experiments: YFH JAP EWJ NAT. Performed the experiments: YFH JAP CPW NAT. Analyzed the data: YFH JAP CPW NAT. Contributed reagents/materials/analysis tools: JAP EWJ NAT. Wrote the paper: YFH JAP CPW EWJ NAT.

References

1. Meuwissen THE, Hayes BJ, Goddard ME (2001) Prediction of Total Genetic Value Using Genome-Wide Dense Marker Maps. Genetics 157: 1819–1829.
2. Heffner EL, Sorrells ME, Jannink J-L (2009) Genomic Selection for Crop Improvement. Crop Sci 49: 1–12.
3. Molnar SJ, Tinker NA, Kaeppler HF, Rines HW (2011) Molecular Genetics of Oat Quality. In: Webster FH, Wood PJ, editors. Oats: Chemistry and Technology. St. Paul, MN: American Association of Cereal Chemists.
4. Oliver RE, Tinker NA, Lazo GR, Chao S, Jellen EN, et al. (2013) SNP Discovery and Chromosome Anchoring Provide the First Physically-Anchored Hexaploid Oat Map and Reveal Synteny with Model Species. PLoS ONE 8: e58068.
5. Massman JM, Jung H-JG, Bernardo R (2013) Genomewide Selection versus Marker-assisted Recurrent Selection to Improve Grain Yield and Stover-quality Traits for Cellulosic Ethanol in Maize. Crop Sci 53: 58–66.
6. Asoro FG, Newell MA, Beavis WD, Scott MP, Tinker NA, et al. (2013) Genomic, Marker-Assisted, and Pedigree-BLUP Selection Methods for β-Glucan Concentration in Elite Oat. Crop Sci 53: 1894–1906.
7. McCouch SR, McNally KL, Wang W, Sackville Hamilton R (2012) Genomics of gene banks: A case study in rice. American Journal of Botany 99: 407–423.
8. Romay M, Millard M, Glaubitz J, Peiffer J, Swarts K, et al. (2013) Comprehensive genotyping of the USA national maize inbred seed bank. Genome Biology 14: R55.
9. Gutierrez-Gonzalez JJ, Wise ML, Garvin DF (2013) A developmental profile of tocol accumulation in oat seeds. Journal of Cereal Science 57: 79–83.
10. Tinker N, Kilian A, Wight C, Heller-Uszynska K, Wenzl P, et al. (2009) New DArT markers for oat provide enhanced map coverage and global germplasm characterization. BMC genomics 10: 39.
11. Elshire R, Glaubitz J, Sun Q, Poland J, Kawamoto K, et al. (2011) A robust, simple genotyping-by-sequencing (GBS) approach for high diversity species. PLoS ONE 6.
12. Poland J, Brown P, Sorrells M, Jannink J-L (2012) Development of high-density genetic maps for barley and wheat using a novel two-enzyme genotyping-by-sequencing approach. PLoS ONE 7.
13. Beissinger TM, Hirsch CN, Sekhon RS, Foerster JM, Johnson JM, et al. (2013) Marker Density and Read Depth for Genotyping Populations Using Genotyping-by-Sequencing. Genetics 193: 1073–1081.
14. Glaubitz JC, Casstevens TM, Lu F, Harriman J, Elshire RJ, et al. (2014) TASSEL-GBS: A High Capacity Genotyping by Sequencing Analysis Pipeline. PLoS ONE 9: e90346.
15. Liu H, Bayer M, Druka A, Russell J, Hackett C, et al. (2014) An evaluation of genotyping by sequencing (GBS) to map the Breviaristatum-e (ari-e) locus in cultivated barley. BMC Genomics 15: 104.
16. Lu F, Lipka AE, Glaubitz J, Elshire R, Cherney JH, et al. (2013) Switchgrass Genomic Diversity, Ploidy, and Evolution: Novel Insights from a Network-Based SNP Discovery Protocol. PLoS Genet 9: e1003215.
17. Saintenac C, Jiang D, Wang S, Akhunov E (2013) Sequence-Based Mapping of the Polyploid Wheat Genome. G3: Genes|Genomes|Genetics 3: 1105–1114.
18. Sonah H, Bastien M, Iquira E, Tardivel A, Légaré G, et al. (2013) An Improved Genotyping by Sequencing (GBS) Approach Offering Increased Versatility and Efficiency of SNP Discovery and Genotyping. PLoS ONE 8: e54603.
19. Spindel J, Wright M, Chen C, Cobb J, Gage J, et al. (2013) Bridging the genotyping gap: using genotyping by sequencing (GBS) to add high-density SNP markers and new value to traditional bi-parental mapping and breeding populations. Theoretical and Applied Genetics 126: 2699–2716.
20. Uitdewilligen JGAML, Wolters A-MA, D'hoop BB, Borm TJA, Visser RGF, et al. (2013) A Next-Generation Sequencing Method for Genotyping-by-Sequencing of Highly Heterozygous Autotetraploid Potato. PLoS ONE 8: e62355.
21. Ward J, Bhangoo J, Fernandez-Fernandez F, Moore P, Swanson J, et al. (2013) Saturated linkage map construction in Rubus idaeus using genotyping by sequencing and genome-independent imputation. BMC Genomics 14: 2.
22. Mascher M, Wu S, Amand PS, Stein N, Poland J (2013) Application of Genotyping-by-Sequencing on Semiconductor Sequencing Platforms: A Comparison of Genetic and Reference-Based Marker Ordering in Barley. PLoS ONE 8: e76925.
23. Poland J, Endelman J, Dawson J, Rutkoski J, Wu S, et al. (2012) Genomic Selection in Wheat Breeding using Genotyping-by-Sequencing. Plant Gen 5: 103–113.
24. Mascher M, Muehlbauer GJ, Rokhsar DS, Chapman J, Schmutz J, et al. (2013) Anchoring and ordering NGS contig assemblies by population sequencing (POPSEQ). The Plant Journal 76: 718–727.
25. Mayer KFX, Waugh R, Langridge P, Close TJ, Wise RP, et al. (2012) A physical, genetic and functional sequence assembly of the barley genome. Nature 491: 711–716.
26. Wight CP, Tinker NA, Kianian SF, Sorrells ME, O'Donoughue LS, et al. (2003) A molecular marker map in 'Kanota' × 'Ogle' hexaploid oat (Avena spp.) enhanced by additional markers and a robust framework. Genome 46: 28–47.
27. Bradbury PJ, Zhang Z, Kroon DE, Casstevens TM, Ramdoss Y, et al. (2007) TASSEL: software for association mapping of complex traits in diverse samples. Bioinformatics 23: 2633–2635.
28. Wu Y, Bhat PR, Close TJ, Lonardi S (2008) Efficient and Accurate Construction of Genetic Linkage Maps from the Minimum Spanning Tree of a Graph. PLoS Genet 4: e1000212.
29. Patterson N, Price AL, Reich D (2006) Population Structure and Eigenanalysis. PLoS Genet 2: e190.
30. R Core Team (2013) R: A language and environment for statistical computing, R Foundation for Statistical Computing, Vienna, Austria, ISBN 3-900051-07-0, URL http://www.R-project.org/
31. Newell MA, Cook D, Tinker NA, Jannink JL (2011) Population structure and linkage disequilibrium in oat (Avena sativa L.): implications for genome-wide association studies. Theoretical and Applied Genetics 122: 623–632.
32. Fraley C, Raftery AE, Murphy TB, Scrucca L (2012) mclust Version 4 for R: Normal Mixture Modeling for Model-Based Clustering, Classification, and Density Estimation. Department of Statistics, University of Washington.
33. Baudry J-P, Raftery AE, Celeux G, Lo K, Gottardo R (2010) Combining Mixture Components for Clustering. Journal of Computational and Graphical Statistics 19: 332–353.
34. Mangin B, Siberchicot A, Nicolas S, Doligez A, This P, et al. (2012) Novel measures of linkage disequilibrium that correct the bias due to population structure and relatedness. Heredity 108: 285–291.
35. Endelman J (2011) Ridge regression and other kernels for genomic selection with R package rrBLUP. Plant Genome 4: 250–255.
36. Sved JA (1971) Linkage disequilibrium and homozygosity of chromosome segments in finite populations. Theoretical Population Biology 2: 125–141.
37. Hill WG, Weir BS (1988) Variances and covariances of squared linkage disequilibria in finite populations. Theoretical Population Biology 33: 54–78.
38. Remington DL, Thornsberry JM, Matsuoka Y, Wilson LM, Whitt SR, et al. (2001) Structure of linkage disequilibrium and phenotypic associations in the maize genome. Proceedings of the National Academy of Sciences 98: 11479–11484.
39. Perrier X, Jacquemoud-Collet JP (2006) DARwin: Dissimilarity Analysis and Representation for Windows. 5.0.158 (2009) ed. Montpellier: CIRAD.
40. Gore MA, Chia J-M, Elshire RJ, Sun Q, Ersoz ES, et al. (2009) A First-Generation Haplotype Map of Maize. Science 326: 1115–1117.
41. Poland JA, Rife TW (2012) Genotyping-by-Sequencing for Plant Breeding and Genetics. Plant Gen 5: 92–102.
42. Marchini J, Howie B (2010) Genotype imputation for genome-wide association studies. Nat Rev Genet 11: 499–511.
43. Oliver R, Lazo G, Lutz J, Rubenfield M, Tinker N, et al. (2011) Model SNP development for complex genomes based on hexaploid oat using high-throughput 454 sequencing technology. BMC genomics 12: 77.
44. Kono TJY, Seth K, Poland JA, Morrell PL (2014) SNPMeta: SNP annotation and SNP metadata collection without a reference genome. Molecular Ecology Resources 14: 419–425.
45. Garris AJ, Tai TH, Coburn J, Kresovich S, McCouch S (2005) Genetic Structure and Diversity in Oryza sativa L. Genetics 169: 1631–1638.
46. Hamblin MT, Close TJ, Bhat PR, Chao S, Kling JG, et al. (2010) Population Structure and Linkage Disequilibrium in U.S. Barley Germplasm: Implications for Association Mapping. Crop Sci 50: 556–566.

47. Vigouroux Y, Glaubitz JC, Matsuoka Y, Goodman MM, Sánchez GJ, et al. (2008) Population structure and genetic diversity of New World maize races assessed by DNA microsatellites. American Journal of Botany 95: 1240–1253.

48. Portyanko V, Hoffman D, Lee M, Holland J (2001) A linkage map of hexaploid oat based on grass anchor DNA clones and its relationship to other oat maps. Genome 44: 249–265.

49. O'Donoughue LS, Sorrells ME, Tanksley SD, Autrique E, Deynze AV, et al. (1995) A molecular linkage map of cultivated oat. Genome 38: 368–380.

14

Genome Duplication and Gene Loss Affect the Evolution of Heat Shock Transcription Factor Genes in Legumes

Yongxiang Lin[1,2☯], Ying Cheng[1☯], Jing Jin[1], Xiaolei Jin[1], Haiyang Jiang[1], Hanwei Yan[1], Beijiu Cheng[1*]

1 Key Lab of Crop Biology of Anhui Province, School of Life Sciences, Anhui Agricultural University, Hefei, Anhui, China, 2 Crop Research Institute, Anhui Academy of Agricultural Sciences, Hefei, Anhui, China

Abstract

Whole-genome duplication events (polyploidy events) and gene loss events have played important roles in the evolution of legumes. Here we show that the vast majority of Hsf gene duplications resulted from whole genome duplication events rather than tandem duplication, and significant differences in gene retention exist between species. By searching for intraspecies gene colinearity (microsynteny) and dating the age distributions of duplicated genes, we found that genome duplications accounted for 42 of 46 Hsf-containing segments in *Glycine max*, while paired segments were rarely identified in *Lotus japonicas*, *Medicago truncatula* and *Cajanus cajan*. However, by comparing interspecies microsynteny, we determined that the great majority of Hsf-containing segments in *Lotus japonicas*, *Medicago truncatula* and *Cajanus cajan* show extensive conservation with the duplicated regions of *Glycine max*. These segments formed 17 groups of orthologous segments. These results suggest that these regions shared ancient genome duplication with Hsf genes in *Glycine max*, but more than half of the copies of these genes were lost. On the other hand, the *Glycine max* Hsf gene family retained approximately 75% and 84% of duplicated genes produced from the ancient genome duplication and recent *Glycine*-specific genome duplication, respectively. Continuous purifying selection has played a key role in the maintenance of Hsf genes in *Glycine max*. Expression analysis of the Hsf genes in *Lotus japonicus* revealed their putative involvement in multiple tissue-/developmental stages and responses to various abiotic stimuli. This study traces the evolution of Hsf genes in legume species and demonstrates that the rates of gene gain and loss are far from equilibrium in different species.

Editor: Marc Robinson-Rechavi, University of Lausanne, Switzerland

Funding: This work was supported by grants from the Special Fund for Agro-scientific Research in the Public Interest (Grant No. 201303001) and the Genetically Modified Organisms Breeding Major Projects (Grant No. 2011ZX08003-002). The funders had no role in study design, data collection and analysis, decision to publish, or preparation of the manuscript.

Competing Interests: The authors have declared that no competing interests exist.

* Email: bjchengahau@163.com

☯ These authors contributed equally to this work.

Introduction

Whole-genome duplication, or polyploidy, is a common phenomenon in the evolution of plants and is particularly widespread in angiosperms [1,2]. Many modern diploid plants have experienced one or more episodes of polyploidy and possess vestiges of multiple rounds of polyploidy [3,4]. Recently, comparisons between legume genomes have revealed that legumes of the large Papilionoideae subfamily (papilionoids) have undergone whole-genome duplication [5]. This older shared polyploidy event is estimated to have occurred 56 to 65 million years ago (Mya) [6,7]. A second, more recent genome duplication event occurred only in the lineage leading to *Glycine* up to 13 Mya [8]. Over time, the genomes of these plants diploidized, accompanied by rearrangements and loss of genes and chromosomal segments; this eliminated much of the evidence of the original duplication. Genome duplication and subsequent fractionation have played key roles in shaping present-day legume genomes [9].

A gene family is a group of similar genes resulting from wholesale or partial gene duplication; the size of a gene family reflects the number of duplicated genes (paralogs) in each species [10]. Whole-genome duplication causes each gene in the genome to be present in two copies. However, after duplication, not all

categories of genes respond the same way to polyploidy; gene loss often occurs independently in different gene families. Some gene families, such as NBS-LRR resistance genes, show rapid rates of turnover among family members, with the loss of major gene lineages in some plant families [11,12]. Alternatively, most genes are retained in other families with highly conserved amino acid sequences, such as transcription factors [13]. In the *Arabidopsis* genome, three whole-genome duplications that have occurred in the past 350 million years have brought about a greater than 90% increase in the number of transcription factor, signal transduction and developmental genes [14]. Genes retained after polyploidy may buffer critical functions, but over time, a gradual erosion of this capacity of duplicated genes may contribute to the cyclicality of genome duplication [15].

Here, we examine the evolution of the heat shock transcription factor (Hsf) family in legume species. Hsfs serve as the terminal components of signal transduction and are the central regulators of the expression of heat shock proteins and other heat shock-induced genes that confer thermotolerance to all eukaryotes [16–21]. Like many other transcription factors, Hsfs have a modular structure [22]. Hsf proteins share a well-conserved DNA binding domain (DBD) at their N termini, an adjacent bipartite oligomerization domain (HR-A/B region) composed of hydrophobic heptad repeats, a nuclear location signal (NLS), a nuclear export signal (NES) and a C-terminal activation domain (AHA motifs) [16,23–26]. In contrast to the small number of Hsf genes found in *Drosophila*, *Caenorhabditis elegans*, yeasts and animals [27–33], the Hsf system is more complex in plants than in any other organisms investigated thus far. To date, on the basis of genome-wide analysis, 21 [34], approximately 25 [35,36] and 25 [37] Hsf genes have been identified in the model plants *Arabidopsis*, rice and maize, respectively. On the basis of sequence divergence, three Hsf classes (A, B and C) and several subgroups are currently recognized [22]. Previous studies have shown that there are no apparent tandem duplications, and no clustered organization, in the Hsf families of several monocot and eudicot species [35,37]. How did the members of this gene family arise, and how are the copy numbers of genes in this family maintained? Multiple rounds of genome duplication and extensive gene loss in different plant lineages may have led to the generation of independent growth and evolution models of Hsf family genes.

In this study, we analyzed the Hsf gene families from six papilionoid legume species for which substantial information about genomes or transcriptomes was available, namely *Lotus japonicus* (birdsfoot trefoil), *Medicago truncatula* (barrel medic), *Cicer arietinum* (chickpea), *Glycine max* (soybean), *Cajanus cajan* (pigeonpea) and *Phaseolus vulgaris* (common bean). The aim of this investigation was to determine which genes were derived from genome duplication, subsequently giving rise to paralogs, which genes descended from speciation events, giving rise to orthologs and which genes have undergone gene loss. In addition to examining the phylogeny of the Hsf family, we performed a comprehensive examination of the legume genome structure anchored by Hsf genes. Furthermore, we searched for gene microsynteny within and between the genomes of the legumes to investigate the evolutionary history of the Hsf regions. Our data show that extensive synteny remains in the homeologous regions within/between legume species. In addition, most Hsf regions can be traced to ancient papilionoid-specific whole-genome duplication or recent *Glycine*-specific whole-genome duplication. However, different rates of gene loss in the Hsf family have occurred along separate lineages of legume. Our results may help facilitate the extrapolation of Hsf gene function from one lineage to another.

Materials and Methods

Data retrieval and sequence analysis

The most recent versions of genome, protein and cDNA sequences of each species were downloaded from the respective genome sequence sites as follows: *L. japonicus* (version 2.5) from the *L. japonicus* Genome Sequencing Project (http://www.kazusa. or.jp/lotus/), *M. truncatula* (version 3.5) from the *M. truncatula* Genome Sequencing Project (http://www.medicagohapmap.org/ ?genome), *G. max* (version 1.01) from the Soybean Genome Sequencing Project (http://www.phytozome.net/soybean.php), *C. cajan* (version 1.0) from the International Crops Research Institute for the Semi-Arid Tropics (http://www.icrisat.org/gt-bt/iipg/ Genome_Manuscript.html), *P. vulgaris* (version 0.9) from the US Department of Energy Joint Genome Institute (http:// www.phytozome.net/commonbean.php), *P. patens* (version 1.6) from the Joint Genome Institute (http://www.phytozome.net/ physcomitrella), *S. moellendorffii* (version 1.0 filtered model 3) from the Joint Genome Institute (http://genome.jgi-psf.org/ Selmo1/Selmo1.download.ftp.html) and *C. reinhardtii* (version 4.3) from the Joint Genome Institute (http://www.phytozome. net/chlamy.php). Although information about the whole genome

of *C. arietinum* is not currently available, the *C. arietinum* transcriptome has been sequenced using next-generation sequence technology [38] and was obtained from the Chickpea Transcriptome Database (http://www.nipgr.res.in/ctdb.html). The downloaded nucleotide and protein sequences of each species were in turn used to build local databases using DNATOOLS software. Published Hsf protein sequences [34,35,39] were used to search the Pfam database [40], and an integrated and exactly conserved Hsf-type DBD domain sequence based on the Hidden Markov Model (HMM) was obtained. The Hsf domain (PF00447) in the Pfam HMM library was then used in BLASTP searches to identify Hsfs from the local databases. For the *C. arietinum* transcriptome, a TBLASTN search was performed, and the identified full-length cDNAs were translated in the correct frame. Only hits returning E-values of less than 0.001 were considered for further analysis. This step was crucial for finding as many similar sequences as possible. Moreover, on the basis of BLASTN search results in the genome databases using the predicted cDNA sequences of Hsf genes, information was obtained about the chromosome locations of these genes. Redundant sequences with different identification numbers and the same chromosome locus were eliminated from the data set. To confirm the presence of both Hsf-type DBD domain and HR-A/B regions in the sequences obtained, the predicted protein sequences of Hsf genes were analyzed in the Pfam HMM database and the SMART tool [41] to find the DBD domain, and proteins without these regions were excluded from the data set. Following this step, the remaining sequences were examined for the HR-A/B regions using the MARCOIL program [42] and the SMART tool, both of which can recognize the coiled-coil structure representing the core of the HR-A/B region; proteins without HR-A/B regions were removed from the data set.

In addition to the DBD and HR-A/B domains, many Hsfs also contain an NLS and an NES domain, and most plant class A Hsfs contain one or several AHA motifs. To identify the NLS domain in the Hsfs, the program PredictNLS [43] (from the website) was used. In addition, the NetNES 1.1 server [44] was used to detect the NES domain in all of the Hsfs. Moreover, since the highly conserved amino acid sequence of AHA motifs has been elaborated previously, and detailed investigations of these motifs have been reported [25,26,34], the AHA motifs could be predicted based on sequence comparisons and their characteristics. Information about the AHA motifs was further verified by alignments with published Hsf sequences.

Multiple sequence alignments and phylogenetic analysis

Multiple sequence alignments using ClustalX (version 1.83) [45] were performed on the N-terminal domains of the Hsfs obtained, including the DBD domains, the HR-A/B regions and parts of the linker between these regions. The alignment was then adjusted manually by Jalview. A phylogenetic tree was constructed with the aligned protein sequences using MEGA (version 4.0) [46] using the NJ method with the following parameters: Poisson correction, pairwise deletion and bootstrap (1,000 replicates; random seed). The Hsf of *Saccharomyces cerevisiae* (ScHsf1), and the Hsfs of *C. reinhardtii*, *S. moellendorffii* and *P. patens*, were used as the outgroup. In order to analyze the classes and subgroups of the legume Hsf families, 25 maize Hsfs (ZmHsfs) [37], 25 rice Hsfs (OsHsfs) [35] and 21 *Arabidopsis* Hsfs (AtHsfs) [34] were included in the phylogenetic analysis by generating a NJ tree (Poisson correction, pairwise deletion and bootstrap = 1,000 replicates). To confirm the robustness of the NJ tree, we built the ML tree using maximum likelihood method (MEGA 6.0; bootstrap = 1,000 replicates, amino acid substitution model, Jones-Taylor-Thornton matrix).

Intraspecies microsynteny analysis

To categorize the expansion of the Hsf gene families, the physical locations of all members of this family were examined in *L. japonicus*, *M. truncatula*, *G. max* and *C. cajan*. Tandem duplication is characterized by multiple gene family members occurring within either the same or neighboring genomic regions. Tandem duplicated genes were defined as genes in any gene pair, T1 and T2, that (1) belong to the same gene family, (2) are located within 100 kb each other, and (3) are separated by zero, one or fewer, five or fewer, or 10 or fewer nonhomologous (not in the same gene family as T1 and T2) spacer genes [47]. A method similar to that of Maher *et al.* [48] and Zhang *et al.* [49] was implemented to identify large-scale duplication events. To classify two Hsf genes as residing within a duplicated block, their neighboring protein-coding genes must be highly similar at the amino acid level. First, all Hsf genes in each family were used as the original anchor points. Next, 15 protein-coding sequences upstream and downstream of each anchor point were compared by pairwise BLASTP analysis to identify duplicated genes between two independent regions. The software then counted the total number of protein-coding genes flanking any anchor point that had the best nonself match (E-$value$ $<10^{-10}$) with a protein-coding gene neighboring another anchor point. When four or more such gene pairs with syntenic relationships were detected, the two regions were considered to have originated from a large-scale duplication event.

Interspecies microsynteny analysis

The analysis of microsynteny across species was based on comparisons of the specific regions containing Hsf genes. Similarly, the Hsf genes of *L. japonicus*, *M. truncatula*, *G. max* and *C. cajan* were set as the anchor points according to their physical locations. The protein-coding sequences assigned to the flanking regions of each Hsf gene in one species were aligned with those in the other species by pairwise comparisons. A syntenic block is defined as the region in which three or more conserved homologs (BLASTP E-$value$ $<10^{-20}$) were located within a 100 kb region between genomes [50].

Duplication event dating and adaptive evolution analysis

The duplicated gene pairs within each duplicated block were used to calculate Ks and to analyze Ka/Ks ratios. Protein sequences of the gene pairs were aligned using MUSCLE [51], and the results were used to guide the codon alignments by PAL2NAL [52]. The generated codon alignments were subjected to computation of Ks and divergence levels (Ka/Ks ratios) using DnaSP software (version 5.10). A sliding window analysis of Ka/Ks ratios was performed with the following parameters: window size, 150 bp; step size, 9 bp.

When dating large-scale duplication events, Ks can be used as the proxy for time. For each pair of duplicated regions, the mean Ks of the flanking conserved genes were calculated, and these values were then translated into divergence time in millions of years assuming a rate of 6.1×10^{-9} substitutions per site per year. The divergence time (T) was calculated as $T=Ks/(2\times6.1\times10^{-9})$ 10^{-6} Mya [53].

Codeml program is available under PAML (phylogenetic analysis maximum likelihood) V. 4.7 software [54]. To further assess whether positive selection acts upon specific sites, six site models that allow ω ratios (Ka/Ks ratios) to vary among sites, as implemented in the program Codeml, were used based on the coding sequences of Hsf genes [55]. These models are the one-ratio model (M0), the nearly neutral model (M1a), the positive-selection model (M2a), the discrete model (M3), the β model (M7)

and the β & ω model (M8). The likelihood ratio tests (LRT) were performed to compare the corresponding models with and without selection (ie, M0 vs M3, M1a vs M2a, and M7 vs M8) [55]. M0–M3 comparison can be used to test whether ω values vary among sites. Both M1a–M2a and M7–M8 comparisons can be used to test positive selection acting on sites [55]. The Bayes empirical Bayes (BEB) were used in the M2a and M8 models to calculate the Bayesian posterior probability (BPP) of the codon sites under a positive selection [56].

Microarray data

The data for evaluating Hsf gene expression in various tissues of *L. japonicus* acquired with the Lotus 52 K Affy chip were obtained from the Lotus Transcript Profiling Resource [57]. The locus names of Hsf genes in the *L. japonicus* Genome Sequencing Project were used to query the corresponding probe set IDs in the GeneChip. The log transformed expression values for the retrieved probe sets were then used to perform cluster analysis by Cluster [58].

Plant material and stress treatments

L. japonicus plants (Miyakojima MG-20) were grown in a greenhouse at $25\pm2°C$ with a 14/10 h (light/dark) photoperiod. Four-week-old seedlings were prepared for abiotic stress treatments. For temperature treatments, the uniform-sized seedlings were transferred to the temperature-controlled growth chambers, which were maintained at $42\pm1°C$ for heat stress and at $4\pm1°C$ for cold stress. For oxidative stress, seedling leaves were sprayed with 10 mM H_2O_2 solution. After each treatment, the leaves of the seedlings were harvested at 0, 1 h and were immediately frozen in liquid nitrogen and kept at $-80°C$ pending the extraction of RNA.

Quantitative real-time PCR

Total RNA was extracted from the collected samples using Trizol reagent (Invitrogen, USA), followed by Dnase I digestion to remove residual genomic DNA contamination. The quality and quantity of the total RNA was measured by electrophoresis on 1% (w/v) agarose gels and examined with a NanoDrop ND-1000 UV-Vis spectrophotometer (NanoDrop Technologies, Inc.). For each sample, the first strand cDNA was reverse transcribed from 1 µg total RNA using the QuantiTect Rev. Transcription Kit (Qiagen, Germany) according to the manufacturer's instructions. Quantitative real-time PCR was conducted on an ABI PRISM 7300 real-time PCR system (Applied Biosystems, USA). All the gene-specific primer sequences for quantitative real-time PCR were designed by Primer Express Version 3.0 software (Applied Biosystems, USA) and are listed in Table S1. Each PCR reaction mixture contained 2.0 µL transcription product, 400 nM primers, and 12.5 µL 2×SYBR Green Master Mix Reagent (Applied Biosystems, USA) in a total volume of 25 µL. The thermal cycle used was as follows: 50°C for 2 min, 95°C for 10 min, 40 cycles of 95°C for 15 s, and 60°C for 1 min. Melting curve analysis was then used to verify the identity of the amplicons and the specificity of the reaction. To normalize the variance among samples, *β-tubulin* was used as an endogenous control. The relative expression of each gene was calculated as the $\Delta\Delta C_T$ value in comparison to unstressed samples (Applied Biosystems, USA). These experiments were independently replicated at least three times for each sample.

Results

Hsf genes form a complex family in legume genomes

Papilionoids represent all major legume crops and model legume species. Most papilionoid species fall into one of two large clades, i.e., the temperate galegoid clade (cool season legumes) and the Millettioid clade (tropical season legumes). To determine the number of full-length Hsf proteins in the six legumes, BLAST and HMM searches were performed against the annotated genomes of *L. japonicus*, *M. truncatula*, *G. max*, *C. cajan* and *P. vulgaris* as well as transcriptome data for *C. arietinum*. A total of 11, 19 and 13 Hsfs were identified in the cool season legumes *L. japonicus*, *M. truncatula* and *C. arietinum*, respectively, while 46, 22 and 29 Hsfs were identified in the tropical season legumes *G. max*, *C. cajan* and *P. vulgaris*, respectively (Table 1, Table S2). To obtain a broader perspective on the evolutionary history of the legume Hsf family, we also searched for Hsf genes in the single-celled green alga *Chlamydomonas reinhardtii*, the lycophyte *Selaginella moellendorffii* and the bryophyte *Physcomitrella patens*, which contain only two, one and seven Hsfs, respectively (Table 1, Table S2). Sequence alignment and domain analysis of the deduced Hsfs showed that the highly structured N-terminal DBD domain of each Hsf is the most conserved region, and the adjacent HR-A/B region, with a heptad pattern of hydrophobic amino acid residues, leads to the formation of a helical coiled-coil structure (Figure S1 and Figure S2; Table S3). These data indicate that the Hsf gene family has expanded in legumes relative to the basal plant taxa analyzed here, and to a greater extent in tropical season legumes than in cool season legumes.

Combining the six legume protein sequences, we constructed a phylogenetic tree using neighbor-joining analysis (Figure 1). The legume Hsfs were grouped into three major classes, A, B and C. Class A and B were further divided into nine (A1–9) and five (B1–5) subgroups with well-supported bootstrap values (Figure S3). Accordingly, previously defined classes and subgroups [22,34] were identified from legume Hsfs (Figure S3).

Furthermore, phylogenetic analysis showed that the Hsf genes from the six legume species could be delineated into 18 well-supported ancient gene lineages (clades 1–18 in Figure 1), and strong amino acid sequence conservation was proven from the short branch lengths at the tips of the clades, indicating close evolutionary relationships among members. In each clade, branches with more than one Hsf gene from the same species are likely to have undergone gene duplication events, whereas the absence of representatives in some species is probably attributable to gene losses. In most cases, Hsf genes from tropical season legumes were more abundant than Hsf genes from cool season legumes. It is worth noting that in almost every clade, at least one extra copy of the Hsfs from *G. max* was present compared with that from the other species. On the other hand, in each clade, the members of different species may have evolved from a common ancestral gene by divergence of the lineage. Therefore, the 18 defined clades provide a framework for inferring parologs and orthologs of Hsfs. The phylogenetic relationships based on the ML tree were largely consistent with these results (Figure S4).

Genome duplication played an important role in the expansion of the Hsf family

To examine the relationship between the genetic divergences within each legume Hsf family and the corresponding expansion patterns, we further surveyed gene duplication events in the legume Hsf families (Figure 2; Table 2). *P. vulgaris* and *C. arietinum* were excluded from this analysis due to the lack of information about the locations of their Hsf genes. We character-

Table 1. The number of Hsfs identified in the legume species and lower plants.

Class	LjHsf	MtHsf	CaHsf	GmHsf	CcHsf	PvHsf	SmHsf	PpHsf	CrHsf
A	9	12	8	26	14	17	0	3	1
B	2	7	4	19	7	11	1	4	1
C	0	0	1	1	1	1	0	0	0
Total	11	19	13	46	22	29	1	7	2

Lj = L. japonicus, Mt = M. truncatula, Ca = C. arietinum, Gm = G. max, Cc = C. cajan, Pv = P. vulgaris, Sm = S. moellendorffii, Pp = P. patens, Cr = C. reinhardtii.

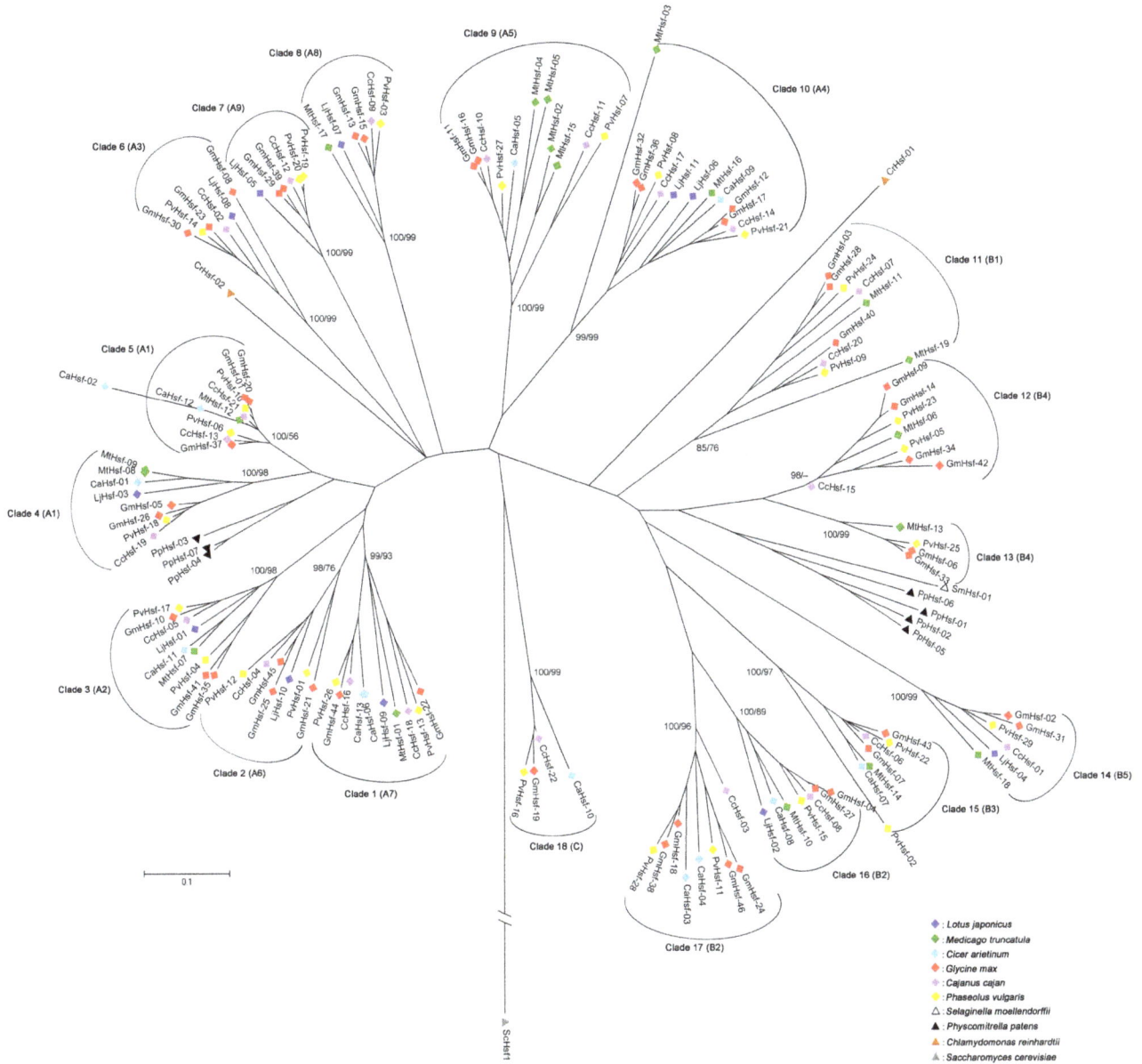

Figure 1. Phylogenetic tree of 140 Hsf proteins from *L. japonicus, M. truncatula, C. arietinum, G. max, C. cajan* **and** *P. vulgaris.* This tree was constructed based on amino acid sequence comparison of the conserved N-terminal regions of Hsfs including the DNA-binding domain, the HR-A/B region and parts of the linker between them, using the neighbor-joining method with 1,000 bootstrap replicates. The Hsf of *Saccharomyces cerevisiae* (ScHsf1) and the Hsfs of *C. reinhardtii, S. moellendorffii* and *P. patens* were used as the outgroup. The colors indicate the species background of the Hsfs. The tree was divided into 18 shared clades (Clades 1–18) according to evolutionary distances. The bootstrap values of both neighbor-joining (NJ) tree (first number; 1000 replicates) and maximum likelihood (ML) tree (second number; 1000 replicates) were shown on the branches leading to each of the clades. The clades were supported by high bootstrap values in neighbor-joining and maximum likelihood analyses. Different subclasses of Hsfs are indicated in brackets. Gene names are presented in Table S2. The scale bar represents 0.1 amino acid changes per site.

ized Hsf paralogs as being cluster or scattered. Chromosomal location analyses showed that the majority of Hsf genes are randomly scattered in the genomes, with tandemly clustered genes occurring in several places (Table S2). Legumes have experienced one or more polyploidy events. Thus, large-scale duplication events may have played an important role in the evolution of the legume Hsf families.

To investigate this possibility, we searched for gene similarity in the Hsfs flanking regions. If four or more of the 15 up- and downstream genes flanking two Hsf genes achieved a best non-self match using BLASTP (*E-value* $<10^{-10}$), we considered these gene

pairs to be conserved and defined these two regions as derived from a large-scale duplication event. We also defined a flexible set as a set of genes in which the flanking regions of an Hsf pair contained two or three conserved genes to avoid the possibility that pairs of Hsf genes resided within more divergent blocks.

We identified three conserved genes flanking the pair *LjHsf-06/LjHsf-11* in *L. japonicus.* Therefore, this pair is considered to have evolved from large-scale duplication, based on our flexible set. In *C. cajan,* one gene pair (*CcHsf-10/CcHsf-11*) was found to have involved large-scale duplication. However, it should be noted that approximately 45% of the Hsf family could not be assigned to

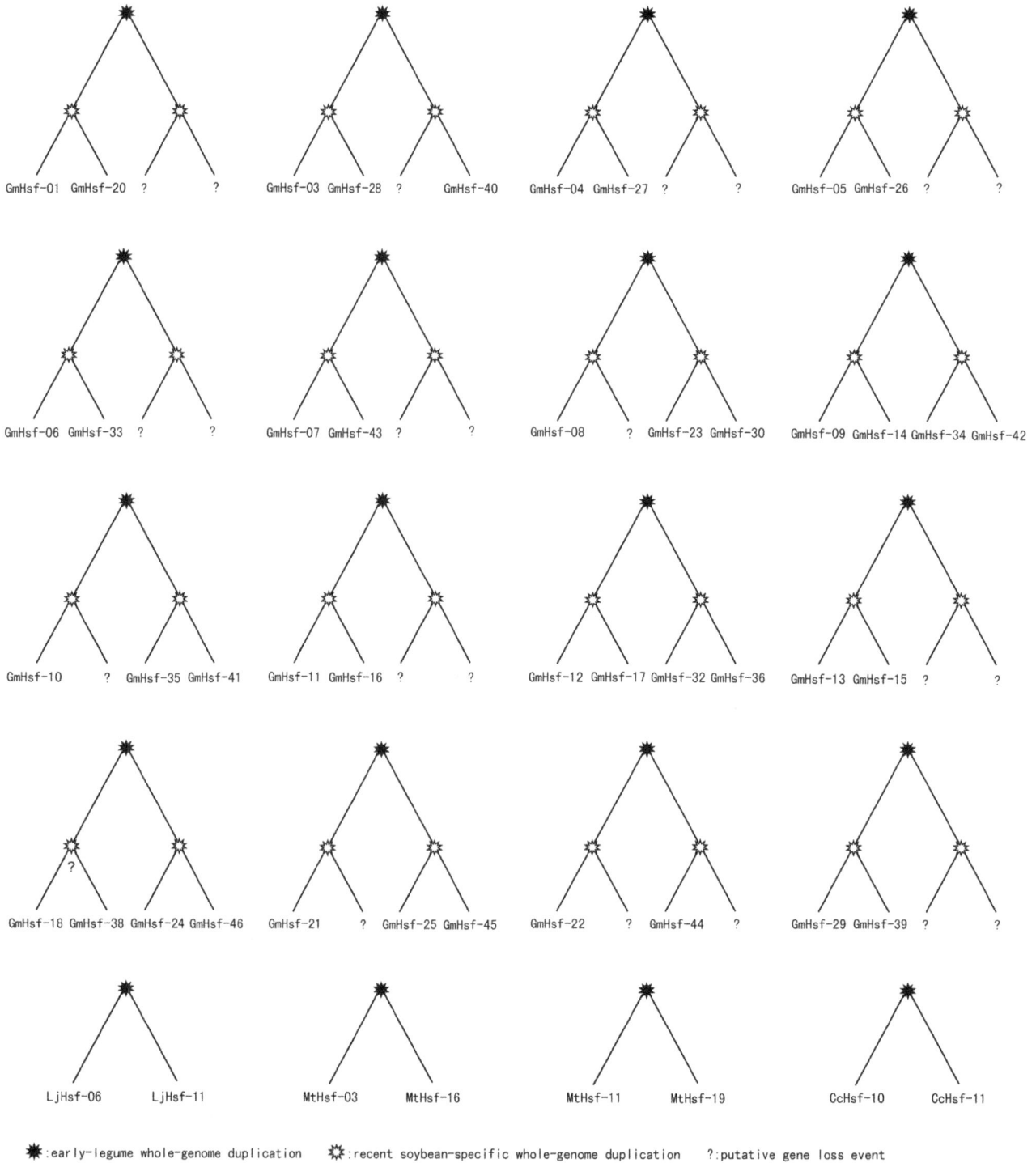

Figure 2. Idealized gene trees of the duplication groups of Hsf genes in *G. max*, *L. japonicus*, *M. truncatula* and *C. cajan*. Each tree represents a duplication group from large-scale gene duplication. As shown in the trees, every Hsf gene of *G. max* was expected to be present in four copies after two rounds of whole-genome duplication (early and recent). Similarly, the number of Hsf genes in *L. japonicus*, *M. truncatula* and *C. cajan* will have doubled after the early-legume whole-genome duplication. The five duplicated gene pairs (*LjHsf-06/LjHsf-11*, *GmHsf-09/GmHsf-34*, *GmHsf-18/GmHsf-24*, *GmHsf-18/GmHsf-46* and *GmHsf-21/GmHsf-45*) were classified in the flexible set. The question marks indicate possible gene loss events. The *GmHsf-18/GmHsf-38* pair could be formed by a segmental duplication that predated the recent whole-genome duplication, and both the *GmHsf-18* and *GmHsf-38* lost homoeologs from the recent whole-genome duplication.

any chromosome. In *M. truncatula*, genes flanking both pairs, *MtHsf-03/MtHsf-16* and *MtHsf-11/MtHsf-19*, were found to be conserved. In addition, two gene pairs (*MtHsf-04/MtHsf-05* and *MtHsf-08/MtHsf-09*) were located near each other on chromosomes 2 and 4, and thus most likely resulted from tandem duplication. Moreover, we found that the DNA sequences for

Table 2. Estimates of the dates for the large scale duplication events in legume species.

Duplicated Hsf gene pairs	Number of conserved flanking protein-coding genes	Ks (mean ± s.d.)	Date (mya)
GmHsf-05 & GmHsf-26	14	0.12±0.02	9.84
GmHsf-25 & GmHsf-45	15	0.12±0.01	9.84
GmHsf-06 & GmHsf-33	16	0.13±0.01	10.66
GmHsf-13 & GmHsf-15	16	0.13±0.02	10.66
GmHsf-04 & GmHsf-27	14	0.14±0.02	11.48
GmHsf-01 & GmHsf-20	16	0.15±0.01	12.30
GmHsf-11 & GmHsf-16	15	0.15±0.02	12.30
GmHsf-24 & GmHsf-46	16	0.15±0.02	12.30
GmHsf-34 & GmHsf-42	12	0.16±0.02	13.11
GmHsf-32 & GmHsf-36	11	0.16±0.01	13.11
GmHsf-12 & GmHsf-17	14	0.16±0.01	13.11
GmHsf-23 & GmHsf-30	7	0.17±0.02	13.93
GmHsf-29 & GmHsf-39	16	0.17±0.03	13.93
GmHsf-03 & GmHsf-28	16	0.18±0.02	14.75
GmHsf-07 & GmHsf-43	6	0.18±0.03	14.75
GmHsf-09 & GmHsf-14	16	0.18±0.03	14.75
GmHsf-35 & GmHsf-41	7	0.18±0.06	14.75
GmHsf-18 & GmHsf-38	5	0.38±0.10	31.15
GmHsf-08 & GmHsf-23	4	0.57±0.02	46.72
GmHsf-10 & GmHsf-41	7	0.58±0.11	47.54
GmHsf-10 & GmHsf-35	6	0.59±0.10	48.36
GmHsf-09 & GmHsf-42	4	0.62±0.04	50.82
GmHsf-21 & GmHsf-25	5	0.62±0.08	50.82
GmHsf-08 & GmHsf-30	4	0.65±0.07	53.28
GmHsf-14 & GmHsf-34	4	0.65±0.06	53.28
GmHsf-22 & GmHsf-44	8	0.65±0.07	53.28
GmHsf-14 & GmHsf-42	5	0.66±0.04	54.10
GmHsf-24 & GmHsf-38	5	0.71±0.10	58.20
GmHsf-38 & GmHsf-46	6	0.72±0.10	59.02
GmHsf-12 & GmHsf-32	6	0.77±0.04	63.11
GmHsf-17 & GmHsf-32	6	0.77±0.06	63.11
GmHsf-12 & GmHsf-36	6	0.79±0.04	64.75
GmHsf-17 & GmHsf-36	6	0.80±0.06	65.57
GmHsf-28 & GmHsf-40	4	0.85±0.17	69.67
GmHsf-03 & GmHsf-40	4	0.91±0.12	74.59
MtHsf-11 & MtHsf-19	5	0.77±0.10	63.05
MtHsf-03 & MtHsf-16	5	0.80±0.03	65.66
CcHsf-10 & CcHsf-11	5	0.74±0.14	60.66

The Hsf gene pairs from flexible sets were not used for calculation.

MtHsf-08 and *MtHsf-09*, as well as their four flanking genes (within an approximately 40-kb region) were completely identical to each other. In *G. max*, no tandem duplication was identified, but 42 out of 46 Hsf genes (approximately 91.3%) were arranged into duplicated chromosomal regions. These 42 genes were classified into 16 duplication groups; each group had two to four members that had conservation between their flanking genes (Figure 2). In the duplication groups of *G. max* Hsfs, the relationships between four putative duplicated gene pairs (*GmHsf-09/GmHsf-34,* *GmHsf-18/GmHsf-24,* *GmHsf-18/*

GmHsf-46 and *GmHsf-21/GmHsf-45*) were judged according to the flexible set.

Assuming that synonymous silent substitutions per site (Ks) occur at a constant rate over time, the conserved flanking protein-coding genes were used to estimate the dates of the large-scale duplication events [48]. In this analysis, the duplicated blocks (excluding the flexible set) were used to date duplication events. The mean Ks values for each duplication event, and the estimated date, are shown in Table 2. The duplicated regions of *G. max* were divided into two groups (except for *GmHsf-18* and *GmHsf-38*) based on the Ks values of paralogs flanking the Hsf pair (Table 2). The Ks values distribution of each pair of genes in duplicated blocks is shown in Figure 3A. In one group, the paralogs flanking 17 Hsf pairs yielded a mean Ks value of 0.155 (the first peak in Figure 3A), corresponding to an event approximately 13 Mya. This estimate is consistent with the timing of a recent *Glycine*-lineage-specific tetraploidization event [8]. In the other group, the paralogs flanking 17 pairs had a mean Ks value of 0.701 (the second peak in Figure 3A), corresponding to an event roughly 57 Mya, concordant with the early-legume duplication that occurred near the origin of the papilionoid lineage [59]. For *GmHsf-18* and *GmHsf-38*, the duplication event was estimated to have occurred approximately 31 Mya, which is between the two rounds of genome duplication. In addition, *GmHsf-38* and *GmHsf-24* or *GmHsf-46* are all related via the ancient genome duplication, but the relationships between *GmHsf-18* and *GmHsf-24* or *GmHsf-46* are uncertain. Therefore, *GmHsf-18* may be the product of a segmental duplication of *GmHsf-38*. From these results, we conclude that two whole-genome duplications played a key role in the expansion of the *G. max* Hsf family.

In the *G. max* Hsf duplicated network, when two duplicated genes from recent duplication could not be found simultaneously, we reasoned that a possible ancient gene loss event occurred. As shown in Figure 2, in the ancestor of *G. max* lineage, ancient genome duplication should have produced at least 32 Hsf genes. However, eight of these lines lacked both copies, which would have been obtained from recent genome duplication, suggesting that approximately 25% of the ancient duplicates were lost over millions of years. Moreover, taking ancient gene losses into account, the number of *G. max* Hsf genes derived from recent genome duplication should be 50. Among these, eight pairs lost one copy of the gene, indicating that only about 16% of recent duplicates were lost. On the contrary, in *L. japonicas*, *M. truncatula* and *C. cajan*, only a few Hsf-containing segments could be matched in duplicated pairs. What is the origin of the remaining Hsf genes in these species?

Massive losses of duplicated Hsf genes in *L. japonicas* and *M. truncatula*

To identify the evolutionary origins and orthologous relationships within the Hsf genes of legumes, Hsf family members were used as anchor genes to study the molecular history of the chromosomal regions in which they reside. Using a stepwise gene-by-gene reciprocal comparison of the regions hosting the Hsf genes, we observed strongly conserved microsynteny among these regions across *L. japonicus*, *M. truncatula*, *G. max* and *C. cajan* (Figure 4 and Figure S5). After this interspecies microsynteny analysis, we were able to assemble 78 out of 86 Hsf-containing genomic segments from these four species into 17 groups (Figure S5 A–Q). We propose that all of the segments within a group descended from a single Hsf-containing segment in the genome of the last common ancestor of the legumes, and thus we refer to these groups as orthologous groups. All of the groups contain at least one cool season legume and one tropical season legume segment with an Hsf gene. A total of 79 Hsfs (11 from *L. japonicus*, 16 from *M. truncatula*, 40 from *G. max* and 12 from *C. cajan*) were present in the 17 orthologous groups of segments. A representative synteny diagram for three of these groups is shown in Figure 5. *L. japonicas-M. truncatula-G. max-C. cajan* microsynteny also allowed us to verify the 18 ancient gene lineages inferred from the phylogenetic analysis. These results demonstrate that there is a one-to-one correspondence between syntenic orthologous groups and ancient gene lineages, except for the ancient gene of clade 18 (class C Hsfs).

Each orthologous group of segments includes between four and 24 orthologous groups of genes (average of 11; including Hsfs) with representation in at least two species. These groups are shown connected by black lines in Figure 5 and Figure S5; most of these groups include genes that obtain 'best hits' in BLASTP searches of entire genomes across species. To estimate the extent of conserved gene content and order, synteny quality was counted for those genes falling into syntenic intervals in *L. japonicus*, *M. truncatula*, *G. max* and *C. cajan*. Synteny quality was calculated as twice the number of matches divided by the total number of genes in both segments; this process discounts gene amplification but counts conservation of genes between species [5]. The average synteny quality of regions orthologous across these four species was 61.32% (Table 3). The lowest synteny quality, 48.97%, was between *M. truncatula* and *G. max* syntenic regions. The *G. max* and *C. cajan* comparison exhibited the highest conservation, 70.91%. These results support the orthology of the segment groups used in this study.

In each orthologous group, high levels of microsynteny were maintained between the members of three legume species (*L. japonicus*, *M. truncatula* and *C. cajan*) and networks of duplicated regions in *G. max*, each anchored by the Hsf gene. Within an orthologous group, segments of different legume species are thought to have shared the ancient legume whole-genome duplication that occurred outside of the papilionoid lineage. In many groups, only one region of *L. japonicus*, *M. truncatula* and *C. cajan* was comparable to homoeologous regions of *G. max*, suggesting that one member of the Hsf gene pair produced from ancient genome duplication was lost in their ancestral lineages. For example, the *MtHsf-17/LjHsf-07/CcHsf-09* anchored regions showed microsynteny with two *G. max* duplicated regions containing *GmHsf-13/GmHsf-15* (Figure 5A). In only a few groups, two duplicated regions of *L. japonicus*, *M. truncatula* or *C. cajan* were syntenic to the *G. max* duplicate regions. In one instance (Figure 5B), where the *LjHsf-06/LjHsf-11* anchored regions were putative duplicated regions in *L. japonicus*, and the *MtHsf-03/MtHsf-16* anchored regions were duplicated regions in *M. truncatula*, these four regions could be aligned with four duplicated regions in *G. max* that contained syntenic counterparts of Hsf genes (*GmHsf-12/GmHsf-17/GmHsf-32/GmHsf-36*). The four *G. max* segments arose from two rounds of whole-genome duplication. Moreover, the Hsf orthologs were usually found in the syntenic regions of the three legume species, and there were no counterparts in one species of *L. japonicus*, *M. truncatula* or *C. cajan*, indicating that two Hsf copies produced from ancient genome duplication were lost in its ancestral lineage. We also uncovered four cases in which the regions containing Hsf orthologs was syntenic between only two legume species. *MtHsf-12* were conserved with those of *G. max* (*GmHsf-01/GmHsf-20*), while the orthologs of *MtHsf-12/GmHsf-01/GmHsf-20* were missing in *L. japonicus* and *C. cajan* (Figure 5C).

Because the 17 orthologous groups of Hsf-containing segments indicate that there are at least 17 Hsf genes in this ancestor, after ancient whole-genome duplication, 34 Hsf genes should have been produced in their progenitor. In nine orthologous groups, only one Hsf-containing region of *L. japonicus* showed microsynteny with the regions of other legumes, and in seven groups, the syntenic intervals anchored by Hsf genes were missing in *L. japonicus*. This indicates that 23 out of 34 ancient duplicated genes (approximately 68%) were lost in the *L. japonicas* Hsf family. Moreover, there were 11 orthologous groups with the single orthologous region in *M. truncatula* and four groups without sharing microsynteny with Hsf anchored regions in *M. truncatula*. This suggests that 19 out of 34 ancient duplicated genes (approximately 56%) were lost in

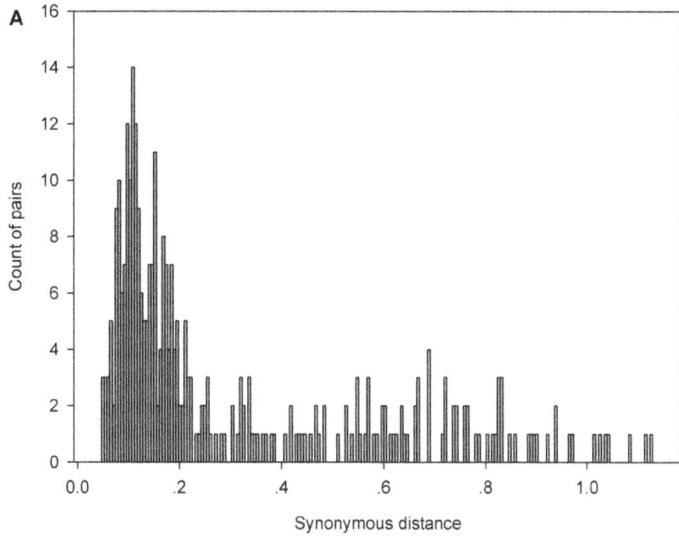

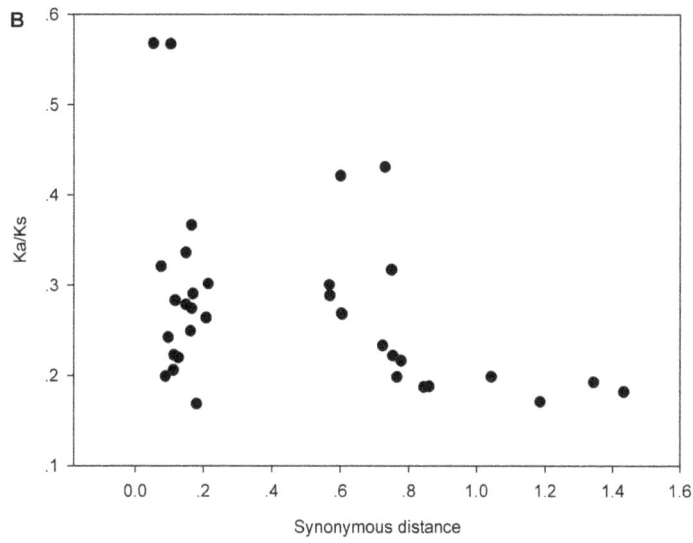

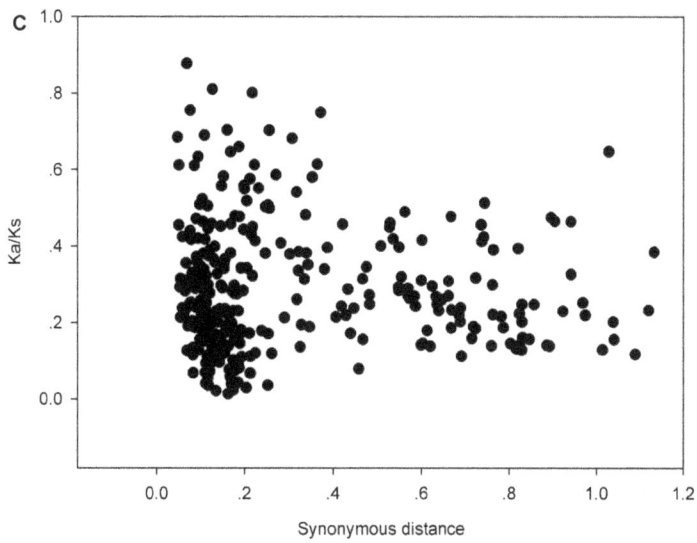

Figure 3. Estimates of Ks and Ka/Ks ratios in pairwise comparisons. (A) Distribution of synonymous distances (Ks) between paralogous genes flanking duplicated Hsf genes in *G. max*. The histogram shows the number of duplicate gene pairs (y-axis) versus synonymous distance between pairs (x-axis). The Ka/Ks ratios of the duplicated Hsf genes (B) and their flanking paralogs (C) in *G. max* are shown in the scatter plots; the y and x axes denote the Ka/Ks ratio and synonymous distance for each pair, respectively.

the *M. truncatula* Hsf family. In *C. cajan*, all 12 mapped Hsf genes were found to possess conserved microsynteny among the species investigated, and Hsf orthologs were found in nine groups located in the single syntenic region compared with other legume species, but 10 other Hsfs in this species could not be localized on the genome, and their regions were not used for comparison. Therefore, the number of duplicated Hsfs that remain in *C. cajan* is uncertain.

Strong purifying selection for Hsf genes in *G. max*

Almost the entire Hsf family of *G. max* has been expanded by two genome duplications. To better understand the evolutionary constraints acting on this gene family, we measured the Ka/Ks ratios for 35 unambiguous pairs of Hsf paralogs (not including paralogs from the flexible set) in the network of duplicated regions of *G. max*. The resulting pairwise comparison data showed that all the paralog pairs have Ka/Ks ratios <1 (Figure 3B), suggesting that the Hsf family has mainly undergone strong purifying

selection, and the Hsf genes are slowly evolving at the protein level. Given the important role of the two rounds of whole-genome duplication in the evolution of the *G. max* Hsf family, the significance of changes in the strength of selection over evolutionary time was also stressed, and the Ka/Ks ratios were sorted into two sets on the basis of the Hsf paralogs that arose from either the recent or earlier whole-genome duplication. The average Ka/Ks ratio for the recent -duplicated Hsfs (0.30) was higher than that of the early-duplicated Hsfs (0.25), but there was no significant difference between these ratios (t-test, P>0.05). Moreover, the variance of the Ka/Ks ratios for the recent-duplicated Hsfs (0.013) was not significantly different from that of the early-duplicated Hsfs (0.006; F-test, P>0.05). This indicates that the younger and older proteins in the Hsf family are under similarly stable evolutionary constraints, which supports the notion that this family is essential for the regulation of cellular processes in *G. max*.

To assess the potential for selection on the regions surrounding Hsfs, pairwise Ka/Ks ratios were also calculated for the duplicated

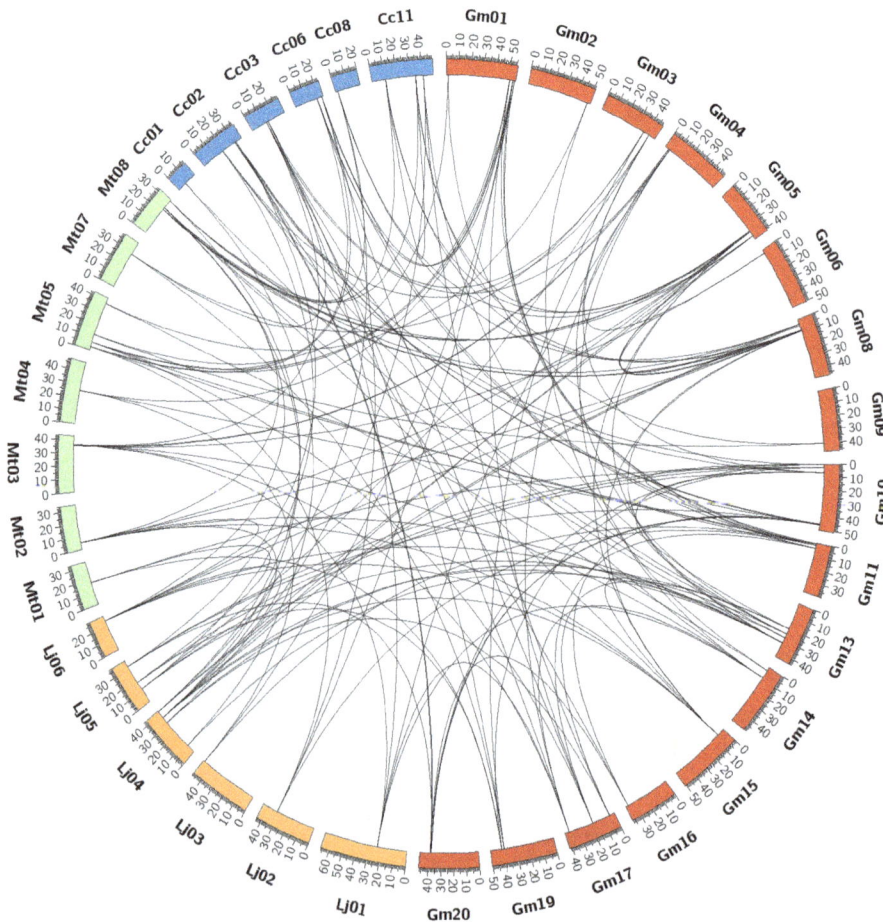

Figure 4. Extensive microsynteny of Hsf regions across *L. japonicus, M. truncatula, G. max* and *C. cajan*. *G. max* chromosomes, labeled Gm, are indicated by red boxes. The *L. japonicus, M. truncatula* and *C. cajan* chromosomes, shown in different colors, are labeled Lj, Mt and Cc, respectively. Numbers along each chromosome box indicate sequence lengths in megabases. The whole chromosomes of these four legumes, harboring Hsf regions, are shown in a circle. Black lines represent the syntenic relationships between Hsf regions.

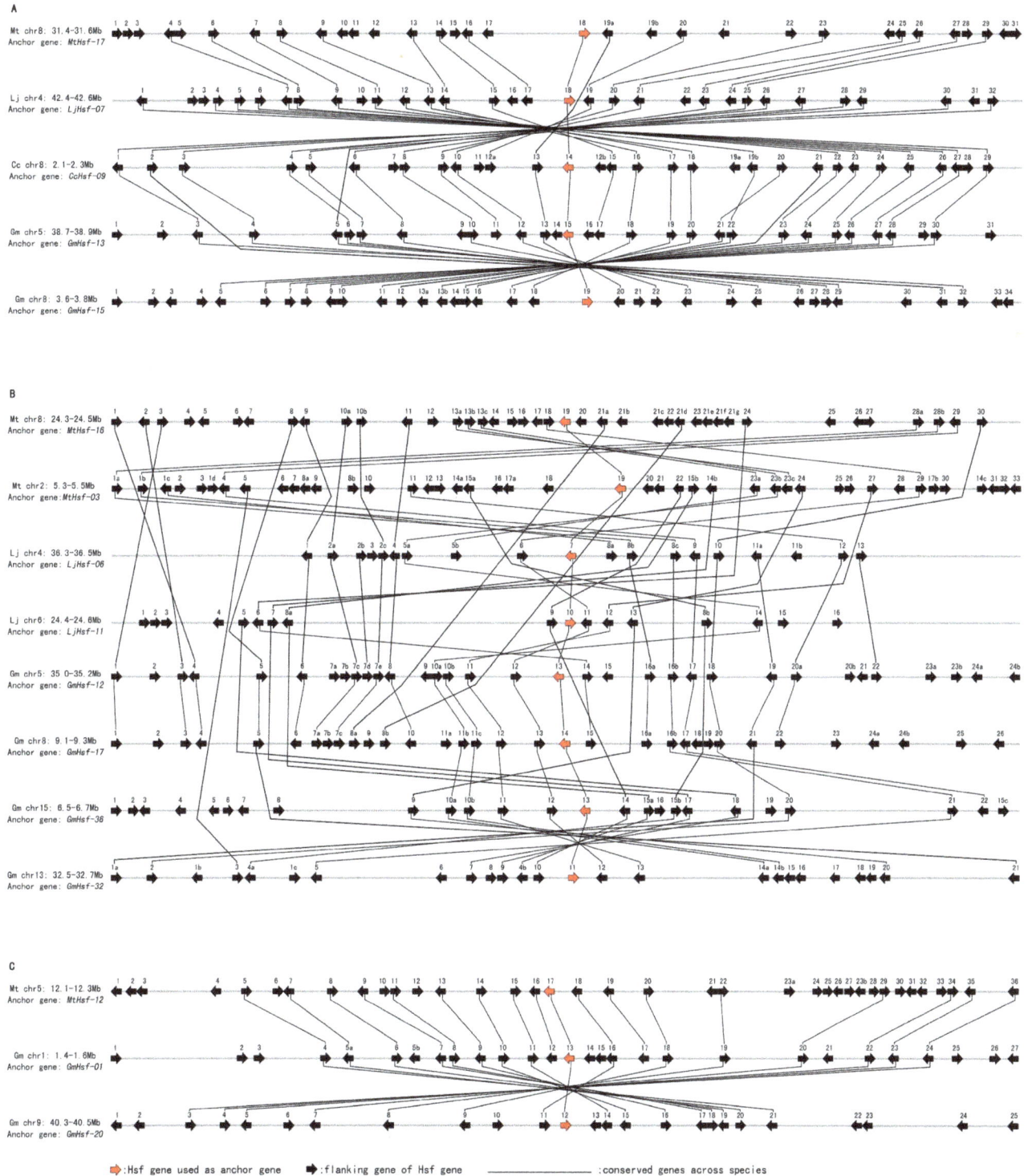

Figure 5. Comparative maps of representative Hsf genes and their flanking genes within syntenic chromosomal intervals across selected legume species. The relative positions of all flanking protein-coding genes were defined by the anchored Hsf genes, highlighted in red. The chromosome segments are shown as gray horizontal lines, with arrows corresponding to individual genes and their transcriptional orientations. All genes are numbered from left to right, in order, for each segment. Where several duplicated genes were present within a region, these genes were given the same number, with the letters a, b, c... appended in order. Conserved gene pairs among the segments are connected with lines. (A) The syntenic chromosomal intervals containing MtHsf-17, LjHsf-07, CcHsf-09, GmHsf-13 and GmHsf-15 across M. truncatula, L. japonicus, C. cajan and G. max. (B) The syntenic chromosomal intervals containing MtHsf-03, MtHsf-16, LjHsf-06, LjHsf-11, GmHsf-12, GmHsf-17, GmHsf-32 and GmHsf-36 across M. truncatula, L. japonicus, and G. max. (C) The syntenic chromosomal intervals containing MtHsf-12, GmHsf-01 and GmHsf-20 across M. truncatula and G. max. The full microsynteny maps of the regions containing Hsf genes within M. truncatula, L. japonicus, C. cajan and G. max are shown in Figure S5.

Table 3. The synteny quality of regions orthologous across *L. japonicus, M. truncatula, G. max* and *C. cajan.*

	L. japonicus	*M. truncatula*	*G. max*	*C. cajan*
L. japonicus				
M. truncatula	58.06%			
G. max	68.00%	48.97%		
C. cajan	70.79%	51.16%	70.91%	

non-Hsf genes (flanking genes) between the duplicated regions containing Hsfs in *G. max*. Interestingly, all Ka/Ks values for 322 pairs of duplicated non-Hsf genes were lower than 1 (Figure 3C), clearly indicating that these genes are evolving under purifying selection. There was no significant difference in average Ka/Ks ratio between the recent-duplicated non-Hsf genes (0.29) and the early-duplicated non-Hsf genes (0.27; t-test, P>0.05). However, the variance of these ratios for the recent-duplicated non-Hsf genes (0.032) was significant greater than that for the early-duplicated ones (0.013; F-test, P<0.01). Duplicated non-Hsf genes have likely evolved in a more "dynamic" regime than that of the Hsf genes.

Since positive selection at a few individual codon sites can be masked by overall strong purifying selection, we performed a sliding-window analysis of Ka/Ks between each pair of Hsf paralogs, which were derived from gene duplication events in *G. max*. As expected from the basic Ka/Ks analysis, sliding window analysis clearly showed that numerous sites/regions are under moderate to strong negative selection (Figure S6). As shown in Figure 6, the conserved domains of Hsfs, such as the DBD domains, HR-A/B regions and NLS motifs, are mainly subjected to strong purifying selection, with Ka/Ks ratios <<1. Moreover, the domains of Hsfs generally had lower Ka/Ka ratios (valleys) than the regions outside of them (peaks), which is consistent with functional constraint being dominant in these domains. There were a few exceptions to the generally low Ka/Ka ratios in domains. For instance, the comparison between *GmHsf-25* and *GmHsf-45* revealed sites with Ka/Ka ratios >>1 in the DBD domain, indicating positive selection in this region.

To further identify possible positive selection acting at specific sites, six site models that allow ω ratios to vary among sites were used based on the coding sequences of *G. max* Hsf family. To detect whether some sites along particular Hsf classes were under positive selection, the hypothesis testing on the class A and B *G. max* Hsfs was also performed by site models. In *G. max* Hsf family, the discrete model M3 fit better than the one-ratio model M0,

suggesting that ω ratios vary among sites (Table S4; LRT, P< 0.01). Both M1a–M2a and M7–M8 comparisons suggested that the most codon sites were under a strong constraint and no reliable positive selection sites were detected in *G. max* Hsf family (Table S4; LRT, P<0.01). In class A and B *G. max* Hsfs, M3 model also appears to be a better fit to the data than the M0 model, and the models M2a and M8 were not significantly better than the null hypothesis models M1a and M7 (Table S4; LRT, P<0.01). Only one positively selected site, listed in Table S4, was detected based on posterior probability in class A Hsfs of *G. max*. The results showed that *G. max* Hsf genes were highly conserved and the majority of sites were dominated by purifying selection.

The expression patterns of Hsf genes in *L. japonicus*

In order to gain insight into the possible functions of Hsf genes, we comprehensively examined information about the expression of all *L. japonicus* Hsf genes using microarray data and quantitative real-time PCR analysis. We first analyzed the expression of *L. japonicus* Hsf genes in nodule, root, stem, leaf and flower from the microarray data (Figure 7). Out of 11 of these genes, the expression data for *LjHsf-05* were not included in the database. The ten remaining genes were expressed in all the tissues investigated, but they exhibited differential patterns in terms of both specificity and expression level. According to their expression profiles, *L. japonicus* Hsf genes can be classified into four types. The transcripts of the first type (*LjHsf-03* and *LjHsf-11*) were highly accumulated in both underground (nodule and root) and aerial (stem, leaf and flower) parts, but the expression level was higher in the underground parts (Figure 7A). In the second type, *LjHsf-04* showed maximum expression in the root and *LjHsf-06* had a similar pattern. However, *LjHsf-06* was much more highly expressed than *LjHsf-04* (Figure 7B). The genes of the third type (*LjHsf-02* and *LjHsf-10*) were expressed preferentially in stem and flower, and *LjHsf-02* showed higher expression than *LjHsf-10* (Figure 7C). The fourth type has four members (*LjHsf-01*,

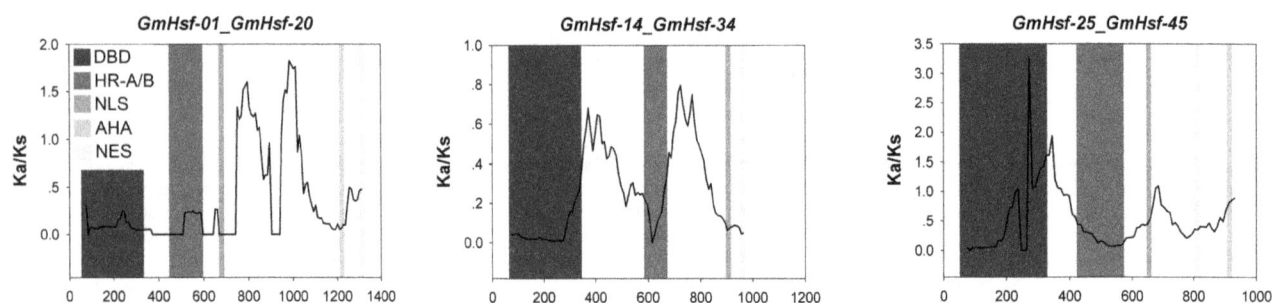

Figure 6. Sliding window plots of representative duplicated Hsf genes in *G. max*. As shown in the key, the gray blocks, from dark to light, indicate the positions of the DBD domain, HR-A/B region, NLS, NES and AHA motifs, respectively. The window size is 150 bp, and the step size is 9 bp. The data for all pairs of duplicated Hsf genes of soybean are shown in Figure S6.

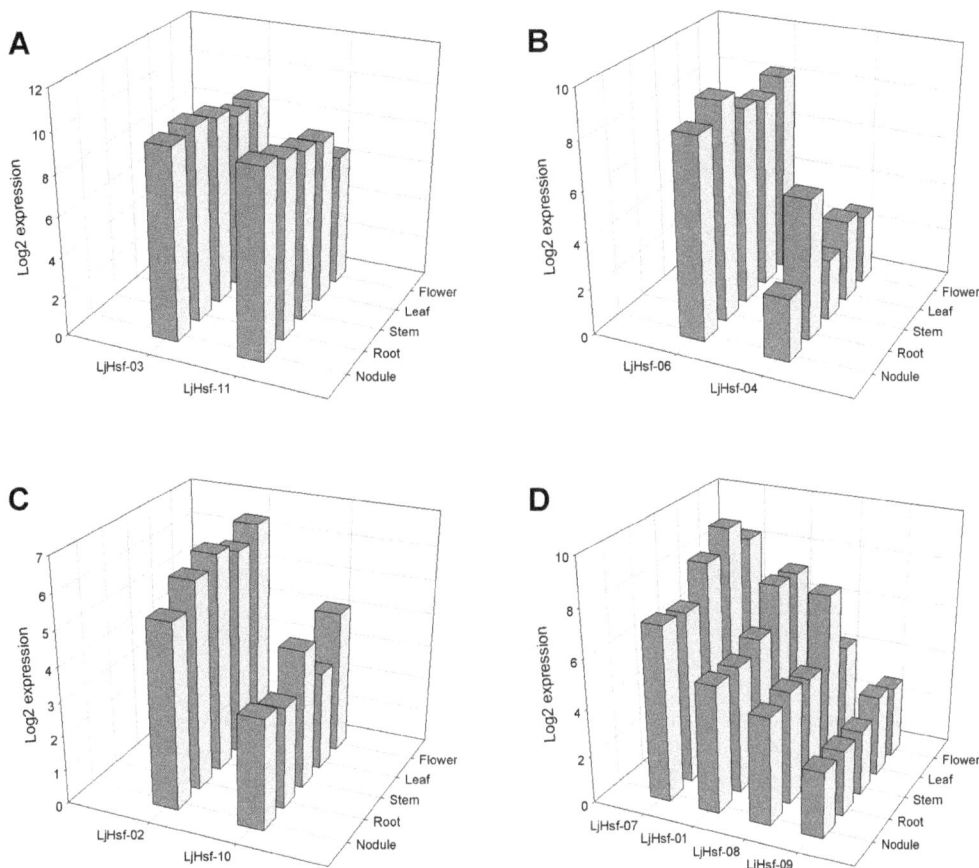

Figure 7. *L. japonicus* **Hsf genes expression in various plant tissues.** The type of tissue (nodule, root, stem, leaf and flower) and the gene name are shown on the y-axis and x-axis, respectively. Hierarchical clustering based on average log signal values in various tissues grouped 10 of the *L. japonicus* Hsf genes into four types (A–D).

LjHsf-07, LjHsf-08 and *LjHsf-09*), the genes predominantly expressed in leaves (Figure 7D). *LjHsf-01* and *LjHsf-09* were also expressed at higher levels in flowers than nodule, root and stem. Although the Hsf genes of the fourth type had similar expression pattern across a range of tissues, their transcript levels were quite diverse. *LjHsf-07* was the most highly expressed gene and *LjHsf-09* the lowest.

Quantitative real-time PCR analysis was then performed to evaluate the response of the *L. japonicus* Hsf gene family to abiotic stress. RNA was isolated from the leaves of 4-week-old *L. japonicus* seedlings subjected to heat, cold and H_2O_2 stress treatment and was used for the experiments. The results showed that a total of ten genes were significantly up- or down-regulated compared to controls (>2 or <0.5) in at least one of the stress conditions examined (Figure 8). Among these genes, most were responsive to more than one stress treatment. Two genes (*LjHsf-04* and *LjHsf-11*) were significantly up-regulated by all three stresses. Three (*LjHsf-01, LjHsf-02* and *LjHsf-09*) were expressed at remarkably high levels in response to both heat and H_2O_2 stress. *LjHsf-08* was induced by heat stress but was suppressed by H_2O_2 stress. *LjHsf-07* was significantly down-regulated upon exposure to heat and H_2O_2 stresses. A few genes were primarily responsive to one stress treatment. *LjHsf-05* and *LjHsf-10* responded specifically to heat stress, while *LjHsf-06* was distinctively up-regulated under H_2O_2 stress. In contrast, *LjHsf-03* showed minor fluctuations during all three stresses. It is worth noting that five genes (*LjHsf-01, LjHsf-02, LjHsf-04, LjHsf-09*

and *LjHsf-10*) were strongly heat-inducible in our experiments, suggesting that they could have important roles in the heat shock regulatory network.

Discussion

In this study, we identified 11, 19 and 13 Hsfs in the cool season legumes *L. japonicus*, *M. truncatula* and *C. arietinum*, respectively, and 46, 22 and 29 Hsfs in the tropical season legumes *G. max*, *C. cajan* and *P. vulgaris*, respectively. Before reconstructing the gene gain/loss history of legume Hsf families, it is necessary to trace the Hsf genes in different legume genomes back to a common ancestor. Phylogenetic trees are quite informative for inferring the number of Hsfs in the most recent common ancestor of the six legume species analyzed in this study [60]. There are 18 well-supported clades representing legume Hsfs (Figure 1), although the representatives of one or two species are missing from some clades. These clades are perceived as shared, and the genes in a shared clade are assumed to be descendants of an ancestral Hsf gene. Therefore, there are at least 18 Hsf genes in the most recent common ancestor of legumes. A large fraction of variability in the clades suggests lineage-specific gene gain and loss. For instance, in almost every clade, Hsfs from *G. max* are present in at least an extra copy compared with the other species, and the extra copy is very close to its potential paralog. These results are consistent with the well-documented fact that *G. max* has undergone an additional WGD not shared with by the other

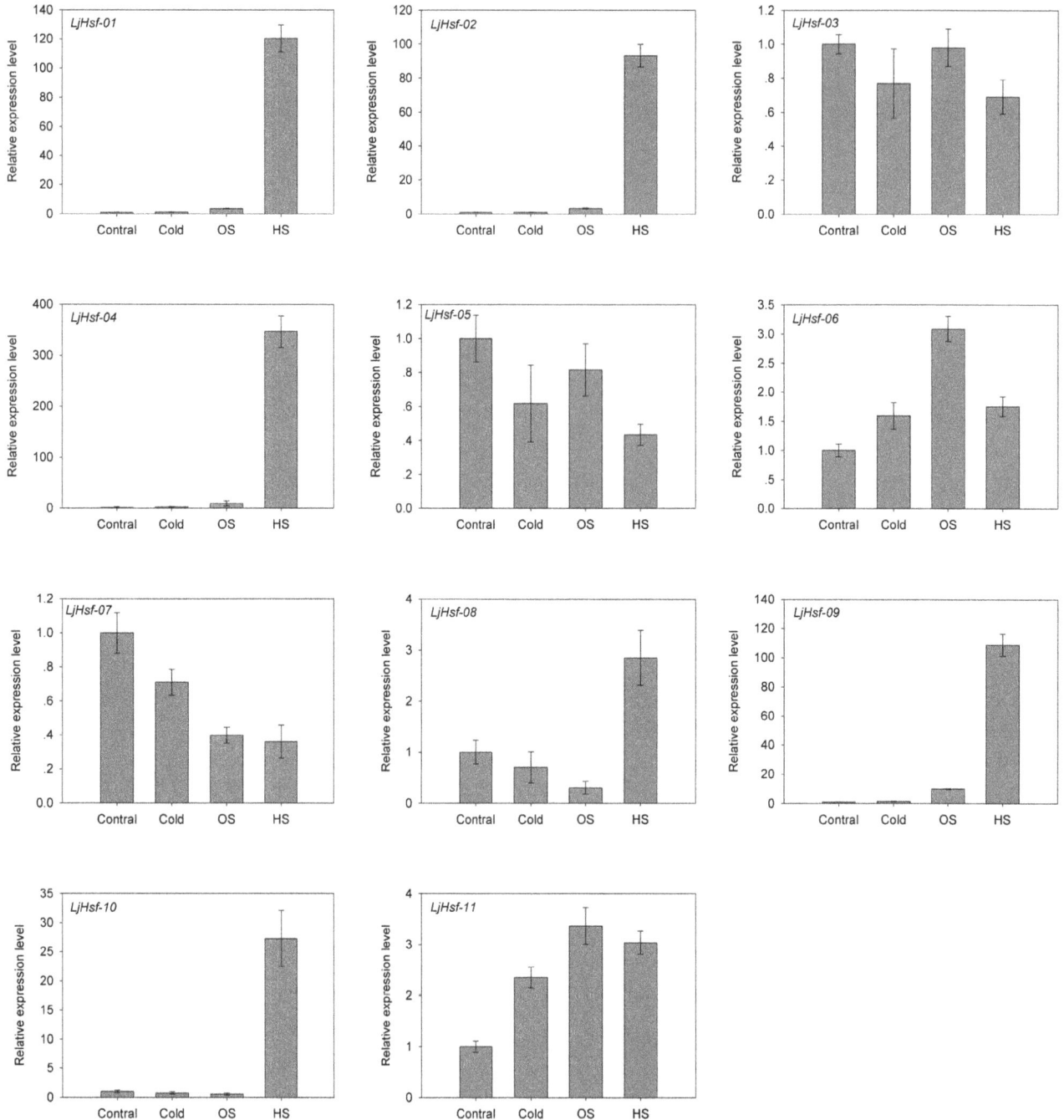

Figure 8. Expression of *L. japonicus* **Hsf genes in response to abiotic stress measured by quantitative real-time PCR.** The mRNA level of each gene in *L. japonicus* seedlings given heat (HS: 42°C), cold (4°C) and oxidative (OS: 10 mM H_2O_2) stress for 1 h was plotted relative to the value obtained for the unstressed contral. Error bars represent standard errors.

three species [8]. However, there are not simply twice as many Hsf genes in *G. max* vs. the other legumes, which indicates that differential gene loss events may have occurred in different species. Furthermore, all of the clades were confirmed by the following interspecies microsynteny analysis (except for the shared clade representing class C Hsfs), indicating most duplicates of class C Hsfs have been lost in legumes. When we compared the number of ancestral genes with those in the extant legumes species, it appeared as though the Hsf family had contracted in most of the cool season legumes and expanded in the tropical season legumes.

For example, compared with the number of ancestral genes, the number of Hsf genes was nearly halved in *L. japonicus* but increased approximately 2.5-fold in *G. max*. These results most likely reflect the complex evolutionary history of the Hsf family in legume species.

In *G. max*, most Hsf genes were assigned to the duplicated segments, and these segments could be divided into two groups based on Ks values. The members of one group were supported with an average Ks value of 0.701, which corresponds to the second peak in Figure 3A. In *L. japonicus*, *M. truncatula* and *C.*

cajan, a few Hsf-containing segments were matched in pairs. The Ks values for these pairs were included in Table 2, such as *MtHsf-03/MtHsf-16* (0.80) and *MtHsf-11/MtHsf-19* (0.77). All of these Ks values from *G. max* and other legumes probably represent the same polyploidy that occurred in all members of Papilionoideae subfamily approximately 59 Mya. Thus, the rate of synonymous substitution in *M. truncatula* appears to be greater than that in *G. max*. In a previous study of *Rxp* regions, higher rates of synonymous substitution were also detected in *M. truncatula* compared with *G. max* [61].

After the shared polyploidy event and following divergence from the other legumes, the lineage leading to present-day *G. max* is known to have undergone a second whole-genome duplication approximately 13 Mya; this duplication did not occur in *L. japonicus*, *M. truncatula* or other legume lineages. Two whole-genome duplications in the ancestor of *G. max* lead to the expectation of a maximum of four homoeologs in *G. max* genome [62,63]. As illustrated in Figure 2, Hsf genes doubled twice in *G. max* and formed two pairs of paralogs, accordingly. In some cases, these four homoeologous Hsf regions were retained. For instance, *GmHsf-12*, *GmHsf-17*, *GmHsf-32* and *GmHsf-36* are located in four homoeologous regions, respectively. More conserved flanking protein-coding genes were identified between the two paralogous Hsf-containing segments derived from the recent duplication event than those from the ancient duplication (Table 2). This suggests that high levels of sequence conservation were maintained between recently duplicated regions in *G. max*, which is similar to previous observations [62,64]. Furthermore, 17 pairs of Hsf genes of *G. max* generated by recent genome duplication were retained, while in eight pairs, one copy was lost (Figure 2). A higher, sharp peak was produced (the first peak in Figure 3A) when the number of gene pairs within these 17 pairs of recent duplicated regions was plotted against Ks values. These results indicate that many younger Hsf genes from recent genome duplications tended to be retained in *G. max* genome. By contrast, it has been demonstrated that massive gene losses occurred after tetraploidization in the maize ancestral lineage, and approximately 50% of the duplicated copies of genes have been removed or severely damaged over the past 12 million years [65,66]. In *Glycine*, the large number of duplicate genes in homeologous regions indicates that the process of diploidization is a slow and ongoing process [67].

In light of the two above-mentioned rounds of genome duplication and the fact that genome duplication should double the number of genes, the *G. max* and other papilionoid legume genomes should show a 2:1 relationship regarding the Hsf-containing regions. Each pair of *G. max* regions should have a corresponding orthologous region in *L. japonicas*, *M. truncatula* and *C. cajan*. In this study, by aligning Hsf-containing regions into paralogous pairs produced by ancient polyploidy, only one pair was detected in *L. japonicas* and *C. cajan*, with two pairs in *M. truncatula*. However, more than 90% of the chromosome regions hosting Hsf genes in *G. max* fell into pairs, triples or quadruples. Further analyses including estimates of the dates of duplications indicated that these duplicated regions arose from two polyploidy events in the *Glycine* lineage. A high degree of microsynteny between the genomes of the two model legumes has previously been found in a comparison of the genomic regions around the apyrase genes [68]. The analysis of microsynteny can help unravel the actual evolutionary relationships between Hsf regions among the legume species by taking advantage of the surrounding genomic sequences. When the Hsf-containing regions in *L. japonicas*, *M. truncatula* and *C. cajan* were compared to those in *G. max*, we found that significant synteny was maintained,

although small insertions/deletions and inversions were observed between regions (Figure 4 and Figure S5). Almost all Hsf-containing regions in *L. japonicas*, *M. truncatula* and *C. cajan* showed close relationships with the orthologous duplicated regions in *G. max*. Syntenic regions are thought to share a common origin derived from ancient legume duplication. In most cases, a single region of *L. japonicus*, *M. truncatula* and *C. cajan* was syntenic to two or three duplicated regions in *G. max*, and in many cases, two paralogons appeared to be missing in *L. japonicus* or *M. truncatula*. This indicates that the duplicated copies of Hsf genes in these genomes have been removed or severely damaged after the genome duplications occurred. Approximately half of the Hsf genes of *C. cajan* were not included in the map-based analyses because these genes could not be localized to the genome owing to their unknown chromosome positions. These results may therefore not hold true for *C. cajan*.

One possible mechanism behind this phenomenon, that many Hsf genes produced by the ancient genome duplication have been lost in *L. japonicus* and *M. truncatula*, is diploidization following the early legume genome duplication. Previous studies have demonstrated that there was substantially less conservation within internal duplications in either *M. truncatula* or *L. japonicus* than in synteny blocks between the two genomes [5]. The fate of these duplicate genes is more likely to be under the control of natural selection (nonrandom loss) than for genes that are not dosage dependant [69].

Duplicated genes undergo a short period of shared "relaxed" selection during their early evolutionary lives, evolving in a neutral way. Following this selection, most paralogs are lost within a few million years, and only a few paralogs are preserved and undergo purifying selection [10,53,70]. Although many gene loss events were found to have occurred in legume Hsf families, two Hsf regions of *L. japonicus* (*LjHsf-06/LjHsf-11*) and the two Hsf regions of *M. truncatula* (*MtHsf-03/MtHsf-16*) were found to have conserved microsynteny to the four Hsf regions of *G. max* (*GmHsf-12/GmHsf-17/GmHsf-32/GmHsf-36*; Figure 5B). This suggests that these sets of Hsf genes may perform a basic, important role in legumes and have remained intact after genome duplications.

The observation that multiple copies of Hsf were retained in *G. max* is reasonable from an evolutionary perspective because Hsfs confer various abiotic and biotic resistance traits to plants [71–77]. The dosage of protein may have increased due to the presence of numerous Hsf genes, thereby leading to a corresponding increase in resistance. Previous studies have demonstrated the advantages of increased dosage of genes involved in plant resistance, for instance, resistance to glyphosate in plants [78]. Furthermore, strong purifying selection was detected between the paralogs produced from the recent genome duplication and the ancient duplication in *G. max*. Purifying selection probably plays a key role in maintaining the long-term stability of biological structures of *G. max* Hsfs by removing deleterious mutations, thus ensuring that gene functions are maintained as long as they are needed.

Moreover, *G. max* is a member of tropical season legumes; natural selection may have played a role in determining the number of duplicates within the Hsf gene family in the tropical season legumes, which are better adapted to more tropical climates. On the contrary, *L. japonicus* and *M. truncatula* are cool season legumes, and they may therefore require fewer Hsf genes. Thus, many copies of these genes may have been lost during the long-term evolutionary process.

Although the legume species differ in genome size, basic chromosome number and ploidy level, comparative genomics can be used for a bridging model with other legume species in view of their close phylogenetic relationships and extensive genome conversation [79]. *L. japonicus* was selected as a model system for gene function characterization because of its small genome size and improving knockout and over-expression techniques, and efforts were made to translate information gained from it into exploration to other legume species. *L. japonicus* maintained only 11 Hsf genes and this number is probably the lowest among the reported Hsf gene families of higher plants. In our study, the orthologs of the 11 Hsf genes were unambiguously defined in the other legumes through microsyntenic analysis. These directed our interest towards further understanding of the expression characteristics of these Hsf genes in different tissues and their responses to abiotic stresses. In several plants, Hsf gene expression has been found to be tissue- and stage-specific [80–82]. Our study revealed that none of the *L. japonicus* Hsf genes examined was expressed only in a particular tissue type, suggesting they play regulatory roles at multiple developmental stages; nevertheless, selected *L. japonicus* Hsf genes exhibited higher levels of expression in particular tissue types, indicating that members of this family might take part in different biological processes in this species. In particular, the member of class B and the member of class A showed similar expression patterns in tissues, e.g. *LjHsf-02* and *LjHsf-10* were expressed at a higher level in stem and flower than other tissues, supporting the assumption that they could co-operate with each other. In *Arabidopsis* and rice, the expression of Hsf genes was strongly induced by heat, cold, salt and osmotic stress [81,83]. In our study, we also found that the Hsf genes of *L. japonicus* were responsive to diverse abiotic stresses, and heat induced their expression more strongly than oxidation and cold. *LjHsf-01* of subclass A2, *LjHsf-02* of subclass B2 and *LjHsf-04* of subclass B1 were strongly and transiently up-regulated by heat shock. Among all Hsf genes in *Arabidopsis*, *HsfA2* was most highly expressed under high temperature conditions and was identified as a key positive regulator of the heat stress response [84,85]. *Arabidopsis HsfB1* and *HsfB2b* repressed the general heat shock response in the absence of excessive heat but were necessary for the development of acquired thermotolerance under heat stress conditions [86]. It is noteworthy that many Hsf genes of *L. japonicus* induced by heat stress were also induced by oxidative stress, and *LjHsf-04* and *LjHsf-11* were induced by all three stresses tested. These findings could support the notion that Hsfs serve as important sensors for H_2O_2 in plants and could be pivotal in linking the heat shock response with other stress-responsive signaling networks [87]. The expression pattern described can provide a basis for identifying the roles of the retained members of the *L. japonicus* Hsf family, and would give clues for studying their syntenic counterparts in other legume species.

Supporting Information

Figure S1 DBD domain alignment of Hsf proteins from *L. japonicus*, *M. truncatula*, *C. arietinum*, *G. max*, *C. cajan*, *P. vulgaris* and three lower plants.

Figure S2 HR-A/B region alignment of Hsf proteins from *L. japonicus*, *M. truncatula*, *C. arietinum*, *G. max*, *C. cajan*, *P. vulgaris* and three lower plants.

Figure S3 NJ phylogeny of *A. thaliana*, *O. sativa*, *Z. mays*, *L. japonicus*, *M. truncatula*, *C. arietinum*, *G. max*, *C. cajan* and *P. vulgaris* Hsf proteins.

Figure S4 ML phylogeny of *L. japonicus*, *M. truncatula*, *C. arietinum*, *G. max*, *C. cajan* and *P. vulgaris* Hsf proteins.

Figure S5 The full syntenic maps of chromosome regions containing Hsf genes across *L. japonicus*, *M. truncatula*, *G. max* and *C. cajan* genes.

Figure S6 Sliding window analysis employed to estimate selective pressures on 35 pairs of duplicated Hsf genes in *G. max*.

Table S1 Primers used in quantitative real-time PCR.

Table S2 Information about Hsfs in *L. japonicus*, *M. truncatula*, *C. arietinum*, *G. max*, *C. cajan*, *P. vulgaris*, *S. moellendorffii*, *P. patens* and *C. reinhardtii*.

Table S3 Domain and motif survey of legume and three lower plant Hsfs.

Table S4 Likelihood ratio tests and parameter estimations for the six site models based on the coding sequences of *G. max* Hsf genes.

Acknowledgments

We thank the members of the Key Lab of Crop Biology of Anhui Province and Dr. Xin Zhang for discussing and editing the manuscript. We are also grateful to Dr. Jia-Xing Yue (Department of Ecology and Evolutionary Biology, Rice University, Houston, USA) for helpful comments on the manuscript.

Author Contributions

Conceived and designed the experiments: YL YC BC. Performed the experiments: YL JJ HY. Analyzed the data: YL XJ HJ. Wrote the paper: YL YC.

References

1. Cui L, Wall PK, Leebens-Mack JH, Lindsay BG, Soltis DE, et al. (2006) Widespread genome duplications throughout the history of flowering plants. Genome Res 16: 738–749.
2. Jaillon O, Aury JM, Noel B, Policriti A, Clepet C, et al. (2007) The grapevine genome sequence suggests ancestral hexaploidization in major angiosperm phyla. Nature 449: 463–467.
3. Blanc G, Wolfe KH (2004) Widespread paleopolyploidy in model plant species inferred from age distributions of duplicate genes. Plant Cell 16: 1667–1678.
4. Adams KL, Wendel JF (2005) Polyploidy and genome evolution in plants. Curr Opin Plant Biol 8: 135–141.
5. Cannon SB, Sterck L, Rombauts S, Sato S, Cheung F, et al. (2006) Legume genome evolution viewed through the Medicago truncatula and Lotus japonicus genomes. Proceedings of the National Academy of Sciences 103: 14959–14964.
6. Fawcett JA, Maere S, Van de Peer Y (2009) Plants with double genomes might have had a better chance to survive the Cretaceous-Tertiary extinction event. Proc Natl Acad Sci U S A 106: 5737–5742.
7. Cannon SB, Ilut D, Farmer AD, Maki SL, May GD, et al. (2010) Polyploidy did not predate the evolution of nodulation in all legumes. PLoS One 5: e11630.
8. Schmutz J, Cannon SB, Schlueter J, Ma J, Mitros T, et al. (2010) Genome sequence of the palaeopolyploid soybean. NATURE 463: 178–183.

9. Young ND, Bharti AK (2012) Genome-enabled insights into legume biology. Annual review of plant biology 63: 283–305.

10. De Grassi A, Lanave C, Saccone C (2008) Genome duplication and gene-family evolution: the case of three OXPHOS gene families. Gene 421: 1–6.

11. Li J, Ding J, Zhang W, Zhang Y, Tang P, et al. (2010) Unique evolutionary pattern of numbers of gramineous NBS–LRR genes. Molecular Genetics and Genomics 283: 427–438.

12. Cheng Y, Li X, Jiang H, Ma W, Miao W, et al. (2012) Systematic analysis and comparison of nucleotide-binding site disease resistance genes in maize. FEBS J 279: 2431–2443.

13. Cannon SB, Mitra A, Baumgarten A, Young ND, May G (2004) The roles of segmental and tandem gene duplication in the evolution of large gene families in Arabidopsis thaliana. BMC Plant Biol 4: 10.

14. Maere S, De Bodt S, Raes J, Casneuf T, Van Montagu M, et al. (2005) Modeling gene and genome duplications in eukaryotes. Proc Natl Acad Sci U S A 102: 5454–5459.

15. Chapman BA, Bowers JE, Feltus FA, Paterson AH (2006) Buffering of crucial functions by paleologous duplicated genes may contribute cyclicality to angiosperm genome duplication. Proc Natl Acad Sci U S A 103: 2730–2735.

16. Baniwal SK, Bharti K, Chan KY, Fauth M, Ganguli A, et al. (2004) Heat stress response in plants: a complex game with chaperones and more than twenty heat stress transcription factors. J Biosci 29: 471–487.

17. Kotak S, Larkindale J, Lee U, von Koskull-Doring P, Vierling E, et al. (2007) Complexity of the heat stress response in plants. Curr Opin Plant Biol 10: 310–316.

18. Morimoto RI (1998) Regulation of the heat shock transcriptional response: cross talk between a family of heat shock factors, molecular chaperones, and negative regulators. Genes Dev 12: 3788–3796.

19. Nover L, Scharf KD (1997) Heat stress proteins and transcription factors. Cell Mol Life Sci 53: 80–103.

20. Schoffl F, Prandl R, Reindl A (1998) Regulation of the heat-shock response. Plant Physiol 117: 1135–1141.

21. Wu C (1995) Heat shock transcription factors: structure and regulation. Annu Rev Cell Dev Biol 11: 441–469.

22. Scharf KD, Berberich T, Ebersberger I, Nover L (2012) The plant heat stress transcription factor (Hsf) family: structure, function and evolution. Biochim Biophys Acta 1819: 104–119.

23. Lutz N (1987) Expression of heat shock genes in homologous and heterologous systems. Enzyme and Microbial Technology 9: 130–144.

24. Peteranderl R, Rabenstein M, Shin YK, Liu CW, Wemmer DE, et al. (1999) Biochemical and biophysical characterization of the trimerization domain from the heat shock transcription factor. Biochemistry 38: 3559–3569.

25. Doring P, Treuter E, Kistner C, Lyck R, Chen A, et al. (2000) The role of AHA motifs in the activator function of tomato heat stress transcription factors HsfA1 and HsfA2. Plant Cell 12: 265–278.

26. Kotak S, Port M, Ganguli A, Bicker F, von Koskull-Doring P (2004) Characterization of C-terminal domains of Arabidopsis heat stress transcription factors (Hsfs) and identification of a new signature combination of plant class A Hsfs with AHA and NES motifs essential for activator function and intracellular localization. Plant J 39: 98–112.

27. Clos J, Westwood JT, Becker PB, Wilson S, Lambert K, et al. (1990) Molecular cloning and expression of a hexameric Drosophila heat shock factor subject to negative regulation. Cell 63: 1085–1097.

28. Hsu AL, Murphy CT, Kenyon C (2003) Regulation of aging and age-related disease by DAF-16 and heat-shock factor. Science 300: 1142–1145.

29. Sorger PK, Pelham HR (1988) Yeast heat shock factor is an essential DNA-binding protein that exhibits temperature-dependent phosphorylation. Cell 54: 855–864.

30. Wiederrecht G, Seto D, Parker CS (1988) Isolation of the gene encoding the S. cerevisiae heat shock transcription factor. Cell 54: 841–853.

31. Xiao X, Zuo X, Davis AA, McMillan DR, Curry BB, et al. (1999) HSF1 is required for extra-embryonic development, postnatal growth and protection during inflammatory responses in mice. EMBO J 18: 5943–5952.

32. Xing H, Wilkerson DC, Mayhew CN, Lubert EJ, Skaggs HS, et al. (2005) Mechanism of hsp70i gene bookmarking. Science 307: 421–423.

33. Fujimoto M, Izu H, Seki K, Fukuda K, Nishida T, et al. (2004) HSF4 is required for normal cell growth and differentiation during mouse lens development. EMBO J 23: 4297–4306.

34. Nover L, Bharti K, Doring P, Mishra SK, Ganguli A, et al. (2001) Arabidopsis and the heat stress transcription factor world: how many heat stress transcription factors do we need? Cell Stress Chaperones 6: 177–189.

35. Guo J, Wu J, Ji Q, Wang C, Luo L, et al. (2008) Genome-wide analysis of heat shock transcription factor families in rice and Arabidopsis. J Genet Genomics 35: 105–118.

36. Xiong Y, Liu T, Tian C, Sun S, Li J, et al. (2005) Transcription factors in rice: a genome-wide comparative analysis between monocots and eudicots. Plant Mol Biol 59: 191–203.

37. Lin YX, Jiang HY, Chu ZX, Tang XL, Zhu SW, et al. (2011) Genome-wide identification, classification and analysis of heat shock transcription factor family in maize. BMC Genomics 12: 76.

38. Garg R, Patel RK, Jhanwar S, Priya P, Bhattacharjee A, et al. (2011) Gene discovery and tissue-specific transcriptome analysis in chickpea with massively parallel pyrosequencing and web resource development. Plant Physiol 156: 1661–1678.

39. Czarnecka-Verner E, Yuan CX, Fox PC, Gurley WB (1995) Isolation and characterization of six heat shock transcription factor cDNA clones from soybean. Plant Mol Biol 29: 37–51.

40. Punta M, Coggill PC, Eberhardt RY, Mistry J, Tate J, et al. (2012) The Pfam protein families database. Nucleic Acids Res 40: D290–301.

41. Letunic I, Doerks T, Bork P (2012) SMART 7: recent updates to the protein domain annotation resource. Nucleic Acids Res 40: D302–305.

42. Delorenzi M, Speed T (2002) An HMM model for coiled-coil domains and a comparison with PSSM-based predictions. Bioinformatics 18: 617–625.

43. Cokol M, Nair R, Rost B (2000) Finding nuclear localization signals. EMBO Rep 1: 411–415.

44. la Cour T, Kiemer L, Molgaard A, Gupta R, Skriver K, et al. (2004) Analysis and prediction of leucine-rich nuclear export signals. Protein Eng Des Sel 17: 527–536.

45. Thompson JD, Gibson TJ, Plewniak F, Jeanmougin F, Higgins DG (1997) The CLUSTAL_X windows interface: flexible strategies for multiple sequence alignment aided by quality analysis tools. Nucleic acids research 25: 4876–4882.

46. Tamura K, Dudley J, Nei M, Kumar S (2007) MEGA4: Molecular Evolutionary Genetics Analysis (MEGA) software version 4.0. Mol Biol Evol 24: 1596–1599.

47. Hanada K, Zou C, Lehti-Shiu MD, Shinozaki K, Shiu SH (2008) Importance of lineage-specific gene and gene pair tandem duplicates in the adaptive response to environmental stimuli. Plant Physiol 148: 993–1003.

48. Maher C, Stein L, Ware D (2006) Evolution of Arabidopsis microRNA families through duplication events. Genome research 16: 510–519.

49. Zhang X, Feng Y, Cheng H, Tian D, Yang S, et al. (2011) Relative evolutionary rates of NBS-encoding genes revealed by soybean segmental duplication. Mol Genet Genomics 285: 79–90.

50. Sato S, Nakamura Y, Kaneko T, Asamizu E, Kato T, et al. (2008) Genome structure of the legume, Lotus japonicus. DNA Res 15: 227–239.

51. Edgar RC (2004) MUSCLE: multiple sequence alignment with high accuracy and high throughput. Nucleic Acids Res 32: 1792–1797.

52. Suyama M, Torrents D, Bork P (2006) PAL2NAL: robust conversion of protein sequence alignments into the corresponding codon alignments. Nucleic acids research 34: W609–W612.

53. Lynch M, Conery JS (2000) The evolutionary fate and consequences of duplicate genes. Science 290: 1151–1155.

54. Yang Z (2007) PAML 4: phylogenetic analysis by maximum likelihood. Mol Biol Evol 24: 1586–1591.

55. Yang Z, Nielsen R, Goldman N, Pedersen AM (2000) Codon-substitution models for heterogeneous selection pressure at amino acid sites. Genetics 155: 431–449.

56. Yang Z, Wong WS, Nielsen R (2005) Bayes empirical bayes inference of amino acid sites under positive selection. Mol Biol Evol 22: 1107–1118.

57. Hogslund N, Radutoiu S, Krusell L, Voroshilova V, Hannah MA, et al. (2009) Dissection of symbiosis and organ development by integrated transcriptome analysis of lotus japonicus mutant and wild-type plants. PLoS One 4: e6556.

58. Eisen MB, Spellman PT, Brown PO, Botstein D (1998) Cluster analysis and display of genome-wide expression patterns. Proc Natl Acad Sci U S A 95: 14863–14868.

59. Bertioli DJ, Moretzsohn MC, Madsen LH, Sandal N, Leal-Bertioli SC, et al. (2009) An analysis of synteny of Arachis with Lotus and Medicago sheds new light on the structure, stability and evolution of legume genomes. BMC Genomics 10: 45.

60. Nam J, Kim J, Lee S, An G, Ma H, et al. (2004) Type I MADS-box genes have experienced faster birth-and-death evolution than type II MADS-box genes in angiosperms. Proceedings of the National Academy of Sciences of the United States of America 101: 1910.

61. Kim KD, Shin JH, Van K, Kim DH, Lee SH (2009) Dynamic rearrangements determine genome organization and useful traits in soybean. Plant Physiol 151: 1066–1076.

62. Innes RW, Ameline-Torregrosa C, Ashfield T, Cannon E, Cannon SB, et al. (2008) Differential accumulation of retroelements and diversification of NB-LRR disease resistance genes in duplicated regions following polyploidy in the ancestor of soybean. Plant Physiol 148: 1740–1759.

63. Shin JH, Van K, Kim DH, Kim KD, Jang YE, et al. (2008) The lipoxygenase gene family: a genomic fossil of shared polyploidy between Glycine max and Medicago truncatula. BMC Plant Biol 8: 133.

64. Van K, Kim DH, Cai CM, Kim MY, Shin JH, et al. (2008) Sequence level analysis of recently duplicated regions in soybean [Glycine max (L.) Merr.] genome. DNA Res 15: 93–102.

65. Lai J, Ma J, Swigonova Z, Ramakrishna W, Linton E, et al. (2004) Gene loss and movement in the maize genome. Genome Res 14: 1924–1931.

66. Messing J, Bharti AK, Karlowski WM, Gundlach H, Kim HR, et al. (2004) Sequence composition and genome organization of maize. Proc Natl Acad Sci U S A 101: 14349–14354.

67. Schlueter JA, Scheffler BE, Schlueter SD, Shoemaker RC (2006) Sequence conservation of homeologous bacterial artificial chromosomes and transcription of homeologous genes in soybean (Glycine max L. Merr.). Genetics 174: 1017–1028.

68. Cannon SB, McCombie WR, Sato S, Tabata S, Denny R, et al. (2003) Evolution and microsynteny of the apyrase gene family in three legume genomes. Mol Genet Genomics 270: 347–361.

69. Blanc G, Wolfe KH (2004) Functional divergence of duplicated genes formed by polyploidy during Arabidopsis evolution. Plant Cell 16: 1679–1691.

70. Lynch M, Conery JS (2003) The evolutionary demography of duplicate genes. J Struct Funct Genomics 3: 35–44.
71. Lohmann C, Eggers-Schumacher G, Wunderlich M, Schoffl F (2004) Two different heat shock transcription factors regulate immediate early expression of stress genes in Arabidopsis. Mol Genet Genomics 271: 11–21.
72. Charng Y, Liu H, Liu N, Chi W, Wang C, et al. (2007) A heat-inducible transcription factor, HsfA2, is required for extension of acquired thermotolerance in Arabidopsis. Plant Physiology 143: 251.
73. Davletova S, Rizhsky L, Liang H, Shengqiang Z, Oliver DJ, et al. (2005) Cytosolic ascorbate peroxidase 1 is a central component of the reactive oxygen gene network of Arabidopsis. The Plant Cell Online 17: 268–281.
74. Sakuma Y, Maruyama K, Qin F, Osakabe Y, Shinozaki K, et al. (2006) Dual function of an Arabidopsis transcription factor DREB2A in water-stress-responsive and heat-stress-responsive gene expression. Proceedings of the National Academy of Sciences 103: 18822.
75. Ogawa D, Yamaguchi K, Nishiuchi T (2007) High-level overexpression of the Arabidopsis HsfA2 gene confers not only increased themotolerance but also salt/osmotic stress tolerance and enhanced callus growth. Journal of experimental botany 58: 3373.
76. Kumar M, Busch W, Birke H, Kemmerling B, Nurnberger T, et al. (2009) Heat shock factors HsfB1 and HsfB2b are involved in the regulation of Pdf1.2 expression and pathogen resistance in Arabidopsis. Mol Plant 2: 152–165.
77. Czarnecka-Verner E, Pan S, Salem T, Gurley WB (2004) Plant class B HSFs inhibit transcription and exhibit affinity for TFIIB and TBP. Plant Mol Biol 56: 57–75.
78. Widholm JM, Chinnala AR, Ryu JH, Song HS, Eggett T, et al. (2001) Glyphosate selection of gene amplification in suspension cultures of 3 plant species. Physiol Plant 112: 540–545.

79. Zhu H, Choi HK, Cook DR, Shoemaker RC (2005) Bridging model and crop legumes through comparative genomics. Plant Physiol 137: 1189–1196.
80. Kotak S, Vierling E, Baumlein H, von Koskull-Doring P (2007) A novel transcriptional cascade regulating expression of heat stress proteins during seed development of Arabidopsis. Plant Cell 19: 182–195.
81. Swindell WR, Huebner M, Weber AP (2007) Transcriptional profiling of Arabidopsis heat shock proteins and transcription factors reveals extensive overlap between heat and non-heat stress response pathways. BMC Genomics 8: 125.
82. Giorno F, Wolters-Arts M, Grillo S, Scharf KD, Vriezen WH, et al. (2010) Developmental and heat stress-regulated expression of HsfA2 and small heat shock proteins in tomato anthers. J Exp Bot 61: 453–462.
83. Mittal D, Chakrabarti S, Sarkar A, Singh A, Grover A (2009) Heat shock factor gene family in rice: genomic organization and transcript expression profiling in response to high temperature, low temperature and oxidative stresses. Plant Physiol Biochem 47: 785–795.
84. Nishizawa A, Yabuta Y, Yoshida E, Maruta T, Yoshimura K, et al. (2006) Arabidopsis heat shock transcription factor A2 as a key regulator in response to several types of environmental stress. The Plant Journal 48: 535–547.
85. Schramm F, Ganguli A, Kiehlmann E, Englich G, Walch D, et al. (2006) The heat stress transcription factor HsfA2 serves as a regulatory amplifier of a subset of genes in the heat stress response in Arabidopsis. Plant molecular biology 60: 759–772.
86. Ikeda M, Mitsuda N, Ohme-Takagi M (2011) Arabidopsis HsfB1 and HsfB2b Act as Repressors of the Expression of Heat-Inducible Hsfs But Positively Regulate the Acquired Thermotolerance. Plant Physiology 157: 1243–1254.
87. Miller G, Shulaev V, Mittler R (2008) Reactive oxygen signaling and abiotic stress. Physiol Plant 133: 481–489.

15

Genomic Distribution of H3K9me2 and DNA Methylation in a Maize Genome

Patrick T. West[1,9], **Qing Li**[1,9], **Lexiang Ji**[2,3], **Steven R. Eichten**[1], **Jawon Song**[4], **Matthew W. Vaughn**[4], **Robert J. Schmitz**[2], **Nathan M. Springer**[1]*

1 Department of Plant Biology, University of Minnesota, Saint Paul, Minnesota, United States of America, **2** Department of Genetics, University of Georgia, Athens, Georgia, United States of America, **3** Institute of Bioinformatics, University of Georgia, Athens, Georgia, United States of America, **4** Texas Advanced Computing Center, University of Texas-Austin, Austin, Texas, United States of America

Abstract

DNA methylation and dimethylation of lysine 9 of histone H3 (H3K9me2) are two chromatin modifications that can be associated with gene expression or recombination rate. The maize genome provides a complex landscape of interspersed genes and transposons. The genome-wide distribution of DNA methylation and H3K9me2 were investigated in seedling tissue for the maize inbred B73 and compared to patterns of these modifications observed in *Arabidopsis thaliana*. Most maize transposons are highly enriched for DNA methylation in CG and CHG contexts and for H3K9me2. In contrast to findings in *Arabidopsis*, maize CHH levels in transposons are generally low but some sub-families of transposons are enriched for CHH methylation and these families exhibit low levels of H3K9me2. The profile of modifications over genes reveals that DNA methylation and H3K9me2 is quite low near the beginning and end of genes. Although elevated CG and CHG methylation are found within gene bodies, CHH and H3K9me2 remain low. Maize has much higher levels of CHG methylation within gene bodies than observed in *Arabidopsis* and this is partially attributable to the presence of transposons within introns for some maize genes. These transposons are associated with high levels of CHG methylation and H3K9me2 but do not appear to prevent transcriptional elongation. Although the general trend is for a strong depletion of H3K9me2 and CHG near the transcription start site there are some putative genes that have high levels of these chromatin modifications. This study provides a clear view of the relationship between DNA methylation and H3K9me2 in the maize genome and how the distribution of these modifications is shaped by the interplay of genes and transposons.

Editor: Beth A. Sullivan, Duke University, United States of America

Funding: The research was supported by a grant from the National Science Foundation (IOS-1237931) to MWV and NMS. This work also used resources or cyberinfrastructure provided by iPlant Collaborative. The iPlant Collaborative is funded by a grant from the National Science Foundation (DBI-0735191; www.iplantcollaborative.org). Start-up funds from the University of Georgia and a research grant from the National Science Foundation (IOS-1339194) to RJS supported aspects of this study. The funders had no role in study design, data collection and analysis, decision to publish, or preparation of the manuscript.

Competing Interests: The authors have declared that no competing interests exist.

* Email: springer@umn.edu

⑨ These authors contributed equally to this work.

Introduction

Cytosine DNA methylation is a chromatin modification involved in many cellular processes including regulating gene expression and silencing repeat sequences and transposons. In plants, DNA methylation occurs in both symmetrical (CG and CHG, where H is A, T, or C) and asymmetrical (CHH) sequence contexts, where DNA methylation in each of the three contexts is thought to be regulated separately [1]. The symmetrical methylation found in CG or CHG contexts can be maintained via methylation of the hemi-methylated molecule generated by DNA replication. In contract, CHH methylation requires continual targeting of *de novo* methyltransferases that do not require hemi-methylation of the target.

Much of our understanding about DNA methylation machinery and mechanisms in plants is based on research in *Arabidopsis thaliana*. In *Arabidopsis*, CG, CHG, and CHH methylation are highly enriched within transposable elements, repeat sequence, and the pericentromeric region [2–3]. CG methylation is largely maintained by the MET1 enzyme [1]. CHG methylation is largely attributed to the CMT3 chromomethylase gene [1]. The proper maintenance of CG and CHG methylation, particularly in heterochromatic regions, requires the chromatin remodeler DDM1 [4–5]. The targeting of chromomethylases involves binding to nucleosomes marked by histone 3 lysine 9 dimethylation (H3K9me2) [6]. There is evidence for interdependence between CHG methylation and H3K9me2 such that if either modification is lost, both show genome wide depletion [7–9]. There are two different pathways for CHH methylation in *Arabidopsis* [1], [5], [9–10]. The RNA-directed DNA methylation (RdDM) pathway requires the DRM methyltransferases and involves PolIV, PolV and other components [1], [9]. The other pathway utilizes the CMT2 methyltransferase and likely requires DDM1 and H3K9me2 methylation [5], [10].

Although the genetic and genomic resources available for *Arabidopsis* have provided substantial opportunities to understand

DNA and histone methylation in plants, the *Arabidopsis* genome may not provide a good model for many of the crop genomes. *Arabidopsis* have very few transposons that are mostly clustered in pericentromeric regions or other heterochromatic knobs [11]. The *Zea mays* genome is ~20 fold larger than the *Arabidopsis* genome and exhibits a complex arrangement of transposons and genes that is observed in many plant species. The majority of maize genes are flanked by transposons [12–14] and the chromatin landscape of maize is much more diverse than that of the segregated *Arabidopsis* genome [15–16]. Profiling of DNA methylation by methyl-filtration sequencing, restriction analyses or methylated DNA immunoprecipitation (meDIP) has revealed that DNA methylation is enriched over transposons and generally lower over genes [14], [16–21]. Several studies have reported whole-genome bisulfite sequencing (WGBS) for maize [22–24]. CG and CHG methylation are highly enriched over transposons and repeat sequences and depleted near genic space [22–23]. CHH methylation on the other hand does not correlate with CG or CHG methylation, is depleted over repeat sequences, and is enriched near the start and end of genes [22]. Cytological analysis of histone modifications in maize revealed that H3K9me2 was localized throughout the chromosome for pachytene chromosomes and was not particularly enriched at pericentromeric or knob heterochromatin [25–26]. However, there is evidence for enrichment of H3K9me2 at transposon sequences in maize that are generally considered to be heterochromatic [27–28].

The relationship between DNA methylation and other chromatin modifications has not been looked at in great detail in maize. Here we combine WGBS with H3K9me2 ChIP-seq to assess the relationship between DNA methylation and H3K9me2 throughout the maize genome. In comparison to *Arabidopsis* we find lower levels of CHH methylation and a distinct relationship of H3K9me2 with CHH methylation patterns. The analysis of different sub-families of transposable elements reveals distinct patterns of CHH and H3K9me2 enrichment for different families. Maize genes contain CHG methylation within the gene body and this is partially attributable to the presence of transposons in introns of >10% of maize genes. The presence of heterochromatin transposons within genes does not appear to restrict expression of these genes. We also identified a subset of maize genes with high levels of CHG methylation or H3K9me2 over the transcription start site (TSS) and find that the majority of these genes are not expressed throughout maize development.

Methods

Bisulfite Sequencing

Genomic DNA was isolated from the third leaf of 14-day after planting seedling from the B73 inbred line. Samples were fragmented and ligated with TruSeq-methylated adapters. Bisulfite conversion was performed on five hundred nanograms of adaptor-ligated DNA using the MethylCode bisulfite conversion kit (Life Technologies) according to manufacturer's guidelines. Converted DNA was split into four reactions and amplified using Pfu Turbo Cx DNA polymerase (Agilent) for four cycles and subsequently pooled. Libraries were sequenced on the HiSeq 2000 (Illumina) for 100 cycles, paired end. Sequencing reads (SRA accession SRP022569) were processed to identify and filter poor 3' quality and incomplete conversion. Sequences were aligned to the B73 reference genome (AGPv2) using the Bismark aligner (v0.7.2; [29]) under the parameters (-n 2, -l 50). Methylated cytosines were extracted from aligned reads using the Bismark methylation extractor under standard parameters. The proportion of CG, CG,

and CHH methylation was determined as weighted methylation levels [30] in 100bp non-overlapping windows across the genome.

H3K9me2 ChIP-seq

H3K9me2 profiling was performed on three replicates of B73 seedling using antibodies specific for H3K9me2 (#07-441) purchased from Millipore (Billerica, USA) according to manufacturer's recommendations as described in Eichten et al. [27]. For ChIP-seq, adapters were ligated to replicates using one of two protocols. In the first, TruSeq-methylated adapters were ligated to the B73 DNA fragments according to the NEBNext DNA Library Prep protocol. In the second, adapters were ligated using the Nextera DNA sample Preparation Kit. Both samples were sequenced to greater than 100 million reads on the HiSeq 2000 (Illumina), single end. Sequencing reads (SRA accession SRP043372) were analyzed using FastQC (http://www.bioinformatics.babraham.ac.uk/projects/fastqc/) to identify and filter poor 3' quality reads. Sequenced reads were aligned to the B73 reference genome (AGPv2) using Bowtie under standard parameters. Duplicate reads from both the NEBNext and Nextera libraries were removed using SAMtools [31] and the samples were merged into one library. The level of H3K9me2 was described as the sum of the intersecting H3K9me2 ChIP-seq reads over 100bp windows across the genome. Intersecting 100 bp windows and H3K9me2 reads were determined using BEDtools [32]. 100 bp windows with significant H3K9me2 were defined as having a sum of reads greater than one standard deviation above the average sum.

RNA-seq and Expression Analysis

RNA isolated from 14-day after planting leaf tissue of the B73 inbred line was prepared for sequencing at the University of Minnesota Genomics Center using the TruSeq library preparation protocol (Illumina). Three independent replicates were included. Libraries were sequenced on the HiSeq 2000. Over 10 million 50bp read pairs were generated for each library. Raw reads (SRA accession SRP018088) were filtered to eliminate poor-quality reads using CASAVA 1.8 (Illumina). High-quality reads were then passed to Trim_glore (http://www.bioinformatics.babraham.ac.uk/projects/trim_galore/) to trim poor bases from 3' end of the sequences, to remove adapters and to filter very short reads resulted from base and adapter trimming. This was run under the pair-end reads mode using standard parameters. Reads that passed quality control were first mapped to the Filtered Gene Set (ZmB73_5a, FGS), and unmapped reads were realigned to the maize reference genome (AGPv2) using TopHat [33] under standard parameters. Only reads that are mapped uniquely to the genome were kept and used to calculate transcript abundance. The number of read pairs that are mapped to each gene were developed using "BAM to Counts" within the iPlant Discovery Environment (www.iplantcollaborative.org). The 'Reads count per kilobase per million mapped' (RPKM) value was calculated and averaged over the three biological replicates to represent the expression level of each filtered gene. Those expression levels were used to group the genes into five categories: not expressed, and four categories with equal number of expressed genes in B73 seedling tissues. The proportion in each of the five categories were determined for genes that were identified to have specific features, e.g., with >1000 bp TE, having high H3K9me2 or CHG in promoters.

Analyzing H3K9me2 and DNA methylation

To analyze the correlation between non-CG methylation (CHG and CHH) and H3K9me2, all the 100 bp windows that have data on all three marks were grouped based on the levels of either CHG or CHH. CHG levels were equally split into 10 groups from 0% to 100%. CHH levels were split into 9 groups, 5 groups from 0 to 5% by an increase of 1%, 3 groups from 5% to 20% by an increase of 5%, and a group of >20%. The CHG and CHH groups were cross-tabulated to give a total of 90 combinations (10 CHG groups * 9 CHH groups), and the average H3K9me2 levels were calculated for each combination and were shown as a heatmap.

The average DNA methylation and H3K9me2 levels for each transposon sub-family was calculated. The classification of transposon sub-families were based on Maize TE Consortium [34] and the study of Eichten et al. [27]. The 100 bp windows that overlap or fall within each transposon sub-family were identified using intersectBed from the BEDtools package. Those windows were used to calculate the average values for both DNA methylation and H3K9me2 using R. We also calculated the mean level of DNA methylation and H3K9me2 in the flanking regions of each transposon sub-family. Briefly, 100 bp windows that overlap with the regions that are 900–1000 bp away from a transposon were identified, and used to get the mean DNA methylation and H3K9me2 for both the upstream and downstream flanking regions.

Analysis of genes with TEs and genes with high CHG/H3K9me2 in promoters

Genes with high CHG or H3K9me2 in their promoter region were identified by assessing the levels of CHG or H3K9me2 in the 100 bp window that overlaps the transcription start site. Genes with greater than 88.5% CHG methylation (top 10% of all CHG values) in the promoter region were defined as having high CHG over the promoter region and genes with greater than 2 standard deviations of H3K9me2 reads above the genome wide average in the promoter region were defined as having high H3K9me2 over the promoter region.

Relative distance line plots

To plot DNA methylation or H3K9me2 levels over transposons and their flanking regions, we first determined the distance between the 100 bp windows and transposons from the Maize TE Consortium (ZmB73_5b). Windows upstream of the transposable elements were given a negative distance value and windows downstream a positive distance value. We then identified the closest transposon to each 100 bp window, and kept those windows that are located within the transposons or the 1000 bp flanking regions on either side. For windows overlapping or within transposons, the normalized distance across the element on a scale of 1 to 1000 was determined. The scaled 1000 bp element, together with 2000 bp flanking regions, were then divided into 60 equal bins, 20 bins each for the 1000 bp upstream region, the scaled 1000 bp element, and the 1000 bp downstream region. The average methylation levels of the bins were then determined and plotted on a line graph in R.

Absolute distance line plots

The absolute distance line plots consist of two parts, the 5' plot and the 3' plot, each of which contains 5 kb genomic segments. The 5' plot contains 2 kb upstream regions of the transcriptional start site (TSS) and 3 kb genic sequences from TSS. The 3' plot contains 2 kb downstream sequences from the transcriptional termination site (TTS) and 3 kb genic sequences from the TTS. For genes that are less than 3 kb, the actual gene size were used, which means less than 5 kb regions will be used. In other words, the further into a gene, the less number of genes will be included in

those plots. To make these plots, the physical distance between genes and nearby 100 bp windows was determined. For the 5' plot, this distance was determined to be the physical distance between the mid-point of the 100 bp window and the TSS. While for the 3' plot, it was calculated as the physical distance between the mid-point of each window and the TTS. Windows that are falling within the respective 5 kb genomic regions of a gene were kept for downstream analysis. For each plot, the 5 kb regions were then divided into 100 equal bins, and the average methylation level for each bin across all genes were calculated using R. Finally, the averaged methylation level was plotted against the center of each bin using R.

Results

To investigate the distribution of DNA methylation and H3K9me2 in the maize genome we performed WGBS and H3K9me2 ChIP-seq on leaf tissue of B73 maize seedlings (Figure S1A). The same tissue was also used to perform RNA-seq in order to compare the distribution of these chromatin modifications relative to gene expression (Figure S1A). On a genome-wide scale, the level of CHH methylation is very low with 1.2% of total CHHs methylated whereas CHG and CG methylation are relatively high with 70.9% and 86.4% methylation respectively (Figure 1A). Similar to maize, the *Arabidopsis* genome (SRA accession SRA035939) also contains more CG and CHG methylation than CHH methylation (Figure 1A). The comparison of maize and *Arabidopsis* reveals ~2-fold higher levels of CG and CHG in the maize genome and ~2-fold lower levels of CHH methylation in maize (Figure 1A). The observation of higher genome-wide CG and CHG methylation in maize may simply reflect the higher transposon content of the maize genome.

The CG, CHG and CHH methylation levels for 100 bp tiles of the maize genome were assessed. The vast majority of regions show less than 10% CHH methylation with only 1.3% of regions exhibiting >10% CHH methylation (Figure 1B). In contrast, the majority of tiles exhibit high levels of CG or CHG methylation, similar to analyses of other maize tissues [22], [23]. However, a small portion (5–10%) of the maize genome exhibits less than 10% CG or CHG methylation. The analysis of 100 bp tiles located in genes or transposons reveals that the majority of 100 bp tiles with low (<10%) CG (72%) or CHG (74%) methylation are located within genes (Figure S1B–C). This supports the utility of the methyl-filtration sequencing that provided targeted sequencing for these unmethylated regions [35]. The levels of CHH methylation are quite low both in genes and within transposons (Figure S1D). The distribution of sequencing depth for 100 bp tiles reveals that genic sequences tend to have much lower levels of H3K9me2 than TEs (Figure S1E).

There is growing evidence that histone modifications, in particular H3K9me2, can play a role in targeting DNA methylation in *Arabidopsis*, especially in the CHG and CHH contexts [6], [10], [36]. We assessed how H3K9me2 levels were associated with CHG and CHH methylation throughout the maize and *Arabidopsis* (SRA accession GSM124393) genomes (Figure 1C–D). Maize exhibits high levels of H3K9me2 whenever there is >20% CHG methylation and less than 10% CHH methylation (Figure 1C). In contrast, H3K9me2 is associated with higher levels of CHH methylation in *Arabidopsis* (Figure 1D).

DNA methylation and H3K9me2 profiles over maize transposons

DNA methylation and H3K9me2 are frequently enriched over transposable element sequences. The profile of DNA methylation

A

B

C

Maize H3K9me2 levels

D

Arabidopsis H3K9me2 levels

Low H3K9me2 ▬▬▬▬▬▬ High H3K9me2

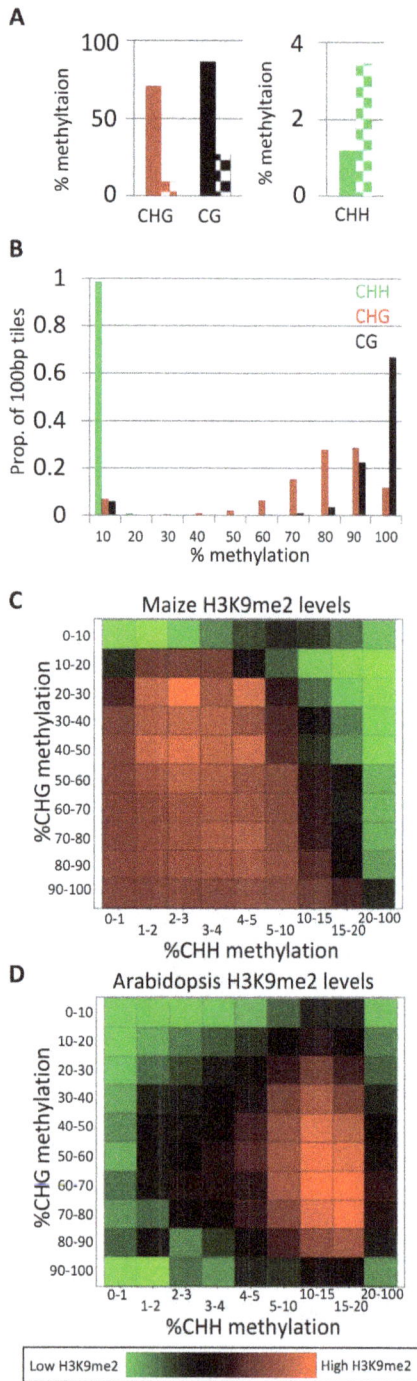

Figure 1. Genome-wide levels of DNA methylation and H3K9me2. (A) Average genome-wide DNA methylation levels in CG, CHG, and CHH sequence contexts. These numbers refer to the average DNA methylation level for all cytosines within this sequence context. Solid bars are maize DNA methylation and dashed bars are *Arabidopsis* DNA methylation (data from [34]). (B) The proportion of 100 bp tiles across the maize genome containing different levels of DNA methylation is shown. (C–D) The average read number for H3K9me2 ChIP-seq was determined for 100 bp tiles in maize (C) or *Arabidopsis* (D) having varying levels of CHG and CHH methylation and is shown using a heat map to illustrate relative enrichment.

and H3K9me2 over maize and *Arabidopsis* DNA transposons or retrotransposons was compared (Figure 2A–C). The maize data are from the current study whereas the *Arabidopsis* DNA methylation data is obtained from Schmitz et al. [37] and the *Arabidopsis* H3K9me2 data were obtained from Stroud et al., [10]. In both maize and *Arabidopsis*, the level of CG and CHG methylation is markedly higher in TEs compared to flanking regions. However, there is more evidence for spreading of this DNA methylation to the flanking regions in maize than in *Arabidopsis* as evidenced by the slope of the profile in maize rather than the sharp drop seen in *Arabidopsis*. In addition, the abundance of CHG methylation is substantially higher for maize than for *Arabidopsis* at both class I (LTR elements - RNA intermediate) and class II (TIR elements - DNA intermediate) transposons. While *Arabidopsis* class II transposons exhibit enrichment for CHH methylation near the TIRs (the bumps in the CHH profile at the beginning and end of TIR elements) this enrichment is not noted in maize. H3K9me2 is enriched over both TIR DNA transposons and LTR retrotransposons in both maize and *Arabidopsis* (Figure 2C). The enrichment of H3K9me2 is more pronounced over LTR elements as compared to TIR elements (Figure 2C).

The profiles of DNA methylation and H3K9me2 were examined for a number of sub-families of maize transposons using the classifications from the maize genome annotation [14], [34]. These include LINE elements, nine sub-types of LTR elements and seven sub-types of TIR elements (full profiles for each class are available in Figure S2). The level of DNA methylation or H3K9me2 was determined within each family and is compared to the average levels observed in flanking regions (1 kb from the elements) or exons (Figure 2D). The levels of CG and CHG methylation are uniformly high for all sub-types of transposons (Figure 2D). The levels of CHH methylation show some unusual trends. The LTR families tend to have quite low levels of CHH methylation but the families that exhibit spreading of heterochromatin to flanking sequences [27] have lower levels of CHH than families that do not exhibit spreading. Some of the TIR families are marked by quite high levels of CHH methylation (Figure 2D) and these same families exhibit elevated CHH in other maize tissues as well [22]. There are also significant differences in the relative levels of H3K9me2 over different families. In general, the LTR families all have quite high levels of H3K9me2 while some of the TIR families have quite low levels of H3K9me2 (Figure 2E). The families with the highest levels of CHH methylation tend to have the least enrichment for H3K9me2.

DNA methylation and H3K9me2 profiles over maize genes

The profile for each type of DNA methylation and H3K9me2 over genes was compared in maize and *Arabidopsis* (Figure 3A–B). The patterns observed in maize and *Arabidopsis* are somewhat similar but there are a number of differences. The level of CG and CHG DNA methylation in the 2 kb upstream of the transcription start site (TSS) or 2 kb downstream of the transcription termination site (TTS) is much lower in *Arabidopsis* than in maize. This is likely a result of many maize genes being flanked by transposons sequences and exemplifies the different chromatin environment surrounding maize genes compared to *Arabidopsis* genes. Maize genes also tend to be closely flanked by regions of elevated CHH methylation, termed CHH islands by Gent et al. [22] and also noted by Regulski et al. [23]. Both *Arabidopsis* and maize genes exhibit increased levels of CG methylation in the middle of the transcribed regions relative to the regions near the

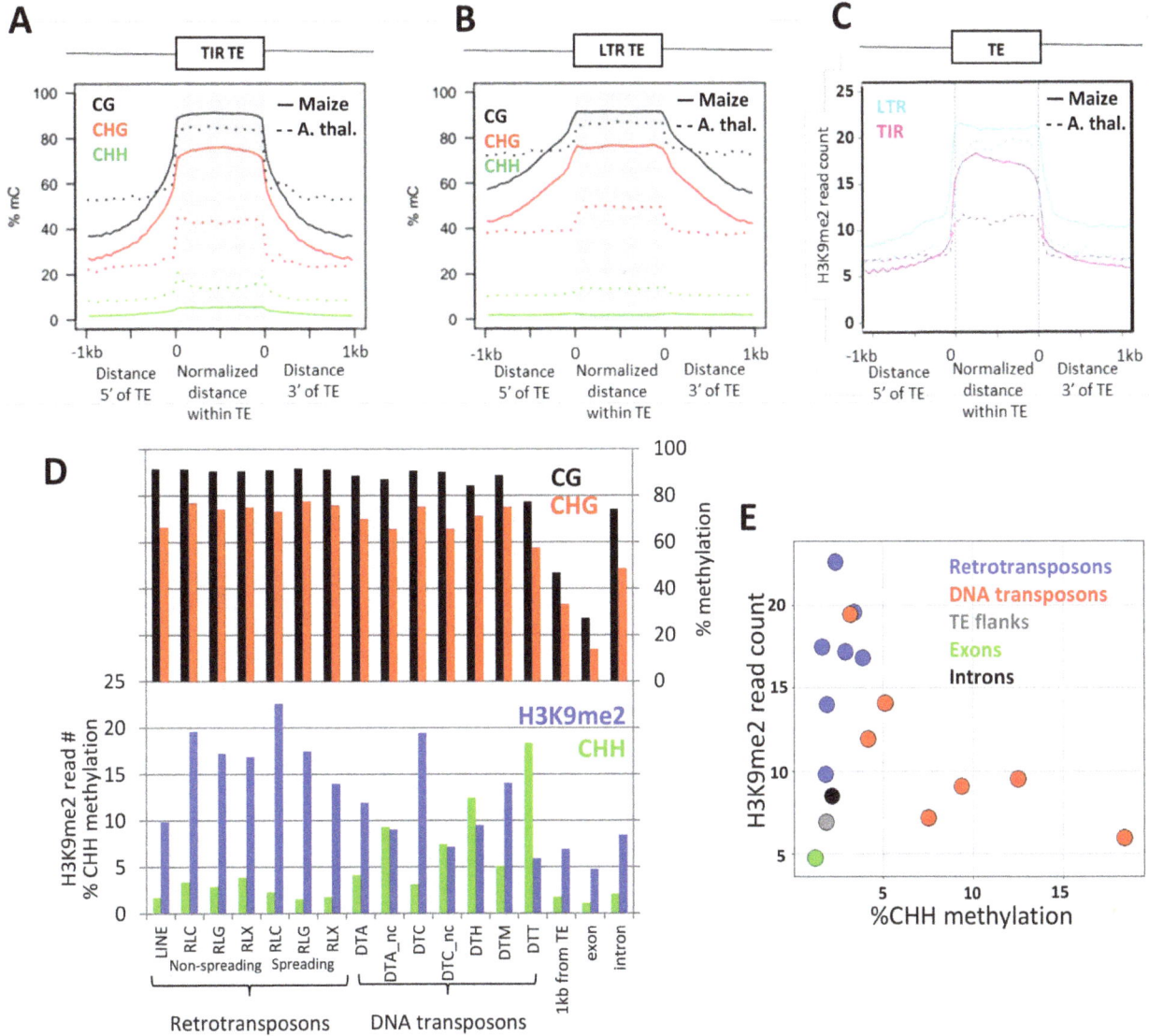

Figure 2. Enrichment of DNA methylation and H3K9me2 over transposable elements. (A–B) Relative distance line plots of DNA methylation over transposable elements in maize and *Arabidopsis*. The plot in A shows the average enrichment for each type of DNA methylation of DNA transposons containing terminal inverted repeats (TIRs). The colors of the lines indicate the context of DNA methylation (black-CG; red-CHG; green-CHH) and type of line indicates the species (solid = maize; dashed = *Arabidopsis*). In (B) similar plots are shown for long terminal repeat (LTR) retrotransposons. (C) Plots of H3K9me2 abundance are shown for DNA transposons (purple) and retrotransposons (blue) in both maize (solid lines) and *Arabidopsis* (dashed lines). (D) The average level of DNA methylation or H3K9me2 is plotted for different sub-classes of transposable elements. The retrotransposons are divided into LINEs, RLG (gypsy-like), RLC (copia-like) and RLX (LTR elements of unknown class). The RLG, RLC and RLX elements are all split according to whether they exhibit evidence for spreading of H3K9me2 or DNA methylation in flanking regions as defined in Eichten et al [27]. The DNA transposons are divided into five major families (DTA (*hAT*), DTC (*CACTA*), DTH (*PIF/Harbinger*), DTM (*Mutator*) and DTT (*Tc1/Mariner*)) and for two of these there are large families of "non-coding" elements that are indicated as "-nc". The last three bars indicate the average levels for each modification 1 kb away from TEs, within exons or within introns. (E) The relative levels of H3K9me2 and CHH methylation are shown for each sub-family of transposons. The levels for TE flanking regions, exons and introns are also shown.

TSS and TTS. Maize gene bodies also contain noticeable CHG methylation whereas this chromatin modification is not observed within *Arabidopsis* gene bodies. The H3K9me2 profiles reveal depletion in the regions immediately preceding or following the TSS and TTS in both maize and *Arabidopsis* (Figure 3B) that could reflect nucleosome free regions.

The profile of DNA methylation over maize genes is influenced by several comparative genomic attributes. The full set of potential annotated maize genes are classified as a working gene set (WGS; n = 110,028) and a subset (n = 39,656) are classified as the filtered

gene set (FGS). The FGS genes are a subset of putative genes with more evidence for functionality (full-length cDNA, homology to coding sequence in other species) whereas the WGS genes may include pseudo-genes, misannotated transposable elements, or gene fragments. The CG and CHG DNA methylation profile for the FGS genes shows much greater reductions in DNA methylation levels at the TSS and TTS (Figure S3A). In contrast, the FGS genes are marked by higher levels of CHH methylation in the regions immediately preceding or following the transcribed region. The H3K9me2 levels are more strongly reduced for FGS

A

B

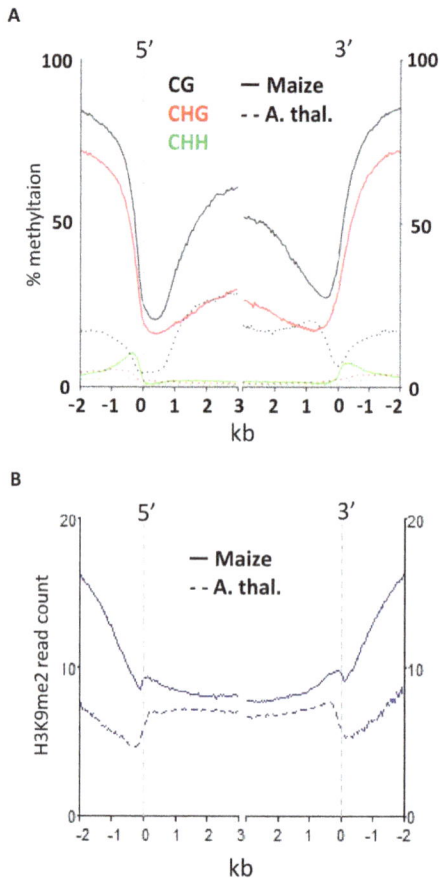

Figure 3. Absolute distance line plots of DNA methylation and H3K9me2 over genic space in maize and *Arabidopsis*. (A) Maize and *Arabidopsis* genes were aligned at the 5′ and 3′ ends and average DNA methylation are plotted for the regions beginning 2 kb from the gene to the regions within the gene. The vertical dashed lines represent the 5′ transcriptional start site (TSS) and 3′ transcriptional termination site (TTS). (B) A similar plot is used to show average levels of H3K9me2 for maize and *Arabidopsis* genes.

genes than WGS genes. The FGS genes can be split into a group with retained syntenic positions relative to sorghum and rice and genes that have inserted into new genomic positions. The inserted genes have much higher levels of CHG and CG methylation both within introns and exon (Figure S3B). The ancient tetraploid nature of the maize genome resulted in many examples of retained paralogs that have been assigned to two subgenomes based on preferential fractionation and expression [38]. However, we did not find evidence for differences in DNA methylation profiles for retained duplicates that are present within both subgenomes (Figure S3C). This suggests that DNA methylation does not play a critical role in distinguishing the sub-genomes but there can be differences in DNA methylation levels at specific pairs of retained duplicates. Similar findings have been reported in maize [21], soybean [39] and brassica [40].

CHG methylation and H3K9me2 within maize gene bodies is due to presence of transposons in maize introns

The presence of CHG methylation within maize gene bodies was unexpected as this chromatin modification is not often found in *Arabidopsis* genes. Separately plotting the levels of CHG for

intron and exon regions reveals that much of this gene-body CHG methylation is derived from introns rather than exons (Figure S3). A recent study noted that a small number of *Arabidopsis* genes contained introns with elevated levels of CHG methylation, often due to the presence of transposons within these introns [41]. We found that 4156 of the 39656 FGS maize genes contain transposons >1,000 bp inserted within introns (examples in Figure 4A and Figure S4). If these transposons within maize genes are masked then we find that the level of CHG and H3K9me2 methylation within maize genes drops substantially (Figure S5A–D).

The transposons insertions within genes were further characterized to understand whether the chromatin of these transposons differed from the chromatin at non-genic insertion (Figure 4B–C). The level of DNA methylation or H3K9me2 for these transposons inserted within genes is similar to transposons located outside of genes and is much higher than the levels observed in exons (Figure 4B–C). The profile of DNA methylation and H3K9me2 for these transposons inserted within genes reveals very stark boundaries between the transposon and the flanking sequence (Figure 4D) providing evidence for precise targeting of these modifications and lack of spreading for the chromatin modifications to flanking exon or intron sequences. The presence of large transposons that are marked by CHG and H3K9me2 within maize FGS genes may pose a problem for gene expression. The relative expression level of each gene was assessed and all expressed genes were assigned to quartiles. The presence of transposons within genes did not result in more examples of genes without expression and was not associated higher or lower expressed genes (Figure 5A). The majority of long TE insertions within genes are class I LTR elements (Figure S5E). A comparison of the frequency for each class of element in the whole genome to the frequency of each class within genes reveals enrichment for LINE elements within genes (Figure S5E).

Genes with elevated CHG or H3K9me2 near TSS

The average levels of CHG and H3K9me2 are quite low near the transcription start site of maize genes (Figure 3A–B). In *Arabidopsis*, the presence of CHG methylation and H3K9me2 over promoter regions is associated with transcriptional silencing [42–43]. Although the average profile shows very low levels of these marks near the TSS there are some genes that exhibit enrichment for CHG methylation and/or H3K9me2 in the 100 bp tile that overlaps the TSS. There are 459 maize FGS genes that have high H3K9me2 (>1 standard deviation above genome-wide average) and 546 FGS genes that had high CHG methylation (>88.5%) in the 100 bp tile that overlaps the TSS (Table S1). Although there is a relatively small overlap in these two sets of genes (Figure 5B) there was evidence that both marks tended to be enriched at most of these genes (Figure 5C). In many cases only one mark met the stringent criteria for discovery but the other mark was also elevated (Figure 5A). In contrast to genes containing transposons, the relative expression of genes with high CHG or H3K9me2 appeared to be strongly depressed relative to all genes (Figure 5A). The expression patterns for these genes were investigated in the developmental atlas representing 60 different tissues or organs of B73 [44]. Many of these genes (45% of high CHG TSSs and 53% of high H3K9me2 TSSs) were not detected in any of the 60 tissues with RNA-seq data. Only 33 of the high CHG TSS genes and 22 of the high H3K9me2 TSS genes exhibit expression levels over 10 FPKM (Fragments Per Kilobase per Million). A small subset of these genes exhibit tissue-specific expression (Figure S6, S7). The genes with elevated CHG or H3K9me2 at the TSS often did not have homology to genes in

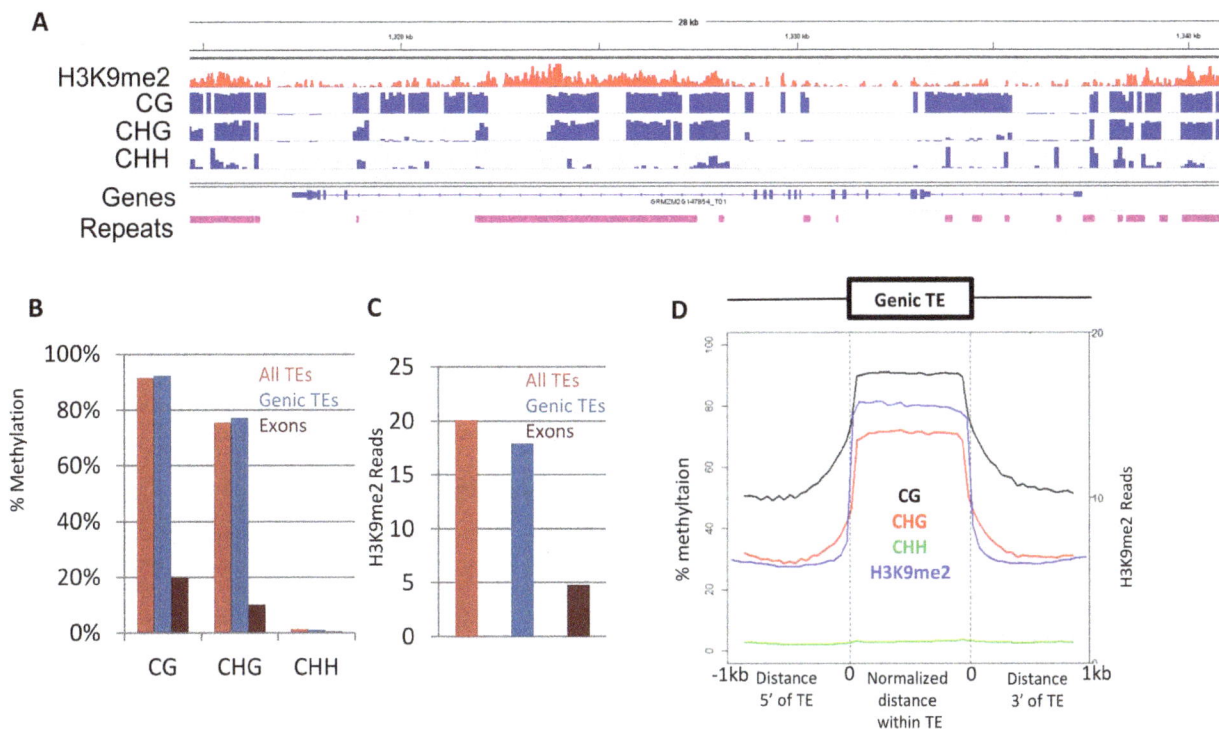

Figure 4. DNA methylation and H3K9me2 levels of genic transposons. (A) An example of a transposable element insertion located within an intron of a maize gene is shown using Integrated Genome Viewer [48]. The H3K9me2 ChIP-seq read count is shown in red. The levels of CG, CHG and CHH methylation (per 100 bp tile) are shown in blue. (B) The average CG, CHG and CHH DNA methylation levels are shown for all TEs (orange), genic TEs (blue) and exons (maroon). (C) The average H3K9me2 read counts are shown for the same regions. (D) The profile of DNA methylation and H3K9me2 over genic TEs is shown.

other grass species (Table S1). Many of these sequences may represent mis-annotated sequences that are not functional genes.

Discussion

Although the genome-wide patterns of DNA methylation and H3K9me2 in maize and *Arabidopsis* are generally similar there are several interesting differences. One of the most notable differences in the pattern of DNA methylation is observed in regions surrounding maize and *Arabidopsis* genes. While *Arabidopsis* genes are generally flanked by regions with low levels of DNA methylation and H3K9me2 maize genes are flanked by regions with elevated levels for these marks. This is likely due to the interspersed organization of genes and transposons within the maize genome [12], [14]. The majority of transposons in the *Arabidopsis* genome are found within percentromeric regions of knob-like heterochromatin structures. In maize, transposons are found throughout the chromosome, interspersed with genes.

The presence of CHG methylation and H3K9me2 within maize gene bodies is somewhat unexpected. One source of CHG methylation and H3K9me2 within maize gene bodies is the presence of transposons within introns. Although long introns containing heterochromatic sequence are common in animal genomes they are relatively rare in *Arabidopsis*. Only ~130 *Arabidopsis* genes contain long introns with elevated levels of CHG methylation [41]. However, these are much more common within the rice genome [41]. Here we show that these long introns, decorated with CHG methylation and H3K9me2, are also present in thousands of maize genes. These introns generally contain transposon insertions. The transposons insertion located within

introns contain levels of CG, CHG and H3K9me2 methylation similar to that observed for transposons located elsewhere in the genome and the levels of these modifications are not influenced by the expression level of the gene itself. The genes containing these insertions show a full range of expression levels similar to that observed for other maize genes suggesting that the presence of a region containing CHG and H3K9me2 does not pose substantial barrier to transcriptional elongation. There is evidence that allelic variation for the insertion of a heavily methylated retrotransposon does not result in differences in transcript abundance at the *Zmet2* locus [45]. In *Arabidopsis*, the *IBM2/ASI1* gene is required for the ability to properly transcribe through introns containing CHG methylation [20], [41]. Orthologs of this gene exist in maize and likely are required for active transcription through introns containing CHG and H3K9me2 methylation.

Maize has lower average levels of CHH methylation than *Arabidopsis*. The CHH methylation that is observed in *Arabidopsis* is found at many different transposons. In *Arabidopsis*, CHH methylation can be due to RdDM-targeting of the DRM enzymes [1] or by the CMT2 gene which seems to be targeted to regions containing H3K9me2 [10] and requires DDM1 [5]. While maize does contain DRM genes [46] and chromomethylases [47], there is no evidence for orthologs of CMT2 in maize [5]. The lack of a CMT2 in maize could explain the lack of elevated CHH methylation levels within retrotransposons that are heavily silenced by H3K9me2. The analysis of H3K9me2 and CHH levels in maize TE families reveals that only one of these two marks is usually enriched in each family (Figure 2D–E). Since a large portion of the maize genome is derived from retrotransposons and these sequences tend to have very low levels of CHH methylation

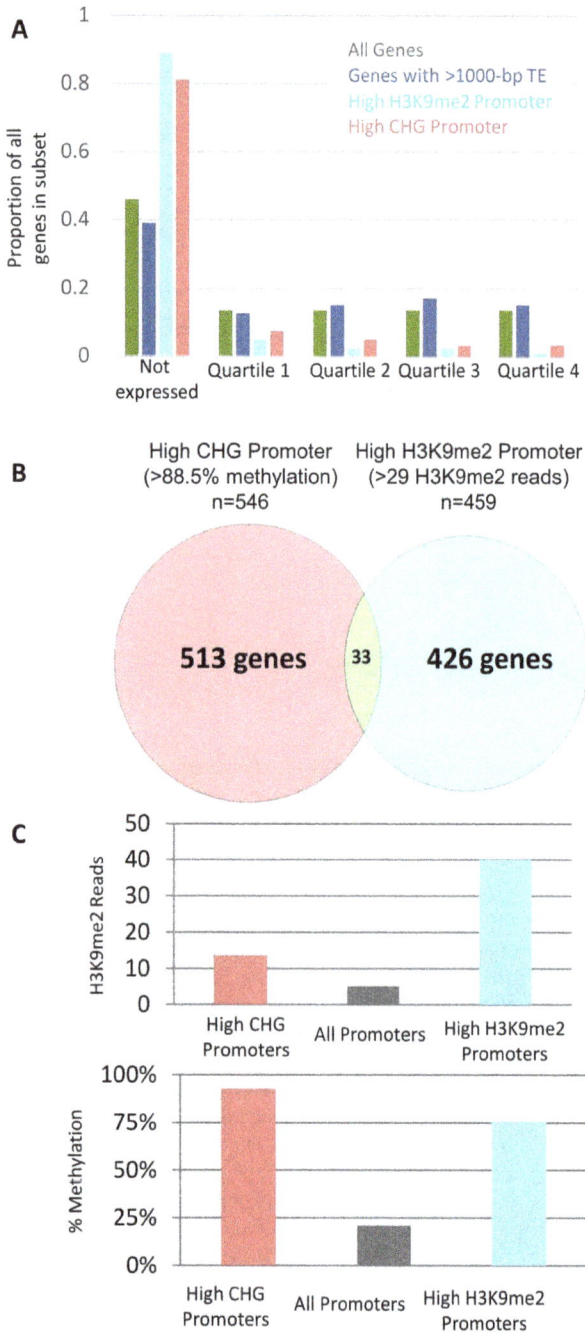

Figure 5. Effect of high CHG methylation or H3K9me2 over promoter on expression. (A) All maize genes were grouped into 5 groups: a subset of genes not expressed and then of the expressed genes, four quartiles of equal size increasing in expression level from 1 to 4 (gray). The relative proportion of genes containing TE insertions (dark blue), genes with high H3K9me2 (light blue) or high CHG in promoters (pink) in these five groups is shown. (B) Genes containing high levels of CHG or H3K9me2 over the transcription start site were identified and the overlap is shown (full list of genes in each category is available in Table S1). (C) Relative level of CHG methylation or H3K9me2 in each subset of promoter regions.

in maize the genome-wide levels of CHH methylation are quite low in maize. The observation that only certain TIR families contain high levels of CHH methylation in maize is intriguing.

These TIR families (DTM and DTC) are enriched for being located near genes (average distance to nearest gene is under 3 kb) compared to the other TIR families (average distance to nearest gen is over 6 kb) which may indicate a preference for insertion in euchromatin. This may allow these elements to by targeted by the RdDM pathway while the other TIR families with insertions in non-genic regions would not be accessible for this pathway and would be silenced by H3K9me2.

Arabidopsis provides an excellent model system for studying the mechanisms that control the distribution of chromatin modifications in plant genomes. However, the relatively simple genome organization in *Arabidopsis* is not common in many plant species. Most plant species, including many crops, contain genomes with more complex organizations and the analysis of the epigenome in these plant species is likely to reveal important differences in the distribution of chromatin modifications.

Supporting Information

Figure S1 Distribution of DNA methylation and H3K9me2 levels. (A) The accession numbers for each of the datasets used in this study is listed. The proportion of 100 bp tiles with varying levels of CG (B), CHG (C) or CHH (D) is shown for all regions, genic regions and TE regions. (E) The distribution of read counts (per 100 bp tile) is shown for all tiles, genic tiles and TE tiles.

Figure S2 Relative distance line plots across sub-families of maize transposable elements. *Zea mays* transposable elements were split by their different sub-families and aligned at their 3′ and 5′ ends. Distance across the transposable elements was normalized to a scale of 1 to 1000. Average percent methylation and H3K9me2 reads at each distance are displayed.

Figure S3 DNA methylation profiles of different types of maize genes. (A) The relative levels of DNA methylation in each context or H3K9me2 ChIP-seq read counts are plotted for all maize genes (black), the filtered gene set (FGS-red) and the working gene set (WGS-green). (B) Maize genes were split by their classification as either syntenic or inserted [35] and aligned at their 3′ and 5′ ends. The average DNA methylation within either exons or introns is shown. (C) The genes were classified as either sub-genome1 or sub-genome 2 [38], aligned at their 3′ and 5′ ends and methylation levels in each context are plotted.

Figure S4 Additional examples of transposable elements located within genes. Genic transposable elements as viewed in Integrated Genomics Viewer (IGV) [48]. H3K9me2 reads are displayed in red; transposable elements in pink; CG, CHG, and CHH methylation are represented as percent methylation across 100 bp tiles.

Figure S5 Absolute distance line plots of DNA methylation and H3K9me2 over genic space in maize and *Arabidopsis*. Maize and *Arabidopsis* genes were aligned at the 5′ and 3′ ends and CG (A), CHG (B) and CHH (C) DNA methylation levels or H3K9me2 read counts (D) are plotted. The vertical dashed lines represent the 5′ and 3′ ends. The regions within genes are classified as introns (red), exons (black) or introns with TEs masked (blue). (E) The proportion of TEs (>1,000 bp) located within maize genes that are annotated as TIR, LINE or LTR elements is shown compared to the proportion of all TEs in the maize genome in each of these three classes.

Figure S6 Clustering of expression levels for genes with high CHG methylation in promoter regions. Many of these genes show very low levels of expression. There are ~80 of these genes with low levels of expression (1-5FPKM) in a large number of tissues. There are only 4 genes with high expression levels (at least 100FPKM). Two of these genes show anther specific expression and the other two exhibit expression in specific leaf tissues.

Figure S7 Clustering of expression levels for genes with high H3K9me2 in promoter regions. The majority of these genes show very low levels of expression. There are 40 of these genes with low levels of expression (1-5FPKM) in a large number of tissues. There are about 10 genes that show high levels of expression (> 100FPKM) in at least one tissue. Four of these genes show anther specific expression and four show endosperm specific expression while the last two have leaf-specific expression.

Table S1 Genes with High H3K9me2 and/or high CHG methylation over transcription start site.

Acknowledgments

The Texas Advanced Computing Center (TACC) at The University of Texas at Austin provided HPC and storage resources. The Minnesota Supercomputing Institute provided access to software and user support for data analyses.

Author Contributions

Conceived and designed the experiments: QL RJS NMS. Performed the experiments: PTW QL SRE. Analyzed the data: PTW QL LJ SRE JS MWV. Contributed reagents/materials/analysis tools: LJ JS MWV. Contributed to the writing of the manuscript: PTW QL RJS NMS.

References

1. Law JA, Jacobsen SE. (2010) Establishing, maintaining and modifying DNA methylation patterns in plants and animals. Nat Rev Genet 11(3): 204–220.
2. Cokus SJ, Feng S, Zhang X, Chen Z, Merriman B, et al. (2008) Shotgun bisulphite sequencing of the arabidopsis genome reveals DNA methylation patterning. Nature 452(7184): 215–219.
3. Lister R, O'Malley RC, Tonti-Filippini J, Gregory BD, Berry CC, et al. (2008) Highly integrated single-base resolution maps of the epigenome in arabidopsis. Cell 133(3): 523–536.
4. Vongs A, Kakutani T, Martienssen RA, Richards EJ. (1993) Arabidopsis thaliana DNA methylation mutants. Science 260(5116): 1926–1928.
5. Zemach A, Kim MY, Hsieh PH, Coleman-Derr D, Eshed-Williams L, et al. (2013) The arabidopsis nucleosome remodeler DDM1 allows DNA methyltransferases to access H1-containing heterochromatin. Cell 153(1): 193–205.
6. Du J, Zhong X, Bernatavichute YV, Stroud H, Feng S, et al. (2012) Dual binding of chromomethylase domains to H3K9me2-containing nucleosomes directs DNA methylation in plants. Cell 151(1): 167–180.
7. Bartee L, Malagnac F, Bender J. (2001) Arabidopsis cmt3 chromomethylase mutations block non-CG methylation and silencing of an endogenous gene. Genes Dev 15(14): 1753–1758.
8. Jackson JP, Lindroth AM, Cao X, Jacobsen SE. (2002) Control of CpNpG DNA methylation by the KRYPTONITE histone H3 methyltransferase. Nature 416(6880): 556–560.
9. Stroud H, Greenberg MV, Feng S, Bernatavichute YV, Jacobsen SE. (2013) Comprehensive analysis of silencing mutants reveals complex regulation of the arabidopsis methylome. Cell 152(1–2): 352–364.
10. Stroud H, Do T, Du J, Zhong X, Feng S, et al. (2014) Non-CG methylation patterns shape the epigenetic landscape in arabidopsis. Nat Struct Mol Biol 21(1): 64–72.
11. Lippman Z, Gendrel AV, Black M, Vaughn MW, Dedhia N, et al. (2004) Role of transposable elements in heterochromatin and epigenetic control. Nature 430(6998): 471–476.
12. SanMiguel P, Tikhonov A, Jin YK, Motchoulskaia N, Zakharov D, et al. (1996) Nested retrotransposons in the intergenic regions of the maize genome. Science 274(5288): 765–768.
13. Messing J, Dooner H. (2006) Organization and variability of the maize genome. Current Opinion in Plant Biology 9(2): 157–163.
14. Schnable PS, Ware D, Fulton RS, Stein JC, Wei F, et al. (2009) The B73 maize genome: Complexity, diversity, and dynamics. Science 326(5956): 1112–1115.
15. Bennetzen JL, Schrick K, Springer PS, Brown WE, SanMiguel P. (1994) Active maize genes are unmodified and flanked by diverse classes of modified, highly repetitive DNA. Genome 37(4): 565–576.
16. Yuan Y, SanMiguel PJ, Bennetzen JL. (2002) Methylation-spanning linker libraries link gene-rich regions and identify epigenetic boundaries in zea mays. Genome Res 12(9): 1345–1349.
17. Rabinowicz PD, McCombie WR, Martienssen RA. (2003) Gene enrichment in plant genomic shotgun libraries. Curr Opin Plant Biol 6(2): 150–156.
18. Palmer LE, Rabinowicz PD, O'Shaughnessy AL, Balija VS, Nascimento LU, et al. (2003) Maize genome sequencing by methylation filtration. Science 302(5653): 2115–2117.
19. Emberton J, Ma J, Yuan Y, SanMiguel P, Bennetzen JL. (2005) Gene enrichment in maize with hypomethylated partial restriction (HMPR) libraries. Genome Res 15(10): 1441–1446.
20. Wang X, Elling AA, Li X, Li N, Peng Z, et al. (2009) Genome-wide and organ-specific landscapes of epigenetic modifications and their relationships to mRNA and small RNA transcriptomes in maize. Plant Cell 21(4): 1053–1069.
21. Eichten SR, Swanson-Wagner RA, Schnable JC, Waters AJ, Hermanson PJ, et al. (2011) Heritable epigenetic variation among maize inbreds. PLoS Genet 7(11): e1002372.
22. Gent JI, Dong Y, Jiang J, Dawe RK. (2012) Strong epigenetic similarity between maize centromeric and pericentromeric regions at the level of small RNAs, DNA methylation and H3 chromatin modifications. Nucleic Acids Res 40(4): 1550–1560.
23. Regulski M, Lu Z, Kendall J, Donoghue MT, Reinders J, et al. (2013) The maize methylome influences mRNA splice sites and reveals widespread paramutation-like switches guided by small RNA. Genome Res 23(10): 1651–1662.
24. Eichten SR, Briskine R, Song J, Li Q, Swanson-Wagner R, et al. (2013) Epigenetic and genetic influences on DNA methylation variation in maize populations. Plant Cell 25(8): 2783–2797.
25. Shi J, Dawe RK. (2006) Partitioning of the maize epigenome by the number of methyl groups on histone H3 lysines 9 and 27. Genetics 173(3): 1571–1583.
26. Jin W, Lamb JC, Zhang W, Kolano B, Birchler JA, et al. (2008) Histone modifications associated with both A and B chromosomes of maize. Chromosome Res 16(8): 1203–1214.
27. Eichten SR, Ellis NA, Makarevitch I, Yeh CT, Gent JI, et al. (2012) Spreading of heterochromatin is limited to specific families of maize retrotransposons. PLoS Genet 8(12): e1003127.
28. Haring M, Offermann S, Danker T, Horst I, Peterhansel C, et al. (2007) Chromatin immunoprecipitation: Optimization, quantitative analysis and data normalization. Plant Methods 3: 11.
29. Krueger F, Andrews SR. (2011) Bismark: A flexible aligner and methylation caller for bisulfite-seq applications. Bioinformatics 27(11): 1571–1572.
30. Schultz MD, Schmitz RJ, Ecker JR. (2012) 'Leveling' the playing field for analyses of single-base resolution DNA methylomes. Trends Genet 28(12): 583–585.
31. Li H, Handsaker B, Wysoker A, Fennell T, Ruan J, et al. (2009) The sequence Alignment/Map format and SAMtools. Bioinformatics 25(16): 2078–2079.
32. Quinlan AR, Hall IM. (2010) BEDTools: A flexible suite of utilities for comparing genomic features. Bioinformatics 26(6): 841–842.
33. Trapnell C, Williams BA, Pertea G, Mortazavi A, Kwan G, et al. (2010) Transcript assembly and quantification by RNA-seq reveals unannotated transcripts and isoform switching during cell differentiation. Nat Biotechnol 28(5): 511–515.
34. Baucom RS, Estill JC, Chaparro C, Upshaw N, Jogi A, et al. (2009) Exceptional diversity, non-random distribution, and rapid evolution of retroelements in the B73 maize genome. PLoS Genet 5(11): e1000732.
35. Whitelaw CA, Barbazuk WB, Pertea G, Chan AP, Cheung F, et al. (2003) Enrichment of gene-coding sequences in maize by genome filtration. Science 302(5653): 2118–2120.
36. Law JA, Du J, Hale CJ, Feng S, Krajewski K, et al. (2013) Polymerase IV occupancy at RNA-directed DNA methylation sites requires SHH1. Nature 498(7454): 385–389. 10.1038/nature12178 [doi].
37. Schmitz RJ, Schultz MD, Lewsey MG, O'Malley RC, Urich MA, et al. (2011) Transgenerational epigenetic instability is a source of novel methylation variants. Science 334(6054): 369–373.
38. Schnable JC, Freeling M, Lyons E. (2012) Genome-wide analysis of syntenic gene deletion in the grasses. Genome Biol Evol 4(3): 265–277.
39. Schmitz RJ, He Y, Valdes-Lopez O, Khan SM, Joshi T, et al. (2013) Epigenome-wide inheritance of cytosine methylation variants in a recombinant inbred population. Genome Res 23(10): 1663–1674.
40. Parkin IA, Koh C, Tang H, Robinson SJ, Kagale S, et al. (2014) Transcriptome and methylome profiling reveals relics of genome dominance in the mesopolyploid brassica oleracea. Genome Biol 15(6): R77.
41. Saze H, Kitayama J, Takashima K, Miura S, Harukawa Y, et al. (2013) Mechanism for full-length RNA processing of arabidopsis genes containing intragenic heterochromatin. Nat Commun 4: 2301.

42. Zhang X, Yazaki J, Sundaresan A, Cokus S, Chan SW, et al. (2006) Genome-wide high-resolution mapping and functional analysis of DNA methylation in arabidopsis. Cell 126(6): 1189–1201.

43. Zhou J, Wang X, He K, Charron JB, Elling AA, et al. (2010) Genome-wide profiling of histone H3 lysine 9 acetylation and dimethylation in arabidopsis reveals correlation between multiple histone marks and gene expression. Plant Mol Biol 72(6): 585–595.

44. Sekhon RS, Briskine R, Hirsch CN, Myers CL, Springer NM, et al. (2013) Maize gene atlas developed by RNA sequencing and comparative evaluation of transcriptomes based on RNA sequencing and microarrays. PLoS One 8(4): e61005.

45. Springer NM, Eichten S, Smith A, Papa CM, Steinway S, et al. (2009) Characterization of a novel maize retrotransposon family SPRITE that shows high levels of variability among maize inbred lines. Maydica 54: 417–428.

46. Cao X, Springer N, Muszynski M, Phillips R, Kaeppler S, et al. (2000) Conserved plant genes with similarity to mammalian de novo DNA methyltransferases. PNAS 97(9): 4979–4984.

47. Papa CM, Springer NM, Muszynski MG, Meeley R, Kaeppler SM. (2001) Maize chromomethylase zea methyltransferase2 is required for CpNpG methylation. Plant Cell 13(8): 1919–1928.

48. Robinson JT, Thorvaldsdottir H, Winckler W, Guttman M, Lander ES, et al. (2011) Integrative genomics viewer. Nat Biotechnol 29(1): 24–26.

16

Chloroplast Genome Differences between Asian and American *Equisetum arvense* (Equisetaceae) and the Origin of the Hypervariable *trnY-trnE* Intergenic Spacer

Hyoung Tae Kim, Ki-Joong Kim*

Division of Life Sciences, School of Life Sciences and Biotechnology, Korea University, Seoul, Korea

Abstract

Comparative analyses of complete chloroplast (cp) DNA sequences within a species may provide clues to understand the population dynamics and colonization histories of plant species. *Equisetum arvense* (Equisetaceae) is a widely distributed fern species in northeastern Asia, Europe, and North America. The complete cp DNA sequences from Asian and American *E. arvense* individuals were compared in this study. The Asian *E. arvense* cp genome was 583 bp shorter than that of the American *E. arvense*. In total, 159 indels were observed between two individuals, most of which were concentrated on the hypervariable *trnY-trnE* intergenic spacer (IGS) in the large single-copy (LSC) region of the cp genome. This IGS region held a series of 19 bp repeating units. The numbers of the 19 bp repeat unit were responsible for 78% of the total length difference between the two cp genomes. Furthermore, only other closely related species of *Equisetum* also show the hypervariable nature of the *trnY-trnE* IGS. By contrast, only a single indel was observed in the gene coding regions: the *ycf1* gene showed 24 bp differences between the two continental individuals due to a single tandem-repeat indel. A total of 165 single-nucleotide polymorphisms (SNPs) were recorded between the two cp genomes. Of these, 52 SNPs (31.5%) were distributed in coding regions, 13 SNPs (7.9%) were in introns, and 100 SNPs (60.6%) were in intergenic spacers (IGS). The overall difference between the Asian and American *E. arvense* cp genomes was 0.12%. Despite the relatively high genetic diversity between Asian and American *E. arvense*, the two populations are recognized as a single species based on their high morphological similarity. This indicated that the two regional populations have been in morphological stasis.

Editor: Marcel Quint, Leibniz Institute of Plant Biochemistry, Germany

Funding: This research was supported by a research grant (#KRF-2010-0011796) from the Korea Research Foundation and by research grants (#062-091-078, #416-111-005) from KEITI to Ki-Joong Kim. A Ph.D. fellowship to HTK was granted by the graduate student education program from NIBR (2012–2013). The funders had no role in study design, data collection and analysis, decision to publish, or preparation of the manuscript.

Competing Interests: The authors have declared that no competing interests exist.

* Email: kimkj@korea.ac.kr

Introduction

Approximately 480 complete chloroplast (cp) genome sequences are currently publicly available (http://www.ncbi.nlm.nih.gov/genome), the majority of which are derived from economically important crop plants. Comparative analysis of chloroplast sequences indicate that genome structure, gene content, and gene order are largely stable in land plant lineages [1]. However, highly rearranged cp genome structures are observed in some land plant lineages, and can be used as molecular markers to elucidate the ancient divergence of specific groups [2–4]. Because of the generally conservative nature of cp genome structure, cp genome data are used most often to address phylogenetic and evolutionary questions at or above the species level. Nevertheless, base substitutions and small indels are seen frequently in cp genomes, even between closely related taxa. [3,5,6].

Cp genome comparative analyses were performed using sequences from seed plants in closely related taxa [7–9]. For example, 72 single-nucleotide polymorphisms (SNPs) and 27 indels were observed when two subspecies of rice (*Oryza sativa* spp. *indica* and *O. sativa* spp. *japonica*) were compared [7], and 32 SNPs were detected between the cp genomes of 17 *Jacobaea vulgaris* individuals [9]. The cp genomes of *Panicum virgatum*, which has different ecotypes in upland and lowland regions, contained 116 SNPs and 46 indels [10], and high variability in indel (3–278) and SNP (6–1000) numbers was noted in the cp genomes of 13 *Gossypium* species [11]. Finally, comparison of intraspecific variation in rare and widespread pines indicated low levels of divergence [8]. These data improved our understanding of cp genome evolution and divergence times of closely related taxa. Recent advances in rapid pyrosequencing techniques provide further opportunities to study population diversification and evolution using large numbers of whole cp genome sequences.

In monilophytes, a sister group to that of seed plants, no comparative cp genome analyses have been performed in intraspecific taxa despite the availability of seven complete cp genome sequences [12–15]. Previous monilophyte cp genome analyses focused on higher taxonomic levels and examined gene content, gene rearrangement, nucleotide substitution, and phylogenetic relationships.

Plant species distribution in different continents is a subject of ongoing interest to botanical researchers. The disjunctive distributions of similar flowering plant species in North America and East Asia have been extensively studied with respect to migration path, migration time, habitat similarity, and phylogenic relationship [16–23]. Single fern species are often distributed throughout

Table 1. The list of taxa used in this study.

Location	Subgenus	Species	GenBank accession #
America	Hippochaete	Equisetum hyemale	KC117177[a]
Asia(Korea)		Equisetum hyemale	KC610090[b]
Asia(China)		Equisetum ramosissimum	HQ658109[b]
America	Equisetum	Equisetum arvense	NC014699[a]
Asia(Korea)		Equisetum arvense	JN968380[a]
Asia(china)		Equisetum arvense	HQ658110[b]

[a]Complete cp genome.
[b]The trnY-trnE region sequence.
The GenBank accession numbers KC610090 and JN968380 were reported in this study and all other sequences are obtained from GenBank.

the two continents. This contrasts with flowering plant distribution, in which the same species rarely occurs in both continents [24,25]. One example is the *Adiantum pedatum* complex of leptosprangiate ferns: molecular data suggest that *A. pedatum* migrated from East Asia to North America through the Bering land-bridge and subsequently migrated from North America to East Asia [25]. Disjunctive distribution within a species is more easily observed in ferns than in flowering plants. Homosporous fern spores are easily dispersible and are able to live independently, facilitating the founding of a new population after migrations up to thousands of kilometers [26].

E. arvense is the most widely distributed fern globally, and is divided into two chemotypes. Plants in the European population do not contain flavonoids, but plants in the Asian and American populations contain luteolin 5-O-glucoside and malonyl ester [27]. However, morphological characteristics do not differ significantly between the three regional *E. arvense* populations and they are recognized as a single species.

With the exception of a common inversion in monilophytes, the American *E. arvense* cp genome structure in the large single-copy (LSC) region resembles the cp genome in moss and hornwort. The *trnY-trnE* intergenic spacer (IGS) region in the American *E. arvense* cp genome has a distinctive length of 5 kb, unlike in other monilophyte cp genomes. The lengths of *trnY-trnE* IGS usually less than I kb in other monilophytes [28]. This unusual *trnY-trnE* IGS length was also detected in *E. ramosissimum*, and this unique characteristic may be attributable to a repetition of the *trnY* anticodon loop [29].

In this study, we investigated the differences between the Asian and American *E. arvense* whole cp genomes. In addition, we analyzed repeat sequences in the *trnY-trnE* IGS region to determine the origin of the repeating unit that forms a hotspot in the cp genomes of genus *Equisetum*. Finally, we used the cp genome differences between Asian and American *E. arvense* to understand correlations between disjunctive distribution and cp genome divergence.

Materials and Methods

Chloroplast genome sequencing and annotation

E. arvense was collected from South Korea (H.-T. Kim, 2009-0413, voucher specimen in Korea University, Seoul (KUS), herbarium) and chloroplasts were separated from fresh leaves using a sucrose step-gradient method [30]. Cp DNAs (PDBK DNA No. 2009-0413) were isolated using 5× lysis buffer [31] and were sequenced using a Genome Sequencer FLX Titanium (Macrogen, Korea). Total FLX read numbers were 123,080, and

the average read length was 337.5 bp. Six large contigs covered 90% of the total cp genome sequence. The *E. arvense* cp genome sequencing strategy was as indicated in Figure S1. Two additional PCR methods were used to fill the 10% gap and low-coverage regions (10%). A long-PCR method was used to fill gaps >3 kb, with conditions as follows: initial denaturation at 94°C, 4 min; (94°C, 15 s; 53–65°C, 30 s; 68°C, 3–10 min)×35 cycles; and post-extension at 72°C, 7 min. A short-PCR method was used to fill gaps <3 kb, with conditions as follows: initial denaturation at 94°C, 4 min; (94°C, 15 s; 50°C, 30 s; 72°C, 2 min)×35 cycles; and post-extension at 72°C, 3 min. PCR products were purified using column-based kits (Qiagen QIAquick PCR purification kit, Hilden, Germany) and sequenced using Big-Dye chemistry (Applied Biosystems, Foster City, CA, USA) and an ABI 3730XL sequencer.

The cp genome sequence of Asian *E. arvense* was assembled using Sequencher 4.7. Genes were annotated using DOGMA [32], and gene locations were determined by NCBI BLAST search (http://blast.ncbi.nlm.nih.gov/Blast.cgi). Secondary tRNA structures and location of tRNA genes were predicted using tRNAscan-SE 1.21 [33] Additional bioinformatic analyses were similar to those described in Kim and Lee [34], Lee et al. [3], and Yi and Kim [35].

Comparison of complete cp genome sequences from Asian and American *E. arvense*

Indels and SNPs between the Asian and American *E. arvense* (NC_014699) cp genomes were detected as follows. The sequences of the two *E. arvense* cp genomes were divided into three functional segments (gene coding regions, introns, and intergenic spacers) and the partitioned sequences were aligned using the MUSCLE program [36]. Positions of indels and SNPs were determined using Geneious 5.6.5 [37]. Indels were divided into two types, A and B. Insertion was observed in the Asian *E. arvense* cp genome sequence for type A indels and in the American *E. arvense* cp genome for type B indels. Insertion event orientation was determined using *E. hyemale* and *Psilotum nudum* cp sequences as references for the outgroup sequence comparison method. A tandem-repeat finder [38] was used to classify indels according to location in 1) mononucleotide repeat regions, 2) tandem-repeat regions, and 3) dispersed repeat regions. The folding structure of the *rrn16* gene was predicted using the Mfold Web Server [39]. The *rrn16* gene sequences from three eusporangiate ferns and *Osmunda cinnamomea* were obtained from GenBank (KF 225592, NC 008829, NC 017006, and NC 003386) and were used for folding structure comparisons.

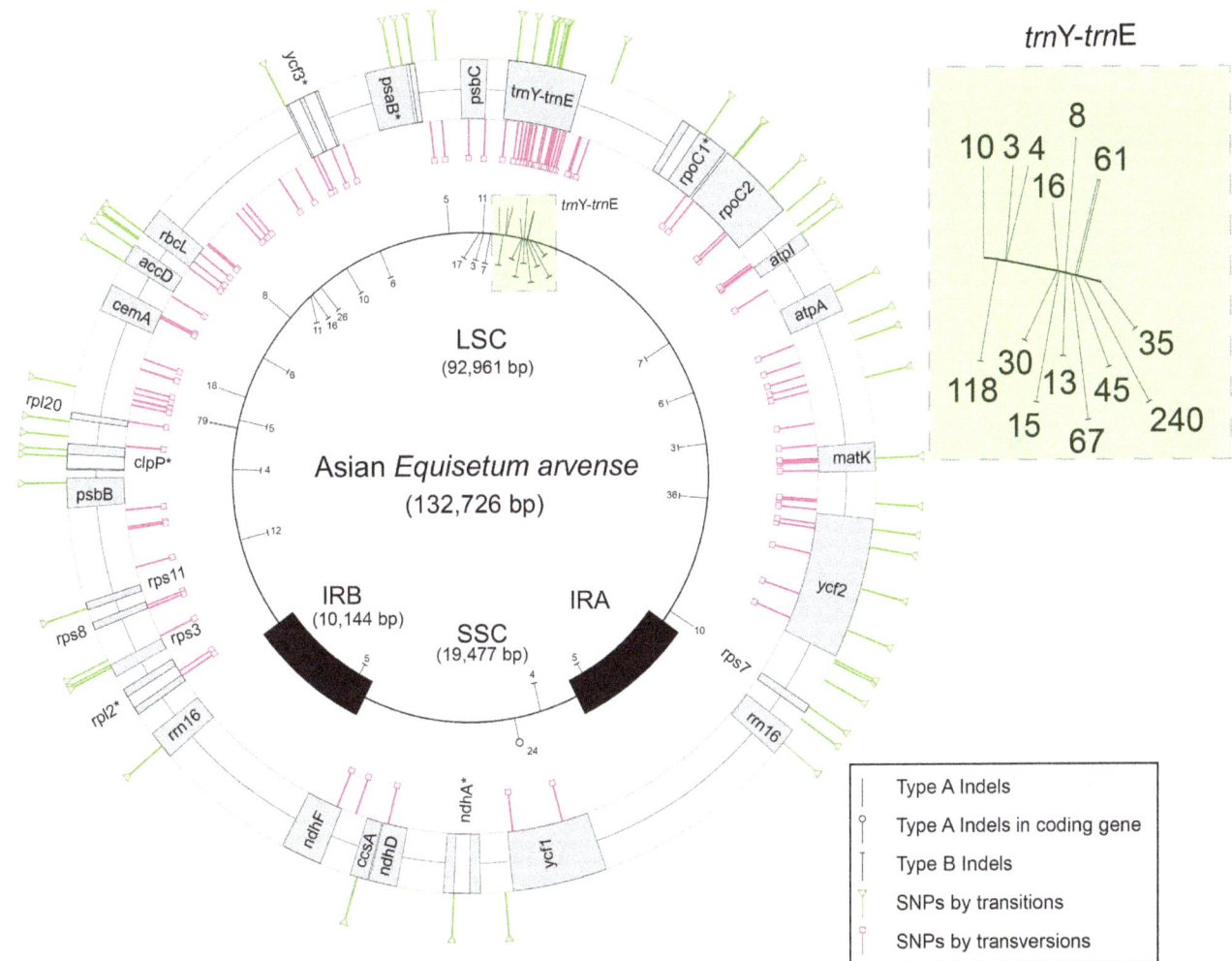

Figure 1. Distribution map of indels and SNPs on the cp genomes of _Equisetum arvense_. This map was generated by comparing Asian and American _E. arvense_ individuals. Indels ≥3 bp in length are indicated on the inner circle and SNPs are indicated on the outer circle. Genes with SNP(s) are listed on the outer circle. The numeric values on the inner circle indicate the indel length at each location. A gene name annotated with an asterisk indicates an intron-containing gene. Detailed locations of SNPs on relevant gene(s) and IGS region(s) are marked on the outer circle. The green box at the right upper corner shows indel locations and indel length on the _trnY-trnE_ IGS region.

Analysis of _trnY-trnE_ IGS sequences between _Equisetum_ species

Six sequences representing the two _Equisetum_ subgenera [40,41] were used to analyze differences between species in the _trnY-trnE_ IGS region. Four sequences were obtained from GenBank and two were generated in this study (Table 1). _E. hyemale_ was collected in South Korea (H.-W. Kim 2007-0543, voucher specimen in KUS herbarium) and genomic DNA (PDBK DNA No. 2007-0543, voucher specimen in KUS herbarium) was extracted from fresh leaves using the CTAB method [42]. DNA was purified using cesium chloride/ethidium bromide gradients. Primers were designed to amplify and sequence the _trnY-trnE_ IGS region of _E. hyemale_ (Forward primer: CAAAGCCAGCGGATT-TACAA, Reverse primer: CCCCATCGTCTAGTGGCCTA) using the cp genome sequence of _E. arvense_ as a reference sequence. The long-PCR method was used to amplify the region. Sequences were assembled with Sequencher 4.7 and the locations of the _trnY_ and _trnE_ genes were confirmed using DOGMA. The six _trnY-trnE_ IGS sequences were aligned using the MUSCLE alignment program [36].

Repeat sequences in _trnY-trnE_ IGS were analyzed using REPuter [43] with a 10 bp minimum length of repeat sequence and a Hamming value of 3 bp. Repeat sequence frequencies were detected using DNA Pattern Search (http://www.geneinfinity. org/sms/sms_DNApatterns.html#). The formation of repeat sequence hairpin structures was confirmed using the Mfold Web Server [39]. Consensus repeat sequences were numbered by position and sequence frequencies at each position were also calculated. Dot-matrix analysis (Serolis dot-plot software version 0.9.9, available from http://www.code10.info) was used to assess the distribution patterns of repeated units and the conservation levels of each repeat unit in the _trnY-trnE_ IGS region.

Results

Length variation caused by insertions/deletions in two _E. arvense_ cp genomes

The complete cp genome sequence of Asian _E. arvense_ was 132,726 bp in length, comprising a 92,961 bp LSC region, a 19,477 bp small single-copy (SSC) region, and two 10,144 bp

Plant Biotechnology: Genetic Modification of Plants

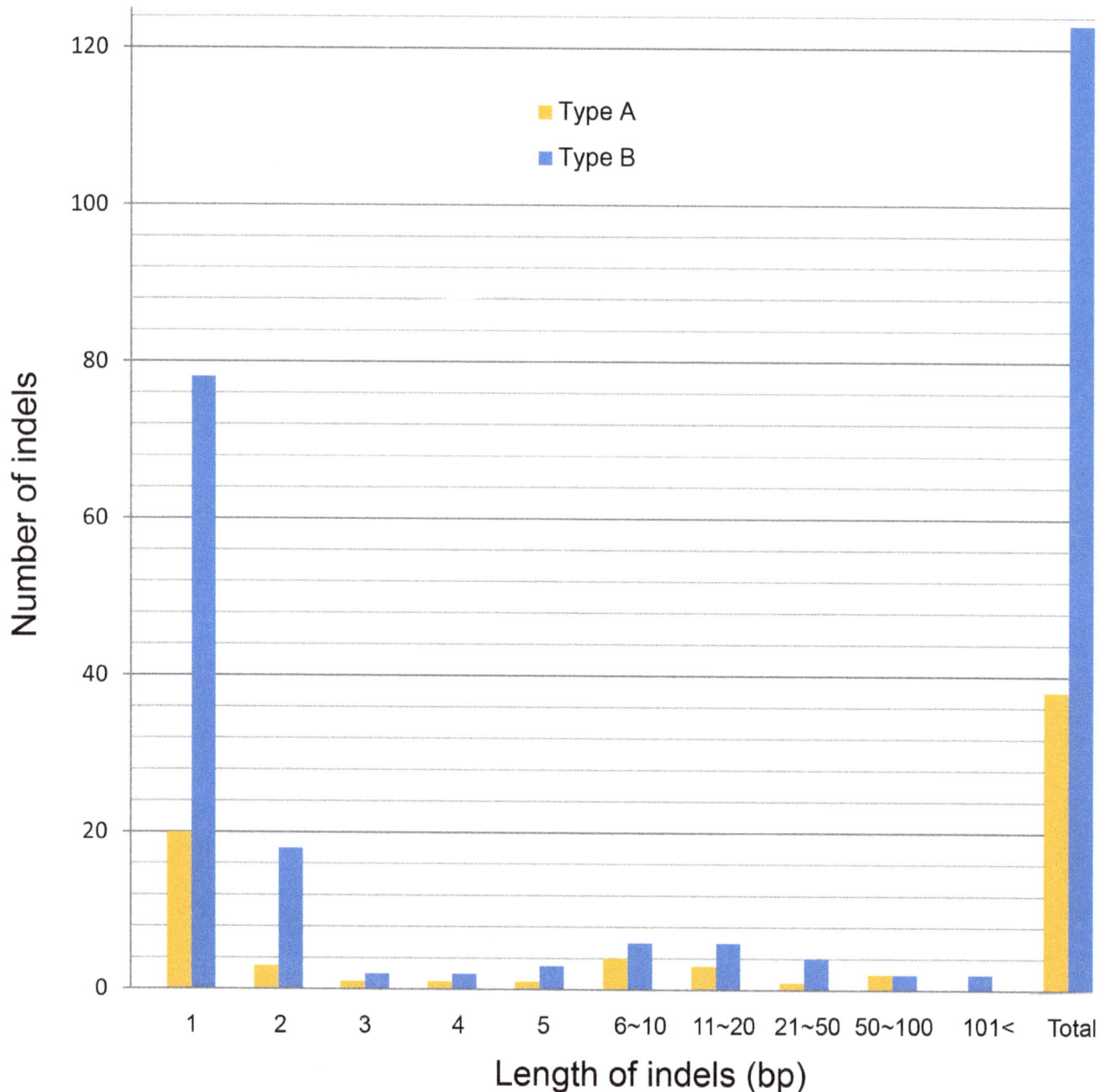

Figure 2. Bar graph showing the number (Y axis) and length (X axis) of indels. Type B indels were predominant over type A indels (see text).

inverted repeats (IRs) (Fig. 1). Comparison of the Asian and American *E. arvense* cp genomes indicated that the Asian *E. arvense* cp genome LSC region was 581 bp shorter, the IR region 5 bp shorter, and the SSC region 8 bp longer than in the American *E. arvense* cp genome (Table 2). Most of the 159 indels (69.2%) occurred in homopolymer regions, and most were type B indels. The remaining indels occurred in tandem-repeat regions and were more than 2 bp in length (15.1%) or were in non-repeat regions (15.7%). Regionally, 142 indels (89.3%) were detected in the LSC region, 13 (8.2%) were detected in the SSC region, and four (2.5%) were detected in the IR regions. The indel number per unit length ratio was 7.7:3.4:1 (LSC:SSC:IR). All indels were found in noncoding regions with the exception of an indel affecting

the *ycf1* gene in the SSC region. An insertion of a 24 bp (ATCAATGCTAGATGTTTCAAAAGT) tandem-repeat unit was observed in the *ycf1* gene in Asian *E. arvense*. Most indels (75%) were 1–2 bp in length; these accounted for 15% of the difference in length between the Asian and American *E. arvense* cp genomes (Fig. 2). Indels ≥3 bp long accounted for 85% of the length difference between the two genomes. The ≥3 bp indels (13 type A and 27 type B) were found throughout the cp genome. A high proportion (35.9%) were concentrated in the *trnY-trnE* IGS region (Fig. 1), with the remainder evenly distributed across the rest of the cp genome.

Table 2. Indel events on the cp genomes of Asian and American *E. arvense*.

	LSC		SSC		IR(×2)		Total	
	Type A	Type B	Type A	Type B	Type A	Type B	Type A	Type B
Indels on homopolymer regions	17	80	1	10	0	1	18	92
Indels on tandem repeat regions	6	15	1	0	0	1	7	17
Indels on other regions	11	13	0	1	0	0	11	14
total	34	108	2	11	0	4	36	123

Single-Nucleotide Polymorphisms in two *E. arvense* cp genomes

In total, 165 SNPs were detected between the Asian and American *E. arvense* genomes (Fig. 1). Of these, 155 (93.9%) were found in the LSC region, eight (4.8%) were located in the SSC region, and two (1.2%) were found in the IR regions (Table 3). The SNP number per unit length ratio was 16.8:4.2:1 (LSC:SSC:IR). Most SNPs (100; 60.6%) were found in IGS, 50 SNPs (30.5%) were located in protein-coding genes, 13 SNPs (7.9%) were in introns, and two SNPs (1.2%) were found in an rRNA gene. Thirty-five SNPs were concentrated in the *trnY-trnE* IGS region. The SNP in the IR region was detected in the *rrn16* gene, and was a unique base substitution reported only in Asian *E. arvense* (Fig. 3). The putative folding structures of *rrn16* due to the SNP are compared in the Figure S2. Of the 50 SNPs found in protein-coding genes, 23 were transitional changes (Ts) and 27 were transversional changes (Tv) (Table 4). Twenty Ts and ten Tv changes were synonymous substitutions (Ks), and three Ts and seven Tv changes were nonsynonymous (Kn). SNPs were detected in 26 of the 84 protein-coding genes. The highest K2P distance was observed for the *matK* gene, which had five Kn changes. The most substitutions were found in the *ycf2* gene, with five Ks and four Kn changes. The Ts/Tv ratio for the whole cp genome was 0.53, while the Ts/Tv ratio for the coding region alone was 0.85. The Ts/Tv ratio in coding regions (0.85) was two-fold higher than the ratio in IGS regions (0.39) and introns (0.41).

Hairpin structures in the *trnY-trnE* IGS in genus *Equisetum*

The *trnY-trnE* IGS was 4,534 bp and 4,991 bp in length in Asian and American *E. arvense*, respectively. This region comprised only 3.4% of the genome but was responsible for 78% of the total variation in length between the two cp genomes. Several secondary-structure-forming repeats were located in the *trnY-trnE* IGS region and had a 19 bp basic unit.

The number of repeats differed between the genomes, with 66 in Asian *E. arvense* and 73 in American *E. arvense*. Most of the repeating units were in one of two different sequence forms, A and B. The A-form consisted of a 7 bp stem region (TATGGAT) and a 5 bp loop region (TTCTT). The same stem region (TATGGAT) was observed in the B-form, but a different loop region was present (TTTAA) (Fig. 4). Eighteen A-form and twenty-three B-form repeats were detected in Asian *E. arvense*, while twenty-eight A-form and twenty-eight B-form repeats were detected in American *E. arvnese*. Indels greater than 10 bp in length occurred mainly in the stem regions of these hairpin structures and contributed to the variation in length between the Asian and American *E. arvense* sequences. In addition, two indels were detected between Chinese and Korean individuals of Asian *E. arvense*, both of which were associated with the A-form repeating unit.

Other species from genus *Equisetum* varied in the length of their *trnY-trnE* IGS region. These species had similar 19 bp repeating units to those in *E. arvense*, but sequence units were slightly modified (Fig. 5). *E. hyemale* and *E. ramosissimum* had 71 and 63 of the 19 bp repeat units, respectively. These repeating units were highly conserved both in structure and sequence, with a higher level of variation observed in the loop region than the stem region (Fig. 5). Stem region sequences were identical at the positions from 5 to 9 in all individual units with the exception of a single A-T pairing of the 5th–15th bases in *E. hyemale*. In addition, the 13th, 14th, and 15th bases paired with the 7th, 6th, and 5th bases, respectively, and showed over 95% identity between/among the four species. The 1st–19th and 17th–3rd pairings showed greater than 90% identity between/among the four species.

Osmunda C— A — G— G—G—C— T — C —A — A—C— C— C —T —G

Angiopteris C— A — G— G—G—C—T — C —A — A—C— C— C —T —G

Mankyua C— A — G— G—G—C—T — C —A — A—C— C— C —T —G

Psilotum C— A — G— G—G—C— T — C —A — A—C— C— C —T —G

American E. arvense C— A — G— G—G—C— T — C —A — A—C— C— C —T —G

Asian E. arvense C— A — G— G—G—C—T — C —A — A—C— C— T —T —G

Figure 3. Aligned SNP sequences of the *rrn16* gene region. All eusporangiate ferns and *Osmunda cinnamomea*, except Asian *E. arvense*, contained identical sequences. C to T substitution was observed in Asian *E. arvense*. Ten *E. arvense* individuals from various locations in Korea, Japan, and China share the C to T substitution.

The 2nd base was most frequently A (83–85%), while the 18th position was usually T (63–69%). C was occasionally found at the 2nd base position (11–12%), and G was sometimes at the 18th base position (25–32%). Three main base-pairings were observed between the 2nd and 18th base positions: A-T had the highest frequency, followed by A-G and C-G pairs. The 10th, 11th, and 12th bases were hypervariable sites, and the main nucleotides at these positions varied according to species. For example, CTT was the main sequence in subgenus *Equisetum*. In subgenus *Hippochaete*, however, ATT was predominant in *E. ramosissimum* but CTT was the main sequence found in *E. hyemale*.

The total sequence lengths in the two species from the subgenus *Hippochaete* were longer than the sequences from two *E. arvense* of subgenus *Equisetum*. We calculated the p-distance and the number of differences in the *trnY-trnE* IGSs between the species (Table 5). The intraspecific p-distances between *E. hyemale* and the two continental individuals of *E. arvense* were both 0.008. The distance between the two Asian individuals was 0.0004. The interspecific p-distances in subgenus *Hippochaete* (*E. hyemale* and *E. ramosissimum*) were in the range 0.031–0.034. The interspecific p-distances between the two subgenera (subgenus *Hippochaete* and subgenus *Equisetum*) were in the range 0.084–0.096. The lowest interspecific p-distance value was observed between Korean *E. arvense* and American *E. hyemale* and the highest value was seen between *E. ramosissimum* and American *E. arvense*.

Dot-plot analysis of Korean *E. arvense* indicated that repeating units were widespread and concentrated in the first half of the *trnY-trnE* IGS region (Fig. 6). A similar pattern was observed with dot-plot analysis of Asian and American *E. arvense*. A few repeat units were found in the second half of the *trnY-trnE* IGS region in *E. arvense* and *E. hyemale*. *E. arvense* and *E. hyemale* had many similar repeat units, and these were located primarily in the 800 to 2,200 bp region. Overall, repeat unit conservation was more prominent in the first half of the IGS than the second half.

Discussion

Molecular evolution of the cp genome in Asian and American *E. arvense*

The Asian and American *E. arvense* cp genomes were of different lengths, and this variation was due to the presence of 159 indels. The majority of the indels were homopolymer length variants; however, these can be produced as artifacts of the 454-sequencing process [44,45]. Nevertheless, we were confident of the accuracy of our complete sequences for three reasons. First, our

pyrosequencing contigs covered the majority of specific sequences with 100× coverage and the majority consensus sequences were derived from numerous overlapping contigs. Second, we manually amplified and sequenced low-coverage regions and found many homo- and heteropolymer repeating regions. Third, the cp genome of the *Olea europaea* complex had approximately 4–8× more 1–2 bp indels than >3 bp indels [46], whereas the ratio of 1–2 bp to >3 bp indels in *E. arvense* was 3:1. If the *E. arvense* indels were identified mainly as a result of 454-pyrosequencing errors, the indel ratio would be biased towards 1–2 bp rather than >3 bp indels. Therefore, we believe that the majority of the homopolymer indels found in *E. arvense* were true indels that reflected the evolutionary history of the two continental individuals.

Relatively few large indels (>3 bp) were previously observed in closely related taxa of plant genera. For example, only seven indels were reported in the cp genomes of the *O. europaea* complex. Nine indels were observed in three species of *Phyllostachys* and 3–21 indels were reported in *Gossypium* species. No indels were detected between the two cp genomes of *Nicotiana tabacum* and *N. sylvestris*. By contrast, 40 indels of >3 bp were seen in Asian and American *E. arvense* despite the recognition of these two continental populations as a single species. In this study, we excluded 14 indels that were identified from the hypervariable *trnY-trnE* region of the *Equisetum* cp genome. Therefore, our data clearly indicated that the variation between the cp genomes of the two *E. arvense* was much higher than the intersubspecific or the interspecific differences observed in various flowering plants. The high intraspecific difference observed between the two continental *E. arvense* was also confirmed by SNP analysis.

SNP variation between Asian and American *E. arvense* was 0.12%. This variation level was higher than intraspecific or interspecific differences previously reported in numerous seed plants, with intraspecific SNP variation being very low in many plant groups. For example, SNP variation was 0.07% between two interecotypes of *Panicum virgatum* [10], and was 0.02% between 17 individuals of *Jacobaea vulgaris* [9]. In addition, SNP variation was 0.05% between *Oryza sativa* subsp. *indica* and *O. sativa* subsp. *japonica* [7] and 0.03–0.07% in the *Olea europaea* complex [46]. The SNP variation between *E. arvense* individuals was higher than interspecific variation observed in several cases. Variation was 0.003% between *Nicotiana tabacum* and *N. sylvestris*, 0.007–0.11% between five *Gossypium* species [11], and 0.02–0.05% between three *Phyllostachys* species [47]. In addition to the substantial indel and SNP differences between Asian and

Table 3. SNPs on the cp genomes of Asian and American *E. arvense*.

	Coding regions		IGSs		Introns		rRNAs		Total	
	Ts	Tv	Ts	Tv	Ts	Tv	Ts	Tv	Ts	Tv
LSC	21	23	28	71	4	9	0	0	53	103
SSC	2	4	0	1	0	0	0	0	2	5
IR(×2)	0	0	0	0	0	0	1	0	2	0
Total	23	27	28	72	4	9	2	0	57	108

Ts and Tv indicate transitional and transversional changes, respectively.

American *E. arvense*, the genetic distance between the two *E. arvense* individuals was higher than the interspecific variation in flowering plants. The high genetic diversity and low morphological diversity indicated that the two continental *E. arvense* populations had high genetic heterogeneity while being in morphological stasis.

A few examples of regionally isolated flowering plant species in similar habitats exhibiting similar morphological characteristics have been reported [19,48]. The two continental populations of *E. arvense* experience similar habitats and have no morphological differences, yet are geographically isolated in Asia and North America. This suggested that the cp genomes would differ considerably as the two *E. arvense* populations were split by disjunctive distribution. The high variance between the cp genomes of the two continental *E. arvense* suggests relatively ancestral divergence if a molecular clock concept is applied.

To estimate the divergence time between Asian and American *E. arvense*, we adopted different SNP data from five data sets. The whole cp SNP data between the two inter-subspecies of *Oryza sativa* (ssp. *indica* and ssp. *japonica*) and two ecotypes of *P. virgatum* were used. The divergence time of the two *O. sativa* inter-subspecies was estimated at 0.4 mya [50], and the divergence time of the two *P. virgatum* eocotypes was estimated at 0.7–1.0 mya [51]. SNP variation between the two *E. arvense* was $2.3\times$ higher than between the two *O. sativa* inter-subspecies and $1.5\times$ higher than between the two *P. virgatum* eocotypes. This suggested that the two *E. arvense* populations diverged 1.0–1.7 mya. We also estimated the divergence time of the two *E. arvense* populations using the partial and whole cp SNP data from the same *Equisetum* genus using a published calibration clock [49]. There are three different data sets available from *Equisetum*. First, the *rbcL* calibration clock suggested that subgenus *Equisetum* and subgenus *Hypochaete* diverged approximately 28.5 ± 5.5 mya. Differences in *rbcL* between the two *E. arvense* were $15\times$ lower than between two subgenera of *Equisetum*. Therefore, the *rbcL* calibration clock suggested that the two *E. arvense* populations diverged approximately 1.9 ± 0.4 mya. Second, the hypervariable *trnY-trnE* intergenic spacer different between the two subgenus *Equisetum* and subgenus *Hypochaete* was 9.1% (Table 5) and they diverged approximately 28.5 ± 5.5 mya. Differences in the hyper-variable *trnY-trnE* intergenic spacer between the two *E. arvense* were $11.4\times$ lower than between two subgenera of *Equisetum*. Therefore, the hypervariable *trnY-trnE* intergenic spacer data suggested that the two *E. arvense* populations diverged approximately 2.5 ± 0.5 mya. Third, the cp SNP difference between the subgenus *Equisetum* (*E. arvense*) and subgenus *Hypochaete* (*E. hyemale*) was 1.44% and they diverged approximately 28.5 ± 5.5 mya as in the *rbcL* calibration. Differences in the whole cp SNP between the two *E. arvense* were $12\times$ lower than between two subgenera of *Equisetum*. Therefore, the whole cp SNP calibration suggested that the two *E. arvense* populations diverged approximately 2.4 ± 0.5 mya. The five independent estimates of divergence time are relatively concordant and fall into an overlapping range. However, we are more confident to the *Equisetum* calibration than the Poaceae calibration because of two reasons. First, the *Equisetum* clock is based upon abundant fossil records. Second, the *Equisetum* calibration uses data from the same lineages, minimizing the lineage bias effect of the molecular clock.

Phylogenetic analysis of genus *Equisetum* estimated the speciation time of *E. arvense* to be 2.588 mya, which lies between the Pliocene and Quaternary periods [49]. *E. arvense* may have migrated from one region to another through the Bering land-bridge after speciation, in a similar manner to the migration that produced the Asian and North American *Adiantum pedatum*

Table 4. Distribution of nucleotide substitutions on coding genes of the cp genomes of Asian and American *E. arvense* populations.

Region	Gene	Ts	Tv	S	N
LSC (21 genes)	*accD*	1	0	1	0
	atpA	1	0	1	0
	atpI	1	0	0	1
	cemA	0	3	1	2
	clpP	1	0	1	0
	matK	1	4	0	5
	psaB	2	0	2	0
	psbB	1	0	1	0
	psbC	0	1	1	0
	rbcL	1	2	2	1
	rpl2	0	1	0	1
	rpl20	0	1	0	1
	rpoC1	1	1	1	1
	rpoC2	3	3	4	2
	rps11	1	0	1	0
	rps14	1	0	1	0
	rps3	1	0	1	0
	rps7	1	0	1	0
	rps8	0	1	0	1
	ycf2	4	5	5	4
	ycf3	0	1	1	0
SSC (5 genes)	*ccsA*	1	0	1	0
	ndhA	1	0	1	0
	ndhD	0	1	0	1
	ndhF	0	1	1	0
	ycf1	0	2	2	0
Total		23	27	30	20

complexes. Japanese *A. pedatum* and northeastern American *A. pedatum* diverged 2.47 mya. Northeastern American *A. pedatum* and Chinese *A. pedatum* subsequently diverged 1.09 mya [25]. The disjunctive distribution of northeastern American *A. pedatum* and Chinese *A. pedatum* through the Bering land-bridge in the mid-Pleistocene is very similar to the scenario for the disjunctive distribution of Asian and American *E. arvense*. Therefore, we believe that *E. arvense* and *A. pedatum* might have migrated during the same period, stimulated by the geological or environmental conditions at that time. The molecular differences therefore accumulated between the two continental *E. arvense* populations since the mid-Pleistocene, 1.9–2.9 mya, with almost no corresponding development of morphological differences.

One rRNA SNP in Asian *E. arvense* was of particular interest because it was shared by all the Asian individuals, suggesting a single origin for all the Asian individuals. *E. arvense* is a member of the eusporangiate ferns, which is the basal ferns lineage. Other eusporangiate ferns (*Mankyua chejuense*, *Psilotum nudum* and *Angiopteris evecta*) and basal leptosporangiate fern (*Osmunda cinamomea*) share the same rRNA sequence as American *E. arvense* (Fig. 3B). The CCCUG sequence in the *rrn16* gene of eusporangiate ferns, including American *E. arvense* and *O. cinamomea*, produces a hairpin structure by pairing with CAGGG (Fig. 3). However, in the *rrn16* gene of Asian *E. arvense*, the

CCCUG sequence was mutated (to CCUUG) and the CCUUG sequence was paired with CAGGG. Therefore, the *rrn16* gene of Asian *E. arvense* had a less stable folding structure than the equivalent American *E. arvense* sequence (Figure S2). The SNP in the *rrn16* gene therefore changed the minimum free energy for RNA secondary structure. We tested for this SNP using PCR-sequencing techniques in ten Asian *E. arvense* individuals from Korea, Japan, and China. All Asian individuals shared this SNP.

Repeat sequence evolution in genus *Equisetum*

Eusporangiate ferns consist of four orders, including Equisetales [52], and the cp genome of one or two species from each order has been sequenced [13,28,53]. However, the *trnY-trnE* IGS expanded up to 5 kb in *Equisetum* and the duplication of a 19 bp repeating unit was responsible for this expansion. The 19 bp repeating unit was not detected in the cp genomes of the other three orders. This indicated that the 19 bp repeating unit might be a unique molecular characteristic that only occurred in the monotypic genus *Equisetum*. The genus *Equisetum* is divided into two subgenera and the 19 bp repeating units with hairpin structures were identified in both. The sequence identity of the 19 bp unit was sufficiently conserved to allow easy recognition. We therefore assumed that the consensus repeating unit originated

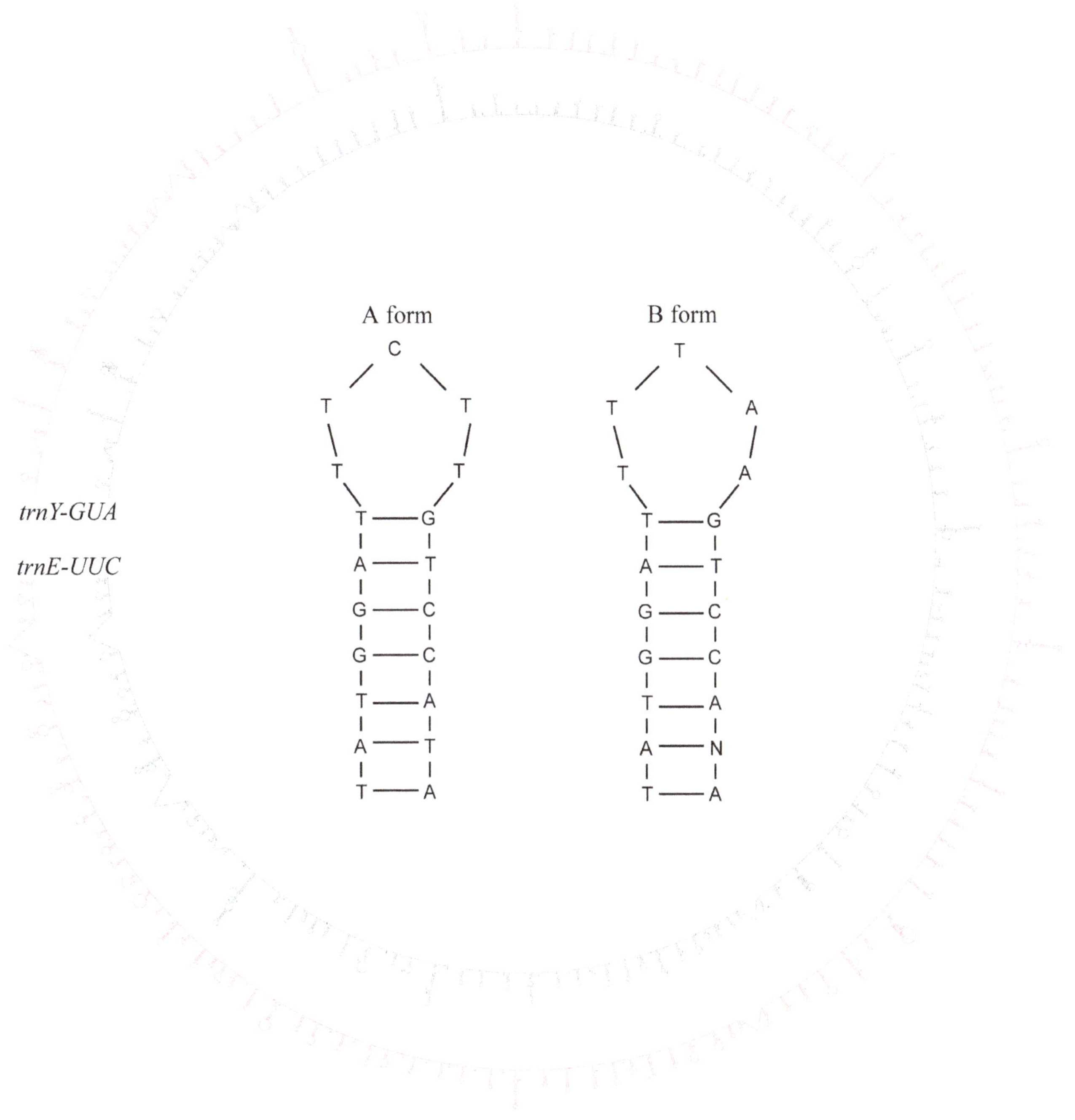

Figure 4. Putative folding structures of *trnY-trnE* IGS sequences in Asian and American *E. arvense*. The inner circle indicates Asian *E. arvense* sequence and the outer circle indicates American *E. arvense* sequence. Both sequences had numerous hairpin structures that were largely distributed evenly across the entire region. The most common two hairpin sequences (A and B forms) are shown inside the two circles. The two forms differed mainly on the loop region sequences. The most frequent base at the N position in the stem of the B form was G or T followed by A or C.

prior to the separation of the two subgenera in genus *Equisetum*. Similar expansion of the *trnY-trnE* IGS (albeit shorter; ~450 bp) was reported in the leptosporangiate fern *Vanderboschia radicans*. In that case, duplication of a 27 bp repeat unit homologous to a *trnY* anticodon was responsible for the expansion of the *trnY-trnE* IGS [29]. A similar duplicated expansion of *trnD-trnY* IGS was responsible for length variation in *Pseudotsuga* species (Gymnosperms) [54]. We compared the *Equisetum trnY-trnE* IGS repeat unit to those in other species but were unable to conclusively determine the origin of the 19 bp repeat unit in *Equisetum trnY-trnE* IGS as sequence identity with *trnY* was low. Gao et al. [29] suggested that a 13 bp repeat in the *Equisetum trnY-trnE* IGS may be derived from partial *trnY* sequences. However, our data do not support this as the extended stem region in our 19 bp consensus repeat unit is substantially different from *trnY* anticodon sequences. Several cp *trn* genes hold similar sequence components and it is therefore difficult to deduce the origin of the *trn* repeat unit. Furthermore, the *trnY* duplication occurs at slightly different

A

Asian **E. arvense**/ American **E. arvense**/ **E. ramosissimum**/ Asian **E. hyemale**

B

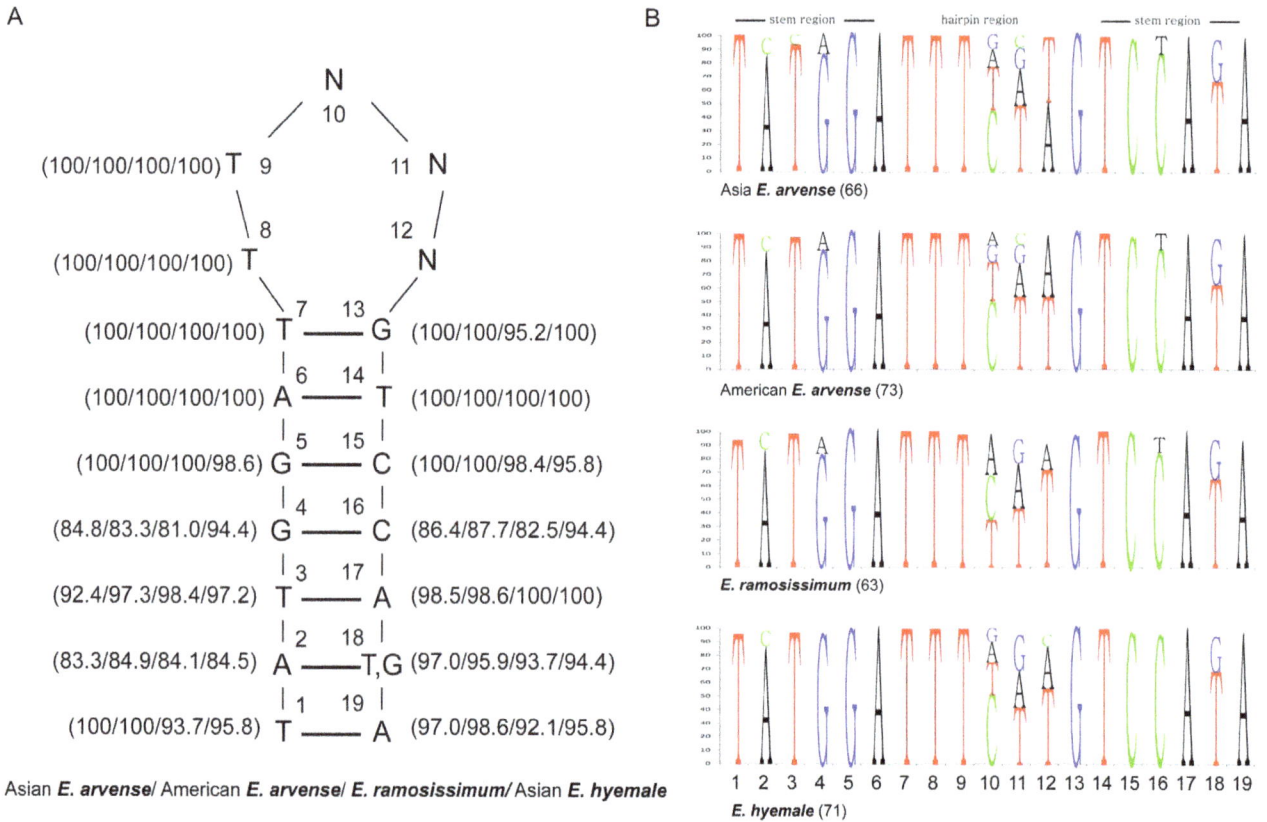

Figure 5. Consensus sequences for the repeated hairpin structure in the *trnY-trnE* **IGS region.** (A) The most common consensus 19 bp sequences and frequencies were as indicated on the hairpin structure. Positions 6, 7, 8, 9, and 14 were invariable and positions 10, 11, and 12 were the most variable sites. (B) Consensus sequences were derived from the 66 repeats of Asian *E. arvense*, the 73 repeats of American *E. arvense*, the 63 repeats of *E. ramosissimum*, and the 71 repeats of *E. hyemale*.

locations in *Pseudotsuga* and *V. radicans*. *Pseudotsuga*, *Equisetum*, and *V. radicans* are not phylogenetically close and no known *trnY-trnE* IGS expansions have occurred in the sister groups of these three lineages. Therefore, we believe that the *trnY-trnE* IGS expansions in *Pseudotsuga*, *Equisetum* and *V. radicans* were independent parallel evolutionary events rather than a homologous synapomorphy. By contrast, the *trnY-trnE* IGS expansion

was discovered in all *Equisetum* species examined in this study and is a single evolutionary event.

Fossils of the genus *Equisetites*, which is the most closely related genera to genus *Equisetum* [55], were discovered worldwide in Europe [56], North America [57], Antarctica [58], China [59], and New Zealand [60] in underlying strata of the post-Mesozoic era. This suggested that the divergence time for genus *Equisetum* was in the Tertiary period. Extant *Equisetum* species diverged

Table 5. No. of nucleotide differences and p-distances of the *trnY-trnE* IGS region in genus *Equisetum*.

	Korean *E. arvense*	Chinese *E. arvense*	American *E. arvense*	Korean *E. hyemale*	American *E. hyemale*	*E. ramosissimum*
Korean *E. arvense* (4534 bp)	-	2	35	351	333	360
Chinese *E. arvense* (4872 bp)	0.0004	-	38	378	369	375
American *E. arvense* (4991 bp)	0.008	0.008	-	397	356	385
Korean *E. hyemale* (5400 bp)	0.090	0.092	0.095	-	44	147
American *E. hyemale* (5645 bp)	0.084	0.089	0.084	0.008	-	159
E. ramosissimum (5000 bp)	0.095	0.095	0.096	0.031	0.034	-

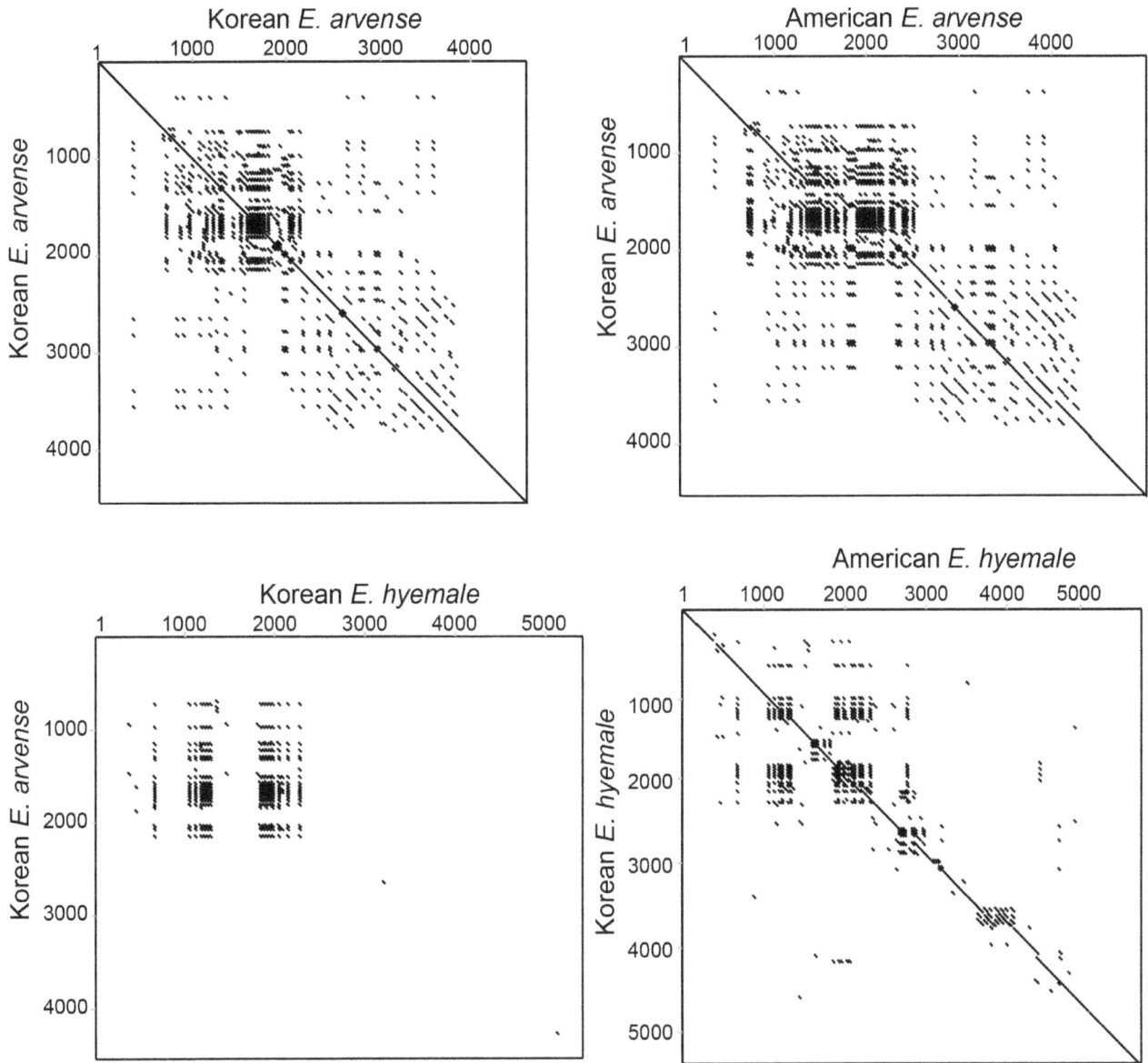

Figure 6. Dot matrix showing the occurrence, conservation, and divergence of the repeating units in the *trnY-trnE* IGS region. The region proximal to the *trnY* gene (1–2,200 bp region) had high density and conservation of repeating sequences compared to the region proximal to the *trnE* gene (>2,200 bp region).

more recently than the Miocene period [49]. Therefore, we reasoned that the repeating units were formed during the Miocene period after or before the formation of genus *Equisetum*, and the repeating units then spread widely and diverged alongside *Equisetum* speciation events. To summarize, the 19 bp repeating unit is a synapomorphic molecular characteristic shared by all living members of *Equisetum*.

The *trnY-trnE* IGS region differed in length between Korean and Chinese *E. arvense* individuals. Chinese *E. arvense* had 76 repeat sequences and a *trnY-trnE* IGS length of 4,872 bp. When repeat sequence numbers and *trnY-trnE* IGS lengths were compared, Chinese *E. arvense* was more superficially similar to American *E. arvense* than to Korean *E. arvense*. However, when the indel and SNP patterns were considered together, Chinese *E. arvense* was found to be more closely related to Korean *E. arvense*

than to American *E. arvense*. Two large indels (241 bp, 97 bp) and two SNPs were detected between Korean *E. arvense* and Chinese *E. arvense*, but 19 indels and 38 SNP differences were noted between the Chinese and American *E. arvense* individuals.

The difference in the number of repeating units observed in Korean *E. arvense* and Chinese *E. arvense* suggested unequal crossover on the tandem-repeat regions. Two unequal crossovers may have generated the large length difference between the Korean and Chinese *E. arvense* individuals. The two large indels, composed of 241 bp and 97 bp, contained six and three 19 bp repeating units, respectively. Small indels of 1–10 bp were not found between Korean and Chinese *E. arvense*. In contrast to the large indels, the large numbers of small indels observed between Korean and American *E. arvense* suggested that the majority of small indels originated as mutation events formed by slipped strand mispairing [61]. A few indels with long length differences were observed between Korean and American *E. arvense*.

Analysis of the 19 bp repeat sequence (TATG-GATTTCTTGTCCATA) (Fig. 5), suggested that the original source sequence was the *trnY-trnE* IGS. The 19 bp sequence found in this study is 6 bp longer than the *trnY* anticodon partial sequence proposed as the repeating unit origin [29]. Additionally, the 19 bp sequence had higher similarities than the 13 bp consensus sequence of the *trnY* anticodon loop region suggested in previous studies [29]. Our 19 bp consensus sequence was based on the wide range of sequence data available for the diverse *Equisetum* taxa. The expected origin sequence of repeating units was commonly abundant not only in subgenus *Equisetum* but also in subgenus *Hippochaete*. In *E. ramosissimum*, of subgenus *Hippochaete*, the repeating unit of TATGGATTTATTGTC-CATA, which differed from the *E arvense* consensus at the 10th position (C to A), was found at a slightly higher frequency than the TATGGATTTCTTGTCCATA sequence.

The 11th and 12th sites, located in the hairpin loop region, showed A to T substitutions. Substitutions, albeit rare, were also observed at the 2nd, 4th, 16th, and 18th sites of the stem region. This indicated that the 19 bp repeating units have mutated constantly and maintained evolutionary heterogeneity in the genus *Equisetum*. Heterogeneity of the repeating unit was confirmed by dot-plot analysis (Fig. 6). The first half of the *trnY-trnE* IGS had more conserved repeating units than the second half. When compared with the length of the *trnY-trnE* IGS in other eusporangiate ferns, it was apparent that many repeats occurred in the second half of *trnY-trnE* IGS, but no apparent homologous repeating sequences were observed between *E. arvense* and *E. hyemale*. Therefore, we assumed that the progression to heterogeneity in the repeating unit proceeded particularly rapidly in the second half of the *trnY-trnE* IGS.

Conclusions

Molecular clock analysis suggested that *E. arvense* migrated between two continents via the Bering land-bridge 1.9–2.9 mya. After migration, its morphological characteristics remained largely unchanged in each region due to adaption to similar habitats, but constant mutational events occurred in the cp genomes. This indicated that the two continental populations of *E. arvense* have been in prolonged morphological stasis while the cp genome sequences in the two regions have changed continuously since population dispersal. The levels of sequence and indel divergence between the two regional cp genomes were far higher than those of closely related interspecific taxa in many seed plants. Two regional genotypes can therefore be recognized. Rigorous comparative analyses of the whole cp genomes from multiple accessions of each continental population, including European populations, are needed to comprehensively address population history and the validity of the species boundary.

References

1. Jansen RK, Ruhlman TA (2012) Plastid Genomes of Seed Plants. In: R. . Bock and V. . Knoop, editors. Genomics of Chloroplasts and Mitochondria. Springer Netherlands. pp. 103–126.
2. Cosner ME, Raubeson LA, Jansen RK (2004) Chloroplast DNA rearrangements in Campanulaceae: phylogenetic utility of highly rearranged genomes. BMC Evol Biol 4: 27. Available: http://www.biomedcentral.com/1471-2148/4/27 Accepted 23 August 2004
3. Lee HL, Jansen RK, Chumley TW, Kim KJ (2007) Gene relocations within chloroplast genomes of Jasminum and Menodora (Oleaceae) are due to multiple, overlapping inversions. Mol Biol Evol 24: 1161–1180.
4. Saski C, Lee SB, Fjellheim S, Guda C, Jansen RK, et al. (2007) Complete chloroplast genome sequences of Hordeum vulgare, Sorghum bicolor and

The *trnY-trnE* IGS is a hypervariable region within the *E. arvense* cp genome, and many indel events and SNPs were concentrated in this region. A unique 19 bp repeating sequence unit that formed a hairpin structure was replicated many times in this IGS region and was responsible for the dynamic sequence evolution of the cp genome. The genus *Equisetum* is a monotypic genus in Equisetales and the repeating units did not exist in other eusporangiate ferns. It was therefore challenging to find the exact origin sequence and to explain the evolutionary paths of repeat unit evolution. A comprehensive study involving additional *Equisetum* species is needed to understand the evolution of repeat units in the genus. However, with the current limited data set, the region showed very different intraspecific, interspecific, and intersubgeneric p-distance values. Therefore, the hypervariable *trnY-trnE* IGS region may be a useful molecular marker to study the evolution and phylogeny of *Equisetum*.

Supporting Information

Figure S1 Sequencing strategy for the *E. arvense* chloroplast genome. The outer blue circle indicates the sequence region generated by next –generation sequencing (NGS). Seven large NGS contigs cover approximately 90% of the genome. The green broken lines indicate the regions sequenced by PCR amplifications and Sanger sequencing. The *trnY-trnE* IGS was amplified by long-range PCR methods. The genome map in the inner circle was generated in OrganellarGenomeDRAW [62] after the completion of sequencing and annotation.

Figure S2 The RNA folding structure differences of *rrn16* gene from two *E. arvense* populations. The gray box regions indicated the folding structure differences due to the SNP(C-U). The sequence -CCCUG- paired with the sequence –CAGGG- and form a hairpin structure in the American *E. arvense* (left). However, the sequence –CCUUG- paired with the sequence –CAAGG- and form a distinct stem structure in the Korean *E. arvense* (left). Two contrasting folding structures are based on the minimum free energy only. Other alternate folding structures are also possible if we consider other factors affecting the secondary structures.

Acknowledgments

We thank two anonymous reviewers for their helpful suggestions for improving the manuscript. All DNA materials used in this study are deposited in the Plant DNA Bank of Korea (PDBK) and are available from the PDBK.

Author Contributions

Conceived and designed the experiments: KJK. Performed the experiments: HTK. Analyzed the data: HTK KJK. Contributed reagents/materials/analysis tools: KJK. Wrote the paper: HTK KJK.

   Agrostis stolonifera, and comparative analyses with other grass genomes. Theor Appl Genet 115: 571–590.
5. Funk HT, Berg S, Krupinska K, Maier UG, Krause K (2007) Complete DNA sequences of the plastid genomes of two parasitic flowering plant species, Cuscuta reflexa and Cuscuta gronovii. BMC Plant Biol 7: 45. Available: http://www.biomedcentral.com/1471-2229/7/45 Accepted 22 August 2007.
6. Yi DK, Lee HL, Sun BY, Chung MY, Kim KJ (2012) The complete chloroplast DNA sequence of Eleutherococcus senticosus (Araliaceae); comparative evolutionary analyses with other three asterids. Mol Cells 33: 497–508.
7. Tang J, Xia H, Cao M, Zhang X, Zeng W, et al. (2004) A comparison of rice chloroplast genomes. Plant Physiol 135: 412–420.

8. Whittall JB, Syring J, Parks M, Buenrostro J, Dick C, et al. (2010) Finding a (pine) needle in a haystack: chloroplast genome sequence divergence in rare and widespread pines. Mol Ecol 19 Suppl 1: 100–114.

9. Doorduin L, Gravendeel B, Lammers Y, Ariyurek Y, Chin AWT, et al. (2011) The Complete Chloroplast Genome of 17 individuals of Pest Species Jacobaea vulgaris: SNPs, microsatellites and barcoding markers for population and phylogenetic studies. DNA Res 18: 93–105.

10. Young HA, Lanzatella CL, Sarath G, Tobias CM (2011) Chloroplast Genome Variation in Upland and Lowland Switchgrass. Plos One 6: e23980. Available: http://www.plosone.org/article/info%3Adoi%2F10.1371%2Fjournal.pone.0023980 Accessed 1 August 2011.

11. Xu Q, Xiong G, Li P, He F, Huang Y, et al. (2012) Analysis of complete nucleotide sequences of 12 Gossypium chloroplast genomes: origin and evolution of allotetraploids. PLoS One 7: e37128. Available: http://www.plosone.org/article/info%3Adoi%2F10.1371%2Fjournal.pone.0037128. Accessed 16 April 2012.

12. Wolf PG, Rowe CA, Sinclair RB, Hasebe M (2003) Complete nucleotide sequence of the chloroplast genome from a leptosporangiate fern, Adiantum capillus-veneris L. DNA Res 10: 59–65.

13. Roper JM, Hansen SK, Wolf PG, Karol KG, Mandoli DF, et al. (2007) The complete plastid genome sequence of Angiopteris evecta (G. Forst.) Hoffm. (Marattiaceae). Am Fern J 97: 95–106.

14. Gao L, Yi X, Yang YX, Su YJ, Wang T (2009) Complete chloroplast genome sequence of a tree fern Alsophila spinulosa: insights into evolutionary changes in fern chloroplast genomes. BMC Evol Biol 9: 130. Available: http://www.biomedcentral.com/1471-2148/9/130 Accepted 11 June 2009.

15. Wolf PG, Der JP, Duffy AM, Davidson JB, Grusz AL, et al. (2011) The evolution of chloroplast genes and genomes in ferns. Plant Mol Biol 76: 251–261.

16. Tiffney BH (1985) Perspectives on the origin of the floristic similarity between eastern Asia and eastern North America. J Arnold Arboretum 66: 73–94.

17. Qiu YL, Chase MW, Parks CR (1995) A chloroplast dna phylogenetic study of the eastern Asia eastern North America disjunct section Rytidospermum of Magnolia (Magnoliaceae). Am J Bot 82: 1582–1588.

18. Xiang QY, Soltis DE, Soltis PS (1998) The eastern Asian and eastern and western North American floristic disjunction: Congruent phylogenetic patterns in seven diverse genera. Mol Phylogenet Evol 10: 178–190.

19. Wen J (1999) Evolution of eastern Asian and eastern north American disjunct distributions in flowering plants. Annu Rev Ecol Syst 30: 421–455.

20. Xiang QY, Soltis DE, Soltis PS, Manchester SR, Crawford DJ (2000) Timing the eastern Asian-eastern north American floristic disjunction: molecular clock corroborates paleontological estimates. Mol Phylogenet Evol 15: 462–472.

21. Wall WA, Douglas NA, Xiang QY, Hoffmann WA, Wentworth TR, et al. (2010) Evidence for range stasis during the latter Pleistocene for the Atlantic Coastal Plain endemic genus, Pyxidanthera Michaux. Mol Ecol 19: 4302–4314.

22. Wen J (2001) Evolution of eastern Asian-eastern north American biogeographic disjunctions: A few additional issues. Int J Plant Sci 162: S117–S122.

23. Li H-L (1952) Floristic relationships between eastern Asia and eastern north America. Trans Am Philoso Soc 42: 371–429.

24. Kato M (1993) Biogeography of ferns - dispersal and vicariance. J Biogeogr 20: 265–274.

25. Lu JM, Li DZ, Lutz S, Soejima A, Yi TS, et al. (2011) Biogeographic disjunction between eastern Asia and north America in the Adiantum Pedatum complex (Pteridaceae). Am J Bot 98: 1680–1693.

26. Wolf PG, Schneider H, Ranker TA (2001) Geographic distributions of homosporous ferns: does dispersal obscure evidence of vicariance? J Biogeogr 28: 263–270.

27. Veit M, Geiger H, Czygan FC, Markham KR (1990) Malonylated flavone 5-O-Glucosides in the barren sprouts of Equisetum arvense. Phytochemistry 29: 2555–2560.

28. Karol KG, Arumuganathan K, Boore JL, Duffy AM, Everett KDE, et al. (2010) Complete plastome sequences of Equisetum arvense and Isoetes flaccida: implications for phylogeny and plastid genome evolution of early land plant lineages. BMC Evol Biol 10: 321. Available: http://www.biomedcentral.com/1471-2148/10/321. Accepted 23 October 2010.

29. Gao L, Zhou Y, Wang ZW, Su YJ, Wang T (2011) Evolution of the rpoB-psbZ region in fern plastid genomes: notable structural rearrangements and highly variable intergenic spacers. BMC Plant Biol 11: 64. Available: http://www.biomedcentral.com/1471-2229/11/64. Accepted 13 April 2011.

30. Palmer JD (1986) Isolation and structural analysis of chloroplast DNA. Methods Enzymol 118: 167–186.

31. Jansen RK, Raubeson LA, Boore JL, dePamphilis CW, Chumley TW, et al. (2005) Methods for obtaining and analyzing whole chloroplast genome sequences. Method Enzymol 395: 348–384.

32. Wyman SK, Jansen RK, Boore JL (2004) Automatic annotation of organellar genomes with DOGMA. Bioinformatics 20: 3252–3255.

33. Lowe TM, Eddy SR (1997) tRNAscan-SE: A program for improved detection of transfer RNA genes in genomic sequence. Nucleic Acids Res 25: 955–964.

34. Kim KJ, Lee HL (2004) Complete chloroplast genome sequences from Korean ginseng (Panax schinseng Nees) and comparative analysis of sequence evolution among 17 vascular plants. DNA Res 11: 247–261.

35. Yi DK, Kim KJ (2012) Complete chloroplast genome sequences of important oilseed crop Sesamum indicum L. Plos One 7: e35872. Available: http://www.

36. Edgar RC (2004) MUSCLE: a multiple sequence alignment method with reduced time and space complexity. BMC Bioinformatics 5: 1–19.

37. Kearse M, Moir R, Wilson A, Stones-Havas S, Cheung M, et al. (2012) Geneious Basic: an integrated and extendable desktop software platform for the organization and analysis of sequence data. Bioinformatics 28: 1647–1649.

38. Benson G (1999) Tandem repeats finder: a program to analyze DNA sequences. Nucleic Acids Res 27: 573–580.

39. Zuker M (2003) Mfold web server for nucleic acid folding and hybridization prediction. Nucleic Acids Res 31: 3406–3415.

40. Hauke RL (1963) A taxonomic monograph of the genus Equisetum subgenus Hippochaete. J. Cramer. Beihefte Nova Hedwigia 8: 1–123.

41. Hauke RL (1978) Taxonomic monograph of Equisetum subgenus Equisetum. Nova Hedwigia 30: 385–455.

42. Doyle JJ, Doyle JL (1987) A rapid DNA isolation procedure for small quantities of fresh leaf tissue. Phytochem Bull 19: 11–15.

43. Kurtz S, Choudhuri JV, Ohlebusch E, Schleiermacher C, Stoye J, et al. (2001) REPuter: the manifold applications of repeat analysis on a genomic scale. Nucleic Acids Res 29: 4633–4642.

44. Wicker T, Schlagenhauf E, Graner A, Close TJ, Keller B, et al. (2006) 454 sequencing put to the test using the complex genome of barley. BMC Genomics 7: 275. Available: http://www.biomedcentral.com/1471-2164/7/275. Accepted 26 October 2006.

45. Gilles A, Meglecz E, Pech N, Ferreira S, Malausa T, et al. (2011) Accuracy and quality assessment of 454 GS-FLX Titanium pyrosequencing. BMC Genomics 12: 245. Available: http://www.biomedcentral.com/1471-2164/12/245. Accepted 19 May 2011.

46. Mariotti R, Cultrera NGM, Diez CM, Baldoni L, Rubini A (2010) Identification of new polymorphic regions and differentiation of cultivated olives (Olea europaea L.) through plastome sequence comparison. BMC Plant Biol 10: 211. Available: http://www.biomedcentral.com/1471-2229/10/211. Accepted 24 September 2010.

47. Zhang YJ, Ma PF, Li DZ (2011) High-throughput sequencing of six bamboo chloroplast genomes: phylogenetic implications for temperate woody bamboos (Poaceae: Bambusoideae). PLoS One 6: e20596. Available: http://www.plosone.org/article/info%3Adoi%2F10.1371%2Fjournal.pone.0020596. Accepted 5 May 2011.

48. Hoey MT, Parks CR (1991) Isozyme Divergence between Eastern Asian, North-American, and Turkish Species of Liquidambar (Hamamelidaceae). Am J Bot 78: 938–947.

49. Des Marais DL, Smith AR, Britton DM, Pryer KM (2003) Phylogenetic relationships and evolution of extant horsetails, equisetum, based on chloroplast DNA sequence data (rbcL and trnL-F). Int J Plant Sci 164: 737–751.

50. Zhu Q, Ge S (2005) Phylogenetic relationships among A-genome species of the genus Oryza revealed by intron sequences of four nuclear genes. New Phytol 167: 249–265.

51. Morris GP, Grabowski PP, Borevitz JO (2011) Genomic diversity in switchgrass (Panicum virgatum): from the continental scale to a dune landscape. Mol Ecol 20: 4938–4952.

52. Smith AR, Pryer KM, Schuettpelz E, Korall P, Schneider H, et al. (2006) A classification for extant ferns. Taxon 55: 705–731.

53. Grewe F, Guo WH, Gubbels EA, Hansen AK, Mower JP (2013) Complete plastid genomes from Ophioglossum californicum, Psilotum nudum, and Equisetum hyemale reveal an ancestral land plant genome structure and resolve the position of Equisetales among monilophytes. BMC Evol Biol 13: 8. Available: http://www.biomedcentral.com/1471-2148/13/8. Accepted 7 January 2013.

54. Hipkins VD, Marshall KA, Neale DB, Rottmann WH, Strauss SH (1995) A mutation hotspot in the chloroplast genome of a conifer (Douglas-Fir, Pseudotsuga) Is caused by variability in the number of direct repeats derived from a partially duplicated transfer rna gene. Curr Genet 27: 572–579.

55. Schaffner JH (1930) Geographic distribution of the species of Equisetum in relation to their phylogeny. Am Fern J: 89–106.

56. Kelber KP, van Konijnenburg-van Cittert JHA (1998) Equisetites arenaceus from the Upper Triassic of Germany with evidence for reproductive strategies. Rev Palaeobot Palyno 100: 1–26.

57. DiMichele WA, Van Konijnenburg-Van Cittert JH, Looy CV, Chaney DS (2005) Equisetites from the early Permian of north central Texas. The Nonmarine Permian 30: 56–59.

58. Cantrill DJ, Hunter MA (2005) Macrofossil floras of the Latady Basin, Antarctic Peninsula. New Zeal J Geol Geop 48: 537–553.

59. Wang ZQ (1996) Recovery of vegetation from the terminal Permian mass extinction in North China. Rev Palaeobot Palyno 91: 121–142.

60. McQueen D (1954) Upper Paleozoic plant fossils from South Island, New Zealand. Trans, roy Soc NZ 82: 231–236.

61. Levinson G, Gutman GA (1987) Slipped-Strand Mispairing - a Major Mechanism for DNA-Sequence Evolution. Mol Biol Evol 4: 203–221.

62. Lohse M, Drechsel O, Bock R (2007) OrganellarGenomeDRAW (OGDRAW): a tool for the easy generation of high-quality custom graphical maps of plastid and mitochondrial genomes. Curr Genet 52: 267–274.

plosone.org/article/info%3Adoi%2F10.1371%2Fjournal.pone.0035872. Accepted 23 March 2012.

17

Characterization of *Ferredoxin-Dependent Glutamine-Oxoglutarate Amidotransferase (Fd-GOGAT)* Genes and Their Relationship with Grain Protein Content QTL in Wheat

Domenica Nigro[1,9], **Antonio Blanco**[1,9], **Olin D. Anderson**[2,9], **Agata Gadaleta**[1*,9]

1 Department of Soil, Plant and Food Sciences, Section of Genetic and Plant Breeding, University of Bari "Aldo Moro", Bari, Italy, 2 Genomics and Gene Discovery Research Unit, Western Regional Research Center, USDA-ARS, Albany, California, United States of America

Abstract

Background: In higher plants, inorganic nitrogen is assimilated via the glutamate synthase cycle or GS-GOGAT pathway. GOGAT enzyme occurs in two distinct forms that use NADH (NADH-GOGAT) or Fd (Fd-GOGAT) as electron carriers. The goal of the present study was to characterize wheat *Fd-GOGAT* genes and to assess the linkage with grain protein content (GPC), an important quantitative trait controlled by multiple genes.

Results: We report the complete genomic sequences of the three homoeologous A, B and D Fd-GOGAT genes from hexaploid wheat (*Triticum aestivum*) and their localization and characterization. The gene is comprised of 33 exons and 32 introns for all the three homoeologues genes. The three genes show the same exon/intron number and size, with the only exception of a series of indels in intronic regions. The partial sequence of the Fd-GOGAT gene located on A genome was determined in two durum wheat (*Triticum turgidum* ssp. *durum*) cvs Ciccio and Svevo, characterized by different grain protein content. Genomic differences allowed the gene mapping in the centromeric region of chromosome 2A. QTL analysis was conducted in the Svevo×Ciccio RIL mapping population, previously evaluated in 5 different environments. The study co-localized the *Fd-GOGAT-A* gene with the marker GWM-339, identifying a significant major QTL for GPC.

Conclusions: The wheat Fd-GOGAT genes are highly conserved; both among the three homoeologous hexaploid wheat genes and in comparison with other plants. In durum wheat, an association was shown between the *Fd-GOGAT* allele of cv Svevo with increasing GPC - potentially useful in breeding programs.

Editor: Tongming Yin, Nanjing Forestry University, China

Funding: The research project was supported by grants from Ministero dell'Istruzione, dell'Universita' e della Ricerca, projects 'PON01_01145-ISCOCEM' and 'PRIN-2010-2011, and by USDA Agricultural Research Service. The funders had no role in study design, data collection and analysis, decision to publish, or preparation of the manuscript.

Competing Interests: The authors have declared that no competing interests exist.

* Email: agata.gadaleta@uniba.it

9 These authors contributed equally to this work.

Introduction

Glutamate is a central molecule in amino acid metabolism in higher plants. The α-amino group of glutamate is directly involved in both the assimilation and dissimilation of ammonia and is transferred to all other amino acids. In addition, both the carbon skeleton and α-amino group form the basis for the synthesis of γ-aminobutyric acid (GABA), arginine, and proline. Glutamate is also the precursor for chlorophyll synthesis in developing leaves [1].

As reviewed by Forde and Lea [2], glutamate synthase (GOGAT) is the key enzyme involved in the *de novo* synthesis of glutamate. It catalyzes the transfer of the amide group of glutamine to 2-oxoglutarate, with the result of two molecules of glutamate yielded.

The history of the discovery of the two enzymes, their structure, and gene regulation has been well documented [3,4]. In plants, GOGAT enzyme occurs in two forms, depending on the electron donor involved in the reaction: it exist as a ferredoxin (Fd) dependent (EC 1.4.7.1), and a NADH dependent (EC 1.4.1.14) form.

Both forms are located in plastids, but, while Fd-dependent enzyme is usually present in high activities in the chloroplasts of photosynthetic tissues, NADH-dependent enzyme is predominantly located in non-photosynthesizing cells. The role of GOGAT enzymes have been well discussed in rice and conifers [5,6]. Several studies showed that GOGAT mutations or gene knock-outs, with a consequent reduced enzyme activity for both forms, seems to be involved in changes in amino acid metabolism

[7,8,9,10]. Only a few studies have reported plant gene isolation and sequencing, probably due togene lengths and structural complexity. For these reasons, the first reported studies on gene sequences described the isolation and sequencing of a full-length cDNA clone for maize *Fd-GOGAT* [11] and a partial cDNA clone for tobacco *Fd-GOGAT* [12]. The first plant complete genomic sequences identified were two genes from *Arabidopsis* [13]. For the *Triticeae* crops (wheat, barley, rye), no complete sequence has been known although partial sequences were reported for barley [14] and fragments for wheat [15].

Recently, NADH-GOGAT genomic sequences has been reported for the A and B genomes of tetraploid durum wheat (*Triticum turgidum ssp. durum*) and for the A, B, and D genomes of hexaploid wheat (*Triticum aestivum*) [16]. Analysis of the gene sequences indicates that all wheat *NADH-GOGAT* genes are composed of 22 exons and 21 introns. A comparative analysis of sequences among di- and mono-cotyledons plants shows both regions of high conservation and of divergence. qRT-PCR performed with the two durum wheat cvs Svevo and Ciccio (characterized by an high and low protein content, respectively) indicates different expression levels of the two *NADH-GOGAT-3A* and *NADH-GOGAT-3B* genes.

The Fd-GOGAT protein is a monomeric enzyme of 140–160 kDa and has been purified from barley leaves as a single polypeptide chain containing iron-sulfur and flavin. [17]. Fd-GOGAT activity has been mapped to the centromeric region of chromosome 2A [18] where we have previously reported a QTL for grain protein content (GPC) [19,20].

GPC partially determines the nutritional value and the baking properties of common wheat (*T. aestivum*) as well as the pasta-making technology characteristics of durum wheat (*T. turgidum* ssp. *durum*). GPC is a typical quantitative trait controlled by a complex genetic system and influenced by environmental factors and management practices (nitrogen and water availability, temperature and light intensity). Breeding for high grain protein concentration has been one of the main goals for wheat breeders for several decades. Simultaneous increases of GPC and grain yield has been difficult to achieve because of a negative correlation between grain yield and GPC [21,22]. This occurs because grain yield is mostly a consequence of starch accumulation, whereas protein accounts for less than 10–15% of the grain dry weight, and any increase in starch accumulation causes a dilution of the protein content if it is not accompanied by an equivalent increase in N accumulation. So far, any environmental factor that affects grain yield also affects GPC.

Recent investigations [23,24,25,26,27,28,29,30,31,32,20] indicated that factors influencing protein concentration in cultivated and wild wheats are located on all chromosomes. Heritability estimates for GPC ranged from 0.41 [33] to 0.70 [31], depending on the genetic material, environment and the computational methods. For this reason, the study of this important character through traditional methods is difficult and time consuming.

In the present work, we determined the DNA sequence of the three homoeologous *Fd-GOGAT* genes in hexaploid wheat, analyzed the exon/intron structure, compared the wheat sequences to other plants, and studied tetraploid durum grain protein content by QTL analysis and identification of candidate genes. In particular, we focused our attention on 2A chromosome where the *Fd-GOGAT* gene is located - identifying and characterizing the genomic sequence in durum wheat and determining its correlation with QTL for grain protein content (GPC).

Results and Discussion

Determination of genomic *Fd-glutamate synthase* (GOGAT) gene sequences

The complete sequences of A, B and D *Fd-GOGAT* genes of hexaploid wheat were obtained by assembling 454 sequences of cv Chinese Spring using a partial barley sequence (NCBI accession S58774; Gene ID: 548298) as the initial query. The Chinese Spring 454 assembly produced one contig comprised of the three A, B and D-genome sequences. Three independent contigs were then obtained by splitting the three sequences and assigned to A, B and D genomes by amplifying genome specific primers [15] on a set of nulli-tetrasomic lines (NTs) for chromosome group 2 of *Triticum aestivum* cv Chinese Spring [34,35]. Examples of the analysis are shown for the A and B genomes in Figure S1 in File S1. The three homoeologous sequences are given in Text S1 in File S1.

Analysis of wheat gene sequences indicates that the wheat *Fd-GOGAT* gene is comprised of 33 exons and 32 introns for all the three homoeologous (Figure 1; Figure S2 in File S1). The sequence for intron 1 and the part of exon 1 containing the 5-coding portion of the exon could only be assigned for the D-genome gene (Figure 1; Text S1 and S2 in File S1). The cause of missing these sequences for the A- and B-genome genes is not known, but among possible causes are randomness of the shotgun sequencing, difficulty in next generation sequencing through the number of homopolymers in this region, or that the sequence is so similar in all three homoeologous that the software and manual assemblies mis-assigned some sequence reads to the D-genome sequence. For the A- and B-genome *Fd-GOGAT* genes, the 5′ extension ended with the final 13 amino acids of the signal peptide (Figure S2 in File S1). The 5′-UTR and the remainder of the coding sequence for the A-genome was found in the Transcriptome Shotgun Assembly (TRA) data at Genbank (Figure S3 in File S1).

The exon/intron borders for all three homoeologous wheat *Fd-GOGAT* genes matched the canonical plant borders (GT...AG) for all 32 identified introns (Figure 1). Consensus exon/introns boundaries were determined using grass EST sequences aligned to the genomic sequence and TRA assemblies. The structure of the *Fd-GOGAT* genes is highly conserved among the three homoeologues. In addition to the same number of exons/introns, no differences were observed in exons size and all three encoded mature enzymes have identical sequence lengths of 1519 amino acids residues (Figure S4 in File S1, Text S2 in File S1). Differences were observed in the intron sequences - in particular differences in length of 1–50 bases in introns 3–5, 8, 14–16, 22–23, 25–26, 28–31; 50–100 bp in introns 6, 12, 13, 21; and more than 500 bases in introns 9, 11, and 19 (Figure S2 in File S1). The three genes vary in length only by intron differences: the *Fd-GOGAT-A* gene is the longest with 15,337 bp, the *Fd-GOGAT-D* is 15,176 bp long and the gene located on chromosome 2B is the shortest at14,750 bp (all determined from the beginning of the sequence encoding the mature protein through the stop codon). The sequences among the three homoeologues are highly conserved – with only seven amino acid residue differences among the three mature polypeptides (Figure S4 in File S1).

Comparison of Fd-GOGAT genes in other species

Conservation of the Fd-GOGAT sequences was further examined by comparing the A-genome wheat Fd-GOGAT mature amino acid sequence with a selection of monocot and dicot sequences available in public databases (Figure 2). The closest match was found with *Brachypodium* Fd-GOGAT (97%), with rice at 94% and maize at 92%. As expected, the dicot

Fd-GOGAT -A

Fd-GOGAT -B

Fd-GOGAT -D

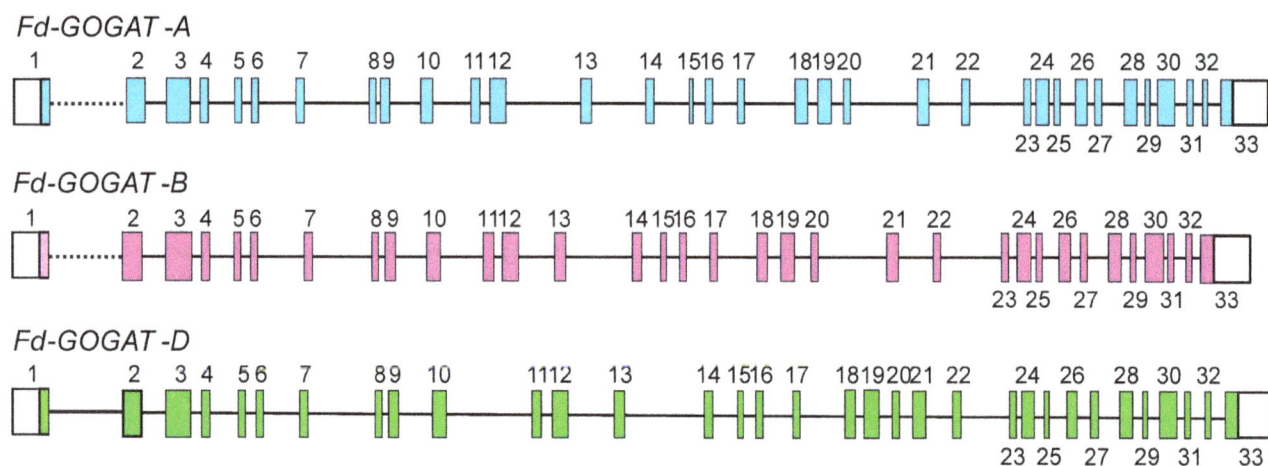

Figure 1. Diagrammatic representation of the structure of *Fd-GOGAT* genes. The three homoeologous Fd-GOGAT genes of wheat cv Chinese Spring are shown. Exons are numbered boxes. Coding sequences are colored, non-coding sequences are uncoloured boxes. Introns are intervening lines. Intron 1 is indicated by the dotted line.

sequences were more divergent, at 82% for *Arabidopsis* and 84% for poplar. The monocots are all of the same length (1519 amino acids), and the dicot examples are of identical length through the first 1501 residues. The only length different, and the major sequence differences between the monocots and dicots is with the signal peptide (Figure S3 in File S1) and residues beyond 1501 (Figure 2).

As noted above, there are two versions of GOGAT in plants – one requiring ferrodoxin and the other NADH. We have previously described the wheat *NADH-GOGAT* genes [16]. Figure 3 compares the amino acids sequences of the three wheat Fd-GOGAT homoeologous polypeptides with other monocots, dicots, and two green algae – the later being the most primitive organismal grouping yet found to contain both GOGAT versions. As expected, the two higher plant groups form two distinct branches for both GOGATs, with the green algae sequences distantly related within each of those two branches – consistent with the evolutionary distances. Less clarity is found when the analysis includes sequences from additional diverse groups; e.g., bacteria, cyanobacteria, Archeae, fungi, arthropods (Figure S5 in File S1).

Glutamate synthase gene mapping

In order to find polymorphisms in the *Fd-GOGAT-A* gene, the "CEL1 technology" was used in the durum wheat cvs Svevo and Ciccio - characterized by high and low grain protein content, respectively.

Allelic variations, SSRs, Indels, and SNPs are the major types of molecular markers that can be developed to detect DNA sequence. In particular, SNPs are nucleotide variations in the DNA sequence of individuals in a population and constitute the most abundant molecular markers in the genome. SNPs are widely distributed throughout genomes [36], vary in occurrence and distribution among species, and are usually more prevalent in the noncoding regions of the genome. There are several methods to discover SNPs; e.g., Sanger sequencing, DGCE (denaturing gradient capillary electrophoresis), denaturating HPLC and enzymatic mismatch cleavage. One of the most efficient and reliable enzymatic cleavage based method is TILLING (Targeting Induced Local Lesion In Genomes) [37,38]. This approach requires a treatment of the amplified DNA with *CEL1* endonuclease, or any of a number of single strand endonucleases, after

heteroduplex formation between the DNA of lines to be investigated. *CEL1* is a glycoprotein from celery, and many green plants. It cuts a DNA heteroduplex that contains a base-substitution or a DNA loop at the 3' most phosphodiester bond of the mismatched nucleotides and produces two complementary sized fragments from the original amplified product (up to ~1,500 bp amplicons). Electrophoresis size-separation on polyacrylamide or agarose gels is required to visualize any cleaved fragment.

A set of primer pairs covering the entire *Fd-GOGAT-A* genomic sequence in (Table 1) was analyzed in the two parental lines Svevo and Ciccio in order to find SNPs. Out of 19 primer combinations amplified and digested with *CEL1*, four of them, (combinations 2, 4, 5 and 19) showed a digestion pattern in the heteroduplex lane but not in the single parental lines digestions, suggesting real differences exists between the two cvs (Figure 4). Mutations were confirmed by sequencing the fragments (see Materials and Method). A partial sequence of *Fd-GOGAT-A* gene was obtained for both cvs Ciccio and Svevo, which allowed finding a total of five SNPs and two indels between the two cultivars. Of the five SNPs detected, 2 were located in intronic regions (one each for intron 5 and intron 31) and 3 were located in exonic region (one for each exon 6, exon 31, and exon 32). Both indels were located in introns, a 2 bp indel in intron 29 and an 8 bp indel in intron 5.

With the objective to genetically map the wheat *Fd-GOGAT-A* gene, a primer pair (Fd-GF, Fd-GR) was designed in one of the genomic region polymorphic between the cvs Ciccio and Svevo and analyzes in the RIL population. The linkage map "Svevo× Ciccio" developed by Gadaleta et al. [39] enriched of new DArT markers, were used for the gene mapping. The Fd-G marker produced, as expected, a single polymorphic fragment of 284 bp present in Svevo and lacking in Ciccio. The fragment was also physically mapped in the centromeric region of chromosome 2A on the bin C-2AS5-0.78 (data not shown). The linkage group identifying the 2A chromosome was of 129,9 cM including 31 markers (12 SSRs, 5 EST-SSRs, and 12DArT) and the *GS2-A2* and *Fd-GOGAT-A* genes (Figure 4A).

Relationship between Fd-GOGAT-A and grain protein content

We focused our attention on the *Fd-GOGAT* gene located on chromosome 2A, where several authors found QTLs for GPC not

```
1          20          40          60          80    x2 x3 100        120    130
CGVGFVANLSNEPSFNVVRDALTALGCMEHRGGCGSDNDSGDGAGLMSGIPWDLFDDWASKEGLAPFERTHTGVGMVFLPQNENSMAEAKAVEKVFTDEGLEVLGWRPVPFNLSVVGPNAKETMPNILQ wheat
CGVGFVANLKNEPSFNIVRDALTALGCMEHRGGCGADNDSGDGAGLMSGIPWDLFNDWASKQGLPPFERTNTGVGMVFLPQNEESMEEAKAVAKVFTDEGLEVLGWRPVPFNLSVVGFAKETMPNIQQ Brachy.
CGVGFVANLKNEPSFNIVRDALTALGCMEHRGGCGADNDSGDGSGLMSGIPWDLFNDWANXQGLPFDRTNTGVGMVFLPQDENSMEEAKAVVAKVFTDEGLEVLGWRTVPFNVSVVGRYAKETMPNIQQ Rice
CGVGFVANLKNMSSFDIVRDALMALGCMEHRGGCGADSDSGDGAGLMSAVPWDLFDDWASKQGLALFDRRNTGVGMVFLPQDEKSMEEAKAATEKVFVDEGLEVLGWRPVPFNVSVVGRNAKETMPNIQQ Maize
CGVGFIANLDNIPSHGVVKDALIALGCMEHRGGCGSDGSGLMSSIPWDFFNVWAKEQSLAPFDKLHTGVGMIFLPQDDTFMQEAKQVIENIFEKEGLQVLGWREVPVNVPIVGKNARETMPNIQQ Arab.
CGVGFIANLENKESHEIVKDALNALSCMEHRGGCGADNDSGDGSGLMTGVPWELFDNXANTQGIASFDKLHTGVGMVFLPKEAQLLNEAKKVIVNIFRQEGLEVLGWRPVPVNTSVVGFYAKETMPNIQQ Soybean
```

```
150         170         190         210         230   x4  250        260
IFVRIAKEDDADDIERELYICRKLIERATKSASWADELYFCSLSSRTIIYKGMLRSEVLGQFYLDLKNELYKSPFAIYHRRFSTNTSPRWPLAQPMRLLGHNGEINTIQGNLNWMRSREATIQSPVWRGR wheat
IFVKVAKEDDADDIERELYICRKLIERAAKSASWADELYFCSLSSRTIIYKGMLRSEVLGQFYLDLQNELYKSPFAIYHRRFSTNTSPRWPLAQPMRLLGHNGEINTIQGNLNWMRSREATIQSPVWRGR Brachy.
IFVKVAKEDNADDIERELYICRKLIERATKSASWADELYFCSLSSRTIVYKGMLRSEILGQFYLDLQNELYKSPFAIYHRRYSTNTSPRWPLAQPMRLLGHNGEINTIQGNLNWMRSREATLQSPVWRGR Rice
IFVKVAKEDNADDIERELYISRKLIERAAKSFSWADELYFCSLSSRTIVYKGMLRSELLGQFYLDLQNELYESPFAIYHRRFSTNTSPRWPLAQPMRLLGHNGEINTIQGNLNWMRSRETTLKSPVWRGR Maize
VFVKIAKEDSTDDIERELYICRKLIERAVATESWGTELYFCSLSNQTIVYKGMLRSEALGLFYLDLQNELYESPFAIYHRRYSTNTSPRWPLAQPMRFLGHNGEINTIQGNLNWMQSREASLKAAVWRGR Arab.
VFVKIVKEENVDDIERELYICRKLIEKAVSSESWGNELYFCSLSNQTIIYKGMLRSEVLGLFYSDLQNDLYKSPFAIYHRRYSTNTSPRWPLAQPMRLLGHNGEINTIQGNLNWMQSREPSLKSPVWRGR Soybean
```

```
280  x5  300         320 x6  340 x7  360 x8  380         390
ENELRPFGDPKASDSANLDSAAELLLRSGRSPAEAMMMLVPEAYKNHPTLSVKYPEVIDFYEYYKGQMEAWDGPALLLFSQGRTVGACLDRNGLRPARYWKTSDGFYVVASEVGVIPMDESKVVMKGRLG wheat
ENEIRPFGDPKASDSANLDNAAELLLRSGRSPAEAMMMLVPEAYKNHPTLSIKYPEVIDFYDYYKGQMEAWDGPALLLFSQGRTVGACLDRNGLRPARYWRTSDGFYVVASEVGVIPMDESKVVMKGRLG Brachy.
EHEIRPFGDPKASDSANLDSTAELLLRSGRSPAEAMMILVPEAYKNHPTLSIKYPEVIDFYDYYKGQMEAWDGPALLLFSQGRTVGACLDRNGLRPARYWRTSDDFYVVASEVGVIPMDESKVVMKGRLG Rice
EHEICPFGDPKASDSANLDSTAELLLRSGRSPAEALMILVPEAYKNHPTLSIKYPEVTDFYDYYKGQMEAWDGPALLLFSQGRTVGATLDRNGLRPARYWRTSDDFYVVASEVGVIPMDESKVVMKGRLG Maize
ENEIRPFGNPRGSDSANLDSAAEIMIRSGRTPEEALMILVPEAYKNHPTLSIKYPEVVDFYDYYKGQMEAWDGPALLLFSQGKTVGACLDRNGLRPARYWRTSDNFYVVASEVGVPVDEAKVTMKGRLG Arab.
ENEIRPFGNPKPGSDSANLDSAAEILIRSGRSPEEAMMILVPEAYKNHPTLSIKYPEVVDFYDYYKGQMEAWDGPALLLFSQGKTVGACLDRNGLRPARYWRTSDNNYVVASEVGVPVDESKVVLKGRLG Soybean
```

```
x9 410         430         450 x10 470         490        x11 520
PGMMITVDLETGGVLENTEVKKNVASAKPYGTWLQESTRSIKPVNFQSSPVMDNETILRHCDAFGYSSEDVQMVIETMASQGKEPTFCMGDDIPLAVLSQKPHMLFDYFKQRFAQVTNPAIDPLREGLVM wheat
PGMMITVDLQTGGVLENTEVKKNVASAKPYGTWLQQSTRSIKPVNFQSSPVMDNETVMRHCDAFGYSSEDVQMVIETMASQGKEPTFCMGDDIPLAVLSQKPHMLFDYFKQRFAQVTNPAIDPLREGLVM Brachy.
PGMMITVDLQTGGVLENTEVKKSVASANPYGSWLQQSTRSIKPVNFQSSVAMDNETVLRHCDAFGYSSEDVQMVIETMASQGKEPTFCMGDDIPLAVLSQKPHMLFDYFKQRFAQVTNPAIDPLREGLVM Rice
PGMMITVDLQTGGVLENTEVKKTVASASPYGTWLQECTRLIKPVNFLSSTIMDNETVLRHCDAFGYSSEDVQMVIESMASQGKEPTFCMGDDIPLAVLSQRPHLLYDYFKQRFAQVTNPAIDPLREGLVM Maize
PGMMIAVDLVNGGVYENTEVKKRISSFNPYGKWIKENSRFLKPVNFKSSTVMENEEILRSCDAFGYSSEDVQMVIESMASQGKEPTFCMGDDIPLAGLSQRPHMLYDYFKQRFAQVTNPAIDPLREGLVM Arab.
PGMMITVDLLGGGVYENTEVKKRVALSSPYGNWIKENLRTLKLLGNFLSASVLDNEAVLRHCDAFGYSSEDVQMVIESMAAQGKEPTFCMGDDIPLAALSQKPHMLFDYFKQRFAQVTNPAIDPLREGLVM Soybean
```

```
540 x12 560         580         600 x13620 640         650
SLEVNIGKRGNILEVGPENADCVTLSSPVLNEGELESLLKDPKLPKVLSTYFNIRKGLDGSLENAIKALCEEADAVRSGSQLLVLSDRSEALEPTRPAVPILLAVGAIHQHLIQNGLRMSASIVADTA wheat
SLEVNIGKRGNILEVGPENADCVALSSPVLNEGELDSLLKDTKLPKVLSTYFSIRKGLDGSLDKAIKALCEEADAAVRSGSQLLVLSDRSEALEPTRPAIPILLAVGAIHQHLIQNGLRMSASIVADTA Brachy.
SLEVNIGKRRNILEVGPENADCVTLSSPVLNEGELESLLNDSKLPKVLSTYFDIRKGLDGSLDKAIKVLCDEADAAVRSGSQLLVLSDRSEALEPTRPAIPILLAVGAIHQHLIQNGLRMSASIVADTA Rice
SLEVNIGKRGNILEVGPENADCVALSSPVLNEGELETLLNDSKLPKVLSTYFDIRKGLDGSLDKTIQALCEEADAAVRSGSQLLVLSDRSEAFEPTRPAIPILLAVGAIHQHLIQNGLRMSASIVADTA Maize
SLEVNIGKRGNILELGPENASCVILSNPVLNEGALEELMKDQYLKPKVLSTYFDIRKGVEGSLQKALVYLCEAADDAVRSGSQLLVLSDRSDRUEPTRPSIPIMLAVGAVHQHLIQNGLRMSASIVADTA Arab.
SLEVNIGKRRNILEIGPENASCVMLSSPVLNEGELESLLKDSYLKPQVLPTFFDITKGIEGSLEKALNKLCEAADEAVRNGSQLLILSDRSEALEPTHPAIPILLAVGTVHQHLIQNGLRMSASIIADTA Soybean
```

```
670 x14 690         710 x15 730 x16 750         770 x17 790
QCFSTHQFACLIGYGASAICPYLALETCRQWRLSNKTVNLMRNGKMPTVTIEQAQRNFIKAVKSGLLKILSKMGISLLSSYCGAQIFEIYGLGQEVVDLAFCGSVSKIGGLTLNELGRETLSFWVRAFSE wheat
QCFSTHQFACLIGYGASAICPYLALETCRQWRLSNKTVNLMRNGKMPTVTIEQAQRNFIKAVKSGLLKILSKMGISLLSSYCGAQIFEIYGLGQEVVDLAFCGSVSKIGGLTLDELGRETLSFWVKAFSE Brachy.
QCFSTHQFACLIGYGASAICPYLALETCRQWRLSNKTVNLMRNGKMPTVTIEQAQRNFIKAVKSGLLKILSKMGISLLSSYCGAQIFEIYGLGQEVVDLAFCGSVSKIGGLTLDELGRETLSFWVKAFSE Rice
QCFSTHHFACLVGYGASAVCPYLALETCRQWRLSNKTVNLMRNGKMPTVTIEQAQRNFIKAVKSGLLKILSKMGISLLSSYCGAQIFEIYGLGQEVVDLAFTGSVSKISGLTFDELARETLSFWVKAFSE Maize
QCFSTHHFACLVGYGASAVCPYLALETCRQWRLSNKTVAFMRNGKIPTVTIEQAQKNYTAVNAGLLKILSKMGISLLSSYCGAQIFEIYGLGQDVVDLAFTGSVSKISGLTFDELARETLSFWVKAFSE Arab.
QCFSTHHFACLIGYGASAVCPYLALETCRQWRLSNKTVNLMRNGKMPTVSIEQAQKNYCKAVKAGLLKILSKMGISLLSSYCGAQIFEVYGLGKEVVDLAFRGSVSKIGGLTFDEVARETLSFWVKAFSE Soybean
```

```
800 x18 820         840 x19 860         880         900 x20 910
DTAKRLENFGFIQSRPCGEFHANNPEMSKLLHKAIREKSDNAYTIYQQHLASRPVNVLRDLVELKSERTPIPIGKVEPATSIVERFCTGGMSLGAISRETHEAIAIAMNRIGGKSNSGEGGEDPIRWSPL wheat
DTAKRLENFGFIQSRPCGEFHANNPEMSKLLHKAIREKSDNAYTIYQQHLASRPVNVLRDLVELKSDRAPIPIGKVEPATSIVERFCTGGMSLGAISRETHEAIAIAMNRIGGKSNSGEGGEDPIRWSPL Brachy.
DTAKRLENFGFIQSRPCGEYHANNPEMSKLLHKAVREKSDNAYTVYQQHLASRPVNVLRDLLELKSDRAPIPIGKVEPATSIVERFCTGGMSLGAISRETHEAIAIAMNRIGGKSNSGEGGEDPIRWSPL Rice
DTAKRLENFGFIQSRPCGEYHANNPEMSKLLHKAIREKRDNAYTVYQQHLASRPVNVLRDLLELKSDRAPIPIGKVESATSIVERFCTGGMSLGAISRETHEAIAIAMNRIGGKSNSGEGGEDPIRWNPL Maize
DTTKRLENFGFIQFRPCGEYHSNNPEMSKLLHKAIREKSETAYAVYQQHLSNRPVNVLRDLLEFKSDRAPIPGKVEPAVAIVQRFCTGGMSLGAISRETHEAIAIAMNRIGGKSNSGEGGEDPIRWKPL Arab.
DTAKRLENFGFIQSRPCGEYHANNPEMSKLLHKAVRQKSQSAFSVYQQYLANRPVNVLRDLFKSDRAPIPVGKVEPASSIVQRFCTGGMSLGAISRETHEAIAIAMNRIGGKSNSGEGGEDPIRWKPL Soybean
```

```
930 x21 950         970         990         1010 x22 1030 1040
TDVVDGYSATLPHLKGLQNGDTATSAIKQVASGRFGVTPTFLVNAEQIEIKIAQGAKPGEGGQLPGKKVSAYIARLRNSKPGVPLISPPPHHDIYSIEDLAQLIFDLHQINPKAKVSVKLVAEAGIGTVA wheat
EDVVDGYSPTLPHLKGLQNGDTATSAIKQVASGRFGVTPTFLVNAEQIEIKIAQGAKPGEGGQLPGKKVSAYIARLRNSKPGVPLISPPPHHDIYSIEDLAQLIYDLHQINPKAKVSVKLVAEAGIGTVA Brachy.
ADVEDGYSPTLPHLKGLQNGDTATSAIKQVASGRFGVTPTFLVNADQIEIKIAQGAKPGEGGQLPGKKVSAYIARLRNSKPGVPLISPPPHHDIYSIEDLAQLIYDLHQINPKAKVSVKLSEAGIGTVA Rice
TDVVDGYSPTLPHLKGLQNGDTATSAIKQVASGRFGVTPTFLVNADQIEIKIAQGAKPGEGGQLPGKKVSAYIARLRNSKPGVPLISPPPHHDIYSIEDLAQLIFDLHQINPNAKVSVKLVAEAGIGTVA Maize
TDVVDGYSPTLPHLKGLQNGDIATSAIKQVASGRFGVTPTFLVNADQLEIKIAQGAKPGEGGQLPGKKVSAYIARLRSSKPGVPLISPPPHHDIYSIEDLAQLIFDLHQINPNAKVSVKLVAEAGIGTVA Arab.
TDVVDGYSPTLPHLKGLQNGDTATSAIKQVASGRFGVTPTFLANADQLEIKIAQGAKPGEGGQLPGKKVSMYIARLRNSKPGVPLISPPPHHDIYSIEDLAQLIFDLHQPKAKVSVKLVAEAGIGTVA Soybean
```

```
x23 1060         1080 x24 1100         1120         1140 x25 1160 1170
SGVSKANADVICISGHDGGTGASPISSIKHAGGPWELGLTETHQTLIQNGLRERVVLRVDGGFRSGLDVLLAAAMGADEYGFGSVAMIATGCVMARICHTNNCPVGVASCREELRARFPGVPGDLVNYFL wheat
SGVSKANADVICISGHDGGTGASPISSIKHAGGPWELGLTETHQTLIQNGLRERVVLRVDGGFRSGLDVLLAAAMGADEYGFGSVAMIATGCVMARICHTNNCPVGVASCREELRARFPGVPGDLVNYFL Brachy.
SGVSKGNADIICISGHDGGTGASPISSIKHAGGPWELGLSETHQTLIQNGLRERVVLRVDGGFRSGLDVLMAAAMGADEYGFGSVAMIATGCVMARICHTNNCPVGVASCREELRARFPGVPGDLVNYFL Rice
SGVAKGNADIICISGHDGGTGASPISSIKHAGGPWELGLTETHQTLIANGLRERVILRVDGGLKSGVDVLMAAAMGADEYGFGSLAMIATGCVMARICHTNNCPVGVASCREELRARFPGVPGDLVNYFL Maize
SGVAKGNADIICISGHDGGTGASPISSIKHAGGPWELGLTESHQTLIENGLRERVILRVDGGFRSGVDVMMAAIMGADEYGFGSVAMIATGCVMARICHTNNCPVGVASCREELRARFPGVPGDLVNYFV Arab.
                                                                                                                              Soybean
```

```
x26 1190         1210 x27 1230         1250 x28 1270         1290 1300
FVAEEVRATLAQLGYEKLDDIIGRTDLLKPKHISLVKTQHIDLAYLLNAGLPKWSSSQIRSQDVHSNGPVLDETILADPEVSDAIENEKEVSKTYPIYNVDRAVCGRVAGAIAKKYGDTGFAGQLNITF wheat
FVAEEVRATLAQLGYEKLDDITGRTDLLKPKHISLVKTQHIDLAYLLNSGLPKWSSSQIRSQDVHSNGPVLDETILADPEVSDAIENEKEVSKTFPIYNVDRAVCGRVAGAIAKKYGDTGFAGQLNITF Brachy.
FVAEEVRATLAQLGFEKLDDIIGRTDILKAKHVSLAKTQHIDLKYLLSAGLPKWSSSQIRSQDVHSNGPVLDETILADPCISDAIENEKEVSKTFQIYNVDRAVCGRVAGVIAKKYGDTGFAGQLNITF Rice
FVAEEVRAALAQLGYEKLDDIIGRTDLLKPKHISLVKTQHIDLGYLLSNAGLPEWSSSQIRSQDVHTNGPVLDETILADPEIADAIENEKEVSKAFQIYNVDRAVCGRVAGVIAKKYGDTGFAGQLNITF Maize
YVAEEVRGILAQLGYNSLDDIIGRTELLLRPRDLSYILSVGTPSLSSTEIRNQEPHTNGPVLDDOILADPLVIDAIENEKVVEKTVKIONVDRAACGRIAGVIAKKYGDTGFAGQLNITF Arab.
YVAEEVRGILAQLGIKKLDDVIGRTDLFQPRDISLAKTQHLDLSYILNVGLPKWSSSEIRNQEPHTNGPVLDDVLLADPEIAYAIENEKVANKTIKIYNIDRAACGRIAGVIAKKYGDTGFAGQLNITF Soybean
```

```
x29 1320         x30 1340         1360         1380         1400 x31 1420 1430
TGSAGQSFGCFLTPGMNVRLVGEANDYVGKGMAGGELVVVPVDDTGFVPEDAAIVGNTCLYGATGGQVFVRGKTGERFAVRNSLGQAVVEGTGDHCCEYMTGGCVVVLGKVGRNVAAGMTGGLAYMLDED wheat
TGSAGQSFGCFLTPGMNVRLVGEANDYVGKGMAGGELVVVPVDDTGFVPEEAAIVGNTCLYGATGGQVFVRGKTGERFAVRNSLGQAVVEGTGDHCCEYMTGGCVVVLGKVGRNVAAGMTGGLAYILDED Brachy.
TGSAGQSFGCFLTPGMNIRLVGEANDYVGKGMAGGELVVVPVEKTGFVPEDAAIVGNTCLYGATGGQVFVRGKAGERFAVRNSLGQAVVEGTGDHCCEYMTGGCVVVLGKAGRNVAAGMTGGLAYILDED Rice
NGSAGQSFGCFLTPGMNIRLVGEANDYVGKGMAGGELVVVPVEKTGFVPEDATIVGNTCLYGATGGQVFVRGKAGERFAVRNSLGQAVVEGTGDHCCEYMTGGCVVVLGKAGRNVAAGMTGGLAYILDED Maize
LGSAGQSFGCFLITPGMNIRLIGESNDYVGKGMAGGEIVVTPVEKIGFVPEEATIVGNTCLYGATGGQIFARGKAGERFAVRNSLAEAVVVEGTGDHCCEYMTGGCVVVLGKVGRNVAAGMTGGLAYLLDED Arab.
TGSAGQSFACFLTPGMNIRLVGEANDYVGKGMAGGEIVVTPVDKTGFEPEDAAIVGNTCLYGATGGQVFVRGRAGERFAVRNSLAEAVVVEGAGDHCCEYMTGGCVVVLGKVGRNVAAGMTGGLAYFLDED Soybean
```

```
x32 1450         1470 x33 1490         1510 1519
DTLVPKVNKEIVKMQRVNAPAGQMQLKGLIEAYVEKTGSTKGAKILSEWEAYLPLFWQLVPPSEEDSPEACAEFERVLARQKTAVQSAK. wheat
DTLVPKVNKEIVKMQRVNAPAGQMQLKGLIEAYVEKTGSVKGAKILSEWEAYLPLFWQLVPPSEEDSPEACAEFERVLARQATAVQSAK. Brachy.
DTLVPKVNKEIVKMQRVNAPAGQMQLKGLIEAYVEKTGSEKGAIILREWEAYLPLFWQLVPPSEEDSPEACAEFERVLAKQATTVQSAK. Rice
DTLVPKVNKEIVKMQRVNAPAGQMQLKGLIEAYVEKTGSIAILREWEAYLPLFWQLVPPSEEDSPEACAEFERVLAKQATTQLSAK. Maize
DTLLPKINREIVKIQRVTAPAGELQLKSLIEAHVEKTGSSKGATILNEWEKYLPLFWQLVPPSEEDTPEASAAYVRTSTGEVTF-QSA. Arab.
NTFIVPKVNGEIVKIQRVSAPVGQMQLKSLIEAHVEKTGSTKGAAILKDWEKYLSLFWQLVPPSEEDTPEANAKYDTTTADQVTY-QSA. Soybean
```

Figure 2. Plant Fd-GOGAT amino acid sequences. Four monocot and two dicot Fd-GOGAT amino acid sequences for the mature protein are aligned with Clustal V: Wheat (A genome; present report), *Brachypodium* (BRADI1G19080), Rice (Os07g46460), Maize (NM_001112223), *Arabidopsis* (CP002688), Soybean. (AK245357).

linked to pleiotropic effects of low productivity in different genetic materials [18,20,40].

The RIL population Ciccio×Svevo was evaluated for grain protein content (GPC) and grain yield components in five field trials in southern Italy. The analysis of variance for GPC revealed highly significant differences between the parental lines Ciccio and Svevo and among the RIL in each of the field trials, suggesting that the RIL population was suitable for studying the putative involvement of the *GOGAT* genes in the control of grain protein content.

QTL analysis reported in Blanco et al. [20] detected ten QTLs for grain protein content on chromosome arms 1AS, 1AL, 2AS (two loci), 2BL, 3BS, 4AL, 4BL, 5AL and 6BS. In order to assess the putative relation between the *Fd-GOGAT-A* genes and GPC we re-analyzed the RIL data with the Inclusive Composite Interval Mapping method [41] in each of the five environments and across environments using the "Ciccio×Svevo" map data [39] enriched with new DArT marker and including the segregation data of the new Fd-GOGAT-A marker.

Among all the putative QTL for GPC in individual environments and across environments, only QTLs with LOD≥3.0 values were considered in the present work. This new QTL analysis

revealed that the *Fd-GOGAT-A* gene mapped in the present work co-localized with a major QTL for GPC. In particular the *Fd-GOGAT-A* gene co-localised with a GPC-QTL detected on chromosome arm 2AS, in the region comprised between the markers *Xgwm372c* and the EST-SSR TC82001 (including the two closer markers *Xgwm339* and *Xgwm95*) significant in two environment and across environments. (Table 2, Figures 4 and 5. The Svevo allele increasing the trait had a positive additive effects ranging from 0.13 to 0.27 with a mean value of 0.24. R^2 value, and the percentage of phenotypic variance explained by the additive effects of the mapped QTL, ranged from 6% to 19.4% between environments and the mean was 19.4 across environments.

Conclusions

The current report describes the characterization of the bread wheat genomic sequence of *Fd-GOGAT* genes and their association with grain protein content (GPC). These gene sequences were useful to study the grain protein content in durum wheat, a quantitative trait controlled by multiple genes and influenced by environmental conditions.

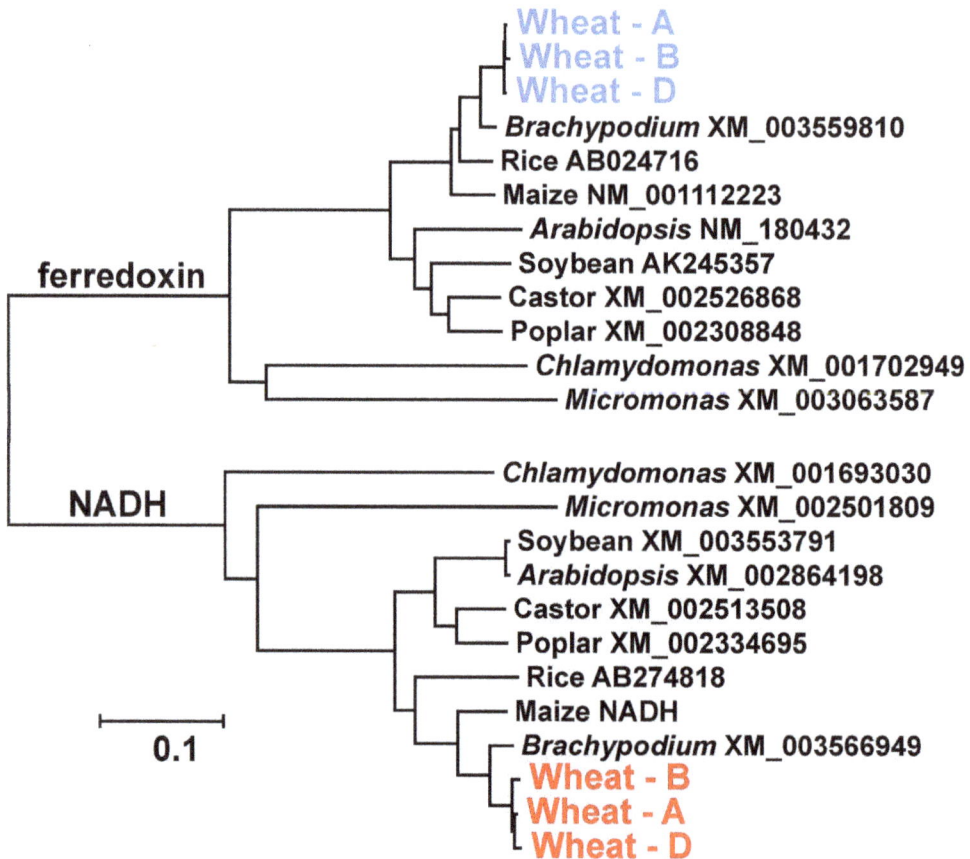

Figure 3. Phylogenetic tree of GOGAT polypeptides. The mature polypeptides for Fd- and NADH-GOGAT from a selection of plants an green algae were aligned with Clustal W and a nearest-neighbour tree generated with MEGA5. All three wheat homoeologues for Fd-GOGAT are in blue and red for the NADH-GOGAT version.

Table 1. *Fd-GOGAT-A1* specific primer name, sequence and PCR condition used for SNPs detection.

Primer name	Forward sequence (5'-3')	Reverse sequence (5'-3')	PCR conditions
FdGOGAT/A_1	CGCCGTCGCTGTTGCCGC	TGCTGGCCCAATCATTAAACAAGT	69°C,
FdGOGAT/A_2	GCTTGTGGTGTTGGGTTTGTC	CTCTCTATCAGCTTTCGGCAG	60°C,
FdGOGAT/A_3	CGGTTCCTTTCAATCTATCAGT	AATCGGATGCTTTAGGGTCAC	58°C,
FdGOGAT/A_4_F	GGAAGCCACAATACAATCTC	CTAAACAAAAGTAAAGCAGG	54°C
DN_FdGOGAT/A_5	TCTATGAATACTACAAAGGT	CAACATAAACAAAACCATCT	54°C,
DN_FdGOGAT/A_6	TGACGGAAGGACGGTAGGGG	GAGGATTGGAAGTTGACAGGCT	63°C
DN_FdGOGAT/A_7	AAACCCTATGGAACTTGGCT	AAATCGCTGCTTGAAATAAT	65°C
DN_FdGOGAT/A_8	TCACAAGGGAAGGAGCCAACAT	TCAGAACGATCGGAAAGCAC	62°C
DN_FdGOGAT/A_9	TACCCTATCAAGTCCTGTCCTG	CAAACTGATGGGTGCTGAAA	58°C
DN_FdGOGAT/A_10	CACGGCCTGCTGTCCCAATAC	TGGTCACTGTGGGCATCTTG	68°C
DN_FdGOGAT/A_11	CATATCTGGCATTGGAAACAT	AGTTTTCCAGCCTCTTTGCG	65°C
DN_FdGOGAT/A_12	CTGGGTCGAGAAACACTATCA	TTGTTGGTAGATGGTGTATGC	58°C
DN_FdGOGAT/A_13	GTCAAAGCTGCTGCACAAAG	TTAATGGCACTTGTGGCGGT	60°C
DN_FdGOGAT/A_14	CAGATGTTGTTGATGGGTATT	CCCTTGTGCAATCTTTATCTCA	58°C
DN_FdGOGAT/A_15	TGCATCTGGACGTTTTGGTG	GCATTTGCCTTAGATACTCCAG	60°C
DN_FdGOGAT/A_16	GGTGTCGGTAAAGCTTGTAG	TGTGTTTCCGTAAGACCAAG	58°C
DN_FdGOGAT/A_17	CTCAATCAAGCATGCTGGGGG	CAAGATCAATGTGCTGCGTTT	62°C
DN_FdGOGAT/A_18	TACGAGCCACATTAGCCCAGT	CACATAATCGTTGGCCTCTC	60°C
DN_FdGOGAT/A_19	GGCAGTCCTTTGGTTGTTTTCT	TCTTCTTCGCTGGGTGGCA	62°C

Each PCR starts with 5 min at 94°C, followed by 35 cycles of 1 min denaturation at 94°C, 1 min annealing at the specific annealing T° reported above, and 2 min elongation at 72°C, and ends with a final elongation of 20 min at 72°C.

The involvement of *Fd-GOGAT-A* gene in the control of GPC was carried out with three actions: isolation of homoeologous allele located on 2A chromosome in two elite durum wheat cultivars with different GPC, gene mapping in a segregant population, and association between the gene and GPC evaluated in five environments. In the present work, we were able to assemble the three complete homoeologous genes from the three hexaploid wheat genomes using as initial query a partial barley sequence to extract and assembly 454 reads from public databases. The three homoeologous genes have the same intron/exon structure with several differences in both intron and exon. We then used a set of aneuploid lines that led us to attribute PCR fragments to the A and B genomes.

In order to quickly screen the gene sequence looking for SNPs between our durum cultivars, we followed a "PCR/*CEL1* strategy" similar to a TILLING approach. In this approach, *CEL1* cleaves at the site of heteroduplex indicating mismatches in the sequences. This allowed us to identify only SNPs between our two cultivars avoiding the sequencing of the complete genes. A total of five SNPs and two indels were found of which an insertion of 8 bp in cv Svevo was used to construct specific markers and map the gene in a segregant population (Svevo×Ciccio). Mapping data of the polymorphic fragment allowed us to identify the locus named *Fd-GOGAT-A*, in the centromeric region of chromosome 2A. QTL analysis performed with CIM (Composite Interval Mapping) confirmed the presence of the marker in a major QTL for grain protein content. Several studies localized QTLs for GPC on chromosomes of group 2 [18,42,43,44]. The QTL analysis carried out in the RIL population Ciccio×Svevo, previously evaluated for grain protein content in five environments, showed that *Fd-GOGAT-A* co-localized with QTLs for GPC on chromosome arms 2AS – the CIM analysis identified a major QTL with a stable effect in two environments and across environments. The genomic differences existing between the two cvs might modify the final predicted protein functionality or might have a key role in the gene regulation and gene expression, determining a different GPC. Further investigation are needed to prove this involvement through genetic transformation and/or sexual complementation.

The influence of the group 2 chromosomes on GPC control was reported in different genetic material have previously indicated the key role of these chromosomes play in the control of GPC. QTLs for GPC on 2A and 2B were firstly reported by Joppa and Cantrell [45] using durum wheat ssp. *dicoccoides* chromosome substitution lines. Prasad et al. [23] reported a protein content QTL on a distal region of the chromosome arm 2AS, while Blanco et al. [40] identified a GPC-QTL on 2AS near the centromere. More recently was found a stable QTL on 2A (*QGPC.usw-A2*) expressed in three environments and a QTL on 2B (*QGpc.usw-B3*) significant in four of the six environments analyzed [31].

The coding sequences of the homoeologous wheat *Fd-GOGAT* genes showed higher conservation among the three than the *NADH-GOGAT* homoeologues previously described [16], with only seven amino acid differences among the three *Fd-GOGAT* homoeologues. The conservation extended in comparisons to other plant Fd-GOGAT amino acid sequences, with only the C-terminal region having major sequence divergence in monocots vs dicots. The two forms of GOGAT are clear when the analysis compares among plants and green algae (Figure 3).

The precise role and location of the two forms of GOGAT are not completely understood, but the NADH form seems involved in development and re-assimilation of ammonia with either the cytosol or amyloplasts [46], while the ferredoxin form in a key component in photosynthesis and nitrogen fixation within the chloroplasts [47]. Both plastid forms are believed to have

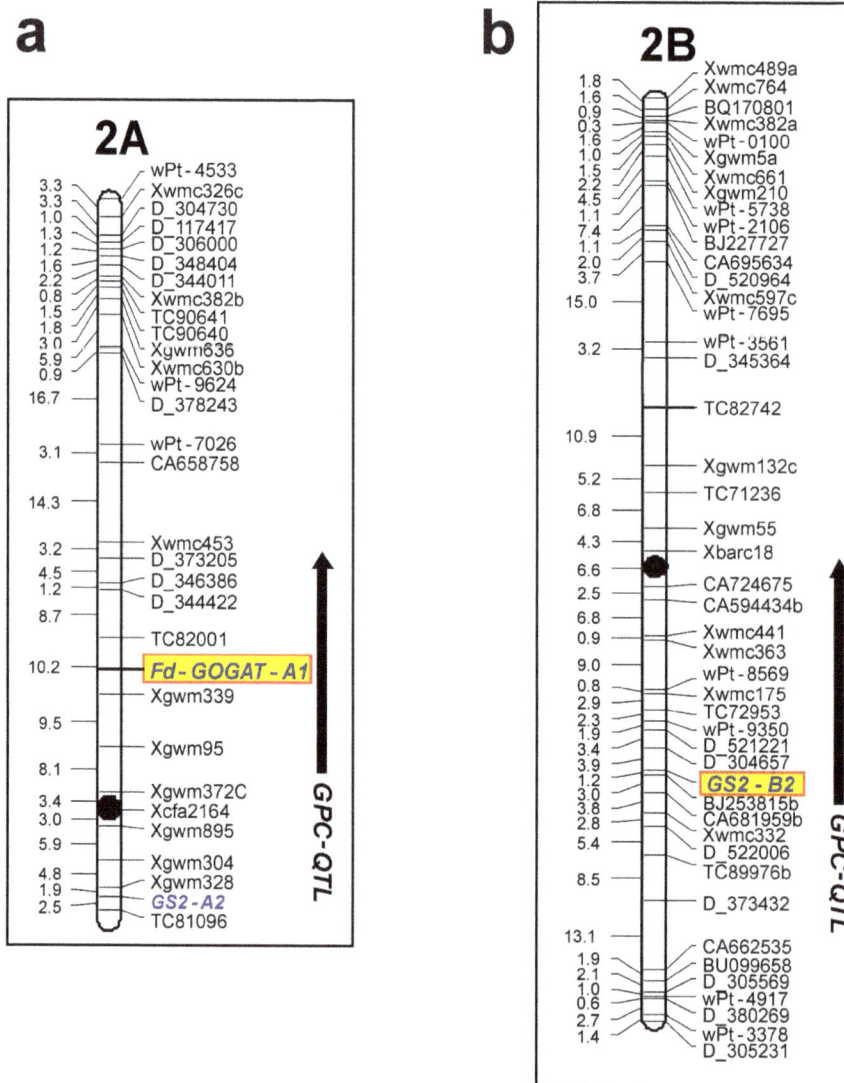

Figure 4. Genetic map position of Fd-GOGAT. The map positions and QTLs for grain protein content are shown in the side panels. *Fd-GOGAT-A* and *GS2-B2* loci, associated with GPC QTLs, respectively on chromosomes 2A and 2B, are highlighted in yellow. Black dots represent centromere. The genetic maps of wheat chromosomes 2A and 2B are from the Svevo×Ciccio RIL mapping population.

Table 2. Additive gene effects of the detected QTL for grain protein content in the 2AS region (flanked by the markers TC82001 and *Xgwm372c* and including *FdGOGAT-A1*) in the Svevo×Ciccio RIL population grown at five environments.

Environment	Effect[a]	LOD	R^2
Valenzano 2006	0,13	1,6	6,4
Foggia 2006	**0,27**	**4,7**	**17,4**
Gaudiano 2006	**0,27**	**3,5**	**13,3**
Valenzano 2007	*0,18*	*2,8*	*10,7*
Foggia 2007	0,20	1,5	6,0
Across environments	**0,24**	**5,3**	**19,4**

[a]Effect: positive additive effects are associated with an increased effect from Svevo allele.
R^2: Percentage of phenotypic variance explained by the additive effects of the mapped QTL.

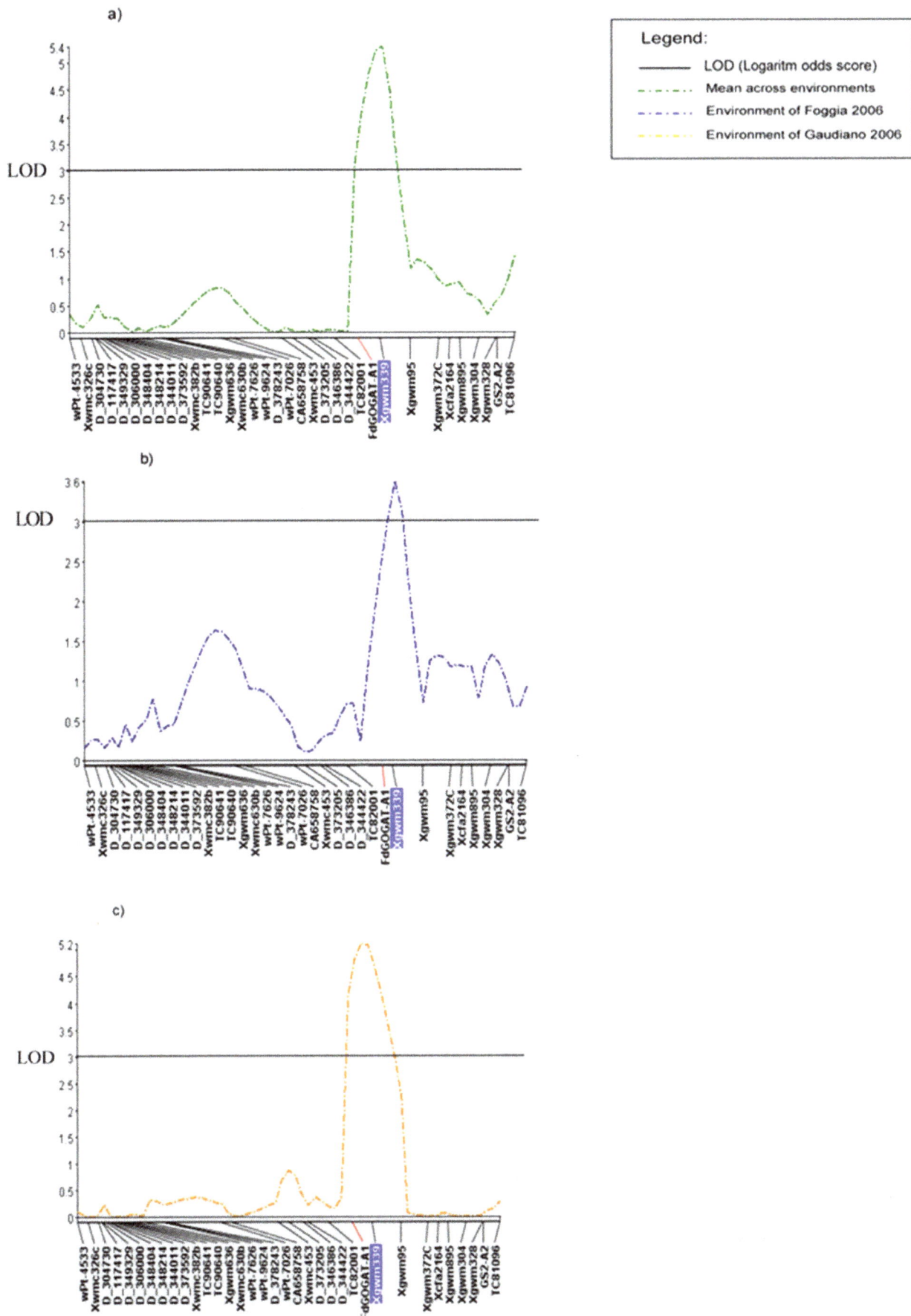

Figure 5. Grain protein QTLs. LOD score scan on chromosome 2A for QTLs associated with grain protein content. The significant scan for QTLs for each environment: A) mean across environments; B) Foggia 2006; C) Gaudiano 2006. The position and the name of molecular markers are shown on the chromosome along the horizontal axis. The LOD score scan was obtained by ICIM with highlights the markers used as cofactors. LOD stands for logarithm of the odds (to the base 10). A LOD score of three or more is generally considered significant - a LOD score of three means the odds are a thousand to one in favour of genetic linkage.

originated from endosymbiotic cyanobacteria [48]. The phylogenetic tree of Figure S5 in File S1 initially suggests cyanobacteria (*Cyanobacterium, Cyanothece, Leptolyngbya*) may possess both forms, but closer examination of the sequences finds different accessions of each of those three species contains only a single gene which can cluster with either the Fd or NADH GOGAT forms (not shown). When combined with the anomalies in annotations and the complex branching in Figure S5 in File S1, a more detailed analysis of GOGAT genes in all phyla is needed to better understand GOGAT gene evolution.

The comparison of a set of plant *Fd-GOGAT* genes (wheat, *Brachypodium*, rice and maize) suggests regions of greater sequence and structure conservation likely related to critical enzymatic functions and metabolic control. The higher identity was observed between wheat and *Brachypodium* sequences, as expected due to the genetic closest between the two species.

Although the two forms of GOGAT, Fd and NADH, catalyze the same reaction, the gene structures and their roles in plant metabolism are not identical. The two forms have detectable conservation in amino acid sequence up to the point where the NADH-GOGAT form encodes a pyridine nucleotide-disulfide oxidoreductase domain at the C-terminus (arrowhead in Figure 2; dot blot in Figure S6 in File S1) – indicating additional metabolic aspect of the *NADH-GOGAT gene*.

In higher plants, ammonium, whether resulting from nitrate assimilation or from secondary sources, is first incorporated into glutamine in a reaction catalysed by glutamine synthetase, and then glutamate synthase (glutamine:oxoglutarate amidotransferase; GOGAT) catalyses the combination of glutamine with 2:oxoglutarate to form two molecules of glutamate, one of which serves as substrate for GS, while the other one is available for transport, storage or further metabolism. These two reactions form a cycle referred to as the GS/GOGAT pathway [49]. Tissues and subcellular localization of both *GS* and *GOGAT* genes, as well as their different expression level during plant growth, resulting in different enzymatic activity along all phenological stages, determine a specific synergy between the two genes [47]. In particular cytosolic GS1 is involved in the pathway with NADH-GOGAT, while plastidic GS2 works preferentially with Fd-GOGAT. This seems to be confirmed in our genetic material, where both genes *Fd-GOGAT-A* and *GS2* were associated with QTLs for GPC. In fact, a *GS2* gene on chromosome 2B (*GS2-B2*) was mapped and found to be involved in GPC control (Figure 4b) [19]. So, from these studies, it's suggested that *Fd-GOGAT-A* gene works in synergy with *GS2-B2* gene. This hypothesis could be confirmed by further analysis such as gene expression analysis.

The present work determined the DNA sequences of the three Fd-GOGAT genes in hexaploid wheat and identified an allele for the increment of GPC in durum wheat and open the way to further investigation using the forward and reverse genetic approaches that have been successfully used to validate the role of *Fd-GOGAT-A* genes for grain production both in rice and maize [43].

Materials and Methods

Plant material

Two durum wheat cultivars (Ciccio and Svevo) were used to investigate the relation of *Fd-GOGAT* genomic sequences to GPC.

The durum cultivars are parents of a mapping population represented by a set of 120 recombinant inbred lines (RILs) [39]. The two parents were chosen for differences in important qualitative and quantitative traits; i.e., grain yield components, grain protein content, yellow pigment, and adaptive traits.

Genomic DNA was isolated from fresh leaves using the method described by Sharp et al. [50] and subsequently purified by phenol-chloroform extraction. DNA amplifications were carried out as described by Gadaleta et al. [39].

Nulli-tetrasomic lines (NTs) of *Triticum aestivum* cv Chinese Spring [34,35] were used to physically localize *Fd-GOGAT* markers to chromosomes. Chinese Spring di-telosomic lines [51] were used for the assignment of markers to each chromosomal arm. Physical location on chromosome bins of each PCR fragment was obtained using a set of wheat deletions lines dividing the A and B genome chromosomes in bins (kindly provided by B. S. Gill, USDA-ARS, Kansas State University) [52].

Genes in cv Chinese spring

To isolate the complete sequences of *Fd-GOGAT* genes in bread wheat, we used the cDNA sequence of a partial barley Fd-GOGAT mRNA (Gene S58774) [14]. This sequence was used as the initial query probe to extract matching 454 genomic sequences of cv Chinese Spring publicly available (http://www.cerealsdb.uk. net/CerealsDB/Documents/DOC_search_reads.php), which were then assembled using the SeqMan module of the Lasergene software (DNAStar, Inc.).

Fd-GOGAT sequences in the Italian durum wheat cvs Ciccio and Svevo

By using Oligo Explorer and Primer3 (http://frodo.wi.mit.edu/primer3/) software, a set of genome specific primer pairs were designed for the distinct *Fd-GOGAT* hexaploid sequences previously obtained: Chinese Spring 2A and 2B *Fd-GOGAT*. A-genome specific primer pairs were used to amplify target DNA from both Ciccio and Svevo parental lines using PCR condition reported in Table 1.

Single PCR fragments were directly purified with EuroGold Cycle Pure Kit and sequenced. Multiple PCR products were first cloned into the pCR4-TOPO vector (Invitrogen, Cloning Kit) following the manufacturer's instructions, and then each fragment was sequenced (http://www.bmr-genomics.it/).

Sequences assembly were carried out using CodonCode Aligner and Geneious software. Gaps and uncertain sequence were resolved by primer walking. Regions of less coverage or ambiguous reads were rechecked with primers designed to cover those regions.

Digestion with CEL1 and revelation fragments

In order to discover and map unknown mutations between the genomic sequences of the two varieties Ciccio and Svevo, Single Nucleotide Polymorphisms (SNPs) were detected using the Surveyor nuclease kit (Transgenomic, Inc.), following manufacture's instruction. This approach requires a treatment of the amplified DNA with *CELI* endonuclease, or any of a number of single strand endonucleases, after heteroduplex formation between the lines to be investigated. Surveyor nuclease cleaves with high specificity at the 3′ side of any mismatch site in both DNA strands, including all base substitutions and insertion/deletions up to at least 12 nucleotides.

Heteroduplex formation, CEL I digestion and gel analysis

An aliquot of each PCR product was used for hybridization to form heteroduplexes between the parental lines, following thermal cycler program: 95°C for 2 min; loop 1 for 8 cycles (94°C for 20 s, 73°C for 30 s, reduce temperature 1°C per cycle, ramp to 72°C at 0.5°C/s, 72°C for 1 min); loop 2 for 45 cycles (94°C for 20 s, 65°C for 30 s, ramp to 72°C at 0.5°C/s, 72°C for 1 min); 72°C for 5 min; 99°C for 10 min; loop 3 for 70 cycles (70°C for 20 s, reduce temperature 0.3°C per cycle); hold at 8°C. After annealing, DNA has been treated with Surveyor nuclease to cleave heteroduplexes by adding 0.2 µl of enzyme and 1.3 µl of Buffer 1X. The digestion step was done at 5°C for 90 minutes and stopped immediately by adding 5 µl of 0.225 EDTA and 2 µl of bromophenol blue loading dye per sample and mixing thoroughly.

PCR fragments of over 1000 bp were analyzed on 3% polyacrylamide gels. The 3% polyacrylamide gel was made with: 15 ml 5×TBE, 120 l dH2O, 11 ml 40% bis-acrylamide, 110 µl TEMED and 1 ml 10% APS. We used 100-lane vertical electrophoresis system (CBS Scientific, Del Mar, CA, USA). The images were analyzed manually on PowerPoint (Microsoft Corp., Seattle, WA, USA).

In order to confirm the polymorphisms within genome specific genes, the heteroduplex hybridization digestion pattern was compared to the ones obtained in each parental lines. Moreover, the PCR product giving a digestion pattern after *CEL1* treatment were re-amplified and sequenced (http://www.bmr-genomics.it/).

Development of Fd-GOGAT specific markers, mapping and correlation with grain protein content

The *Fd-GOGAT* sequences of the two cvs Svevo and Ciccio were aligned using ClustalW from EBI website to identify polymorphisms. The marker Fd-G (Forward 5′-GCAAAACAAC-CAGGGCACATA-3′, Reverse 5′-TAGCTCCCTTCCCCAA-TACAT-3′) for *Fd-GOGAT-A*, was designed with Oligo Explorer software in the polymorphic regions. The polymorphic marker was mapped in the "Svevo×Ciccio" mapping population. The observed segregation ratio for the marker was tested by chi-square analysis for deviation from the expected 1:1 ratio. The linkage analysis was performed by JoinMap v. 4.0 [53] and the Kosambi mapping function was used to calculate map distances [54].

Grain protein content (GPC) and yield components were evaluated in the RIL population "Svevo×Ciccio" in five different environments (Valenzano 2006, Gaudiano 2006, Foggia 2006, Valenzano 2007, Foggia 2007). QTL analysis was performed following the procedure indicated by Blanco et al. [20].

DNA sequence analysis

DNA sequences were analyzed using the Seqman and Megalign modules of the Lasergene software (DNAstar, Inc.), and MEGA5 [55].

References

1. Yaronskaya E, Vershilovskaya I, Poers Y, Alawady AE, Averina N, et al. (2006) Cytokinin effects on tetrapyrrole biosynthesis and photosynthetic activity in barley seedlings. Planta 224: 700–709.
2. Forde BG, Lea PJ (2007) Glutamate in plants: metabolism, regulation, and signalling. J Exp Bot 58: 2339–2358.
3. Lea PJ, Miflin BJ (2003) Glutamate synthase and the synthesis of glutamate in plants. Plant Physiol Bioch 41: 555–564.

Supporting Information

File S1 Supporting figures and text. Figure S1. Chromosome mapping of the Fd-GOGAT genes on chromosomes 2A and 2B. Genome specific markers were amplified from durum cultivars and hexaploid wheat cv Chinese Spring genetic stocks. A) The A-genome specific marker amplified in cvs Svevo, Ciccio, and Chinese Spring and sets of Chinese Spring nulli-tetrasomic deletion lines for chromosome group 2. The 350 bp fragment was absent in the nulli-2A-tetra-2B line, as indicated by arrows and confirming the localization on chromosome 2A. B) The B-genome specific marker amplified in cvs, Ciccio, Svevo, Chinese Spring and sets of nullitetrasomic deletion lines for chromosome group 2. The 450 bp fragment was absent in the nulli-2B-tetra-2A line, as indicated by arrows and confirming the localization on chromosome 2B. **Figure S2. Alignment of wheat FD-GOGAT genes.** Alignment of wheat cv Chinese Spring A, B, and D genome *Fd-GOGAT* genes from the beginning of exon 2 through the stop codons. Exons are indicated by red brackets and exon number above the alignments. The blue line indicates the signal peptide/mature polypeptide boundary. The stop codons are boxed in red. **Figure S3. Fd-GOGAT signal sequences.** The signal sequences encoded by six plant *Fd-GOGAT* genes are aligned with Clustal V: Wheat (A genome; present report and GAJL01283868), Brachypodium (BRA-DI1G19080), Rice (Os07g46460), Maize (NM_001112223), *Arabidopsis* (CP002688), Soybean. (AK245357). The red vertical line indicates the exon 1/exon 2 junction. The blue line indicates the end of the signal peptide. **Figure S4. Alignment of wheat Fd-GOGAT polypeptides.** Alignment of the three hexaploid wheat Fd-GOGAT polypeptides from the A, B, and D genomes. Differences in amino acid sequence are highlighted in yellow. **Figure S5. Phylogenetic tree of Fd- and NADH-GOGAT proteins from diverse species.** A selection of available GOGAT amino acid sequences from diverse phyla through genera were aligned with Clustal W and a phylogenetic tree formed by nearest-neighbor analysis. The wheat Fd-GOGAT homoeologues are in blue and red for the NADH-GOGAT version. Annotated sequences from other organisms have the same color coding as wheat. Unannotated sequences are in black. **Figure S6. Comparison of Fd-GOGAT and NADH-GOGAT amino acid sequences.** The amino acid sequences of the mature Fd-GOGAT from the A-genome is compared to the mature NADH-GOGAT from the A-genome by dot blot. Matching criteria was 80 match in a 5 amino acid residue window. **Text S1. Fasta file of the three cv Chinese Spring Fd-GOGAT genes. Text S2. Fasta file of the three cv Chinese Spring Fd-GOGAT mature proteins.**

Author Contributions

Conceived and designed the experiments: DN AB OA AG. Performed the experiments: DN OA AG. Analyzed the data: DN AB OA AG. Contributed reagents/materials/analysis tools: DN AB OA AG. Contributed to the writing of the manuscript: DN AB OA AG.

4. Suzuki A, Knaff DB (2005) Glutamate synthase: structural, mechanistic and regulatory properties, and role in the amino acid metabolism. Photosynth Res 83: 191–217.
5. Tabuchi M, Abiko T, Yamaya T (2007) Assimilation of ammonium ions and reutilization of nitrogen in rice. J Exp Bot 58: 2319–2327.
6. Cánovas FM, Avila C, Cantón FR, Cañas RA, De La Torre F (2007) Ammonium assimilation and amino acid metabolism in conifers. J Exp Bot 58: 2307–2318.

7. Leegood RC, Lea PJ, Adcock MD, Hausler RE (1995) The regulation and control of photorespiration. J Exp Bot 46: 1397–1414.

8. Ferrario-Méry S, Hodges M, Hirel B, Foyer CH (2002a) Photorespiration-dependent increases in phosphoenolpyruvate carboxylase, isocitrate dehydrogenase and glutamate dehydrogenase in transformed tobacco plants deficient in ferredoxin-dependent glutamine-alpha-ketoglutarate aminotransferase. Planta 214: 877–888.

9. Ferrario-Méry S, Valadier MH, Godefroy N, Miallier D, Hodges M, et al. (2002b) Diurnal changes in ammonia assimilation in transformed tobacco plants expressing ferredoxin-dependent glutamate synthase mRNA in the antisense orientation. Plant Sci 163: 59–67.

10. Lancien M, Martin M, Hsieh M-H, Leustek T, Goodman H, et al. (2002) Arabidopsis glt-T mutant defines a role for NADH-GOGAT in the non-photorespiratory ammonium assimilation pathway. Plant J 29: 347–358.

11. Sakakibara H, Kawabata S, Takahashi H, Hase T, Sugiyama T (1992) Molecular cloning of the family of glutamine synthetase genes from maize: expression of genes for glutamine synthetase and ferredoxin-dependent glutamate synthase in photosynthetic and non-photosynthetic tissues. Plant Cell Physiol 33: 49–58.

12. Zehnacker C, Becker TW, Suzuki A, Carrayol E, Caboche M, et al. (1992) Purification and properties of tobacco ferredoxin-dependent glutamate synthase, and isolation of corresponding cDNA clones. Light-inducibility and organ-specificity of gene transcription and protein expression. Planta 187: 266–274.

13. Coschigano KT, Melo-Oliveira R, Lim J, Coruzzi GM (1998) Arabidopsis gls mutants and distinct Fd-GOGAT genes: implications for photorespiration and primary nitrogen assimilation. Plant Cell 10: 741–752.

14. Avila C, Márquez AJ, Pajuelo P, Cannell ME, Wallsgrove RM, et al. (1993) Cloning and sequence analysis of a cDNA for barley ferredoxin-dependent glutamate synthase and molecular analysis of photorespiratory mutants deficient in the enzyme. Planta 189: 475–83.

15. Boisson M, Mondon K, Torney V, Nicot N, Laine A-L, et al. (2005) Partial sequences of nitrogen metabolism genes in hexaploid wheat. Theor Appl Genet 110: 932–940.

16. Nigro D, Gu YQ, Huo N, Marcotuli I, Blanco A, et al. (2013) Structural analysis of the wheat genes encoding NADH-dependent glutamine-2-oxoglutarate amidotransferases genes and correlation with grain protein content. PLoS ONE 8(9): e73751. doi:10.1371/journal.pone.0073751.

17. Márquez AJ, Avila C, Forde BG, Wallsgrove RM (1988) Ferredoxin-glutamate synthase from barley leaves: rapid purification and partial characterization. Plant Physiol Bioch 26: 645–651.

18. Bernard SM, Blom-Møller AL, Dionisio G, Kichey T, Jahn TP, et al. (2008) Gene expression, cellular localisation and function of glutamine synthetase isozymes in wheat (Triticum aestivum L.) Plant Mol Biol 67, 89–105.

19. Gadaleta A, Nigro D, Giancaspro A, Blanco A (2011) The glutamine synthetase (GS2) genes in relation to grain protein content of durum wheat. Functional and integrative genomics 11: 665–670.

20. Blanco A, Mangini G, Giancaspro A, Giove S, Colasuonno P, et al. (2012) Relationships between grain protein content and grain yield components through quantitative trait locus analyses in a recombinant inbred line population derived from two elite durum wheat cultivars. Mol Breeding 30, 79–92.

21. Lawlor DW (2002) Carbon and nitrogen assimilation in relation to yield: mechanisms are the key to understanding production systems. J Exp Bot 53: 789–99.

22. Triboi E, Triboi-Blondel AM (2002) Productivity and grain or seed composition: a new approach to an old problem. Eur J Agron 16: 163 186.

23. Prasad M, Varshney RK, Kumar A, Balyan HS, SharmaPC, et al. (1999) A microsatellite marker associated with a QTL for grain protein content on chromosome arm 2DL of bread wheat. Theor Appl Genet 99: 341–345.

24. Khan IA, Procunier JD, Humphreys DG, Tranquilli G, Schlatter AR, et al. (2000) Development of PCR-based markers for a high grain protein content gene from Triticum turgidum ssp. dicoccoides transferred to bread wheat. Crop Sci 40: 518–524.

25. Zanetti S, Winzeler M, Feuillet C, Keller B, Messmer M (2001) Genetic analysis of bread-making quality in wheat and spelt. Plant Breeding 120: 13–19.

26. Campbell KG, Finney PL, Bergman CJ, Anderson JA, Giroux MJ, et al. (2001) Quantitative trait loci associated with milling and baking quality in a soft×hard wheat cross. Crop Sci 41: 1275–1285.

27. Groos C, Robert N, Bervas E, Charmet G (2003) Genetic analysis of grain protein content, grain yield and thousand-kernel weight in bread wheat. Theor Appl Genet 106: 1032–1040.

28. Prasad M, Kumar N, Kulwal PL, Roder MS, Balyan HS, et al. (2003) QTL analysis for grain protein content using SSR markers and validation studies using NILs in bread wheat. Theor Appl Genet 106: 659–667.

29. Zhang L, Spiertz JHJ, Zhang S, Li B, Werf WVD (2008) Nitrogen economy in relay intercropping systems of wheat and cotton. Plant Soil 303: 55–68.

30. Raman R, Milgate AW, Imtiaz M, Tan M-K, Raman H, et al. (2009) Molecular mapping and physical location of major gene conferring seedling resistance to Septoriatritici blotch in wheat. Mol Breeding 24: 153–164.

31. Suprayogi Y, Pozniak CJ, Clarke FR, Clarke JM, Knox RE, et al. (2009) Identification and validation of quantitative trait loci for grain protein concentration in adapted Canadian durum wheat populations. Theor Appl Genet 119: 437–448.

32. Sun Y, Wang J, Crouch JH, Xu Y (2010) Efficiency of selective genotyping for genetic analysis of complex traits and potential applications in crop improvement. Mol Breeding 26: 493–511.

33. Kramer T (1979) Environmental and genetic variation for protein content in winter wheat (Triticum aestivum L.) Euphytica 28: 209–218.

34. Sears ER (1954) The aneuploids of common wheat. University of Missouri, College of Agriculture, Agricultural Experiment Station Bulletin 572: 1–58.

35. Sears ER (1966) Nullisomic-tetrasomic combinations in hexaploid wheat. In: R Riley and K. R Lewis eds. Chromosome manipulations and plant genetics. Oliver & Boyd, Edinburgh. pp. 29–45.

36. Halushka MK, Fan JB, Bentley K, Hsie L, Shen N, et al. (1999) Patterns of single-nucleotide polymorphisms in candidate genes for blood-pressure homeostasis. Nature Genet 22: 239–247.

37. Till BJ, Burtner C, Comai L, Henikoff S (2004a) Mismatch cleavage by single-strand specific nucleases. Nucleic Acids Res 32: 2632–2641.

38. Till BJ, Reynolds SH, Weil C, Springer N, Burtner C, et al. (2004b) Discovery of induced point mutations in maize genes by TILLING. BMC Plant Biol 4–12.

39. GadaletaA, Giancaspro A, Giove SL, Zacheo S, Mangini G, et al. (2009) Genetic and physical mapping of new EST-derived SSRs on the A and B genome chromosomes of wheat. Theor Appl Genet 118: 1015–1025.

40. Blanco A, Simeone R, Gadaleta A (2006) Detection of QTLs for grain protein content in durum wheat. Theor Appl Genet 112: 1195–1204.

41. Li H, Ye G, Wang J (2007) A modified algorithm for the improvement of composite interval mapping. Genetics 175: 361–374.

42. Habash DZ, Bernard S, Schondelmaier J, Weyen J, Quarrie SA (2007) A genetic study of nitrogen use in hexaploid wheat in relation to N utilisation, development and yield. Theor Appl Genet 114: 403–419.

43. Fontaine J, Ravel C, Pageau K, Heumez E, Dubois F, et al. (2009) A quantitative genetic study for elucidating the contribution of glutamine synthetase, glutamate dehydrogenase and other nitrogen-related physiological traits to the agronomic performance of common wheat. Theor Appl Genet 119: 645–662.

44. Bordes J, Ravel C, Jaubertie JP, Duperrier B, Gardet O, et al. (2011) Use of a global wheat core collection for association analysis of flour and dough quality traits. J Cereal Sci 54: 137–147.

45. Joppa LR, Cantrell RG (1990) Chromosomal location of genes for grain protein content of wild tetraploid wheat. Crop Sci 30: 1059–1064.

46. Trapp GB, Plank DW, Gantt JS, Vance CP (1999) NADH-glutamate synthase in alfalfa root nodules. Immunocytochemical localization. Plant Physiol 119: 829–837.

47. Tobin AK, Yamaya T (2001) Cellular compartmentation of ammonium assimilation in rice and barley. J Exp Bot 52: 591–604.

48. Nakayama T, Archibald JM (2012) Evolving a photosynthetic organelle. BMC Biology 10: 35. doi:10.1186/1741-7007-10-35.

49. Miflin BJ, Lea PJ (1980) Ammonia assimilation. In: Miflin BJ, ed. The biochemistry of plants, Vol. 5. Toronto: Academic Press.

50. Sharp PJ, Kreis M, Shewry PR, Gale MD (1988) Location of β-amylase sequences in wheat and its relatives Theor Appl Genet 75: 286–290.

51. Sears ER, Sears LMS (1978) The telocentric chromosomes of common wheat. Proceedings of the 5th International wheat genetics symposium. New Delhi: Indian Society of Genetics and Plant Breeding. pp. 389–407.

52. Endo TR, Gill BS (1996) The deletion stocks of common wheat. J Hered, 87: 295–307.

53. Van Ooijen JW, Voorrips RE (2001) JoinMap version 3.0: software for the calculation of genetic linkage maps. Wageningen: Plant Research International.

54. Kosambi DD (1944) The estimation of map distances from recombination values. Ann Eugenic 12: 172–175.

55. Tamura K, Peterson D, Peterson N, Stecher G, Nei M, et al. (2011) MEGA5: molecular evolutionary genetics analysis using maximum likelihood, evolutionary distance, and maximum parsimony methods. Mol Biol Evol 28: 2731–2739.

18

Transcriptome/Degradome-Wide Discovery of MicroRNAs and Transcript Targets in Two *Paulownia australis* Genotypes

Suyan Niu[1,2], Guoqiang Fan[1,2]*, Enkai Xu[1,2], Minjie Deng[1,2], Zhenli Zhao[1,2], Yanpeng Dong[1,2]

1 Institute of Paulownia, Henan Agricultural University, Zhengzhou, Henan, P.R. China, 2 College of Forestry, Henan Agricultural University, Zhengzhou, Henan, P.R. China

Abstract

MicroRNAs (miRNAs) are involved in plant growth, development, and response to biotic and abiotic stresses. Most of the miRNAs that have been identified in model plants are well characterized, but till now, no reports have previously been published concerning miRNAs in *Paulownia australis*. In order to investigate miRNA-guided transcript target regulation in *P. australis*, small RNA libraries from two *P. australis* (diploids, PA2; and autotetraploids, PA4) genotypes were subjected to Solexa sequencing. As a result, 10,691,271 (PA2) and 10,712,733 (PA4) clean reads were obtained, and 45 conserved miRNAs belonging to 15 families, and 31 potential novel miRNAs candidates were identified. Compared with their expression levels in the PA2 plants, 26 miRNAs were up-regulated and 15 miRNAs were down-regulated in the PA4 plants. The relative expressions of 12 miRNAs were validated by quantitative real-time polymerase chain reaction. Using the degradome approach, 53 transcript targets were identified and annotated based on Gene Ontology and Kyoto Encyclopedia of Genes and Genomes pathway analysis. These targets were associated with development, stimulus response, metabolism, signaling transduction and biological regulation. Among them, 11 targets, including TCP transcription factors, auxin response factors, squamosa promoter-binding-like proteins, scarecrow-like proteins, L-type lectin-domain containing receptor kinases and zinc finger CCCH domain-containing protein, cleaved by four known miRNA family and two potentially novel miRNAs were found to be involved in regulating plant development, biotic and abiotic stresses. The findings will be helpful to facilitate studies on the functions of miRNAs and their transcript targets in *Paulownia*.

Editor: Keqiang Wu, National Taiwan University, Taiwan

Funding: This work was supported by the National Natural Science Foundation of China (Grant nos. 30271082, 30571496, U1204309 to GQF) (http://www.nsfc.gov.cn/); Outstanding Talents Project of Henan Province (Grant no. 122101110700 to GQF); Transformation Project of the National Agricultural Scientific and Technological Achievement of China (Grant no. 2012GB2D000271 to GQF); and Science and Technology Innovation Team Project of Zhengzhou City (Grant no. 121PCXTD515 to GQF). The funders had no role in study design, data collection and analysis, decision to publish, or preparation of the manuscript.

Competing Interests: The authors have declared that no competing interests exist.

* Email: zlxx64@126.com

Introduction

MicroRNAs (miRNAs), a class of endogenous 21–24 nucleotide (nt), single-stranded, non-coding RNAs, modulate the gene expression at the transcription and post-transcriptional levels [1,2]. In plants, emerging data demonstrate that miRNAs play vital regulatory roles in a wide range of biological processes, including regulation of plant growth, development, signal transduction [3–6], and response to biotic and abiotic stresses via interactions with their specific target mRNAs [7–10]. Thereby, understanding of miRNAs functions in plant requires the recognition of their target genes. Initially, based on the either perfect or near-perfect sequence complementarity between miRNAs and their target mRNAs, the plant miRNA transcript targets have been identified via computational prediction [11,12]. However, computational target prediction method is very challenging because of the existence of a higher mismatch in miRNA-target mRNA pairing [13]. With the development of high-throughput sequencing technology, recently, a transcriptome wide experimental method combining high-throughput miRNA profiling with degradome sequencing analysis, has been successfully developed to identify miRNA-directed mRNA cleavage at a large scale [14,15]. Using this strategy, so many plant species have been studied, for instance, *Arabidopsis* [15], *Oryza sativa* [16], maize [17], *Brassica napus* [18], *Medicago truncatula* [19], *Fragaria ananassa* [20], grapevine [21], *Taxus* [22], *Populus trichocarpa* [23], *Populus euphratica* [24], *Trifoliate orange* [25], *Paulownia tomentosa* [26], and *Paulownia fortunei* [27]. These discoveries have triggered detailed biological studies on gene regulation with the result that the number of reported plant miRNAs and their transcript targets has been increasing rapidly.

Paulownia is a genus of deciduous tree species in the family Paulowniaceae, related to and sometimes included in the family Scrophulariaceae. It is a very adaptable, extremely fast-growing woody plant, and in managed plantations, it can be harvested for saw timber in as little as 5 years. *Paulownia* can not only provide wood for a variety of products including timber, fuelwood, herbal medicines, boxes, clogs, musical instruments, and surfboards, but can also benefit the environment and ecology; for example, it can increase crop production when used for intercropping on farmland, and prevent soil erosion [28,29]. Because of its specific characteristics and economic value, in its native China, *Paulownia* is popular for reforestation, intercropping on farmland, and roadside and garden planting. To enlarge the germplasm and breed new varieties, in recent years, several autotetraploid *paulownia* seedlings were induced successfully using colchicines [30–33]. The autotetraploid cultivars of *paulownia* contains two sets of the same chromosomes, and compared with their diploid progenitors, the autotetraploids have better timber quality and improved stress resistance [34,35], which makes understanding the molecular mechanisms underlying the differences of diploids and autotetraploids imperative. To our knowledge, the different paulownia species have genetic diversity and differentiation, although miRNAs in the *P. tomentosa* [26] and *P. fortunei* species have been identified [27], there have been no detailed reports of miRNAs in *P. australis*. In the current study, we used Solexa sequencing and degradome technology to analyze four sequencing libraries constructed from the seedlings of the two *P. australis* genotypes to identify conserved and novel miRNAs and to investigate the potential roles of their transcript targets.

Materials and Methods

Plant culture and treatment

Plantlets of two *P. australis* genotypes (diploids, PA2; and autotetraploids, PA4) were cultured *in vitro* on medium as described previously [30,36]. All the samples were grown at 25°C under a 16/8 h (light/dark) photoperiod for 30 days, and at least three parallel samples were prepared for each genotype. After culturing, samples were collected from nine different seedlings of each genotype and subsequently mixed, frozen immediately in liquid nitrogen, and stored at −86°C until further use.

Construction and sequencing of the small RNA library

Total RNA was isolated from the leaves of the PA2 and PA4 samples using TRIzol reagent (Invitrogen, Carlsbad, CA, USA) following the manufacturer's instructions. Two small RNA (sRNA) libraries were constructed, one from the PA2 sample and the other from the PA4 sample, using methods described elsewhere [24,37]. Briefly, bands in the 18–30 nt size range were purified by electrophoretic separation on a 15% denaturing polyacrylamide gel and then successively ligated with 5′ and 3′ adapters. After reverse transcription and PCR, the amplified products were subjected to Solexa sequencing (Illumina, San Diego, CA, USA) at the Beijing Genomics Institute, Shenzhen, China. The data used in this publication have been deposited in the NIH Short Read Archive database (http://www.ncbi.nlm.nih.gov/sra) with accession number SRP041440 and SRP041442 (Alias: PRJNA245384 and PRJNA245383).

Identification of conserved and novel miRNAs in the *P. australis* libraries

Low-quality reads, poly(A) reads, oversized insertions, reads shorter than 18 nt, and adaptor contaminated reads were filtered out from the raw reads to yield the clean reads. The length distribution of the 18–30 nt clean reads was analyzed and the reads were mapped to the sequences in the *P. australis* UniGene database (http://www.ncbi.nlm.nih.gov/sra; Accession Number SRP032321) using SOAP [38]. Perfectly matched reads were analyzed by running Blastall (http://www.ncbi.nlm.nih.gov/staff/tao/URLAPI/blastall/) against the Rfam (http://www.sanger.ac.uk/software/Rfam) and GenBank databases (http://www.ncbi.nlm.nih.gov/) to discard tRNAs, rRNAs, snRNAs, snoRNAs, and other ncoRNAs. The remaining reads were annotated by aligning

them against the sequences in the miRBase database 19.0 (http://www.mirbase.org/) to identify known miRNAs. When a read in the *P. australis* libraries shared homology with fewer than two mismatches with a miRNA sequence in miRBase, it was considered to be an evolutionarily conserved miRNA.

The MIREAP software (https://sourceforge.net/projects/mireap/) was used to detect novel miRNAs in the two libraries by predicting the stem-loop structures and estimating the minimal folding free energy of the unannotated sRNAs (no matches in miRBase) that mapped to the *P. australis* UniGene sequences. The stem-loop structures of the candidate precursor miRNAs (pre-miRNAs) were predicted using Mfold (http://mfold.rna.albany.cdu/?q0mfold) [39]. The strict criteria used to annotate the candidate miRNAs were as described by Meyers *et al.* [40].

Differential expression analysis of miRNAs in the two *P. australis* libraries

To identify the differentially expressed miRNAs between the PA2 and PA4 libraries, the abundance of the miRNAs in the two libraries were normalized to one million, regardless of the total number of miRNAs in each sample. Then, statistical analysis was performed based on a Poisson distribution. Finally, when the fold-change was greater than 1.0 or less than −1.0 and the *P*-values were less than 0.05, the miRNA was considered to be significantly differentially expressed. The fold-changes and *P*-values [41,42] were calculated from the normalized expression using the following formulas:

Normalized expression = (actual miRNA reads/total count of clean reads) × 1,000,000

Fold-change = log2 (PA4 normalized reads/PA2 normalized reads)

P-values were calculated as,

$$p(x|y) = \left(\frac{N_2}{N_1}\right) \frac{(x+y)!}{x!y!(1+\frac{N_2}{N_1})^{(x+y+1)}}$$

$$C(y \leq y_{min}|x) = \sum_{y=0}^{y \leq y_{min}} p(y|x)$$

$$D(y \geq y_{max}|x) = \sum_{y \geq y_{max}}^{\infty} p(y|x)$$

Degradome library construction, data analysis, and target identification

Two degradome libraries were constructed from the leaves of the PA2 and PA4 plants based on a protocol published previously [14,15]. Briefly, using T4 RNA ligase (Takara, Dalian, China), poly(A) enriched RNA was ligated to a 5′RNA adapter containing a *Mme*I recognition site. Reverse transcription was performed to generate first-strand cDNA, followed by PCR enrichment, and digestion using the *Mme*I restriction enzyme (NEB, Ipswich, MA, USA). A double-stranded DNA adapter was then ligated to the digested products using T4 DNA ligase (NEB, Ipswich, MA, USA) and gel purified for PCR amplification. The resulting PCR products were sequenced using an Illumina HiSeqTM 2000 system. Low-quality sequences and adapters were removed from the raw reads of the degradome libraries, then the remaining reads were aligned to the *P. australis* UniGene database (http://www. ncbi.nlm.nih.gov/sra; Accession Number SRP032321) using SOAP software (http://soap.genomics. org.cn/) to define the coverage rate. The PAIRFINDER software (version 2.0) [22] was used to identify the miRNA-mediated cleaved fragments. Alignments with scores not exceeding four and no mismatches at the cleavage site (between the 10th and 11th nucleotides) were considered to be the potential miRNA targets. Furthermore, t-plots were built according to the distribution of signatures (and abundances) along these transcripts to help analyze the miRNA targets and RNA degradation patterns. Based on the locations of the target genes in the *P. australis* UniGene sequences, putative target genes were selected manually and subsequently mapped to the previously identified genes that were annotated according to the annotations of their homologous sequences in the GenBank Nr and Nt databases, and in the Swiss-Prot database using BLASTX searches with an E-value cutoff of 10^{-5}.

Verification of miRNAs and their targets by quantitative real-time polymerase chain reaction

Quantitative real-time polymerase chain reaction (qRT-PCR) was used to validate and measure the expressions of selected miRNAs and their targets obtained from the Solexa sequencing and degradome analysis. The qRT-PCR was performed according to the protocol described previously [43]. For the qRT-PCR, total RNA were isolated from the leaves of two individual PA2 and PA4 plants at two developmental stages (30-day-old *in vitro* plantlets, and two-year-old saplings from field). Three biological replicates of each stage were used. The qRT-PCRs were performed in triplicate on a CFX96TM Real-Time PCR System (Bio-Rad, Hercules, CA, USA) using a SuperScript III platinum SYBR Green one-step qRT-PCR kit (Invitrogen, Carlsbad, CA, USA) according to the manufacturer's instructions. The PCR conditions were 50°C for 3 min, 95°C for 5 min, then 40 cycles of 95°C for 15 s, 55°C for 30 s, and 40°C for 10 min. Specific stem-loop primers and other primers for the miRNAs were designed based on the mature miRNA sequence. The U6 was used as an endogenous control. All the primers used for the qRT-PCR are listed in Table S1. The primers for the target genes were designed using Beacon Designer, version 7.7 (Premier Biosoft International, Ltd., Palo Alto, CA, USA), and the 18S rRNA of Paulownia was chosen as an endogenous reference gene for normalization. The relative expression level of a miRNA or target gene was calculated according to the method described previously [44].

Results

Analysis of sRNAs in the two *P. australis* libraries

A total of 13,895,340 and 13,537,466 raw reads were obtained in the PA2 and PA4 sRNA libraries, respectively. After removing low-quality reads, poly(A) reads, oversized insertions, reads shorter than 18-nt, and adaptor contaminated reads, we obtained 10,691,271 (PA2) and 10,712,733 (PA4) clean reads that represented 2,006,153 (PA2) and 2,418,971 (PA4) unique reads with lengths ranging from 18 to 30-nt. The majority of clean reads, 63.24% (PA2) and 68.79% (PA4), ranged from 20 to 24-nt in length, among which the 24-nt long reads were the most abundant, followed by the 21-nt long reads (Figure S1). The abundances of the 24-nt and 21-nt long sRNAs in the PA4 library were 6.48% and 0.04% more, respectively, than their abundances in the PA2 library. This atypical situation has also been reported in other hardwood species, such as *Liriodendron chinense* [45].

The clean reads were used to query the *P. australis* UniGene database (http://www.ncbi.nlm.nih.gov/sra), the non-coding RNAs sequences deposited in GenBank (http://www.ncbi.nih. gov/Genbank), the Rfam database (http://rfam.sanger.ac.uk/), and miRBase 19.0 (http://microrna.sanger.ac.uk/sequences). These searches allowed us to assign annotations to each sRNA sequence, including rRNA, tRNA, snRNA, snoRNA, and known miRNA (Table S2). As a result, 848,652 (PA2) and 931,989 (PA4) known miRNA reads were detected. Finally, 1,809,478 (PA2) and 2,308,527 (PA4) unique unannotated sRNAs remained for predicting potentially novel miRNAs.

Identification of conserved miRNAs in *P. australis*

To detect conserved miRNAs in the two libraries, the unique reads were compared with the known miRNAs in miRBase 19.0, allowing two mismatches. In the two libraries, a total of 45 conserved miRNAs belonging to 15 miRNA families were found to share high identity with known plant miRNAs (Table 1 and Table S3). Among these 45 miRNAs, 40 and 43 were identified in PA2 and PA4, respectively; 38 of them were expressed in both libraries, two were found only in the PA2 library, and five occurred only in the PA4 library. Most of the conserved miRNAs (91.67%) were 21-nt long and the remaining miRNAs were 20 or 23-nt long (Table 1 and Table S3). This result is consistent with the current understanding that canonical miRNAs are 21-nt long, while canonical small interfering RNAs (siRNAs) are 24-nt long [46]. Among the 15 miRNA families, the miR166 family was dominant in both libraries, accounting for 89.44% (PA2) and 86.35% (PA4) of all conserved miRNA reads, followed by the miR159 family (Table 1). Several miRNA families, such as miR156, miR396, miR397, miR398, and miR482, had moderate expression levels in the two libraries, while other miRNA families showed very low levels of expression, with fewer than 100 reads, in both libraries. Moreover, different members in the same miRNA family displayed significantly different expression levels. For instance, members of the miR166 family varied in abundance from 50,932 to 563,946 reads in the two libraries (Table 1).

Identification of novel miRNAs in the *P. australis* sRNA libraries

The characteristic stem-loop structure of pre-miRNA was employed to predict novel miRNAs [40]. We found that 31 of the potential pre-miRNAs met this requirement; 15 pre-miRNAs exhibited the 3p:5p miRNA pair, providing more evidence that they were novel miRNAs, while the other 16 miRNAs were also considered to be potentially novel miRNAs (Table 2 and Table S4). Among the 31 candidate novel miRNAs, 10 miRNAs were common between the two libraries, while 10 and 11 were specific to PA2 and PA4, respectively. The length of the mature miRNAs varied from 20 to 23 nt, with the majority being 21-nt long. The mature miRNA sequences were localized inside the stem-loop structures, with almost half in either the 3p or 5p arms. We observed that the average pre-miRNA length was 151 nt, and these precursors had minimal folding free energies that ranged from -142.5 kcal mol^{-1} to -30.8 kcal mol^{-1} with an average of -56.3 kcal mol^{-1}. In this study, compared with the abundance of the conserved miRNAs, the majority of novel miRNAs had relatively low expressions, and only five (PA2) and six (PA4) of the novel miRNA candidates had more than 1,000 reads (Table 2 and Table S4).

Table 1. Conserved miRNAs identified from *P. australis*.

Family	miRNA	expression		fold-change (log2PA4/PA2)	P-value	MiRNA*expression	
		PA2	PA4			PA2	PA4
MIR169	pas-miR169a-3p	252	142	-0.83	2.27E-08	180	87
	pas-miR169b-3p	244	0	-11.16	2.77E-74	173	0
	pas-miR169c-3p	244	0	-11.16	2.77E-74	173	0
	pas-miR169d	18	26	0.53	2.35E-01	0	0
	pas-miR169e	13	26	0.53	2.35E-01	0	0
MIR159	pas-miR159-3p	117706	141387	0.26	0	115	83
	pas-miR319a-3p	1312	85	-3.95	0	16	0
MIR408	pas-miR408a-3p	1475	6263	2.08	0	50	208
	pas-miR408b-3p	1475	6263	2.08	0	50	208
MIR396	pas-miR396a	7879	6831	-0.21	1.86E-18	142	153
	pas-miR396b	2E08	3441	0.29	2.11E-15	120	263
	pas-miR396c-3p	0	30	8.13	9.61E-10	13	11
	pas-miR396d-3p	0	30	8.13	9.61E-10	13	11
MIR397	pas-miR397a	675	6543	3.27	0	16	368
	pas-miR397b	673	6601	3.28	0	16	368
MIR398	pas-miR398a-3p	532	3071	2.53	0	44	1042
	pas-miR398b-3p	532	3071	2.53	0	44	1042
	pas-miR398c-3p	532	3071	2.53	0	44	1042
MIR166	pas-miR166a-3p	178621	189900	0.09	3.20E-72	1809	1883
	pas-miR166b-3p	180817	192950	0.09	1.90E-82	141	332
	pas-miR166c-3p	560961	563946	0.03	2.62E-29	1689	1755
	pas-miR166d-3p	422623	424471	0.00	2.78E-01	373	236
	pas-miR166e-3p	50932	51265	0.01	4.71E-01	490	455
MIR160	pas-miR160a	718	1445	1.01	8.76E-56	0	1
	pas-miR160b	30	40	0.41	2.38E-01	0	0
	pas-miR160c	30	40	0.41	2.38E-01	0	0
	pas-miR160d	30	40	0.41	2.38E-01	0	0
	pas-miR160e	30	40	0.41	2.38E-01	0	0
	pas-miR160f	30	40	0.41	2.38E-01	0	0
MIR156	pas-miR156a	2251	3231	0.51	5.59E-39	0	0
	pas-miR156b	2251	3221	0.51	4.75E-39	46	69
	pas-miR156c	1349	2284	0.76	1.94E-54	278	493
	pas-miR156d	1349	2284	0.76	1.94E-54	278	493
MIR164	pas-miR164	396	284	-0.48	1.51E-05	85	40

Table 1. Cont.

Family	miRNA	expression		fold-change (log2PA4/PA2)	P-value	MiRNA*expression	
		PA2	PA4			PA2	PA4
MIR167	pas-miR167	509	864	0.76	8.46E-22	18	27
MIR168	pas-miR168a	1893	2713	0.52	2.06E-33	143	236
	pas-miR168b	1895	2712	0.51	3.58E-33	137	185
MIR2118	pas-miR2118a-3p	109	136	0.32	8.78E-02	0	0
	pas-miR2118b-3p	109	136	0.32	8.78E-02	0	0
MIR482	pas-miR482a-3p	4727	6806	0.52	5.00E-83	0	0
	pas-miR482b-3p	4529	4723	0.06	5.48E-02	0	0
	pas-miR482c-3p	4728	6804	0.52	8.48E-83	0	0
MIR171	pas-miR171a	0	10	6.54	9.87E-04	0	0
	pas-miR171b	0	8	6.22	3.94E-03	0	0
	pas-miR171c	0	8	6.22	3.94E-03	0	0

Expression patterns of conserved and novel miRNAs

The differential expression analysis was performed based on the normalized read counts for each identified miRNA. A total of 41 miRNAs showed statistically significant changes (fold change ≥ 1.0 or ≤ -1.0, and P-values ≤ 0.05) in their relative abundance between the two libraries. Among them, 26 miRNAs were up-regulated (13 conserved miRNAs and 13 novel miRNAs; fold-change ≥ 1.0) and 15 were down-regulated (three conserved miRNAs and 12 novel miRNAs; fold-change ≤ -1.0) in the PA4 library compared with in the PA2 library. The expression levels of some miRNAs changed significantly. For example, the expression levels of pas-miR169b-3p, pas-miR169c-3p, pas-miR396c-3p, pas-miR396d-3p, pas-miR171a, pas-miR171b, and pas-miR171c increased or decreased by about 5-fold in the PA4 library (Table 1). We also detected 21 novel miRNAs in the PA4 library that had 5-fold greater or lesser expression levels than in the PA2 library (Table 2).

Identification of miRNAs transcript targets in *P. australis* by using degradome analysis

To better understand the physiological functions and biology processes that these miRNAs may be involved in during the development of *P. australis*, transcript target identification was performed based on the degradome approach. Using PAIRFIN-DER, a total of 53 targets were predicted to be cleaved by 11 of the conserved miRNA families and three novel miRNA candidates (Table S5). The target transcripts were pooled and grouped into three categories (I–III) according to their relative abundances [14,15] (Figure 1). Among these identified targets, 32 (37 cleavage sites) belonged to category I, and 20 (34 cleavage sites) and six (six cleavage sites) targets belonged to categories II and III, respectively (Table S5). We then performed a Gene Ontology (GO) analysis to assign functional annotations to the predicted target genes, as described previously [47]. More than 80% of the target genes were annotated as being involved in regulation of biological processes, and the GO annotations for the predicted target genes are shown in Figure 2. The Kyoto Encyclopedia of Genes and Genomes (KEGG) Pathway database was used to further classify the miRNA target genes. The KEGG Pathway annotation showed that the target genes were involved in metabolic pathways, plant-pathogen interaction, pyruvate metabolism, carbon fixation in photosynthetic organisms, purine metabolism, RNA polymerase, pyrimidine metabolism, nitrogen metabolism, plant hormone signal transduction, cellular metabolism, and disease. (Table S5). The prediction and annotation of the miRNA target genes may provide some new insights into how *P. australis* miRNAs regulate gene expression in this plant.

Validation of the candidate miRNAs and their targets by qRT-PCR

To verify the existence and expression levels of the miRNAs determined from the high-throughput Solexa sequencing, 12 miRNAs with different expression levels were randomly selected for qRT-PCR analysis. As shown in Figure S2, the expression patterns of the miRNAs obtained by qRT-PCR showed similar trends to their expression patterns in the two libraries as determined by Solexa sequencing. Compared with their expressions in PA2, the expression levels of pas-miR156c, pas-miR398a-3p, pas-miR408a-5p, and pas-mir22-3p were up-regulated in PA4 in the 30-day plantlets and in the 2-year saplings, pas-miR319a-3p were on the contrary. Further, the expressions of pas-miR160a, pas-miR167, pas-miR171a, pas-miR397a, pas-mir1, and pas-mir14 in PA4 were up-regulated in the 30-day plantlets, and

Table 2. Novel miRNAs identified from *P. australis*.

miRNA	expression		Fold-change	P-value	MiRNA* expression	
	PA2	PA4	(log2PA4/PA2)		PA2	PA4
pas-mir1	116	422	1.86	7.02E-42	106	74
pas-mir2	43	0	−8.65	1.09E-13	0	0
pas-mir3	8	3	−1.42	1.45E-01	6	5
pas-mir4a	6	0	−5.81	1.55E-02	0	0
pas-mir4b	6	0	−5.81	1.55E-02	0	0
pas-mir5a	7116	6078	−0.23	5.38E-20	208	223
pas-mir5b	7116	6078	−0.23	5.38E-20	208	223
pas-mir6a-3p	2125	4098	0.94	1.51E-139	67	71
pas-mir6b-3p	2125	4098	0.94	1.51E-139	67	71
pas-mir6c-3p	2125	4098	0.94	1.51E-139	67	71
pas-mir7-3p	696	0	−12.67	1.51E-210	6	0
pas-mir8a-3p	29	0	−8.08	4.57E-01	0	0
pas-mir8b-3p	29	11	−1.4	5.48E-02	0	0
pas-mir9	8	0	−6.23	3.87E-03	0	0
pas-mir10-3p	10	0	−6.55	9.66E-04	0	0
pas-mir11-3p	17	0	−7.31	7.49E-06	0	0
pas-mir12	5	0	−5.55	3.11E-02	0	0
pas-mir13-3p	934	1506	0.69	5.10E-31	1128	1403
pas-mir14	10	28	1.48	3.45E-03	2	1
pas-mir15a	345	0	−11.66	9.86E-105	107	0
pas-mir16a-3p	0	19	7.47	1.95E-06	0	0
pas-mir16b-3p	0	19	7.47	1.95E-06	0	0
pas-mir16c-3p	0	19	7.47	1.95E-06	0	0
pas-mir17-3p	0	8	6.22	3.94E-03	0	7
pas-mir18-3p	0	11	6.68	4.94E-04	0	0
pas-mir19-3p	0	126	10.2	1.34E-38	0	0
pas-mir20-3p	0	54	8.98	5.87E-17	0	2
pas-mir21a	0	12	6.81	2.47E-04	0	9
pas-mir21b	0	12	6.81	2.47E-04	0	9
pas-mir22-3p	0	11	6.68	4.94E-04	0	0
pas-mir23-3p	0	28	8.03	3.84E-09	0	0

down-regulated in the PA4 in the 2-year saplings, while the expressions of pas-mir3 were the opposite to this. Thus, with the development of the plants, the expression levels of some of the miRNAs showed different trends. Six miRNAs (pas-miR160a, pas-miR167, pas-miR319a-3p, pas-miR398a,-3p pas-miR408a-3p, pas-mir1) had the same expression trend between PA2 and PA4 in the two stages. Furthermore, to confirm the reliability of degradome sequencing technology and the potential correlation between miRNAs and their transcript targets, 12 genes from two *P. australis* genotypes were also selected for qRT-PCR assays. The results showed that except the targets of auxin response factor (ARF) ARF8 (CL4211.Contig3) and scarecrow-like protein (SCL) SCL15 (CL10503.Contig1), the expression levels of the rest targets were inversely correlated with these of the corresponding miRNAs (Figure 3). During the different developmental stages, pas-miR160a, pas-miR167, pas-miR171 and pas-mir1 at a relatively higher level in the PA4 than in the PA2 at the 30-day plantlets stage, and lower level at the 2-year saplings stage, while its

transcript targets, CL3173.Contig7, CL11603.Contig1, CL6407.Contig9, CL11078.Contig2, CL11078.Contig3, and Unigene9061 expressed in the reverse way as expected, and these coding for proteins are members of the auxin response factors ARF10, ARF18 and ARF6, the scarecrow-like proteins SCL6 and SCL22, and serine/threonine protein kinase, respectively. The expression levels of pas-miR319a-3p in PA4 were significantly lower than in the PA2 at two treatment stages, while the reverse was true for the transcription factor TCP4 (CL9103.Contig3) (Figure 3). Moreover, a reverse trend was noted between pas-miR156c and its target genes coding for squamosa promoter-binding-like protein (SPL) SPL6 and SPL12 (CL11428.Contig2 and CL5129.Contig2), and between pas-mir22-3p and its target gene coding for zinc finger CCCH domain-containing protein 53 (CL1197.Contig2) in the PA4 as compared to the PA2 (Figure 3). These results indicated that the miRNA and transcript target expression patterns were very complex and varied during the growth and development of *P. australis*. Furthermore, the possible

A

Unigene11776_All

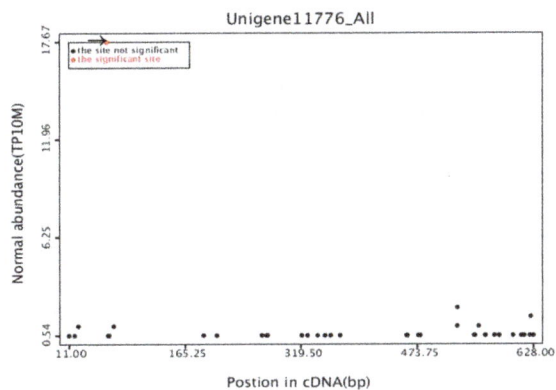

NBS-LRR class disease resistance

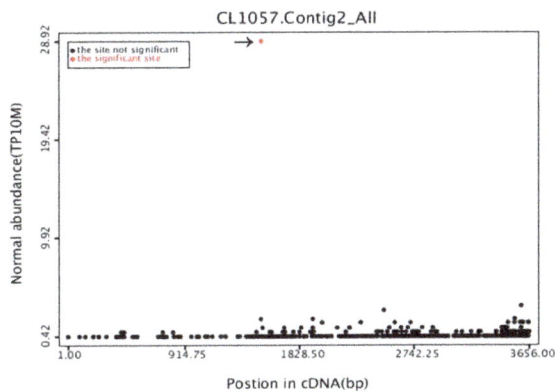

```
5' AGTTGGCATGGGAGGTATTGGTAAGACTACT 3'  Unigene11776_All
     |||||x|||||||x|x||||||x|||
3' ----AACCCTACCCACCTTAACCTTT-------------- 5'  pas-miR482a-3p
```

CL1057.Contig2_All

Homeobox_leucine zipper protein

```
5' TGCCTGGAATGAAGCCTGGTCCGGATTCCAT 3'  CL1057.Contig2_All
     ||||||||||||||o x x
3' -----------CCTTACTTCGGACCAGGCTCT--------- 5'  pas-miR166c
```

B

CL5129.Contig3_All

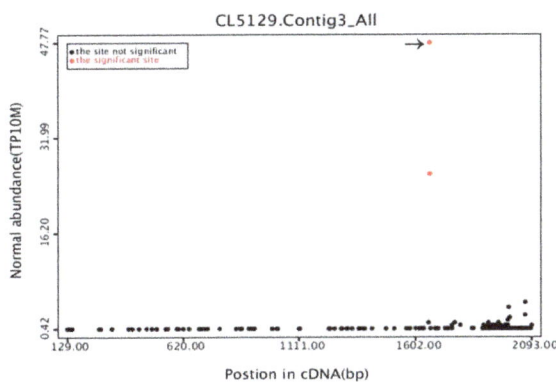

Squamosa promoter-binding-like protein 12

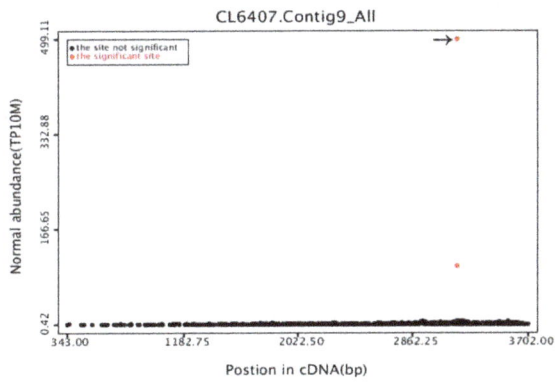

```
5' CGTCGTGCTCTCTCTCTTCTGTCAAGTAATT 3'  CL5129.Contig2_All
     ||||||||x||||||||||||||
3' -------CACGAGTGAGAGAAGACAGT------------ 5'  pas-miR156d
```

CL6407.Contig9_All

Auxin response factor 6

```
5' CTTGAGATCAGGCTGGCAGCTTGTATTTGTA 3'  CL6407.Contig9_All
     x ||||||x||||||||||x x
3' ------ ATCTAGTACGACCGTCGAAGT------- 5'  pas-miR167
```

C

CL11907.Contig1_All

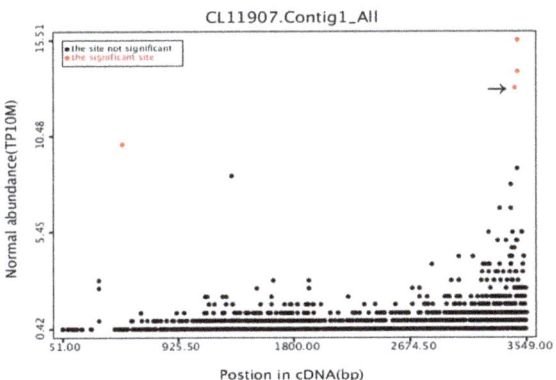

Protein argonaute 1

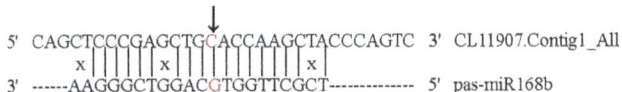

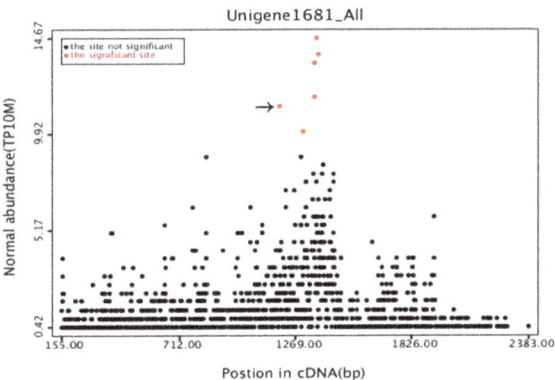

```
5' CAGCTCCCGAGCTGCACCAAGCTACCCAGTC 3'  CL11907.Contig1_All
     x|||||x|||||||||x|||||
3' ------AAGGGCTGGACGTGGTTCGCT-------------- 5'  pas-miR168b
```

Unigene1681_All

Cytosolic NADP-malic enzyme

```
5' GGAAGAAACACGTAAAAAGATTTGGCTGGTA 3'  Unigene1681_All
     |||o|||xxo|||||o
3' -------------------GTGCGTTTTTAGGAACCGAT----- 5'  pas-miR169e
```

Figure 1. Target plots (t-plots) of miRNA targets in different categories confirmed by degradome sequencing. (A) T-plot (top) and miRNA: mRNA alignments (bottom) for two category I targets, Unigene11776_All and CL1057.Contig2_All transcripts. The arrow indicates signatures consistent with miRNA-directed cleavage. The solid lines and dot in miRNA: mRNA alignments indicate matched RNA base pairs and GU mismatch, respectively, and the red letter indicates the cleavage site. (B) CL5129.Contig3_All and CL6407.Contig9_All, a category II target for pas-miR156d and pas-miR167. (C) CL11907.Contig1_All and Unigene1681_All, a category III target for pas-miR168b and pas-miR169e.

roles of these miRNAs in the genome duplication changes from diploid to autotetraploid in these plants were revealed.

Discussion

Variations in plant morphology and physiology resulting from genome duplication have occurred in many plants, such as *Triticum*, *Gossypium*, *Spartina*, *Tragopogon*, *Brassica* and *Solanum* [48,49]. Generally speaking, genome duplication has led to the production of fast-growing, high-quality plants [50,51]. The autotetraploid *P. australis*, which contains two sets of the same chromosomes, displayed apparent alterations in morphology, growth development, physiology, and gene expression when compared with their diploid counterparts [34,35]. However, the mechanisms for these changes are poorly understood. MiRNAs are a class of endogenous small RNAs that have been involved in many processes, including growth, development, and resistance to stress and disease, by their ability to regulate gene expression in plants [52–54]. To understand the functions of miRNAs in diploid and autotetraploid *P. australis*, in this present study, we used Solexa sequencing and degradome approaches to construct two sRNA libraries and two degradome libraries from the PA2 and PA4 plants to identify conserved and novel miRNAs and their transcript targets. A total of 45 conserved miRNAs belonging to 15 miRNA families and 31 potential novel miRNA candidates along with 53 transcript targets were identified across the PA2 and PA4 libraries. Most of the identified conserved miRNA families are also conserved in other plant species, including *Populus tomentosa* [53], *Populus euphratica* [55], *Oryza sativa* [3], and *Arabidopsis thaliana* [56]. In the PA2 and PA4 libraries, the expression

patterns varied dramatically among the different miRNA families, and different members in the same miRNA family also displayed significantly different expression levels. For instance, the read number varied from 26 (miR171 family) to 1,422,532 (miR166 family) (Table 1), and members of the miR166 family varied in abundance from 50,932 to 563,946 reads (Table 1). Moreover, the majority of novel miRNAs had relatively low expressions, and only five (PA2) and six (PA4) of the novel miRNA candidates had more than 1,000 reads (Table 2). Our results are in accordance with previous reports that novel miRNAs were often represented in relatively lower levels than conserved miRNAs [57,58]. Furthermore, it is possible that the low-expression novel miRNAs may play particular functions in specific tissues, during developmental stages, or under various growth conditions. Whether these low-expression miRNAs are expressed at higher levels in other tissues or developmental stages, or are regulated by environmental stresses, remain to be investigated.

Previous studies have shown that many of the genes appeared to be methylated in tetraploid Paulownia plants specifically after genome duplication by the methylation-sensitive amplified polymorphism analysis [59]. DNA methylation has been reported to be involved in inducing gene silencing, which can restart or change the genes expression levels [60]. Salmon et al. [61] found that significant changes in DNA methylation patterns could explain the morphological plasticity and larger ecological amplitude of Spartina allopolyploids. Indeed, we found that the expression levels of many of the differentially expressed miRNAs in the PA4 library were not increased by more than two-fold compared with their expressions in the PA2 library. However, the expressions of

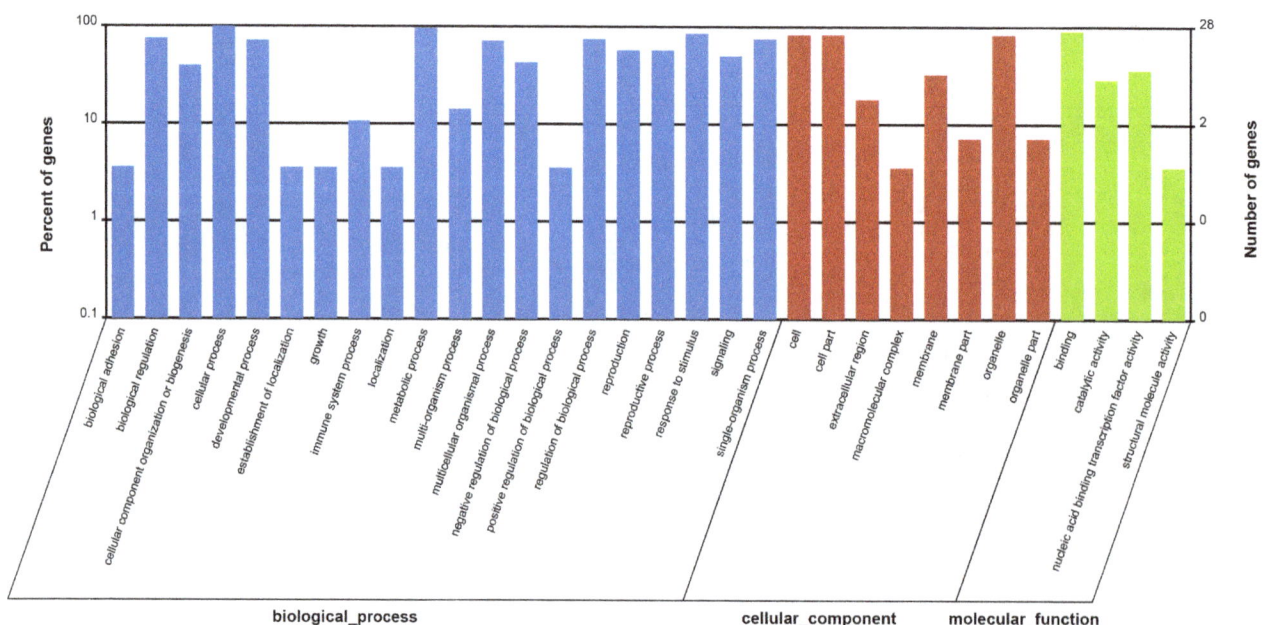

Figure 2. Gene Ontology analysis of miRNA targets in *P. australis*.

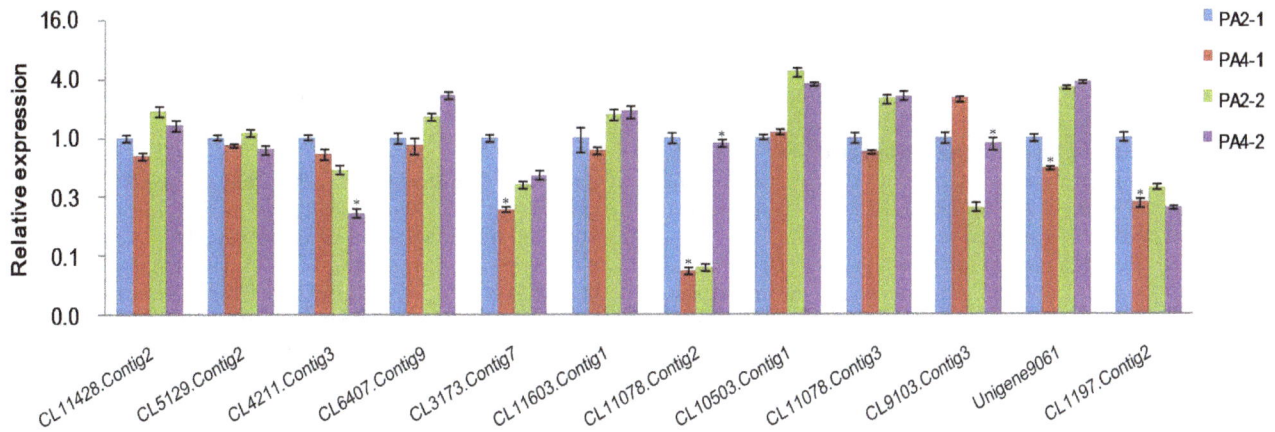

Figure 3. Relative expression levels of the target genes in *P. australis*. PA2-1, 30-day-old diploid in vitro plantlets; PA4-1, 30-day-old autotetraploid in vitro plantlets; PA2-2, two-year-old diploid saplings; PA4-2, two-year-old autotetraploid saplings. CL11428.Contig2 (SPL6) and CL5129.Contig2 (SPL12) targeted by pas-miR156c, CL3173.Contig7 (ARF10) and CL11603.Contig1 (ARF18) targeted by pas-miR160a, CL4211.Contig3 (ARF8) and CL6407.Contig9 (ARF6) targeted by pas-miR167, CL11078.Contig2 (SCL6), CL10503.Contig1 (SCL15) and CL11078.Contig3 (SCL22) targeted by pas-miR171a, CL9103.Contig3 (Transcription factor TCP4) targeted by pas-miR319a-3p; Unigene9061 (Serine/threonine protein kinase) targeted by pas-mir1; CL1197.Contig2 (Zinc finger CCCH domain-containing protein 53) targeted by pas-mir22-3p. Three independent biological replicates were performed. Values are means ± SD (n = 3).The expression levels of targets were normalized to 18SrRNA. The normalized miRNA levels in the PA2-1 were arbitrarily set to 1. *: Statistically significant differences between PA2 and PA4 under the same developmental stages (p-value was less than 0.05)

about half of these miRNAs were significantly different in the two libraries. Some of the miRNAs were expressed at similar level in the PA4 and PA2 libraries (Table 2). These findings are similar to those reported previously in *P. tomentosa*, *P. fortunei* and tetraploid *Arabidopsis thaliana* lines [26,27,62], suggesting that the genome merger in the PA4 plants lead to nonadditive expression of the miRNA primary transcripts and miRNA target genes. Furthermore, the expression patterns of the differentially expressed miRNAs and transcript targets at different development stages were validated by qRT-PCR. The result showed that the differentially expressed miRNAs caused different expression levels in their transcript targets. Interestingly, we also found the expression levels of two target genes (CL4211.Contig3 and CL10503.Contig1) were inconsistent with those of their corresponding miRNAs (pas-miR167 and pas-miR171a). The similar phenomena were also observed in *P. tomentosa*, *Phalaenopsis Aphrodite* and cotton [26,63,64], indicating that the other mechanisms of regulating expression of the target genes exist. Above all, these results imply that the miRNAs with significantly varied expressions in the PA2 and PA4 *P. australis* plants are probably involved in the epigenetic changes of PA4 plants; however, the relation between the miRNA expression patterns and genome duplication may be more complex than we first thought.

To understand the biological functions of miRNAs, it is necessary to identify their transcript targets. In the present study, to avoid false-positive predictions of miRNA transcript targets in *P. australis*, we identified the 53 transcript targets for 11 miRNA families and three novel miRNA candidates by degradome sequencing, which opens up a new avenue for high-throughput validation of splicing targets [14,15]. The target genes predicted for the conserved miRNAs in *P. australis* were similar or functionally related to validated plant miRNA targets, which were annotated as being involved in diverse physiological processes. For instance, pas-miR156 targeted the SPL protein family, which can affect diverse developmental processes such as leaf development, shoot maturation, phase change, and flowering in plants [65]; and pas-miR167 targeted ARF6 and ARF 8, which belong to a class of

transcription factors known to control multiple processes in plants, including the regulation of gynoecium and stamen maturation, and seed dispersal [66,67]. Thus, our results support the idea that conserved miRNAs take part in essential physiological processes in plants.

The analysis of the target genes identified for the differentially expressed miRNAs revealed that some of the target genes may play important roles in plant morphology and physiology. We found that the expression level of pas-miR319a-3p decreased by about 3-fold in PA4 compared with its level in PA2, and pas-miR319a-3p was predicted to target TCP transcription factors, which are plant-specific transcription factors that have been shown to participate in specifying plant morphological traits, such as organ border delimitation [68], cell division and proliferation [69], flower and leaf shape, and shoot outgrowth [70]. In *Arabidopsis*, the TCP transcription factors have been related to control of the morphology of shoot lateral organs and formation of the shoot meristem-dependent regulation of the expression of boundary-specific genes [71]. The putative transcription factor SCL is a number of GRAS protein family that are involved in several aspects of plant growth and development, including control of asymmetric cell division, maintenance of stem cell status, and induction of the regeneration of the root tip after laser ablation, and SCL expression has been associated with auxin distribution in root apical meristem [72–74]. In this study, three SCL genes were predicted to be targeted by the miR171 family, which was up-regulated in the PA4 plants. This result suggested that SCL may be involved in the formation of adventitious root and the other aerial organs of *Paulownia*. L-type lectin-domain containing receptor kinases (lecRKs) and zinc finger CCCH domain-containing protein were predicted to be targeted by pas-miR1 and pas-miR22-3p, respectively, suggesting that, besides their possible involvement in plant development, some of the miRNA target genes identified in this study could play fundamental roles in biotic and abiotic stresses. A previous study showed that LecRKs were most likely plasma membrane proteins, and were probably involved in mediating protein-protein interactions with a wide range of functions such as recognition of oligosaccharide or lipochitooligosaccharide signals, linking ABA-signaling, and response to salt, drought, cold stress, wounding, and disease in plants [75,76]. Zinc finger CCCH domain-containing protein is a kind of RNA-binding protein and many studies have shown that it may be regulated by abiotic or biotic stresses, and could have regulatory functions in mRNA processing [77–80], thus supporting the possible roles for pas-miR1 and pas-miR22-3p in the adaptive response of PA4 to abiotic stress.

Based on the GO analysis of the targets of the identified miRNAs in the two *P. australis* genotypes plants, the target genes were separated into the biological process, cellular component, and molecular function. Some of the genes were annotated as being involved in biological regulation, cellular process, developmental process, response to stimulus, metabolic process, cell, cell part, organelle components, binding, and catalytic activity (Figure 2). Thus, our results suggest that the transcript targets might be closely related to the observed differences of phenotype (environmental adaptations) and resistance to biotic and abiotic stresses between the PA4 and PA2 plants. The functional role of the differentially expressed miRNAs will be the focus of future investigations. No target genes were predicted for many of the miRNAs identified in the two *P. australis* sRNA libraries, and a few of the predicted target genes were annotated as being of unknown function and hypothetical genes. Careful analysis of these potential targets will contribute further to our understanding of the role of miRNAs in *P. australis*.

In summary, miRNA target genes were identified using a degradome approach that included functional annotation and pathway analyses based on the GO and KEGG databases. Some of the transcript targets regulated by the differentially expressed miRNAs were related to the physiology and environmental adaptations. Our results suggest that the significantly varied expression miRNAs in the two *P. australis* genotypes are probably involved in the epigenetic changes of PA4 plants, the correlation between the miRNA expression pattern and genome duplication may be more complex than we first thought. Taken together, this study provides beneficial information for elucidating the miRNA-mediated regulation of transcript targets in *P. australis* and related species.

Supporting Information

Figure S1   Length distribution of sRNAs in *P. australis*. (A) Size distribution of total sequences. (B) Size distribution of unique sequences.

Figure S2   Results from qRT-PCR of miRNAs in *P. australis*. PA2-1, 30-day-old diploid in vitro plantlets; PA4-1, 30-day-old autotetraploid in vitro plantlets; PA2-2, two-year-old diploid saplings; PA4-2, two-year-old autotetraploid saplings. Three independent biological replicates were performed. Values are means $+$ SD (n $-$ 3). The expression levels of miRNAs were normalized to U6. The normalized miRNA levels in the PA2-1 were arbitrarily set to 1. *: Statistically significant differences between PA2 and PA4 under the same developmental stages (p-value was less than 0.05).

Author Contributions

Conceived and designed the experiments: GQF SYN EKX MJD ZLZ YPD. Performed the experiments: GQF SYN EKX MJD ZLZ YPD. Analyzed the data: GQF SYN EKX MJD ZLZ YPD. Contributed reagents/materials/analysis tools: GQF SYN EKX MJD ZLZ YPD. Contributed to the writing of the manuscript: SYN GQF.

References

1. Vaucheret H (2006) Post-transcriptional small RNA pathways in plants: mechanisms and regulations. Genes Dev 20: 759–771.
2. Meng Y, Shao C, Wang H, Chen M (2011) The regulatory activities of plant microRNAs: a more dynamic perspective. Plant Physiol 157: 1583–1595.
3. Sunkar R, Zhou X, Zheng Y, Zhang W, Zhu JK (2008) Identification of novel and candidate miRNAs in rice by high throughput sequencing. BMC Plant Biol 8: 25.
4. Bartel DP (2004) MicroRNAs: genomics, biogenesis, mechanism, and function. Cell 116: 281–297.
5. Voinnet O (2009) Origin, biogenesis, and activity of plant microRNAs. Cell 136: 669–687.
6. Mallory A, Vaucheret H (2006) Functions of microRNAs and related small RNAs in plants. Nat Genet 38: 31–36.
7. Sunkar R, Chinnusamy V, Zhu J, Zhu JK (2007) Small RNAs as big players in plant abiotic stress responses and nutrient deprivation. Trends Plant Sci 12: 301–309.
8. Shukla LI, Chinnusamy V, Sunkar R (2008) The role of microRNAs and other endogenous small RNAs in plant stress responses. Biochim Biophys Acta 1779: 743–748.
9. Chuck G, Candela H, Hake S (2009) Big impacts by small RNAs in plant development. Curr Opin Plant Biol 12: 81–86.
10. Zhou Z, Huang S, Yang Z (2008) Bioinformatic identification and expression analysis of new microRNAs from Medicago truncatula. Biochem Biophys Res Commun 374: 538–542.
11. Wang X, Reyes J, Chua N, Gaasterland T (2004) Prediction and identification of Arabidopsis thaliana microRNAs and their mRNA targets. Genome Biol 5: R65.
12. Adai A, Johnson C, Mlotshwa S, Archer-Evans S, Manocha V, et al. (2005) Computational prediction of miRNAs in Arabidopsis thaliana. Genome Res 15: 78–91.
13. Jones-Rhoades M, Bartel D (2004) Computational identification of plant microRNAs and their targets, including a stress-induced miRNA. Mol Cell 14: 787–799.
14. Addo-Quaye C, Eshoo TW, Bartel DP, Axtell MJ (2008) Endogenous siRNA and miRNA targets identified by sequencing of the Arabidopsis degradome. Curr Biol 18: 758–762.
15. German MA, Pillay M, Jeong DH, Hetawal A, Luo S, et al. (2008) Global identification of microRNA-target RNA pairs by parallel analysis of RNA ends. Nat Biotechnol 26: 941–946.
16. Zhou M, Gu L, Li P, Song X, Wei L, et al. (2010) Degradome sequencing reveals endogenous small RNA targets in rice (Oryza sativa L. ssp. indica). Front Biol 5: 67–90.
17. Shen Y, Jiang Z, Lu S, Lin H, Gao S, et al. (2013) Combined small RNA and degradome sequencing reveals microRNA regulation during immature maize embryo dedifferentiation. Biochem Biophys Res Commun 441: 425–430.
18. Zhou Z, Song J, Yang Z (2012) Genome-wide identification of Brassica napus microRNAs and their targets in response to cadmium. J Exp Bot 63: 4597–4613 doi: 4510.1093/jxb/ers 4136.
19. Zhou Z, Zeng H, Liu Z, Yang Z (2012) Genome-wide identification of Medicago truncatula microRNAs and their targets reveals their differential regulation by heavy metal. Plant Cell Environ 35: 86–99.

20. Xu X, Yin L, Ying Q, Song H, Xue D, et al. (2013) High-throughput sequencing and degradome analysis identify miRNAs and their targets involved in fruit senescence of Fragaria ananassa. PLoS One 8: e70959.

21. Pantaleo V, Szittya G, Moxon S, Miozzi L, Moulton V, et al. (2010) Identification of grapevine microRNAs and their targets using high-throughput sequencing and degradome analysis. Plant J 62: 960–976.

22. Hao DC, Yang L, Xiao PG, Liu M (2012) Identification of Taxus microRNAs and their targets with high-throughput sequencing and degradome analysis. Physiol Plant 146: 388–403.

23. Shuai P, Liang D, Zhang Z, Yin W, Xia X (2013) Identification of drought-responsive and novel Populus trichocarpa microRNAs by high-throughput sequencing and their targets using degradome analysis. BMC Genomics 14: 233.

24. Li B, Qin Y, Duan H, Yin W, Xia X (2011) Genome-wide characterization of new and drought stress responsive microRNAs in Populus euphratica. J Exp Bot 62: 3765–3779.

25. Zhang JZ, Ai XY, Guo WW, Peng SA, Deng XX, et al. (2012) Identification of miRNAs and their target genes using deep sequencing and degradome analysis in trifoliate orange [Poncirus trifoliata L. Raf]. Mol Biotechnol 51: 44–57.

26. Fan G, Zhai X, Niu S, Ren Y (2014) Dynamic expression of novel and conserved microRNAs and their targets in diploid and tetraploid of Paulownia tomentosa. Biochimie 102: 68–77.

27. Niu S, Fan G, Zhao Z, Deng M, Dong Y (2014) High-throughput sequencing and degradome analysis reveal microRNA differential expression profiles and their targets in Paulownia fortunei. Plant Cell Tiss Org DOI 10.1007/s11240-014-0546-9.

28. Chirko CP, Gold MA, Nguyen PV, Jiang J (1996) Influence of direction and distance from trees on wheat yield and photosynthetic photon flux density in a Paulownia and wheat intercropping system. Forest Ecol Manag 83: 171–180.

29. Bayliss K, Saqib M, Dell B, Jones M, Hardy GSJ (2005) First record of 'Candidatus Phytoplasma australiense' in Paulownia trees. Australas Plant Path 34: 123–124.

30. Fan G, Wei Z, Yang Z (2009) Induction of autotetraploid of Paulownia australis and its in vitro plantlet regeneration. J Northwest Sci-Tech Univ 37: 83–90.

31. Fan G, Cao Y, Zhao Z, Yang Z (2007) Induction of Autotetraploid of Paulownia fortunei. Sci Silv Sin 43: 31–35.

32. Fan G, Yang Z, Cao Y, Q. Z (2007) Induction of Autotetraploid of Paulownia tomentosa. Plant physiol Commun 43: 109–111.

33. Fan G, Zhai X, Wei Z, Yang Z (2010) Induction of autotetrapioid from the somatic cell of Paulownia tomentosa×Paulownia fortunei and its in vitro plantley regeneration. J Northeast Forestry Univ 38:22–26.

34. Zhang X, Zhai X, Fan G, Deng M, Zhao Z (2012) Observation on Microstructure of Leaves and Stress Tolerance Analysis of Different Tetraploid Paulownia. J Henan Agric Univ 46: 646–650.

35. Zhai X, Zhang X, Zhao Z, Deng M, Fan G (2012) Study on Wood Physical Properties of Tetraploid Paulownia fortunei. J Henan Agric Univ 46: 651–654.

36. Fan G, Zhai X, Liu X (2002) Callus induction from Paulownia plant leaves and their plantlet regenerations. Sci Silv Sin 38: 29–35.

37. Hafner M, Landgraf P, Ludwig J, Rice A, Ojo T, et al. (2008) Identification of microRNAs and other small regulatory RNAs using cDNA library sequencing. Methods 44: 3–12.

38. Li R, Li Y, Kristiansen K, Wang J (2008) SOAP: short oligonucleotide alignment program. Bioinformatics 24: 713–714.

39. Zuker M (2003) Mfold web server for nucleic acid folding and hybridization prediction. Nucleic Acids Res 31: 3406–3415.

40. Meyers BC, Axtell MJ, Bartel B, Bartel DP, Baulcombe D, et al. (2008) Criteria for annotation of plant MicroRNAs. Plant Cell 20: 3186–3190.

41. Audic S, Claverie JM (1997) The significance of digital gene expression profiles. Genome Res 7: 986–995.

42. Man MZ, Wang X, Wang Y (2000) POWER_SAGE: comparing statistical tests for SAGE experiments. Bioinformatics 16: 953–959.

43. Chen C, Ridzon DA, Broomer AJ, Zhou Z, Lee DH, et al. (2005) Real-time quantification of microRNAs by stem-loop RT-PCR. Nucleic Acids Res 33: e179.

44. Livak KJ, Schmittgen TD (2001) Analysis of relative gene expression data using real-time quantitative PCR and the 2(-Delta Delta C(T)). Methods 25: 402–408.

45. Wang K, Li M, Gao F, Li S, Zhu Y, et al. (2012) Identification of Conserved and Novel microRNAs from Liriodendron chinense Floral Tissues. PLoS one 7: e44696.

46. Chen X (2009) Small RNAs and their roles in plant development. Ann Rev Cell Dev 25: 21–44.

47. Morin RD, O'Connor MD, Griffith M, Kuchenbauer F, Delaney A (2008) Application of massively parallel sequencing to microRNA profiling and discovery in human embryonic stem cells. Genome Res 18: 610–621.

48. Cifuentes M, Grandont L, Moore G, Chevre AM, Jenczewski E (2010) Genetic regulation of meiosis in polyploid species: new insights into an old question. New Phytol 186: 29–36.

49. Chen ZJ (2007) Genetic and epigenetic mechanisms for gene expression and phenotypic variation in plant polyploids. Annu Rev Plant Biol 58: 377–406.

50. Leitch AR, Leitch IJ (2008) Genomic plasticity and the diversity of polyploid plants. Science 320: 481–483.

51. Adams K (2007) Evolution of duplicate gene expression in polyploid and hybrid plants. J hered 98: 136–141.

52. Vashisht D, Nodine MD (2014) MicroRNA functions in plant embryos. Biochem Soc Trans 42: 352–357.

53. Ren Y, Chen L, Zhang Y, Kang X, Zhang Z, et al. (2012) Identification of novel and conserved Populus tomentosa microRNA as components of a response to water stress. Funct Integr Genomics 12: 327–339.

54. Zhao M, Ding H, Zhu J, Zhang F, Li W (2011) Involvement of miR169 in the nitrogen-starvation responses in Arabidopsis. New Phytol 190: 906–915.

55. Li B, Yin W, Xia X (2009) Identification of microRNAs and their targets from Populus euphratica. Biochem Bioph Res Co 388: 272–277.

56. Liu HH, Tian X, Li YJ, Wu CA, Zheng CC (2008) Microarray-based analysis of stress-regulated microRNAs in Arabidopsis thaliana. RNA 14: 836–843.

57. Mao W, Li Z, Xia X, Li Y, Yu J (2012) A combined approach of high-throughput sequencing and degradome analysis reveals tissue specific expression of microRNAs and their targets in cucumber. PloS one 7: e33040.

58. Chi X, Yang Q, Chen X, Wang J, Pan L, et al. (2011) Identification and characterization of microRNAs from peanut (Arachis hypogaea L.) by high-throughput sequencing. PLoS One 6: e27530.

59. Zhang X, Fan G, Zhao Z, Cao X, Zhao G, et al. (2013) Analysis of Diploid and Its Autotetraploid Paulownia tomentosa× P. fortunei with AFLP and MSAP. Sci Silv Sin 10: 026.

60. Shen H, He H, Li J, Chen W, Wang X, et al. (2012) Genome-wide analysis of DNA methylation and gene expression changes in two Arabidopsis ecotypes and their reciprocal hybrids. Plant Cell 24: 875–892.

61. Salmon A, Ainouche ML, Wendel JF (2005) Genetic and epigenetic consequences of recent hybridization and polyploidy in Spartina (Poaceae). Mol Ecol 14: 1163–1175.

62. Ha M, Lu J, Tian L, Ramachandran V, Kasschau K D., et al. (2009) Small RNAs serve as a genetic buffer against genomic shock in Arabidopsis interspecific hybrids and allopolyploids. Proc Natl Acad Sci USA 106: 17835–17840.

63. An F, Chan M (2012) Transcriptome-wide characterization of miRNA-directed and non-miRNA-directed endonucleolytic cleavage using Degradome analysis under low ambient temperature in Phalaenopsis aphrodite subsp. formosana. Plant Cell Physiol 53: 1737–1750. doi:1710.1093/pcp/pcs1118.

64. Wei M, Wei H, Wu M, Song M, Zhang J, et al. (2013) Comparative expression profiling of miRNA during anther development in genetic male sterile and wild type cotton. BMC Plant Biol 13: 66.

65. Rhoades MW, Reinhart BJ, Lim LP, Burge CB, Bartel B, et al. (2002) Prediction of plant microRNA targets. Cell 110: 513–520.

66. Nagpal P, Ellis CM, Weber H, Ploense SE, Barkawi LS, et al. (2005) Auxin response factors ARF6 and ARF8 promote jasmonic acid production and flower maturation. Development 132: 4107–4118.

67. Kwak P, Wang Q, Chen X, Qiu C, Yang Z (2009) Enrichment of a set of microRNAs during the cotton fiber development. BMC Genomics 10: 457.

68. Weir I, Lu J, Cook H, Causier B, Schwarz-Sommer Z, et al. (2004) CUPULIFORMIS establishes lateral organ boundaries in Antirrhinum. Development 131: 915–922.

69. Kosugi S, Ohashi Y (2002) DNA binding and dimerization specificity and potential targets for the TCP protein family. Plant J 30: 337–348.

70. Cubas P (2004) Floral zygomorphy, the recurring evolution of a successful trait. Bioessays 26: 1175–1184.

71. Koyama T, Furutani M, Tasaka M, Ohme-Takagi M (2007) TCP transcription factors control the morphology of shoot lateral organs via negative regulation of the expression of boundary-specific genes in Arabidopsis. Plant Cell 19: 473–484.

72. Sabatini S, Beis D, Wolkenfelt H, Murfett J, Guilfoyle T, et al. (1999) An auxin-dependent distal organizer of pattern and polarity in the Arabidopsis root. Cell 99: 463–472.

73. Sánchez C, Vielba JM, Ferro E, Covelo G, Solé A, et al. (2007) Two SCARECROW-LIKE genes are induced in response to exogenous auxin in rooting-competent cuttings of distantly related forest species. Tree Physiol 27: 1459–1470.

74. Wysocka-Diller JW, Helariutta Y, Fukaki H, Malamy JE, Benfey PN (2000) Molecular analysis of SCARECROW function reveals a radial patterning mechanism common to root and shoot. Development 127: 595–603.

75. Bouwmeester K, Govers F (2009) Arabidopsis L-type lectin receptor kinases: phylogeny, classification, and expression profiles. J Exp Bot 60: 4383–4396.

76. Singh P, Zimmerli L (2013) Lectin receptor kinases in plant innate immunity. Front Plant Sci 4: 124.

77. Li J, Jia D, Chen X (2001) HUA1, a regulator of stamen and carpel identities in Arabidopsis, codes for a nuclear RNA binding protein. Plant Cell 13: 2269–2281.

78. Berg JM, Shi Y (1996) The galvanization of biology: a growing appreciation for the roles of zinc. Science 271: 1081–1085.

79. Wang D, Guo Y, Wu C, Yang G, Li Y, et al. (2008) Genome-wide analysis of CCCH zinc finger family in Arabidopsis and rice. BMC Genomics 9: 44.

80. Peng X, Zhao Y, Cao J, Zhang W, Jiang H, et al. (2012) CCCH-type zinc finger family in maize: genome-wide identification, classification and expression profiling under abscisic acid and drought treatments. PLoS One 7: e40120.

19

Unique Features of a Japanese '*Candidatus* Liberibacter asiaticus' Strain Revealed by Whole Genome Sequencing

Hiroshi Katoh[1]*, Shin-ichi Miyata[1], Hiromitsu Inoue[2], Toru Iwanami[1]

1 NARO Institute of Fruit Tree Science, Tsukuba, Ibaraki, Japan, 2 Kuchinotsu Citrus Research Station, NARO Institute of Fruit Tree Science, Minami-shimabara, Nagasaki, Japan

Abstract

Citrus greening (huanglongbing) is the most destructive disease of citrus worldwide. It is spread by citrus psyllids and is associated with phloem-limited bacteria of three species of α-Proteobacteria, namely, '*Candidatus* Liberibacter asiaticus', '*Ca.* L. americanus', and '*Ca.* L. africanus'. Recent findings suggested that some Japanese strains lack the bacteriophage-type DNA polymerase region (DNA pol), in contrast to the Floridian psy62 strain. The whole genome sequence of the pol-negative '*Ca.* L. asiaticus' Japanese isolate Ishi-1 was determined by metagenomic analysis of DNA extracted from '*Ca.* L. asiaticus'-infected psyllids and leaf midribs. The 1.19-Mb genome has an average 36.32% GC content. Annotation revealed 13 operons encoding rRNA and 44 tRNA genes, but no typical bacterial pathogenesis-related genes were located within the genome, similar to the Floridian psy62 and Chinese gxpsy. In contrast to other '*Ca.* L. asiaticus' strains, the genome of the Japanese Ishi-1 strain lacks a prophage-related region.

Editor: Paul Jaak Janssen, Belgian Nuclear Research Centre SCK•CEN, Belgium

Funding: The National Agriculture and Food Research Organization Institute of Fruit Tree Science provided funding for this study. The grant number is 199998. The funders had no role in study design, data collection and analysis, decision to publish, or preparation of the manuscript.

Competing Interests: The authors have declared that no competing interests exist.

* Email: katohh@affrc.go.jp

Introduction

Citrus greening (huanglongbing) is a devastating citrus disease that affects crops around the world. The disease was first noted in China in the early 20th century [1]. Three species of phloem-limited, gram-negative bacteria in the genus '*Candidatus* Liberibacter' are associated with greening. '*Ca.* L. africanus' is mainly present in Africa [2]; '*Ca.* L. americanus' is found in Brazil [3]. A third species, '*Ca.* L. asiaticus' is particularly widespread in Asian countries as well as in Sao Paulo, Brazil and Florida, USA. '*Ca.* L. asiaticus' is transmitted by phloem-feeding insect vectors, the Asian citrus psyllid *Diaphorina citri* [4] and the African citrus psyllid *Trioza erytreae* [5]. A new Liberibacter species, '*Ca.* L. solanacearum', was recently associated with the emerging 'zebra chip' disease of potatoes in the U.S. and tomatoes in New Zealand [6].

Little is known about the genetic diversity of '*Ca.* L. asiaticus'; the bacteria are difficult to culture, although some successes have been reported [7,8,9]. Diversity studies of '*Ca.* L. asiaticus' have been restricted to the 16S/23S rRNA genes, the *omp* gene region, the *rplKAJL-rpoBC*, *nus*G-*rpl*K operon sequence, or bacteriophage-type DNA polymerase region (DNA pol) [10–20]. However, the complete genomic sequence of the pathogenic '*Ca.* L. asiaticus' Floridian strain "psy62" (1.23 Mb) [21] has been determined, thus enabling genome-wide analysis. In fact, Chen et al. characterized variation in "*Ca.* L. asiaticus" strains by using one repeat unit (AGACACA) [22]. From the whole-genome sequence, we selected 25 simple sequence repeat loci, including one repeat unit reported by Chen et al. [22] and successfully

differentiated '*Ca.* L. asiaticus' strains using these SSR loci [23,24]. Zhou et al. identified two hypervariable genes in the prophage regions of the psy62 genome [25]. Morgan et al. improved real-time PCR detection of '*Ca.* L. asiaticus' from citrus and psyllid hosts by using the prophage gene [26]. The whole-genome sequencing of '*Ca.* L. asiaticus' Floridian psy62 strain significantly advanced the study of diversity in this species.

Zhang et al. [27] reported two highly related, circular bacteriophage-type genes associated with '*Ca.* L. asiaticus', named SC1 and SC2. Both were found integrated into the '*Ca.* L. asiaticus' Floridian UF506 strain genome as prophages [27]. SC1 was apparently a fully functional, temperate phage with a lytic cycle that was seemingly activated when its host bacterium was present in plants but not when in psyllids [27]. SC2 replicates as an excision plasmid when its '*Ca.* L. asiaticus' host is present in either plants or psyllids [27]. These findings suggest the bacteriophage-type genes are important for infection and virulence expression. However, most of the Japanese '*Ca.* L. asiaticus' strains lack the bacteriophage-type DNA polymerase gene [18,19]. In Floridian UF506, the bacteriophage-type DNA polymerase gene is flanked by SC1 and SC2. Thus, absence of the bacteriophage-type DNA polymerase gene in Japanese strains suggests they also lack SC1 and SC2. Thus, the Japanese strains have unique genomic features.

In contrast to '*Ca.* L. asiaticus' Floridian strains psy62 and UF506, the whole genome sequence of a Japanese '*Ca.* L. asiaticus' strain lacking the bacteriophage-type DNA polymerase gene has not been reported. Recently, the complete genome sequence of the Chinese '*Ca*. L. asiaticus' strains gxpsy [28] and A4 [29] were reported, although the latter remains in the draft form. Both Chinese '*Ca*. L. asiaticus' strains also contained the bacteriophage-type DNA polymerase gene. The results encouraged us to perform whole-genome sequencing of a Japanese strain lacking this gene. Duan et al. [21] obtained a complete circular '*Ca*. L. asiaticus' Floridian psy62 strain genome by metagenomic analysis of DNA extracted from a single '*Ca*. L. asiaticus'-infected psyllid. We used a similar method to obtain the complete genome of the uncultured '*Ca*. L. asiaticus' Japanese strain Ishi-1.

Materials and Methods

Bacterial strains

Japanese '*Ca*. L. asiaticus' strain Ishi-1 was used throughout the study. The strain was originally found in local citrus of unidentified cultivars on Ishigaki Island, Okinawa prefecture, Japan. The infected scion was sent to the NARO Institute of Fruit Tree Science (NIFTS) with permission from the plant quarantine office of Japan, and kept in the isolated greenhouse after grafting on rough lemon (*Citrus jambhiri* Lush) rootstocks. The strain Ishi-1

induced severe symptoms on rough lemon, yuzu (*Citrus junos* Tanaka, Figure 1) and other citrus cultivars.

Psyllid treatment

All experiments using live individuals of *D. citri* were performed in insect-proof growth chambers at 25°C with a 16L:8D photoperiod at the Kuchinotsu Citrus Research Station, NIFTS (Otsu 954, Kuchinotsu, Minamishimabara, Nagasaki 859–2501, Japan). Healthy fifth instars of psyllids were transferred to an HLB-affected rough lemon tree (*Citrus jambhiri*, approximately 40 cm in height) with a high titer of '*Ca*. L. asiaticus' bacteria. After acquisition feeding for 20 days on the infected plant, nine emerged adults were reared individually for 20 days on healthy *Citrus junos* seedlings for incubating the HLB bacteria, and they were stored at −50°C.

DNA extraction and quantitative real-time PCR

Total DNA was purified from the entire body of single psyllids using the DNeasy Blood and Tissue Kit (Qiagen, Tokyo, Japan) and a plastic homogenizer pestle (As One, Tokyo, Japan)

Figure 1. Foliar symptoms on Yuzu (*Citrus junos* Tanaka) induced by '*Ca*. L. asiaticus' Japanese Ishi-1. Severe yellowing on the leaves of a Yuzu plant kept in a closed chamber at the NARO Institute of Fruit Tree Science.

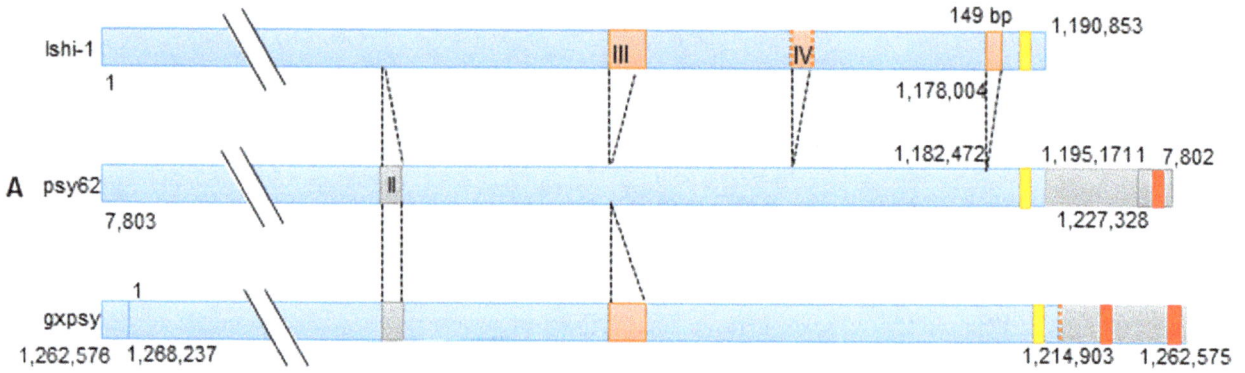

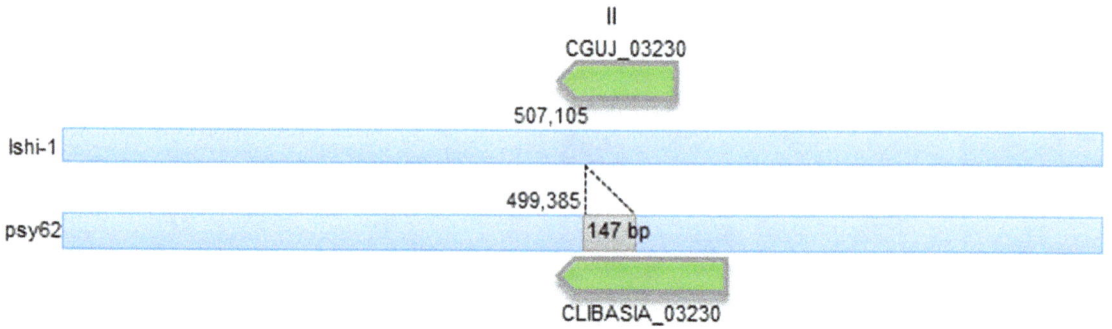

CGUJ_03230   MNFRIAMLISFLASGCVAHALLTKKIESDTDSRHEKATISLSAHDKEGSKHTMNAEFSVPKNDE-------------------
CLIBASIA_03230   MNFRIAMLISFLASGCVAHALLTKKIESDTDSRHEKATISLSAHDKEGSKHTMNAEFSVPKNDEKYTISSLTKKIESDTDFR
WSI_02190   MNFRIAMLISFLASGCVAHALLTKKIESDTDSRHEKATISLSAHDKEGSKHTMNAEFSVPKNDEKYTISSLTKKIESDTDFR
                 ****

CGUJ_03230   --------------------------------KYTISACASDDKGNKSTLCVECPSPSTPGQYDLNHCAECENTTSKGLCP
CLIBASIA_03230   REKATISLSAHDKEGSKHTMNAKEFSVPKNDEYTISACASDDKGNKSTLCVECPSPSTPGQYDLNHCAECENTTSKGLCP
WSI_02190   REKATISLSAHDKEGSKHTMNAEFSVPKNDEKYTISACASDDKGNKSTLCVECPSPSTPGQYDLNHCAECENTTSKGLCP
                                                 ****

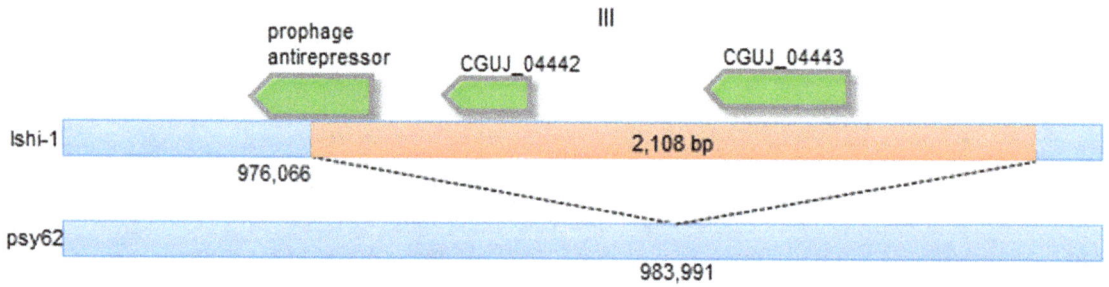

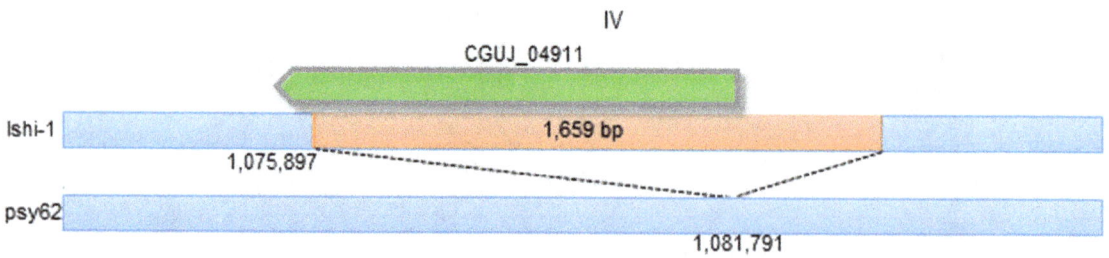

Figure 2. Whole-genome comparison of '*Ca.* L. asiaticus' Floridian psy62 and Japanese Ishi-1. A, Schematic linear alignment between '*Ca.* L. asiaticus' Floridian psy62 and Japanese Ishi-1. Orange/gray boxes (designated I, II, III, and IV) represent four large insertion/deletion domains in '*Ca.* L. asiaticus' Japanese Ishi-1. Other In/Del and SNP variants are ignored. Vertical dotted lines in domain IV in the Ishi-1 box indicate the unclear insertion borders. The number by each box indicates the nucleotide position of each strain. **B, C, and D,** Enlarged maps of domains II, III, and IV in Figure 3 A. Green arrows indicate CDS. B, Deduced amino acid sequences of the hypothetical protein at CLIBASIA_03230 of psy62, WSI_02190 of gxpsy, and CGUJ_03230 of Ishi-1 aligned by CLUSTAL W [48] and identical residues are indicated with asterisks. Databank accession numbers are CP001677 for psy62 [21], AP014595 for Ishi-1, and CP004005 for gxpsy [28].

according to the manufacturer's instructions, and eluted in 150 µL.

Individual psyllid DNA samples were analyzed for '*Ca.* L. asiaticus' populations by quantitative real-time PCR analysis as described by Inoue et al. [30]. Samples containing copies of '*Ca.* L. asiaticus' genomic DNA were selected by real-time PCR (data not shown). Whole-genome amplification was performed with Illustra GenomiPhi V2 (GE Healthcare, Buckinghamshire, England) according to the manufacturer's instructions. DNA concentration was estimated with the Qubit 2.0 instrument and the Qubit dsDNA HS Assay (Life Technologies, Invitrogen, California).

Genome sequencing and mapping

Sequencing was performed at the Bio Dragon Genomics Center (Takara Bio Co. Ltd. Mie, Japan). DNA libraries with 300~350 bp inserts were constructed according to manufacturer's instructions (Illumina GaIIx platform) and 75-bp paired-end reads were generated on an Illumina HiSeqTM 2000 platform. Reads were mapped to the '*Ca.* L. asiaticus' Floridian psy62 genome using BWA [31] and Bowtie [32]. Mapping results were visualized with Integrative Genomics Viewer (IGV) version 2.3 [33].

Polymerase chain reaction for whole genome mapping confirmation

After initial genome mapping results were obtained, ambiguous sequences were determined by PCR amplification and conventional sequencing on an ABI 3130×*l* instrument. Total DNA was extracted from the leaf midrib tissue of citrus trees infected with the '*Ca.* L. asiaticus' Japanese Ishi-1 strain. Total DNA was extracted with the DNeasy plant minikit (Qiagen, Valencia, CA) according to manufacturer's instructions with minor modifications: approximately 0.2 g of the leaf midrib was placed in 400 µL AP1 buffer in a mortar and ground with a pestle until the leaf midrib became a fine green liquid.

Many In/Dels and SNPs were found by mapping the sequence reads of Ishi-1 to the complete sequence of the pathogenic '*Ca.* L. asiaticus' Floridian psy62 (1.23 Mb) strain, and primers were designed from the surrounding sequences (Primer3, http://frodo. wi.mit.edu/primer3/) (Table S1). Other primers were selected from Duan et al. [21]. PCR was performed with the Gene Amp PCR System 9700 (Applied Biosystems, Foster City, CA) in 20-µl reactions containing 1 µl DNA template, 0.1 µM each primer, 200 µM dNTPs, 1× PCR buffer, and 2.5 units of *Ex Taq* DNA polymerase Hot Start Version (TaKaRa, Shiga, Japan) under the following cycling conditions: initial denaturation at 92°C for

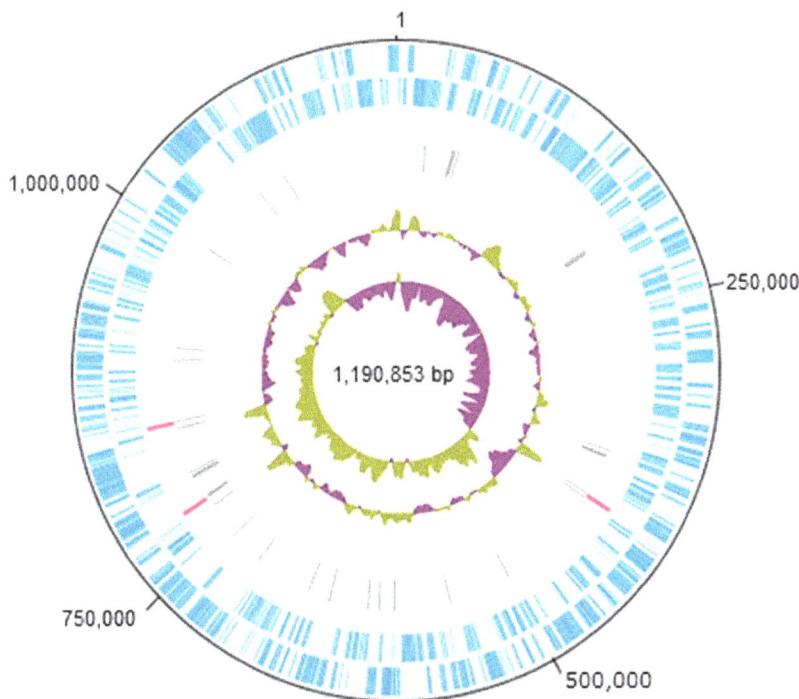

Figure 3. Schematic representation of the genome of '*Ca.* L. asiaticus' Japanese Ishi-1. Circular representation of the 1.19 Mbp genome. The tracks from the outmost circles represent (1) Forward CDS (blue) and (2) Reverse CDS (blue); (3) six copies of the rRNA operon (16S, 23S and 5S) (pink); (4) tRNA (gray): (5)% G+C content (yellow-green, purple), and (6) GC skew [(G−C/(G+C))] (yellow-green, purple).

Plant Biotechnology: Genetic Modification of Plants

Table 1. Comparison of the whole genome among three strains of 'Ca. Liberibacter asiaticus' and 'Ca. L. solanacearum.'

Features	'Ca. L. asiaticus'			'Ca. L. solanacearum'
	Ishi-1[a]	psy62[b]	gxpsy[c]	CIso-ZC1[d]
Size(bp)	1,190,853	1,227,328	1,268,237	1,258,278
GC%[e]	36.3	36.5	36.5	35.2
rRNA operons	13	9	6	9
tRNA	44	44	44	45
RBS	975	1022	1078	1093
CDS	1075	1134	1165	1192
hypothetical protein	313	358	368	409

[a]Accession number AP014595.
[b]Accession number CP001677 [21].
[c]Accession number CP004005 [28].
[d]Accession number NC_014774 [37].
[e]GC contents were calculated using GENETYX ver. 11.

2 min; 35 cycles of denaturation at 92°C for 30 s, annealing at 54°C for 30 s, and extension for 1 min/kb of the desired product at 72°C. Long-range PCR of products above 3.0 kbp was performed with Tks Gflex DNA polymerase (TaKaRa). Each 50-μl reaction contained 1 μl DNA template, 0.1 μM each primer, 2× Gflex PCR buffer (Mg^{2+}, dNTP plus), and 1.25 U Tks Gflex DNA polymerase. Cycling conditions were as follows: 30 cycles of denaturation at 98°C for 10 s, annealing at 52°C for 15 s, and extension for 30 s/kb of the desired product at 68°C.

DNA sequences were aligned using GENETYX-windows ver. 11 (Software Development, Tokyo, Japan), and homology analysis was performed as recommended by the DNA Data Bank of Japan (http://www.ddbj.nig.ac.jp/Welcome-j.html).

Gene prediction and functional annotation

Gene prediction and functional annotation were performed with the Microbial Genome Annotation Pipeline (MiGAP) (http://www.migap.org/index.php/en, [34]). Detection of tRNAs and rRNA was performed with tRNAScan-SE 1.23 (http://lowelab.ucsc.edu/tRNAscan-SE/, [35]) and RNAmmer 1.2 (http://www.cbs.dtu.dk/services/RNAmmer/, [36]).

Results

Whole genome re-sequencing

Sequencing yielded 2,721,927,150 bp of DNA from 36,292,362 pair-end reads of 75 bp. As a result of mapping using BWA, Bowtie against the psy62 strain reference, reads of 14.6% of the 2.7 Gbp were mapped. Coverage to the reference was 96.9%. The

sequence reads of Ishi-1 were not mapped near the nucleotide position of 0 to 7,803 and after nucleotide position 1,195,171 of the linear genomic map of psy62, indicating that Ishi-1 lacks a large genomic fragment. PCR amplification of the Ishi-1 template with the LJ754r and LJ764f primers, which are separated by about 35 kbp on the psy62 sequence, yielded a 2.6-kbp product. Sequence analysis showed that the 2.6 kbp fragment filled the 35-kbp gap in the Ishi-1genome. The results clearly demonstrated that Ishi-1 lacks about 33 kbp, corresponding to both ends of the linear genomic map of psy62 (Figure 2). Likewise, by mapping the candidate SNPs/InDels and PCR verification, the draft genome of Ishi-1 was obtained.

In order to confirm the whole genome sequence of Ishi-1 strain, six primers published by Duan [21] were selected, and 128 new primers were designed for conventional and long PCR (Table S1). All amplicons generated from these primers were directly sequenced on an ABI 3130×l. In total, we re-sequenced over 40,000 bp by Sanger sequencing. These efforts generated a circular chromosome sequence consisting of 1,190,853 bp (Figure 3).

General features of the 'Ca. L. asiaticus' Japanese Ishi-1 genome

The calculated GC content of the Ishi-1 genome is 36.32%, similar to other 'Ca. L. asiaticus' strains (Table 1). After annotation, the newly confirmed CDS regions were compared to those of other 'Ca. L. asiaticus' strains. Then, the tRNAs and rRNA of Japanese Ishi-1 were compared to those of Floridian

Table 2. Base substitution, insertion, deletion mutation and repeat number at respective SSR motif between 'Ca. L. asiaticus' Japanese Ishi-1 and Floridian psy62 strain.

	Ishi-1	psy62	SSR motif	Ishi-1	psy62
Base substitution	291	–	TACAGAA	14	8
Insertion (one base insertion)	66 (58)	–	AGACACA	8	5
Deletion (One base deletion)	56 (38)	–	TTTG	9	14
			TTTTAA	5	3
			AGA	6	5

Table 3. List of CDS encoded in the large 33 kbp fragment that is retained by Floridian psy62, UF506 and Chinese gxpsy strains, but not by Japanese Ishi-1 strain of 'Ca. L. asiaticus.'

psy62			UF506			gxpsy		
product	locus tag	Locus location in genome[a]	product	locus tag	Locus location in genome[b]	product	locus tag	Locus location in genome[c]
hypothetical protein[d]	CLIBASIA_00005	36..407	hypothetical protein	SC1_gp185 SC2_gp185	76449..76820 36401..36772	hypothetical protein	WSI_05545	1214938..1215309
hypothetical protein	CLIBASIA_00010	497..820	hypothetical protein	SC1_gp190 SC2_gp190	76910..77233 36862..37185	hypothetical protein	WSI_05550 WSI_05750	1215399..1215722 1255266..1255589
hypothetical protein	CLIBASIA_00015	948..2114	hypothetical protein	SC1_gp195 SC2_gp195	77361..78527 37313..38479	hypothetical protein	WSI_05555 WSI_05755	1215850..1217016 1257717..1256883
prophage antirepressor	CLIBASIA_00020	2285..3073	putative Bro-N family phage antirepressor	SC1_gp200 SC2_gp155	78698..79486 32107..32292	prophage antirepressor	WSI_05560 WSI_05760	1217187..1217978 1257054..1257845
hypothetical protein	CLIBASIA_00025	3091..3741	hypothetical protein	SC1_gp205 SC2_gp205	79504..80154 39456..40106	hypothetical protein	WSI_05565 WSI_05765	1217996..1218646 1257863..1258513
putative DNA polymerase from bacteriophage origin	CLIBASIA_00030	3745..5772	DNA polymerase A	SC1_gp210 SC2_gp210	80158..82185 40110..42137	putative DNA polymerase from bacteriophage origin	WSI_05570 WSI_05770	1218650..1220677 1258517..1260544
VRR-NUC domain-containing protein	CLIBASIA_00035	5769..6080	endonuclease	SC1_gp215 SC2_gp215	82182..82493 42134..42445	VRR-NUC domain-containing protein	WSI_05575 WSI_05775	1220674..1220985 1260541..1260852
SNF2 related protein	CLIBASIA_00040	6065..6727	SNF2 Dead box helicase	SC2_gp220	42430..43815	SNF2 related protein	WSI_05580 WSI_05780	1220970..1222355 1260837..1262222
DNA ligase, NAD-dependent	CLIBASIA_00050	7442..7801	DNA ligase	SC2_gp225 SC1_gp225	4811..5170 43808..44167	DNA ligase, NAD-dependent	WSI_05360 WSI_05585 WSI_05785	1183651..1184010 1222348..1222707 1262215..1262574
guanylate kinase	CLIBASIA_05525	1195911..1196264	hypothetical protein	SC1_gp235	44890..45261	guanylate kinase	WSI_05595	1223448..1223780
hypothetical protein	CLIBASIA_05530	complement(1196268..1196741)	hypothetical protein	SC1_gp005	complement(45265..45738)	hypothetical protein	WSI_05600	complement(1223784..1224257)
hypothetical protein	CLIBASIA_05531	complement(1196738..1197010)	hypothetical protein	SC1_gp010	complement(45735..46007)	hypothetical protein	WSI_05410	complement(1188796..1189068)
hypothetical protein	CLIBASIA_05538	complement(1197003..1199762)	hypothetical protein	SC1_gp025	complement(46325..48448)	hypothetical protein	WSI_05610	complement(1224519..1226903)
hypothetical protein	CLIBASIA_05545	complement(1199769..1202363)	hypothetical protein	SC1_gp030	complement(48455..51049)	hypothetical protein	WSI_05615	complement(1226910..1229504)
hypothetical protein	CLIBASIA_05550	complement(1202360..1203796)	hypothetical protein	SC1_gp035	complement(51046..52482)	hypothetical protein	WSI_05620	complement(1229501..1230937)
hypothetical protein	CLIBASIA_05555	complement(1203814..1205937)	hypothetical protein	SC1_gp045	complement(52500..54623)	hypothetical protein	WSI_05625	complement(1230955..1233078)
hypothetical protein	CLIBASIA_05560	complement(1205934..1206449)	hypothetical protein	SC1_gp050	complement(54620..55123)	hypothetical protein	WSI_05630	complement(1233075..1233596)

Table 3. Cont.

psy62			UF506			gxpsy		
product	locus tag	Locus location in genome[a]	product	locus tag	Locus location in genome[b]	product	locus tag	Locus location in genome[c]
hypothetical protein	CLIBASIA_05565 CLIBASIA_05570	complement(1205934..1206449) complement(1210209..1210451)	hypothetical protein	SC1_gp060	complement(55116..59138)	hypothetical protein	WSI_05635	complement(1233577..1237620)
hypothetical protein	CLIBASIA_05575	complement(1210476..1212212)	hypothetical protein	SC1_gp080	complement(59163..60899)	hypothetical protein	WSI_05640	complement(1237645..1239381)
hypothetical protein	CLIBASIA_05580	complement(1212205..1212732)	hypothetical protein	SC1_gp085	complement(60892..61419)	hypothetical protein	WSI_05645	complement(1239374..1239901)
hypothetical protein	CLIBASIA_05585	complement(1212732..1213763)	putative major capsid protein	SC1_gp090 SC2_gp090	complement(61419..62450) complement(22025..22945)	hypothetical protein	WSI_05455	complement(1200642..1201673)
hypothetical protein	CLIBASIA_05590	complement(1213776..1214480)	hypothetical protein	SC1_gp095	complement(62463..63167)	hypothetical protein	WSI_05655	complement(1240893..1241594)
hypothetical protein	CLIBASIA_05595	complement(1214491..1214820)	hypothetical protein	SC1_gp100	complement(63178..63507)	hypothetical protein	WSI_05660	complement(1241605..1241934)
head-to-tail joining protein, putative	CLIBASIA_05600	complement(1214813..1216483)	putative phage-related head-to-tail joining protein	SC1_gp105	complement(63500..65170)	head-to-tail joining protein, putative	WSI_05665	complement(1241927..1243597)
hypothetical protein	CLIBASIA_05605	complement(1216480..1216812)	hypothetical protein	SC1_gp110	complement(65167..65499)	hypothetical protein	WSI_05670	complement(1243594..1243926)
putative phage terminase, large subunit	CLIBASIA_05610	complement(1216885..1218420)	putative phage terminase, large subunit	SC1_gp115	complement(65572..67107)	putative phage terminase, large subunit	WSI_05680	complement(1244335..1245870)
hypothetical protein	CLIBASIA_05615	complement(1218677..1218793)				hypothetical protein	WSI_05450 WSI_05685	complement(1206133..1206249) complement(1246127..1246243)
hypothetical protein	CLIBASIA_05620	complement(1218955..1219443)	hypothetical protein	SC1_gp120 SC2_gp120	complement(67642..68157) complement(27687..28202)	hypothetical protein	WSI_05690	complement(1246405..1246893)
hypothetical protein	CLIBASIA_05625	complement(1220547..1221164)	hypothetical protein	SC1_gp125	complement(69236..69853)	hypothetical protein	WSI_05695	complement(1247997..1248614)
hypothetical protein	CLIBASIA_05630	1221334..1221936	hypothetical protein	SC1_gp130	70104..70625	hypothetical protein	WSI_05700	1248784..1249386
hypothetical protein	CLIBASIA_05635	1221998..1222390	hypothetical protein	SC1_gp135 SC2_gp135	70687..71079 30619..31011	hypothetical protein	WSI_05500 WSI_05705	1209271..1209663 1249448..1249840
hypothetical protein	CLIBASIA_05640	1222526..1222732	hypothetical protein	SC1_gp140 SC2_gp140	71215..71421 31147..31353	hypothetical protein	WSI_05505 WSI_05710	1209800..1210006 1249977..1250183
hypothetical protein	CLIBASIA_05645	1222725..1222937	hypothetical protein	SC1_gp145 SC2_gp145	71414..71626 31346..31558	hypothetical protein	WSI_05510 WSI_05715	1209999..1210211 1250176..1250388
intrrupted gp228, phage associated protein	CLIBASIA_05650	1222969..1223301	hypothetical protein[e]	SC1_gp150 SC2_gp150	71658..72062 31590..31994	intrrupted gp229, phage associated protein	WSI_05515 WSI_05720	1210243..1210575 1250420..1250752
hypothetical protein	CLIBASIA_05655	complement(1223555..1223866)	hypothetical protein	SC1_gp160 SC2_gp160	complement(72591..72968) complement(32523..32900)	hypothetical protein	WSI_05725	complement(1251006..1251317)

Table 3. Cont.

psy62			UF506			gxpsy		
product	locus tag	Locus location in genome[a]	product	locus tag	Locus location in genome[b]	product	locus tag	Locus location in genome[c]
P4 family phage/plasmid primase	CLIBASIA_05660	complement(1223914..1226283)	phage associated primase	SC1_gp165	complement(73016..75388)	P4 family phage/plasmid primase	WSL_05730	complement(1251365..1253734)
hypothetical protein	CLIBASIA_05665	complement(1226284..1226673)	hypothetical protein	SC1_gp170 SC2_gp170	complement(75389..75778) complement(35321..35710)	hypothetical protein	WSI_05535 WSI_05735	complement(1213869..1214258) complement(1253735..1254124)
hypothetical protein	CLIBASIA_05670	complement(1226691..1226897)	hypothetical protein	SC1_gp175 SC2_gp175	complement(75796..76002) complement(35728..35934)	hypothetical protein	WSI_05540 WSI_05740	complement(1214276..1214482) complement(1254142..1254348)
hypothetical protein	CLIBASIA_05675	complement(1226894..1227157)	hypothetical protein	SC2_gp180	complement(35931..36194)			

[a]Data are based on the genome sequence of 'Ca. L. asiaticus' Floridian psy62 strain. The accession number is CP001677 [21].
[b]Data are based on the genome sequence of 'Ca. L. asiaticus' Floridian UF506 strain. The accession number is HQ377374 [27].
[c]Data are based on the genome sequence of 'Ca. L. asiaticus' Chinese gxpsy strain. The accession number is CP004005 [28].
[d]Products on the same line indicates that the deduced amino acid sequences has a huge similarity to one of 'Ca. L. asiaticus' Floridian psy62 strain on the far left.
[e]Underline revealed that the deduced amino acid sequences showed about 80% similarity to one of 'Ca. L. asiaticus' Floridian psy62 strain on the far left.

psy62. These analyses revealed 1,075 coding sequences and 975 ribosome binding sites. We also found 44 tRNA genes that were shared with the Floridian psy62 and Chinese gxpsy strains, as well as 13 rRNA operons and 313 hypothetical proteins (Table 1). Our comparison of 'Ca. L. asiaticus' Japanese Ishi-1 and Floridian psy62 revealed 291 base substitutions (Table 2). We also confirmed 122 in/del loci (Table 2). The five SSR loci were also polymorphic between the two strains (Table 2).

Bacteriophage-type polymerase and other genes

As described above, the biggest difference in Japanese Ishi-1 is the absence of the 33-kbp fragment (Figure 3 A). In psy62, this fragment encodes 40 CDS, including the bacteriophage-type polymerase gene between the SC1 and SC2 genes in the prophage region (Table 3). Most of the 40 CDS are shared between psy62, UF506, and Gxpsy (Table 3). None of these 40 CDS, including the bacteriophage-type polymerase gene, were found elsewhere in the genome of Ishi-1. Thus, Ishi-1 lacks the bacteriophage-type DNA polymerase gene found in Floridian strains psy62 and UF506, and the Chinese gxpsy strain ([21], [27], [28], shown by a vertical red line in Figure 3A). Another bacteriophage-type DNA polymerase is encoded in the middle of the linear schematic representation of the psy62 genome (shown by a vertical yellow line in Figure 3A). This bacteriophage-type DNA polymerase is also encoded in the corresponding region of the Ishi-1 genome (Figure 3A). Thus, it became clear that Ishi-1 carries a single bacteriophage-type DNA polymerase gene, whereas psy62 has two. In contrast, Chinese gxpsy and Floridian UF506 carry three bacteriophage-type DNA polymerase genes (WSI_05345, 05570, 05770, UF506_015, SC1_gp210, SC2_gp210).

Absence of the 33 kbp-fragment means other genes are also missing from the Ishi-1 genome. For example, Ishi-1 carries two NAD-dependent DNA Ligase genes (CGUJ_05395, 05515), whereas psy62 carries three (CLIBASIA_00050, 05395, 05515)—as does Chinese gxpsy (WSI_05360, 05585, 05785). In addition, the putative phage terminase, large subunit, exists in a single copy (CGUJ_05470) in the genome of Ishi-1, but as two copies in psy62 (CLIBASIA_05470, 05610), and Chinese gxpsy carries three (WSI_05315, 05475, 05680). Furthermore, the genome of Ishi-1 does not contain a full-length P4 family phage/plasmid primase gene; psy62 carries one (CLIBASIA_05660) and the Chinese gxpsy genome carries two (WSI_05530, 05730).

Characteristics of 'Ca. L. asiaticus' Japanese Ishi-1 strain marked by large In/Del variations

Several large In/Dels are shown in the simplified schematic presentation of the genome (Figure 3 A). The 147-bp deletion at nucleotide positions 507106 through 507252 of the Floridian psy62 strain was detected in the genome of 'Ca. L. asiaticus' Japanese Ishi-1 (Figure 3 B). This deletion reduced the hypothetical protein sequence at CGUJ_03230 by 49 amino acids in comparison to CLIBASIA_03230 of psy62 and WSI_02190 of Chinese gxpsy (Figure 3 B). In contrast, the 2,108 bp insertion between nucleotide positions 983990 and 983991 (Figure 3 C), an **untranslated region** in psy62, was detected in Ishi-1. This insertion carries a prophage anti-repressor at CGUJ_04441, and two hypothetical proteins at CGUJ_04442 and CGUJ_04443 were newly confirmed. The deduced amino acid sequence of the prophage anti-repressor at CGUJ_04441 was identical to that of Chinese gxpsy (WSI_04270), but different from those of Floridian psy62 and UF506. The hypothetical protein at CGUJ_04442 was also identical to that of Chinese gxpsy (WSI_04275). The hypothetical protein at CGUJ_04443 locus shared 99% amino

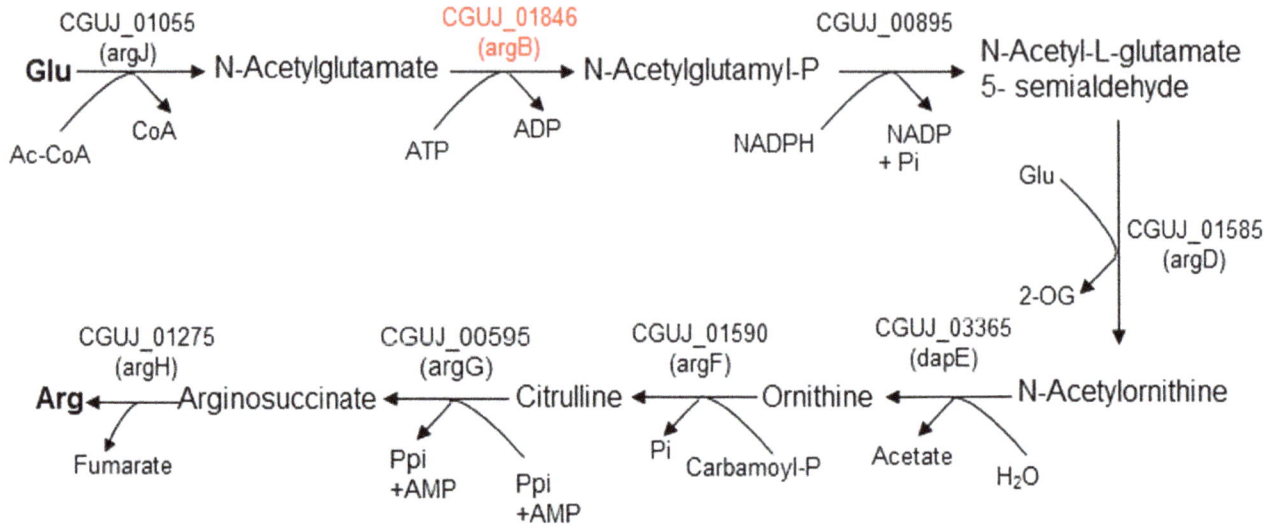

Figure 4. Analysis of the 'Ca. L. asiaticus' Japanese Ishi-1 arginine biosynthesis pathway. The typical prokaryotic arginine biosynthesis pathway. The NAGK family of enzymes catalyze the second step of arginine biosynthesis and are known as argB in many bacteria [44,45,46,47]. The argB that was not encoded by 'Ca. L. asiaticus' Floridian psy62 but encoded by Japanese Ishi-1 and Chinese gxpsy is indicated in red letters.

acid sequence identity with the putative WSI_04280 in Chinese gxpsy. These two hypothetical proteins shared no identity with any of the hypothetical proteins from Floridian psy62 and UF506.

Another insertion around nucleotide position 1081791 (psy62) was detected in Ishi-1 (Figure 3 D), encoding a hypothetical protein at CGUJ_04911 within the 1,660-bp span. The deduced amino acid sequence of the hypothetical protein shares 48% identity with the hypothetical protein CKC_03455 from 'Ca. L. solanacearum' CLso-ZC1, a pathogen of zebra chip. Ishi-1 also carries a 149 bp-long insertion that correspond to the nucleotide positions 1182471 and 1182472 of psy62 (Figure 3A). No reading frames were found in the insertion.

Other In/Del and non-synonymous SNPs affecting annotation of 'Ca. L. asiaticus' Japanese Ishi-1

Lin et al. noted the absence of a full-length N-acetylglutamate kinase (NAGK) in the genome of 'Ca. L. asiaticus' Floridian psy62, although it is present in 'Ca. L. solanacearum' CLso-ZC1 [37]. However, Japanese Ishi-1 (CGUJ_01846) and Chinese gxpsy (WSI_01760, [28]) encode identical full-length NAGK. Within the three 'Ca. L. asiaticus' strains, psy62 lacks an adenine between 406695 and 406696, thus truncating the sequence. The presence of an NAGK coding sequence indicates that Ishi-1 has a complete

pathway for the production of arginine from glutamine, unlike psy62 (Figure 4).

Because of a single base insertion, Ishi-1 has two copies of the malic enzyme gene at CGUJ_00080 and CGUJ_00081, while 'Ca. L. asiaticus' Floridian psy62 (CLIBASIA_00080) and Chinese gxpsy (WSI_00005) each carry a single copy.

The genome of Ishi-1 also encodes a non-heme ferritin-like protein (CGUJ_03035), just like psy62 [38,39]. This ferritin-like protein is also found in 'Ca. L. solanacearum' [37], but is absent from the genomes of all other Rhizobiaceae. The ferritin superfamily of proteins includes several diverse members that are typically involved in iron storage and detoxification [40,41,42]. Lin et al. hypothesized that this ferritin-like protein may play a critical role in the survival and/or virulence of 'Ca. L. solanacearum' and 'Ca. L. asiaticus' [37]. The genome of 'Ca. L. asiaticus' Chinese gxpsy also encodes a ferritin-like protein (WSI_2370). These sequences were identical but for a single amino acid substitution in Japanese Ishi-1 (Figure S1). In contrast, Floridian UF506 does not encode ferritin, indicating that this protein is dispensable.

Lin et al. reported that the 'Ca. L. solanacearum' genome encodes three known proteins involved in DNA replication and repair, all of which are absent from 'Ca. L. asiaticus' Floridian

Table 4. Presence of deduced amino acid sequence related DNA replication.

	LexA		DnaE		RadC	
	locus tag	nucleotide position	locus tag	nucleotide position	locus tag	nucleotide position
Ishi-1	-	-	CGUJ_03631	789771..793445	CGUJ_03976	ccomplement(866926..867195)
psy62	-	-	-	-	-	874850..875119
gxpsy	-	-	WSI_03515	782257..785931	WSI_03815	complement(859383..859676)
CIso-ZC1	CKC_02355	507699..508370	CKC_05200	complement(1116198..1119878)	CKC_04675	986531..987244

-: not identified.

psy62: LexA, DnaE, and RadC [37]. The genome Japanese Ishi-1 does not encode LexA, but does encode DnaE at CGUJ_03631 and RadC at CGUJ_03976. Chinese gxpsy also encodes DnaE and RadC, at WSI_03515 and WSI_03815, respectively (Table 4). However, a nucleotide sequence identical to that encoding RadC on the other 'Ca. L. asiaticus' strains is also carried in psy62 (Table 4), while LexA and DnaE are absent. This difference might be an annotation error.

The hypothetical protein at CGUJ_03991 was newly confirmed because of a one-base substitution in the untranslated region of the psy62 genome. The hypothetical proteins shared no similarity to other proteins of 'Ca. L. asiaticus'. These proteins are listed in Table S2.

Discussion

As described previously, 'Ca. L. asiaticus' Japanese Ishi-1 lacks a bacteriophage-type DNA polymerase gene [19]. Our study showed that Ishi-1 lacks a large fragment of about 33 kbp that contains the bacteriophage-type DNA polymerase gene. It is noteworthy that this strain is found only in Japan; despite having the smallest genome of all 'Ca. L. asiaticus' strains, Floridian UF506 carries the large 33-kbp fragment [27]. We suggest the large 33-kbp fragment is associated with neither pathogenicity nor transmissibility, because Ishi-1 induced severe symptoms on citrus and propagated to a high titer in the vector insect. This is in sharp contrast to the discussions of Zhang et al. [27] regarding UF506, where the SC1 and SC2 genes flanking the bacteriophage-type DNA polymerase gene are suspected to be important for infection and virulence expression. It is likely that Ishi-1 carries different virulence factors. Most Japanese strains also lack the bacterio-phage-type DNA polymerase gene [19]. Thus, the large 33-kbp fragment encoding the bacteriophage-type DNA polymerase gene may be absent from other Japanese strains, although confirmation by sequencing is needed. Another bacteriophage-type DNA polymerase is encoded in the middle of the Floridian psy62 and Japanese Ishi-1 genomes (Figure 3A), while Chinese gxpsy and Floridian UF506 carry two additional polymerases. These differences also suggest Ishi-1 (and perhaps other Japanese strains) are distinct from other strains from the US and China.

Zhou et al. identified two related and hypervariable genes (hyv_I and hyv_{II}) in the large 33-kbp fragment of the psy62 genome [25]. Although all DNA samples were obtained from symptomatic tissue and tested positive by 16S rRNA gene-based real-time PCR, neither the hyv_I nor the hyv_{II} gene was amplified from eight Indian citrus DNA samples and six Philippine psyllid DNA samples using the same primer sets [25]. These 14 strains likely lack the large 33-kbp fragment as Japanese Ishi-1 does. Thus, 'Ca. L. asiaticus' lacking the large fragment are not limited to Japan but are widespread in South Asia, pending confirmation by genome

sequencing. We conclude it is best not to use primer sets specific to the large 33 kbp fragment for PCR detection of 'Ca. L. asiaticus', because some strains might escape detection.

Two malic enzyme genes were identified in the genome sequence of Ishi-1 in contrast to 'Ca. L. asiaticus' Floridian psy62 and Chinese gxpsy. The malic enzyme of the microaerophilic protist *Entamoeba histolytica* decarboxylates malate to pyruvate [43]. In 'Ca. L. asiaticus', a phloem-limited bacterium [2], the enzyme might play a similar role to that of *E. histolytica*. However, both malic enzyme genes of Ishi-1 are shorter than those of other 'Ca. L. asiaticus' strains. The Ishi-1 enzymes might not maintain their original function, suggesting a possible limitation of the Ishi-1 fermentation pathway.

In conclusion, whole genome sequencing of Japanese 'Ca. L. asiaticus' strain Ishi-1 revealed unique genomic features and suggested novel expression of virulence and establishment in host plant as well as distinct molecular evolution. We hope this study will advance our understanding of 'Ca. L. asiaticus' and facilitate efforts to control this devastating disease in the citrus industry.

Supporting Information

Figure S1 Comparison of deduced amino acid sequences of ferroxidase from 'Ca. L. asiaticus'. Ferroxidase sequences were aligned by CLUSTAL W [48], and identical residues are indicated with asterisks. Databank accession numbers are CP001677 for psy62 [21], AP014595 for Ishi-1, and CP004005 for gxpsy [28], respectively.

Table S1 Conventional and long-distance PCR primers used for 'Ca. L. asiaticus' Japanese Ishi-1 strain sequence confirmation and gap closure in this study.

Table S2 List of CDS in the genome of 'Candidatus Liberibacter asiaticus' Japanese Ishi-1 strain that revealed no deduced amino acid sequence similarity to those of other 'Ca. L. asiaticus' strains.

Acknowledgments

We thank Ms. H. Hatomi for providing meticulous technical assistance to this work.

Author Contributions

Conceived and designed the experiments: TI. Performed the experiments: HK SM HI. Analyzed the data: HK SM. Contributed reagents/materials/analysis tools: SM HI TI. Contributed to the writing of the manuscript: HK SM HI TI.

References

1. Zhao X (1981) Citrus yellow shoot disease (Huanglongbing)—A review. Proc Int Soc Citriculture 1: 466–469.
2. Jagoueix S, Bové JM, Garnier M (1994) The phloem-limited bacterium of greening disease of citrus is a member of the α subdivision of the Proteobacteria. Int J Syst Bacteriol 44: 379–386.
3. Teixeira D (2005) First report of a Huanglongbing-like disease of citrus in São Paulo State, Brazil, and association of a new liberibacter species, 'Candidatus Liberibacter americanus', with the disease. Plant Dis 89: 107.
4. Halbert SE, Manjunath KL (2004) Asian citrus psyllids (Sternorrhyncha: Psyllidae) and greening disease of citrus: a literature review and assessment of risk in Florida. Fla Entomol 87: 330–353.
5. Bové JM (2006) Huanglongbing: a destructive, newly-emerging, century-old disease of citrus. J Plant Pathol 88: 7–37.
6. Liefting LW, Perez-Egusquiza ZC, Clover GRG, Anderson JAD (2008) A new 'Candidatus Liberibacter' species in *Solanum tuberosum* in New Zealand. Plant Dis 92: 1474.
7. Garnett HM (1984) Isolation of the greening organism. Citrus Subtrop Fruit J 611: 4–5.
8. Davis MJ, Mondal SN, Chen H, Rogers ME, Brlansky RH (2008) Co-cultivation of 'Candidatus Liberibacter asiaticus' with Actinobacteria from Citrus with Huanglongbing. Plant Dis 92: 1547–1550.
9. Sechler A, Schuenzel EL, Cooke P, Donnua S, Thaveechai N, et al. (2009) Cultivation of 'Candidatus Liberibacter asiaticus', 'Ca. L. africanus', and 'Ca. L. americanus' associated with Huanglongbing. Phytopathology 99: 480–486.
10. Bastianel C, Garnier-Semancik M, Renaudin J, Bové JM, Eveillard S (2005) Diversity of "Candidatus Liberibacter asiaticus," based on the *omp* gene sequence. Appl Environ Microbiol 71: 6473–6478.

11. Ding F, Deng X, Hong N, Zhong Y, Wang G, et al. (2009) Phylogenetic analysis of the citrus Huanglongbing (HLB) bacterium based on the sequences of 16S rDNA and 16S/23S rDNA intergenic regions among isolates in China. Eur J Plant Pathol 124: 495–503.

12. Furuya N, Matsukura K, Tomimura K, Okuda M, Miyata S, et al. (2010) Sequence homogeneity of the ψserA-trmU-tufB-secE-nusG-rplKAJL-rpoB gene cluster and the flanking regions of 'Candidatus Liberibacter asiaticus' isolates around Okinawa Main Island in Japan. J Gen Plant Pathol 76: 122–131.

13. Furuya N, Truc NTN, Iwanami T (2011) Recombination-like sequences in the upstream region of the phage-type DNA polymerase in 'Candidatus Liberibacter asiaticus.' J Gen Plant Pathol 77: 295–298.

14. Jagoueix S, Bové JM, Garnier M (1997) Comparison of the 16S/23S ribosomal intergenic regions of "Candidatus Liberobacter asiaticum" and huanglongbing (greening) disease. Int J Syst Bacteriol 47: 224–227.

15. Miyata S, Kato H, Davis R, Smith MW, Weinert M, et al. (2011) Asian common strain of 'Candidatus Liberibacter asiaticus' are distributed in Northeast India, Papua New Guinea and Timor-Leste. J Gen Plant Pathol 77: 43–47.

16. Planet P, Jagoueix S, Bové JM, Garnier M (1997) Detection and characterization of the African citrus greening Liberibacter by amplification, cloning and sequencing of the rplKAJL-rpoBC operon. Curr Microbiol 30: 137–141.

17. Subandiyah S, Iwanami T, Tsuyumu S, Ieki H (2000) Comparison of 16S rDNA and 16S/23S intergenic region sequences among citrus green organisms in Asia. Plant Dis 84: 15–18.

18. Tomimura K, Miyata S, Furuya N, Kubota K, Okuda M, et al. (2009) Evaluation of Genetic Diversity among 'Candidatus Liberibacter asiaticus' isolates collected in Southeast Asia. Phytopathology 99: 1062–1069.

19. Tomimura K, Furuya N, Miyata S, Hamashima A, Torigoe H, et al. (2010) Distribution of Two Distinct Genotypes of Citrus Greening Organism in the Ryukyu Islands of Japan. Jpn Agric Res Q 44: 151–158.

20. Villechanoux S, Garnier M, Laigret F, Renaudin J, Bové JM (1993) The genome of the non-cultured, bacterial-like organism associated with citrus greening disease contains the nusG-rplKAJLrpoBC gene cluster and the gene for a bacteriophage type DNA polymerase. Curr Microbiol 26: 161–166.

21. Duan YL, Zhou DG, Hall W, Li H, Doddapaneni H, et al. (2009) Complete genome sequence of citrus Huanglongbing bacterium, 'Candidatus Liberibacter asiaticus' obtained through metagenomics. Mol Plant-Microbe Interact 22: 1011–1020.

22. Chen J, Deng X, Sun X, Jones D, Irey M, et al. (2010) Guangdong and Florida populations of 'Candidatus Liberibacter asiaticus' distinguished by a genomic locus with short tandem repeats. Phytopathology 100: 567–572.

23. Katoh H, Subandiyah S, Tomimura K, Okuda M, Su HJ, et al. (2011) Differentiation of "Candidatus Liberibacter asiaticus" isolates by Variable Number of Tandem Repeat Analysis. Appl Environ Microbiol 77: 1910–1917.

24. Katoh H, Davis R, Smith MW, Weinert M, Iwanami T (2012) Differentiation of Indian, East Timorese, Papuan and Floridian 'Candidatus Liberibacter asiaticus' isolates on the basis of simple sequence repeat and single nucleotide polymorphism profiles at 25 loci. Ann Appl Biol 160: 291–297.

25. Zhou L, Powell CA, Hoffman MT, Li W, Fan G, et al. (2011) Diversity and plasticity of the intracellular plant pathogen and insect symbiont "Candidatus Liberibacter asistics" as revealed by hypervariable prophage genes with intragenic tandem repeats. Appl Environ Microbiol 77: 6663–6673.

26. Morgan JK, Zhou L, Li W, Shatters RG, Keremane M, et al. (2012) Improved real-time PCR detection of 'Candidatus Liberibacter asiaticus' from citrus and psyllid hosts by targeting the intragenic tandem-repeat of its prophage genes. Mol Cell Probes 26: 90–98.

27. Zhang A, Flores-Cruz Z, Zhou L, Kang B, Fleites LA, et al. (2011) 'Ca. Liberibacter asiaticus' carries an excision plasmid prophage and a chromosomally integrated prophage that becomes lytic in plant infections. Mol Plant-Microbe Intract 24: 458–468.

28. Lin H, Han CS, Liu B, Lou B, Bai X, et al. (2013) Complete Genome Sequence of a Chinese Strain of 'Candidatus Liberibacter asiaticus'. Genome Announc 1: E00184–13.

29. Zheng Z, Deng X, Chen J (2014) Whole-Genome Sequence of 'Candidatus Liberibacter asiaticus' from Guangdong, China. Genome Announc 2 (2), e00273–14.

30. Inoue H, Ohnishi J, Ito T, Tomimura K, Miyata S, et al. (2009) Enhanced proliferation and efficient transmission of Candidatus Liberibacter asiaticus by adult Diaphorina citri after acquisition feeding in the nymphal stage. Ann Appl Biol 155: 29–36.

31. Li H, Durbin R (2009) Fast and accurate short read alignment with Burrows-Wheeler transform. Bioinformatics 25: 1754–1760.

32. Langmead B, Trapnell C, Pop M, Salzberg SL (2009) Ultrafast and memory-efficient alignment of short DNA sequences to the human genome. Genome Biol 10: R25.

33. Robinson JT, Thorvaldsdottir H, Winckler W, Guttman M, Lander ES, et al. (2011) Integrative genomics viewer. Nat Biotech 29: 24–26.

34. Sugawara H, Ohyama A, Mori H, Kurokawa K (2009) Microbial genome annotation pipeline (MiGAP) for diverse users. abstr S-001, p 1–2. Abstr. 20th Int. Conf. Genome Informatics, Kanagawa, Japan.

35. Lowe TM, Eddy SR (1997) tRNAscan-SE: a program for improved detection of transfer RNA genes in genomic sequence. Nucl Acids Res 25: 955–964.

36. Lagesen K, Hallin P, Rødland EA, Stærfeldt H-H, Rognes T, et al. (2007) RNAmmer: consistent and rapid annotation of ribosomal RNA genes. Nucl Acids Res 35: 3100–3108.

37. Lin H, Lou B, Glynn JM, Doddapaneni H, Civerolo EL, et al. (2011) The complete genome sequence of 'Candidatus Liberibacter solanacearum', the bacterium associated with potato zebra chip disease. PLoS ONE 6: E19135.

38. Reindel S, Anemüller S, Sawaryn A, Matzanke BF (2002) The DpsA homologue of the archaeon Halobacterium salinarum is a ferritin. Biochimica et Biophysica Acta (BBA) - Proteins & Proteomics 1598: 140–146.

39. Zeth K, Offermann S, Essen L-O, Oesterhelt D (2004) Iron-oxo clusters biomineralizing on protein surfaces: Structural analysis of Halobacterium salinarum DpsA in its low- and high-iron states. Proc Natl Acad Sci USA 101: 13780–13785.

40. Andrews SC (2010) The ferritin-like superfamily: Evolution of the biological iron storeman from a rubrerythrin-like ancestor. Biochimica et Biophysica Acta (BBA) - General Subjects 1800: 691–705.

41. Carrondo MA (2003) Ferritins, iron uptake and storage from the bacterioferritin viewpoint. EMBO J 22: 1959–1968.

42. Zhao G, Ceci P, Ilari A, Giangiacomo L, Laue TM, et al. (2002) Iron and hydrogen peroxide detoxification properties of DNA-binding protein from starved cells: A ferritin-like DNA-binding protein of Escherichia coli. J Biol Chemistry 277: 27689–27696.

43. Field J, Rosenthal B, Samuelson J (2000) Early lateral transfer of genes encoding malic enzyme, acetyl-CoA synthetase and alcohol dehydrogenases from anaerobic prokaryotes to Entamoeba histolytica. Mol Microbiol 38: 446–455.

44. Fernández-Murga ML, Gil-Ortiz F, Llácer JL, Rubio V (2004) Arginine biosynthesis in Thermotoga maritima: Characterization of the Arginine-Sensitive N-Acetyl-L-Glutamate Kinase. J Bacteriol 186: 6142–6149.

45. Hass D, Holloway BW, Schambõck A, Leisinger T (1977) The genetic organization of arginine biosynthesis in Pseudomonas aeruginosa. Mol Gen Genet 154: 7–22.

46. Ikeda M, Mitsuhashi S, Tanaka K, Hayashi M (2009) Reengineering of a Corynebacterium glutamicum L-Arginine and L-Citrulline Producer. Appl Environ Microbiol 75: 1635–1641.

47. Picard FJ, Dillon JR (1989) Cloning and organization of seven arginine biosynthesis genes from Neisseria gonorrhoeae. J Bacteriol 171: 1644–1651.

48. Thompson JD, Higging DG, Gibson TJ (1994) CLUTALW: improving the sensitivity of progressive multiple sequence alignment through sequence weighting, position-specific gap penalties and weight matrix choice. Nucleic Acids Res 22:4673–4680.

Permissions

All chapters in this book were first published in PLOS ONE, by The Public Library of Science; hereby published with permission under the Creative Commons Attribution License or equivalent. Every chapter published in this book has been scrutinized by our experts. Their significance has been extensively debated. The topics covered herein carry significant findings which will fuel the growth of the discipline. They may even be implemented as practical applications or may be referred to as a beginning point for another development.

The contributors of this book come from diverse backgrounds, making this book a truly international effort. This book will bring forth new frontiers with its revolutionizing research information and detailed analysis of the nascent developments around the world.

We would like to thank all the contributing authors for lending their expertise to make the book truly unique. They have played a crucial role in the development of this book. Without their invaluable contributions this book wouldn't have been possible. They have made vital efforts to compile up to date information on the varied aspects of this subject to make this book a valuable addition to the collection of many professionals and students.

This book was conceptualized with the vision of imparting up-to-date information and advanced data in this field. To ensure the same, a matchless editorial board was set up. Every individual on the board went through rigorous rounds of assessment to prove their worth. After which they invested a large part of their time researching and compiling the most relevant data for our readers.

The editorial board has been involved in producing this book since its inception. They have spent rigorous hours researching and exploring the diverse topics which have resulted in the successful publishing of this book. They have passed on their knowledge of decades through this book. To expedite this challenging task, the publisher supported the team at every step. A small team of assistant editors was also appointed to further simplify the editing procedure and attain best results for the readers.

Apart from the editorial board, the designing team has also invested a significant amount of their time in understanding the subject and creating the most relevant covers. They scrutinized every image to scout for the most suitable representation of the subject and create an appropriate cover for the book.

The publishing team has been an ardent support to the editorial, designing and production team. Their endless efforts to recruit the best for this project, has resulted in the accomplishment of this book. They are a veteran in the field of academics and their pool of knowledge is as vast as their experience in printing. Their expertise and guidance has proved useful at every step. Their uncompromising quality standards have made this book an exceptional effort. Their encouragement from time to time has been an inspiration for everyone.

The publisher and the editorial board hope that this book will prove to be a valuable piece of knowledge for researchers, students, practitioners and scholars across the globe.

List of Contributors

Guillaume Martin, Franc-Christophe Baurens, Céline Cardi and Angélique D'Hont
CIRAD (Centre de coopération Internationale en Recherche Agronomique pour le Développement), UMR AGAP, Montpellier, France

Jean-Marc Aury
Genoscope, Evry, France

Brian M. Forde, Nouri L. Ben Zakour, Mitchell Stanton-Cook, Minh-Duy Phan, Makrina Totsika, Kate M. Peters, Mark A. Schembri and Scott A. Beatson
Australian Infectious Diseases Research Centre, School of Chemistry & Molecular Biosciences, The University of Queensland, Queensland, Australia

Kok Gan Chan
Division of Genetics and Molecular Biology, Institute of Biological Sciences, Faculty of Science, University of Malaya, Kuala Lumpur, Malaysia

Mathew Upton
Plymouth University Peninsula Schools of Medicine and Dentistry, Plymouth, United Kingdom

Martin Mascher and Uwe Scholz
Department of Cytogenetics and Genome Analysis, Leibniz Institute of Plant Genetics and Crop Plant Research (IPK), Corrensstrabe 3, Stadt Seeland, Germany

Nina Gerlach and Marcel Bucher
Botanical Institute, Cologne Biocenter, Cluster of Excellence on Plant Sciences (CEPLAS), University of Cologne, Zülpicherstrasse 47b, Cologne, Germany

Manfred Gahrtz and Thomas Dresselhaus
Cell Biology and Plant Biochemistry, Biochemie-Zentrum Regensburg, University of Regensburg, Universitätsstrabe 31, Regensburg, Germany

Marcus Lechner and Roland K. Hartmann
Institut für Pharmazeutische Chemie, Philipps-Universität Marburg, Marburg, Germany

Maribel Hernandez-Rosales
Bioinformatics Group, Department of Computer Science, Universität Leipzig, Leipzig, Germany Interdisciplinary Center for Bioinformatics, Universität Leipzig, Leipzig, Germany Max Planck Institute for Mathematics in the Sciences, Leipzig, Germany, Departamento de Ciência da Computação, Instituto de Ciências Exatas, Universidade de Brasília, Brasília, Brasil

Daniel Doerr, Annelyse Thévenin and Jens Stoye
Genome Informatics, Faculty of Technology, Bielefeld University, Bielefeld, Germany Institute for Bioinformatics, Center for Biotechnology, Bielefeld University, Bielefeld, Germany

Nicolas Wieseke
Faculty of Mathematics and Computer Science University of Leipzig, Leipzig, Germany

Sonja J. Prohaska
Computational EvoDevo Group, Department of Computer Science, Universität Leipzig, Leipzig, Germany

Peter F. Stadler
Bioinformatics Group, Department of Computer Science, Universität Leipzig, Germany Interdisciplinary Center for Bioinformatics, Universität Leipzig, Leipzig, Germany Max Planck Institute for Mathematics in the Sciences, Leipzig, ermany Institute for Theoretical Chemistry, University of Vienna, Vienna, Austria Center for non-coding RNA in Technology and Health, University of Copenhagen, Frederiksberg, Denmark The Santa Fe Institute, Santa Fe, New Mexico, United States of America RNomics Group, Fraunhofer Institut for Cell Therapy and Immunology, Leipzig, Germany

Adnan Niazi and Erik Bongcam-Rudloff
Department of Animal Breeding and Genetics, SLU Global Bioinformatics Centre, Swedish University of Agricultural Sciences, Uppsala, Sweden

Shahid Manzoor
Department of Animal Breeding and Genetics, SLU Global Bioinformatics Centre, Swedish University of Agricultural Sciences, Uppsala, Sweden University of the Punjab, Lahore, Pakistan

Shashidar Asari, Sarosh Bejai and Johan Meijer
Department of Plant Biology, Linnéan Center for Plant Biology, Uppsala Biocenter, Swedish University of Agricultural Sciences, Uppsala, Sweden

Zihao Xia, Yongqiang Li, Ling Chen, Shuai Li, Tao Zhou and Zaifeng Fan
State Key Laboratory of Agro-biotechnology and Ministry of Agriculture Key Laboratory for Plant Pathology, China Agricultural University, Beijing, China

Jun Peng
State Key Laboratory of Agro-biotechnology and Ministry of Agriculture Key Laboratory for Plant Pathology, China Agricultural University, Beijing, China
Ministry of Agriculture Key Laboratory of Integrated Pest Management on Tropical Crops, Environmental and Plant Protection Institute, Chinese Academy of Tropical Agricultural
Sciences, Haikou, Hainan, China

Monica A. Kehoe, Brenda A. Coutts, Bevan J. Buirchell and Roger A. C. Jones
School of Plant Biology and Institute of Agriculture, Faculty of Science, University of Western Australia, Crawley, WA, Australia
Crop Protection and Lupin Breeding Branches, Department of Agriculture and Food Western Australia, Perth, WA, Australia

Yankun Wang, Pu Chu, Qing Yang, Shengxin Chang and Jianmei Chen
State Key Laboratory of Crop Genetics and Germplasm Enhancement, Nanjing Agricultural University, Nanjing, Jiangsu, China

Maolong Hu
Institute of Economic Crop, Jiangsu Academy of Agricultural Sciences, Nanjing, Jiangsu, China

Rongzhan Guan
State Key Laboratory of Crop Genetics and Germplasm Enhancement, Nanjing Agricultural University, Nanjing, Jiangsu, China
Nanjing Agricultural University, Jiangsu Collaborative Innovation Center for Modern Crop Production, Nanjing, Jiangsu, China

Mårten Lind, Malin Elfstrand and Jan Stenlid
Department of Forest Mycology and Plant Pathology, Swedish University of Agricultural Sciences, Uppsala, Sweden

Thomas Källman, Jun Chen, Xiao-Fei Ma and Martin Lascoux
Department of Ecology and Genetics, Evolutionary Biology Centre, Uppsala University, Uppsala, Sweden

Jean Bousquet
Institute for Systems and Integrative Biology, Université Laval, Québec City, Québec, Canada

Michele Morgante and Giusi Zaina
Dipartimento di Scienze Agrarie e Ambientali, Universita di Udine, Udine, Italy

Bo Karlsson
Skogforsk, Svalöv, Sweden

Mehmet Cengiz Baloglu
Kastamonu University, Faculty of Engineering and Architecture, Department of Genetics and Bioengineering, Kastamonu, Turkey

Vahap Eldem
Istanbul University, Faculty of Science, Department of Biology, Istanbul, Turkey

Mortaza Hajyzadeh and Turgay Unver
Cankırı Karatekin University, Faculty of Science, Department of Biology, Cankiri, Turkey

Thierry Candresse
INRA, UMR 1332 Biologie du Fruit et Pathologie, CS 20032, 33882 Villenave d'Ornon Cedex, France
Université de Bordeaux, UMR 1332 Biologie du Fruit et Pathologie, CS 20032, 33882 Villenave d'Ornon Cedex, France

Denis Filloux, Charlotte Julian, Serge Galzi, Guillaume Fort, Pauline Bernardo, Jean-Heindrich Daugrois, Emmanuel Fernandez and Philippe Roumagnac
CIRAD, UMR BGPI, Campus International de Montferrier-Baillarguet, 34398 Montpellier Cedex-5, France

Brejnev Muhire and Darren P. Martin
Computational Biology Group, Institute of Infectious Diseases and Molecular Medicine, University of Cape Town, Cape Town, South Africa

Arvind Varsani
School of Biological Sciences and Biomolecular Interaction Centre, University of Canterbury, Christchurch, New Zealand

Department of Plant Pathology and Emerging Pathogens Institute, University of Florida, Gainesville, Florida, United States of America
Electron Microscope Unit, Division of Medical Biochemistry, Department of Clinical Laboratory Sciences, University of Cape Town, Observatory, South Africa

Yung-Fen Huang, Charlene P. Wight and Nicholas A. Tinker
Eastern Cereal and Oilseed Research Centre, Agriculture and Agri-Food Canada, Ottawa, Ontario, Canada

Jesse A. Poland
Department of Plant Pathology, Kansas State University, Manhattan, Kansas, United States of America

Eric W. Jackson
General Mills Crop Biosciences, Manhattan, Kansas, United States of America

Yongxiang Lin
Key Lab of Crop Biology of Anhui Province, School of Life Sciences, Anhui Agricultural University, Hefei, Anhui, China
Crop Research Institute, Anhui Academy of Agricultural Sciences, Hefei, Anhui, China

Ying Cheng, Jing Jin, Xiaolei Jin, Haiyang Jiang, Hanwei Yan and Beijiu Cheng
Key Lab of Crop Biology of Anhui Province, School of Life Sciences, Anhui Agricultural University, Hefei, Anhui, China

Patrick T. West, Qing Li, Steven R. Eichten and Nathan M. Springer
Department of Plant Biology, University of Minnesota, Saint Paul, Minnesota, United States of America

Lexiang Ji
Department of Genetics, University of Georgia, Athens, Georgia, United States of America
Institute of Bioinformatics, University of Georgia, Athens, Georgia, United States of America

Jawon Song and Matthew W. Vaughn
Texas Advanced Computing Center, University of Texas-Austin, Austin, Texas, United States of America

Robert J. Schmitz
Department of Genetics, University of Georgia, Athens, Georgia, United States of America

Hyoung Tae Kim and Ki-Joong Kim
Division of Life Sciences, School of Life Sciences and Biotechnology, Korea University, Seoul, Korea

Domenica Nigro, Antonio Blanco and Agata Gadaleta
Department of Soil, Plant and Food Sciences, Section of Genetic and Plant Breeding, University of Bari "Aldo Moro", Bari, Italy

Olin D. Anderson
Genomics and Gene Discovery Research Unit, Western Regional Research Center, USDA-ARS, Albany, California, United States of America

Suyan Niu, Guoqiang Fan, Enkai Xu, Minjie Deng, Zhenli Zhao and Yanpeng Dong
Institute of Paulownia, Henan Agricultural University, Zhengzhou, Henan, P.R. China
College of Forestry, Henan Agricultural University, Zhengzhou, Henan, P.R. China

Hiroshi Katoh, Shin-ichi Miyata and Toru Iwanami
NARO Institute of Fruit Tree Science, Tsukuba, Ibaraki, Japan

Hiromitsu Inoue
Kuchinotsu Citrus Research Station, NARO Institute of Fruit Tree Science, Minami-shimabara, Nagasaki, Japan

Index

www.ingramcontent.com/pod-product-compliance
Lightning Source LLC
Chambersburg PA
CBHW061250190326
41458CB00011B/3635